Birkhäuser

Progress in Nonlinear Differential Equations and Their Applications

Volume 85

For further volumes:
http://www.springer.com/series/4889

Djairo G. de Figueiredo • João Marcos do Ó
Carlos Tomei
Editors

Analysis and Topology in Nonlinear Differential Equations

A Tribute to Bernhard Ruf on the Occasion of his 60th Birthday

Birkhäuser

and extremals of a suitable trace Sobolev embedding in $\mathbb{R}^N_+$. For more details, we refers to [1] where the authors studied some properties on the solutions of (P_∞).

Next, if we denote by $\mathcal{H}(z)$ the mean curvature of the boundary at the point $z \in \partial\Omega$, we have.

Theorem 1.2. *Assume that $N \geq 4$ and let u_ε be the least energy solution of (P_ε) obtained in Theorem 1.1. If $z_\varepsilon \in \partial\Omega_\varepsilon$ is a point where u_ε achieves its maximum value then*

$$\mathcal{H}(\varepsilon z_\varepsilon) \to \max_{z\in\partial\Omega} \mathcal{H}(z), \ \ as\ \varepsilon \to 0.$$

Moreover, there are positive constants $\gamma = \gamma(p, N)$ and $\tilde{\gamma} = \tilde{\gamma}(p, N)$ such that:

(i) *the associated critical value $C_p(\Omega_\varepsilon)$ can be estimated as*

$$C_p(\Omega_\varepsilon) = C_p(\mathbb{R}^N_+) - \varepsilon\gamma \max_{z\in\partial\Omega} \mathcal{H}(z) + o(\varepsilon), \ \ as\ \varepsilon \to 0; \tag{1.8}$$

(ii) *the best constant $S_p(\Omega_\varepsilon)$ can be estimated as*

$$S_p(\Omega_\varepsilon) = S_p(\mathbb{R}^N_+) - \varepsilon\tilde{\gamma} \max_{z\in\partial\Omega} \mathcal{H}(z) + o(\varepsilon), \ \ as\ \varepsilon \to 0. \tag{1.9}$$

Finally, we will study problem (P_ε) for large ε, for which the main result can be stated as follows.

Theorem 1.3. *There exists $\varepsilon^* > 0$ such that for each $\varepsilon > \varepsilon^*$, $u \equiv 1$ is the unique positive solution of (P_ε).*

Remark 1.4. *In the light of Theorem 1.3, the ground state of (P_ε) is the constant function $u_\varepsilon \equiv 1$, and the best constant $S_p(\Omega_\varepsilon) = S_p(\Omega_\varepsilon) = \left(\varepsilon^{1-N}|\partial\Omega|\right)^{1-\frac{2}{p}}$ as $\varepsilon \to \infty$, which implies that $S_p(\Omega_\varepsilon) \to 0$ as $\varepsilon \to \infty$, since $p > 2$.*

Some related sharp inequalities involving Sobolev trace imbedding are given by Bonder–Rossi [3], Adimurthi–Yadava [2], Escobar [7], and the references therein. See also [2, 12]. Problems with nonlinear boundary conditions appear in a natural way when one considers the Sobolev trace embedding, see for example [5], where existence and qualitative behavior of solutions were investigated. When $p = 2_*$, Adimurthi–Yadava in [2, see proof of Theorem 2] proved that (P_ε) does not have solution for ε large. We quote that in their approach they used strongly the extremal function w of the critical Sobolev imbedding $H^1(\mathbb{R}^n_+) \hookrightarrow L^{2_*-1}(\mathbb{R}^{n-1})$. They also obtain similar existence result for ε small.

2. Existence of extremal

As we quote in the introduction, the existence of extremal is equivalent the existence of least energy solutions to (P_ε). For that we study critical points of the associated functional to (P_ε),

$$I_{\Omega_\varepsilon}(u) := \frac{1}{2}\int_{\Omega_\varepsilon} |\nabla u|^2 \,\mathrm{d}z + \frac{1}{2}\int_{\partial\Omega_\varepsilon} u^2 \,\mathrm{d}\sigma - \frac{1}{p}\int_{\partial\Omega_\varepsilon} (u^+)^p \,\mathrm{d}\sigma,$$

defined on the Hilbert space $H^1(\Omega_\varepsilon)$ endowed with inner product

$$\langle u, v\rangle = \int_{\Omega_\varepsilon} [\nabla u \nabla v + uv]\ dx$$

and the induced norm $\|u\|_{H^1(\Omega_\varepsilon)} := \langle u, u\rangle^{1/2}$. The energy functional I_{Ω_ε} is well defined and C^1 with

$$I'_{\Omega_\varepsilon}(u)\varphi = \int_{\Omega_\varepsilon} \nabla u \nabla \varphi\ dz + \int_{\partial\Omega_\varepsilon} u\varphi\ d\sigma - \int_{\partial\Omega_\varepsilon} (u^+)^{p-1}\varphi\ d\sigma, \quad \varphi \in H^1(\Omega_\varepsilon). \tag{2.1}$$

Consequently, the weak solutions of (P_ε) are critical points of I_{Ω_ε} and conversely (see [13]).

Lemma 2.1. *The functional I_{Ω_ε} satisfies the following conditions:*

(i) *for each $u \in H^1(\Omega_\varepsilon)$ such that the trace of u_+ is not identically zero on $\partial\Omega_\varepsilon$, we have $\lim_{s\to\infty} I_{\Omega_\varepsilon}(su) = -\infty$.*

(ii) *there exist ρ, $\alpha > 0$, such that $I_{\Omega_\varepsilon}(u) \geq \alpha$ if $\|u\|_{H^1(\Omega_\varepsilon)} = \rho$.*

Proof. Let $u \in H^1(\Omega_\varepsilon)$ such that the trace of u_+ is not identically zero on $\partial\Omega_\varepsilon$. From

$$I_{\Omega_\varepsilon}(su) := \frac{s^2}{2}\left\{\int_{\Omega_\varepsilon} |\nabla u|^2\ dz + \int_{\partial\Omega_\varepsilon} u^2\ d\sigma\right\} - \frac{s^p}{p}\int_{\partial\Omega_\varepsilon} (u^+)^p\ d\sigma,$$

we see that (i) holds, because $p > 2$. By inequality (1.4) we get

$$I_{\Omega_\varepsilon}(u) \geq C_1\|u\|^2_{H^1(\Omega_\varepsilon)} - C_2\|u\|^p_{H^1(\Omega_\varepsilon)},$$

which implies (ii), and this completes the proof. □

Lemma 2.2. *I_{Ω_ε} satisfies the Palais–Smale condition.*

Proof. Let (u_k) in $H^1(\Omega_\varepsilon)$ be a (PS)-sequence for the functional I_{Ω_ε}, that is, $|I_{\Omega_\varepsilon}(u_k)| \leq C$ and $I'_{\Omega_\varepsilon}(u_k) \to 0$. Since

$$\begin{aligned}\left(\frac{1}{2} - \frac{1}{p}\right)\left[\int_{\Omega_\varepsilon} |\nabla u_k|^2\ dz + \int_{\partial\Omega_\varepsilon} u_k^2\ d\sigma\right] &= I_{\Omega_\varepsilon}(u_k) - \frac{1}{p}I'_{\Omega_\varepsilon}(u_k)u_k \\ &\leq C_1 + C_2\|u_k\|_{H^1(\Omega_\varepsilon)},\end{aligned}$$

using inequality (1.4), we get $\|u_k\|^2_{H^1(\Omega_\varepsilon)} \leq C'_1 + C'_2\|u_k\|_{H^1(\Omega_\varepsilon)}$, which implies that (u_k) is bounded. Thus, up to a subsequence, we can assume that $u_k \rightharpoonup u$ in $H^1(\Omega_\varepsilon)$ and $u_k \to u$ in $L^p(\partial\Omega_\varepsilon)$. Now observe that

$$\begin{aligned}&\int_{\Omega_\varepsilon} |\nabla(u_k - u)|^2\ dz + \int_{\partial\Omega_\varepsilon} (u_k - u)^2\ d\sigma \\ &\quad = (I'_{\Omega_\varepsilon}(u_k) - I'_{\Omega_\varepsilon}(u))(u_k - u) + \int_{\partial\Omega_\varepsilon} ((u_k^+)^{p-1} - (u^+)^{p-1})(u_k - u)\ d\sigma.\end{aligned} \tag{2.2}$$

By Hölder's inequality, we have $\int_{\partial\Omega_\varepsilon}((u_k^+)^{p-1} - (u^+)^{p-1})(u_k - u)\ d\sigma \to 0$. Since $u_k \rightharpoonup u$ and $I'_{\Omega_\varepsilon}(u)(u_k - u) \to 0$, we obtain $(I'_{\Omega_\varepsilon}(u_k) - I'_{\Omega_\varepsilon}(u))(u_k - u) \to 0$. Thus, I_{Ω_ε} satisfies the (PS) condition. □

Since $w(r,0) \to 0$ as $r \to \infty$ we get $-w_t(r,0)-Aw(r,0) = w(r,0)(w^{p-2}(r,0)-(A+1)) < 0$, $\quad r \geq r_0$. Thus, we can choose $A > 0$ such that $\varphi_3(r,0) = (-w_t(r,0) - Aw(r,0))_+ = 0$, $\quad \forall\, r \geq 0$, where $\varphi_3 = (-w_t - Aw)_+$. It follows from (3.7), that

$$\int_{\mathbb{R}^N_+} \nabla(-w_t - Aw)\nabla\varphi_3 dz = \int_{\mathbb{R}^{N-1}} \frac{\partial(-w_t - Aw)}{\partial\eta}\varphi_3 dx = 0.$$

Notice that $\varphi_3 \equiv 0$ on the set $\{w_t > 0\}$. Thus, we have

$$0 = \int_{\mathbb{R}^N_+\cap\{w_t<0\}} \nabla(-w_t - Aw)\nabla\varphi_3 dz = \int_{\mathbb{R}^N_+} |\nabla((w_t)_- - Aw)_+|^2 dz,$$

which implies

$$(w_t)_- \leq Aw, \quad \text{in } \overline{\mathbb{R}^N_+}. \tag{3.8}$$

From (3.4), (3.6), and (3.8) we get the desired result. □

4. Upper estimate for $C_p(\Omega_\varepsilon)$

In order to prove that the minimax solution u_ε is nonconstant for ε sufficiently small, we will obtain an upper bound estimate to the minimax level $C_p(\Omega_\varepsilon)$ using the characterization given in (2.4). In order to avoid technicalities we assume from now on that Ω is a domain strictly convex. Let w be a positive solution of (P_∞) and fix $z_0 \in \partial\Omega$. After a translation and rotation of the coordinate system we may assume that z_0 is the origin and the inner normal to Ω at z_0 is pointing in the direction of the positive t-axis. On the other hand, there exists a C^2 function $G : B_{r_0} \to \mathbb{R}$ defined in a ball $B_{r_0} = \{x = (x_1, \ldots, x_{N-1}) \in \mathbb{R}^{N-1} : |x| < r_0\}$, such that $G(0) = 0, \nabla G(0) = 0$. Since Ω is strictly convex we consider the following cylinder in $\mathbb{R}^N$:

$$\mathcal{U} = \{(x,t) \in \mathbb{R}^N : |x| \leq r_0 \quad \text{and} \quad 0 \leq t \leq t_0\},$$

where $t_0 = \min_{|x|=r_0} G(x) > 0$. Notice that

$$\begin{aligned} \partial\Omega \cap \mathcal{U} &= \{(x,t) : t = G(x)\} \\ \Omega \cap \mathcal{U} &= \{(x,t) : t > G(x)\}. \end{aligned} \tag{4.1}$$

Using the minimax characterization of $C_p(\Omega_\varepsilon)$ given in (2.4) with $v_0(x,t) = w(\varepsilon(x,t) - z_0)$ we obtain the following estimate.

Proposition 4.1. *There exists a positive constant γ, depending on N and p, such that*

$$C_p(\Omega_\varepsilon) \leq C_p(\mathbb{R}^N_+) - \varepsilon\gamma \max_{z\in\partial\Omega} \mathcal{H}(z) + o(\varepsilon), \quad \text{as} \quad \varepsilon \to 0. \tag{4.2}$$

We split the proof of Proposition 4.1 into two lemmas. We set

$$g(x) = \langle D^2G(0)x, x\rangle, \quad x \in \mathbb{R}^{N-1},$$

and

$$R_1(\varepsilon) := \int_{\Omega_\varepsilon} |\nabla w|^2 \, dz - \int_{\mathbb{R}^N_+} |\nabla w|^2 \, dz,$$

$$R_2(\varepsilon) := \int_{\partial\Omega_\varepsilon} \left(\frac{w^2}{2} - \frac{w^p}{p}\right) d\sigma - \int_{\mathbb{R}^{N-1}} \left(\frac{w^2}{2} - \frac{w^p}{p}\right) dx.$$

Choosing $s_\varepsilon > 0$ such that $\max_{s>0} I_{\Omega_\varepsilon}(sv_0) = I_{\Omega_\varepsilon}(s_\varepsilon v_0)$ we can state.

Lemma 4.2. *The following estimates hold as* $\varepsilon \to 0$,

$$R_1(\varepsilon) = -\varepsilon \int_{\mathbb{R}^{N-1}} |\nabla w(x,0)|^2 g(x) \, dx + o(\varepsilon),$$

$$R_2(\varepsilon) = \varepsilon \int_{\mathbb{R}^{N-1}} w_t^2(x,0) g(x) \, dx + o(\varepsilon).$$

Moreover, $s_\varepsilon^{p-2} = 1 + O(\varepsilon)$.

Proof. Let $\mathcal{U}_\varepsilon = \varepsilon^{-1}\mathcal{U}$ that is, $\mathcal{U}_\varepsilon = \{(x,t) \in \mathbb{R}^N : |\varepsilon x| \le r_0 \quad \text{and} \quad 0 \le \varepsilon t \le t_0\}$, where $t_0 = \min_{|x|=r_0} G(x) > 0$. Now observe that

$$\begin{aligned} -R_1(\varepsilon) &= \int_{\mathbb{R}^N_+ \setminus \Omega_\varepsilon} |\nabla w|^2 \, dz \\ &= \int_{\mathbb{R}^N_+ \setminus \mathcal{U}_\varepsilon} |\nabla w|^2 \, dz + \int_{\mathcal{U}_\varepsilon \setminus (\Omega_\varepsilon \cap \mathcal{U}_\varepsilon)} |\nabla w|^2 \, dz - \int_{\Omega_\varepsilon \cap (\mathbb{R}^N_+ \setminus \mathcal{U}_\varepsilon)} |\nabla w|^2 \, dz \\ &= A_1(\varepsilon) + A_2(\varepsilon) + A_3(\varepsilon). \end{aligned}$$

By Proposition 3.1, there is a positive constant $C = C(N)$ such that

$$\begin{aligned} A_1(\varepsilon) := \int_{\mathbb{R}^N_+ \setminus \mathcal{U}_\varepsilon} |\nabla w|^2 \, dz &\le C \int_{\mathbb{R}^N_+ \setminus \mathcal{U}_\varepsilon} \frac{e^{-2c_2|x|}}{(1+t^2)^{N-2}} \, dz \\ &\le C \int_{t_0\varepsilon^{-1}}^{+\infty} \frac{1}{t^{2N-4}} \, dt \int_0^{r_0\varepsilon^{-1}} e^{-2c_2 r} r^{N-2} \, dr \\ &\quad + C \int_0^{+\infty} \frac{1}{1+t^2} \, dt \int_{r_0\varepsilon^{-1}}^{+\infty} e^{-2c_2 r} r^{N-2} \, dr \\ &\le C \int_{t_0\varepsilon^{-1}}^{+\infty} \frac{1}{t^{2N-4}} \, dt + C \int_{r_0\varepsilon^{-1}}^{+\infty} e^{-2c_2 r} r^{N-2} \, dr \\ &= o(\varepsilon). \end{aligned} \tag{4.3}$$

Since $(\Omega_\varepsilon \cap (\mathbb{R}^N_+ \setminus \mathcal{U}_\varepsilon)) \subset \mathbb{R}^N_+ \setminus \mathcal{U}_\varepsilon$, the last estimate implies

$$A_2(\varepsilon) := \int_{\Omega_\varepsilon \cap (\mathbb{R}^N_+ \setminus \mathcal{U}_\varepsilon)} |\nabla w|^2 \, dz = o(\varepsilon).$$

Setting $D_\varepsilon = \{(x,t) : |\varepsilon x| \le r_0, t_0 \le \varepsilon t \le G(\varepsilon x)\} \subset \mathbb{R}^N_+ \setminus \mathcal{U}_\varepsilon$, we have

$$
\begin{aligned}
A_3(\varepsilon) &:= \int_{\mathcal{U}_\varepsilon \setminus (\Omega_\varepsilon \cap \mathcal{U}_\varepsilon)} |\nabla w|^2 \,\mathrm{d}z \\
&= \int_{|\varepsilon x| \le r_0} \int_0^{G(\varepsilon x)\varepsilon^{-1}} |\nabla w(x,t)|^2 \,\mathrm{d}t \,\mathrm{d}x - \int_{D_\varepsilon} |\nabla w|^2 \,\mathrm{d}z \\
&= \int_{|\varepsilon x| \le r_0} \int_0^{G(\varepsilon x)\varepsilon^{-1}} |\nabla w(x,t)|^2 \,\mathrm{d}t \,\mathrm{d}x + o(\varepsilon).
\end{aligned}
$$

Applying the mean value theorem there exists $c \in (0,t)$ such that $|\nabla w(x,t)|^2 = |\nabla w(x,0)|^2 + 2\langle \nabla w(x,c), \nabla w_t(x,c)\rangle t$, which together with the fact that $G(x) = g(x) + o(|x|^2)$ implies

$$
\begin{aligned}
A_3(\varepsilon) &= \varepsilon \int_{|\varepsilon x| \le r_0} |\nabla w(x,0)|^2 g(x) \,\mathrm{d}x + o(\varepsilon) \\
&= \varepsilon \int_{\mathbb{R}^{N-1}} |\nabla w(x,0)|^2 g(x) \,\mathrm{d}x - \varepsilon \int_{|\varepsilon x| \ge r_0} |\nabla w(x,0)|^2 g(x) \,\mathrm{d}x + o(\varepsilon).
\end{aligned}
$$

As in estimate (4.3) one has $\int_{|\varepsilon x| \ge r_0} |\nabla w(x,0)|^2 g(x) \,\mathrm{d}x = o(\varepsilon)$. Thus,

$$
R_1(\varepsilon) = -\varepsilon \int_{\mathbb{R}^{N-1}} |\nabla w(x,0)|^2 g(x) \,\mathrm{d}x + o(\varepsilon). \tag{4.4}
$$

To estimate $R_2(\varepsilon)$ we write $R_2(\varepsilon) = I_2(\varepsilon) - I_p(\varepsilon)$ where

$$
2I_2(\varepsilon) = \int_{\partial\Omega_\varepsilon} w^2 \,\mathrm{d}\sigma - \int_{\mathbb{R}^{N-1}} w^2 \,\mathrm{d}x \quad \text{and} \quad pI_p(\varepsilon) = \int_{\partial\Omega_\varepsilon} w^p \,\mathrm{d}\sigma - \int_{\mathbb{R}^{N-1}} w^p \,\mathrm{d}x.
$$

If $\Gamma_\varepsilon = \partial\Omega_\varepsilon \cap \mathcal{U}_\varepsilon$ we have

$$
2I_2(\varepsilon) = \int_{\Gamma_\varepsilon} w^2 \,\mathrm{d}\sigma - \int_{|\varepsilon x| \le r_0} w^2(x,0) \,\mathrm{d}x + \int_{\partial\Omega_\varepsilon \setminus \Gamma_\varepsilon} w^2 \,\mathrm{d}\sigma - \int_{|\varepsilon x| \ge r_0} w^2(x,0) \,\mathrm{d}x.
$$

It follows from the exponential decay of $w(x,t)$ in the variable x that

$$
\int_{|\varepsilon x| \ge r_0} w^2(x,0) \,\mathrm{d}x = o(\varepsilon). \tag{4.5}
$$

Setting $\tilde{\Omega_\varepsilon} = \Omega_\varepsilon \setminus (\Omega_\varepsilon \cap \mathcal{U}_\varepsilon)$ it follows from [3, Lemma 1.3] that

$$
\int_{\partial\Omega_\varepsilon \setminus \Gamma_\varepsilon} w^2 \,\mathrm{d}\sigma \le \int_{\partial\tilde{\Omega}_\varepsilon} w^2 \,\mathrm{d}\sigma \le S(\tilde{\Omega}_\varepsilon) \|w\|^2_{H^1(\tilde{\Omega}_\varepsilon)},
$$

where $S(\tilde{\Omega}_\varepsilon)$ is bounded independent of ε. Thus, using the same approach as in the proof to estimate $R_1(\varepsilon)$, we obtain

$$
\int_{\partial\Omega_\varepsilon \setminus \Gamma_\varepsilon} w^2 \,\mathrm{d}\sigma = o(\varepsilon). \tag{4.6}
$$

By estimates (4.5) and (4.6) it follows that

$$\begin{aligned} 2I_2(\varepsilon) &= \int_{\Gamma_\varepsilon} w^2 \, \mathrm{d}\sigma - \int_{|\varepsilon x| \le r_0} w^2(x,0) \, \mathrm{d}x + o(\varepsilon) \\ &= \int_{|\varepsilon x| \le r_0} \Big(w^2(x, \varepsilon^{-1} G(\varepsilon x)) \sqrt{1 + |\nabla G(\varepsilon x)|^2} - w^2(x,0) \Big) \, \mathrm{d}x + o(\varepsilon). \end{aligned} \tag{4.7}$$

Considering the function $f_\varepsilon(r) = w^2(x, r\varepsilon^{-1}G(\varepsilon x))\sqrt{1 + r^2|\nabla G(\varepsilon x)|^2}$ one can see that

$$f(1) - f(0) = w^2(x, \varepsilon^{-1}G(\varepsilon x))\sqrt{1 + |\nabla G(\varepsilon x)|^2} - w^2(x,0).$$

By the mean value theorem we obtain $0 \le r_\varepsilon \le 1$ such that

$$\begin{aligned} 2I_2(\varepsilon) = 2 \int_{|\varepsilon x| \le r_0} & w(x, r_\varepsilon \varepsilon^{-1} G(\varepsilon x)) w_t(x, r_\varepsilon \varepsilon^{-1} G(\varepsilon x)) \\ & \times \sqrt{1 + r_\varepsilon^2 |\nabla G(\varepsilon x)|^2} (\varepsilon^{-1} G(\varepsilon x)) \, \mathrm{d}x + o(\varepsilon). \end{aligned}$$

Since $\varepsilon^{-1}G(\varepsilon x)) = \varepsilon g(x) + o(\varepsilon)$ we get

$$\begin{aligned} I_2(\varepsilon) &= \varepsilon \int_{\mathbb{R}^{N-1}} w(x, r_\varepsilon \varepsilon^{-1} G(\varepsilon x)) w_t(x, r_\varepsilon \varepsilon^{-1} G(\varepsilon x)) \\ &\qquad \times \sqrt{1 + r_\varepsilon^2 |\nabla G(\varepsilon x)|^2} g(x) \chi_{\{|\varepsilon x| \le r_0\}} \, \mathrm{d}x + o(\varepsilon) \\ &= \varepsilon \int_{\mathbb{R}^{N-1}} w(x,0) w_t(x,0) g(x) \, \mathrm{d}x + o(\varepsilon). \end{aligned}$$

A similar argument, we obtain the following estimate to $I_p(\varepsilon)$,

$$I_p(\varepsilon) = \varepsilon \int_{\mathbb{R}^{N-1}} w^{p-1}(x,0) w_t(x,0) g(x) \, \mathrm{d}x + o(\varepsilon).$$

Thus, we concluded that

$$\begin{aligned} R_2(\varepsilon) &= \varepsilon \int_{\mathbb{R}^{N-1}} \left[w(x,0) - w^{p-1}(x,0) \right] w_t(x,0) g(x) \, \mathrm{d}x + o(\varepsilon) \\ &= \varepsilon \int_{\mathbb{R}^{N-1}} w_t^2(x,0) g(x) \, \mathrm{d}x + o(\varepsilon). \end{aligned} \tag{4.8}$$

The estimate for s_ε is now obvious in view of the estimates founded above. Indeed,

$$\begin{aligned} s_\varepsilon^{p-2} &= \frac{\int_{\Omega_\varepsilon} |\nabla w|^2 \, \mathrm{d}z + \int_{\partial\Omega_\varepsilon} w^2 \, \mathrm{d}\sigma}{\int_{\partial\Omega_\varepsilon} w^p \, \mathrm{d}\sigma} \\ &= \frac{R_1(\varepsilon) + \int_{\mathbb{R}^N_+} |\nabla w|^2 \, \mathrm{d}z + \int_{\mathbb{R}^{N-1}} w^2 \, \mathrm{d}x + 2I_2(\varepsilon)}{pI_p(\varepsilon) + \int_{\mathbb{R}^{N-1}} w^p \, \mathrm{d}x}. \end{aligned}$$

Since w is a solution of (P_∞), it follows that $s_\varepsilon^{p-2} = 1 + O(\varepsilon)$, which completes the proof. □

Proof of Proposition 4.1. As we are supposing that $z_0 = 0$, by estimate (2.4) with $v_0 = w(\varepsilon(x,t))$ we get

$$\begin{aligned} C_p(\Omega_\varepsilon) \leq \Bigg\{ &\frac{s_\varepsilon^2}{2}\Big(\int_{\mathbb{R}^N_+} |\nabla w|^2 \, dz + \int_{\Omega_\varepsilon} |\nabla w|^2 \, dz - \int_{\mathbb{R}^N_+} |\nabla w|^2 \, dz\Big) \\ &+ \frac{s_\varepsilon^2}{2}\Big(\int_{\mathbb{R}^{N-1}} w^2 \, dx + \int_{\partial\Omega_\varepsilon} w^2 \, d\sigma - \int_{\mathbb{R}^{N-1}} w^2 \, dx\Big) \\ &- \frac{s_\varepsilon^p}{p}\Big(\int_{\mathbb{R}^{N-1}} w^p \, dx + \int_{\partial\Omega_\varepsilon} w^p \, d\sigma - \int_{\mathbb{R}^{N-1}} w^p \, dx\Big)\Bigg\}, \end{aligned}$$

which together with the estimates obtained in Lemma 4.2 implies

$$\begin{aligned} C_p(\Omega_\varepsilon) \leq &\left\{ \frac{s_\varepsilon^2}{2}\left(\int_{\mathbb{R}^N_+} |\nabla w|^2 \, dz + \int_{\mathbb{R}^{N-1}} w^2 \, dx\right) - \frac{s_\varepsilon^p}{p}\int_{\mathbb{R}^{N-1}} w^p \, dx \right\} \\ &+ \frac{R_1(\varepsilon)}{2} + R_2(\varepsilon) + o(\varepsilon). \end{aligned}$$

Since w is a least energy solution of (P_∞), using once again Lemma 4.2 and Lemma 8.1 in the Appendix, we get

$$\begin{aligned} C_p(\Omega_\varepsilon) &\leq C_p(\mathbb{R}^N_+) - \varepsilon \int_{\mathbb{R}^{N-1}} \left(\frac{|\nabla w(x,0)|^2}{2} - w_t^2(x,0)\right) g(x) \, dx + o(\varepsilon) \\ &= C_p(\mathbb{R}^N_+) - \varepsilon \mathcal{H}(z)\gamma + o(\varepsilon) \end{aligned}$$

where (see Lemma 8.2)

$$\gamma = \int_{\mathbb{R}^{N-1}} \left(\frac{|\nabla w(x,0)|^2}{2} - w_t^2(x,0)\right) |x|^2 \, dx > 0 \tag{4.9}$$

and this completes the proof of Proposition 4.1. □

5. L^∞ estimates for solutions of (P_ε)

Next we adapt the Nash–Moser iterative methods to obtain L^∞ estimate for weak solutions of (P_ε) with uniform bounded energy.

Proposition 5.1. *There exists $\varepsilon_o > 0$ and a positive constant $C = C(\Omega, p, N)$ such that for all nonnegative solution u_ε of (P_ε) with $\varepsilon \in (0, \varepsilon_o)$, we have*

$$1 < \sup_{\overline{\Omega}} u_\varepsilon(\varepsilon^{-1}z) \leq C. \tag{5.1}$$

Proof. Let z_ε be such that $u_\varepsilon(z_\varepsilon) = \max_{\overline{\Omega}_\varepsilon} u_\varepsilon(z)$. It follows from Hopf's lemma that

$$0 < \frac{\partial u_\varepsilon}{\partial \eta}(z_\varepsilon) = u_\varepsilon^{p-1}(z_\varepsilon) - u_\varepsilon(z_\varepsilon),$$

which implies the first inequality in (5.1) because $p > 2$. The second inequality in (5.1) follows by one well-known Moser iteration method. □

If $A \subset \mathbb{R}^N$ is an open set we use the notation

$$J_A(u) = \frac{1}{2}\int_A |\nabla u|^2 dz + \frac{1}{2}\int_{\partial A} u^2 d\sigma - \frac{1}{p}\int_{\partial A} |u|^p d\sigma.$$

Lemma 5.2. *The function u_ε decay uniformly at infinity, namely, given $\eta > 0$ there exists an $R > 0$ such that $u_\varepsilon(z) < \eta$, if $|z - z^\varepsilon| > R$, where z^ε denotes any maximum point of u_ε in $\overline{\Omega}_\varepsilon$.*

Proof. By contradiction let us assume that for some $\eta > 0$, there are sequences $\varepsilon_k \to 0$ and $z^k \in \overline{\Omega}_{\varepsilon_k}$ (for short we denote Ω_{ε_k}, u_{ε_k} by $\overline{\Omega}_k$ and u_k respectively) such that

$$|z^{\varepsilon_k} - z^k| \to +\infty \quad \text{and} \quad u_k(z^k) \geq \eta. \tag{5.2}$$

We claim that

$$2C_p(\mathbb{R}^N_+) \leq \liminf J_{\Omega_k}(u_k) \tag{5.3}$$

which is a contradiction, because Proposition 4.1 implies

$$\limsup J_{\Omega_k}(u_k) \leq C_p(\mathbb{R}^N_+).$$

Thus it only remains to prove (5.3). Since (u_k) is uniformly bounded in $C^{1,\alpha}(\overline{\Omega}_k)$ we may assume up to a subsequence that $u_k(z^{\varepsilon_k} + z) \to u(z)$ uniformly over compacts subsets of $\mathbb{R}^N_+$ and u satisfies

$$\begin{cases} \Delta u = 0 & \text{in } \mathbb{R}^N_+, \\ \dfrac{\partial u}{\partial \eta} + u = u^p & \text{on } \mathbb{R}^{N-1}. \end{cases} \tag{5.4}$$

Since $u(0) = \lim_{k\to\infty} u_k(z^{\varepsilon_k}) \geq \eta$ we have that $u \geq 0$ and by the maximum principle $u > 0$. Since u is a nontrivial solution of (5.4), we have $C_p(\mathbb{R}^N_+) \leq J_{\mathbb{R}^N_+}(u) = J_{B_R}(u) + J_{B_R^c}(u)$. Now, using that $\lim_{R\to\infty} J_{B_R^c}(u) = 0$ and $\lim_{k\to\infty} J_{B_R}(u_k) = J_{B_R}(u)$, given $\delta > 0$, for all R sufficiently large we get

$$\lim_{k\to\infty} J_{B_R(z^{\varepsilon_k})\cap\Omega_k}(u_k) \geq \frac{C_p(\mathbb{R}^N_+)}{2} - \delta.$$

Similarly,

$$\lim_{n\to\infty} J_{B_R(z^k)\cap\Omega_k}(u_k) \geq \frac{C_p(\mathbb{R}^N_+)}{2} - \delta.$$

Let us consider $R > 0$ and a smooth cut-off function η_R^k with $0 \leq \eta_R^k \leq 1$ and $|\nabla \eta_R^k| \leq C$, where C is independent of R and k such that

$$\eta_R^k = 0 \quad \text{on} \quad B_{R-1}(z^{\varepsilon_k}) \cap B_{R-1}(z^k),$$

and

$$\eta_R^k \equiv 1 \quad \text{on} \quad \mathbb{R}^N_+ \setminus (B_{R-1}(z^{\varepsilon_k}) \cup B_{R-1}(z^k)).$$

Taking $w_k = \eta_R^k u_k$ as a test function to $J'_{\varepsilon_k}(u_k) = 0$ we obtain

$$\begin{aligned} 0 = J'_{\Omega_k}(u_k)w_k &= J'_{\Omega_k \cap A_R}(u_k)w_k + J'_{\Omega_k \setminus (B_1 \cup B_2)}(u_k)w_k \\ &= J'_{\Omega_k \cap A_R}(u_k)w_k + 2J_{\Omega_k \setminus (B_1 \cup B_2)}(u_k) - 2J_{\Omega_k \setminus (B_1 \cup B_2)}(u_k) + J'_{\Omega_k \setminus (B_1 \cup B_2)}(u_k)w_k \\ &= E_k - \int_{\partial\Omega_k \cap A_R} u_k^p \eta_R^k d\sigma + 2J_{\Omega_k \setminus (B_1 \cup B_2)}(u_k) + \left(\frac{2}{p} - 1\right) \int_{\partial\Omega_k \setminus (B_1 \cap B_2)} u_k^p d\sigma \end{aligned}$$

where

$$E_k = \int_{\Omega_k \cap A_R} \nabla u_k \nabla w_k dx + \int_{\partial\Omega_k \cap A_R} u_k^2 \eta_R^k d\sigma.$$

Since $p > 2$ we conclude that $0 \leq E_k + 2J_{\Omega_k \setminus (B_1 \cup B_2)}(u_k)$. Now, notice that $E_k \to 0$. Thus, $J_{\Omega_k \setminus (B_1 \cup B_2)}(u_k) \geq -\delta$. On the other hand,

$$J_{\Omega_k}(u_k) \quad = J_{\Omega_k \cap B_1}(u_k) + J_{\Omega_k \cap B_2}(u_k) + J_{\Omega_k \setminus (B_1 \cup B_2)}(u_k) \geq 2C_p(\mathbb{R}_+^N) - \delta - \delta,$$

which implies estimate (5.3) and this completes the proof. □

In order to establish the polynomial decay of u_ε we use $U(x) = (1 + |x|^2)^{\frac{2-N}{2}}$ solution of

$$-\Delta U = N(N-2)U^{\frac{N+2}{N-2}} \quad \text{in} \quad \mathbb{R}^N$$

to build a suitable test function for our arguments in the next result.

Lemma 5.3. *Let $u_\varepsilon \in C^\infty(\Omega_\varepsilon) \cap C^{1,\beta}(\overline{\Omega}_\varepsilon)$ be a positive solution of (P_ε). Then, there is a positive constant C_0 independent of ε such that*

$$u_\varepsilon(z) \leq \frac{C_0}{(1 + |z|^2)^{(N-2)/2}}, \quad \forall\, z \in \overline{\Omega}_\varepsilon.$$

Proof. Let us consider the function $W_\varepsilon(z) = u_\varepsilon(z)/U(z)$ in Ω_ε. We claim that W_ε is uniformly bounded. For this, notice that W_ε is a solution to the problem

$$\begin{cases} -\Delta W_\varepsilon - \displaystyle\sum_{i=1}^N b_i(z) \frac{\partial W_\varepsilon}{\partial z_i} + a(z) W_\varepsilon = 0 & \text{in } \Omega_\varepsilon, \\ \dfrac{\partial W_\varepsilon}{\partial \eta} + g_1(z) W_\varepsilon(z) - g_2(z) W_\varepsilon^{p-1} = 0 & \text{on } \partial\Omega_\varepsilon, \end{cases} \tag{5.5}$$

where

$$b_i(z) = \frac{2}{U(z)} \cdot \frac{\partial U(z)}{\partial z_i} = -\frac{2(N-2)z_i}{1 + |z|^2} \ (i = 1, \dots, N), \quad a(z) = \frac{N(N-2)}{(1 + |z|^2)^2}, \ z \in \Omega_\varepsilon$$

$$|g_1(z)| = \left| 1 + \frac{1}{U(z)} \cdot \frac{\partial U}{\partial \eta}(z) \right| \leq \left(1 + \frac{(N-2)|z|}{1 + |z|^2} \right) \quad \text{and} \quad g_2(z) = U^{p-2}(z).$$

One can see that there exists $C > 0$ independent of ε such that

$$\|a\|_{L^\infty(\Omega_\varepsilon)}, \|b_i\|_{L^\infty(\Omega_\varepsilon)}, \|g_1\|_{L^\infty(\Omega_\varepsilon)} \|g_2\|_{L^\infty(\Omega_\varepsilon)} \leq C,$$

for all $i = 1, \dots, N$. Assume by contradiction that there is a sequence $z_\varepsilon \in \overline{\Omega}_\varepsilon$ such that $W_\varepsilon(z_\varepsilon) \to +\infty$. By the weak maximum principle we may assume that $z_\varepsilon \in \partial\Omega_\varepsilon$ for all $\varepsilon > 0$. So, we define $M_\varepsilon = W_\varepsilon(z_\varepsilon)$. We have two cases to consider:

Case 1: (z_ε) is bounded. In this case let us consider

$$\widetilde{W}_\varepsilon(z) = \frac{W_\varepsilon(z_\varepsilon + M_\varepsilon^\alpha z)}{M_\varepsilon}, \quad z \in \widetilde{\Omega}_\varepsilon := M_\varepsilon^{-\alpha}(\Omega_\varepsilon - z_\varepsilon), \quad \alpha = \frac{2-p}{2}.$$

Since $\|\widetilde{W}_\varepsilon\|_{L^\infty(\Omega_\varepsilon)} \leq C$ (independent of ε), using the regularity result due to Lieberman [10], one can prove that

$$\|\widetilde{W}_\varepsilon\|_{C^{1,\beta}(\widetilde{\Omega}_\varepsilon)} \leq C, \tag{5.6}$$

for some $0 < \beta < 1$ and C positive constant independent of ε. Therefore, straightening out the boundary in a neighborhood of z_ε, one can prove that $\widetilde{\Omega}_\varepsilon \to \mathbb{R}^N_+$ as $\varepsilon \to 0$. Using (5.6), the Arzelà–Ascoli theorem and the diagonal argument we obtain a nonnegative function $\widetilde{W} \in C^{1,\beta/2}(\overline{\mathbb{R}^N_+})$ such that

$$\lim_{\varepsilon \searrow 0} \widetilde{W}_\varepsilon(z) = \widetilde{W}(z) \geq 0 \quad \text{and} \quad \widetilde{W}(0) = 1. \tag{5.7}$$

Since (z_ε) is bounded we can assume that $\lim_{\varepsilon\searrow 0} z_\varepsilon = 0 \in \partial\Omega$. Then, one readily deduces on any compact subset of $\overline{\mathbb{R}^N_+}$ that

$$\begin{aligned} &\lim_{\varepsilon\to 0} a(z_\varepsilon + M_\varepsilon^\alpha z) = a(0), \quad \lim_{\varepsilon\to 0} b_i(z_\varepsilon + M_\varepsilon^\alpha z) = b_i(0), \quad i = 1, \ldots, N, \\ &\lim_{\varepsilon\to 0} g_1(z_\varepsilon + M_\varepsilon^\alpha z) = g_1(0), \quad \lim_{\varepsilon\to 0} g_2(z_\varepsilon + M_\varepsilon^\alpha z) = g_2(0). \end{aligned} \tag{5.8}$$

It follows from (5.5)–(5.8), that the limit function $\widetilde{W} \geq 0$ satisfies the following limit problem

$$\begin{cases} -\Delta \widetilde{W} = 0 & \text{in } \mathbb{R}^N_+, \\ \dfrac{\partial \widetilde{W}(x,0)}{\partial t} = -\widetilde{W}^{p-1}(x,0) & \text{on } \mathbb{R}^{N-1}. \end{cases}$$

Since $2 < p < 2_* - 1$, we obtain by Theorem 1.2 in Hu [9] that $\widetilde{W} = 0$, which is a contradiction with the fact that $\widetilde{W}(0) = 1$.

Case 2: There exists a sequence $z_k = z_{\varepsilon_k}$ such that $|z_k| \to \infty$. In this case let us consider

$$v_k(z) = \frac{W_k(z_k + z)}{M_k} \qquad z \in \widetilde{\Omega}_k = \Omega_{\varepsilon_k} - z_{\varepsilon_k},$$

where $M_k = W_k(z_k) = \frac{u_k(z_k)}{U(z_k)}$. Notice that $M_k \to \infty$ and $\|v_k\|_{L^\infty(\widetilde{\Omega}_k)} \leq 1$. Moreover, v_k satisfies

$$\begin{cases} -\Delta v_k - \displaystyle\sum_{i=1}^N b_i(z_k + z)\frac{\partial v_k}{\partial z_i} + a(z_k + z)v_k = 0 & \text{in } \widetilde{\Omega}_k, \\ \dfrac{\partial v_k}{\partial \eta} + g_1(z_k + z)v_k(z) - u_k^{p-2}(z_k + z)v_k(z) = 0 & \text{on } \partial\widetilde{\Omega}_k. \end{cases} \tag{5.9}$$

Since $u_k^{p-2}(z_k + z) \to 0$, similarly one can see that $v_k \to v \in C^{1,\beta/2}(\overline{\mathbb{R}^N_+})$,

$$\lim_{n\to\infty} v_k(z) = v(z) \geq 0, \quad v(0) = 1 \tag{5.10}$$

and

$$\begin{cases} -\Delta v = 0 & \text{in } \mathbb{R}^N_+, \\ \dfrac{\partial v}{\partial \eta} = -v & \text{on } \mathbb{R}^{N-1}. \end{cases}$$

On the other hand, using the Hopf lemma at $z_0 = 0$ we get $0 < \frac{\partial v}{\partial \eta} = -v < 0$, which is a contradiction, and this conclude the proof. □

In order to obtain the exponential decay of u_ε let us consider the auxiliary function $v(x,t) = \varphi_0(x)\psi_0(t)$ where

$$\varphi_0(x) := e^{-\alpha(|x|)} \quad \text{and} \quad \psi_0(t) := \left(\frac{1}{1+t^2}\right)^{\frac{N-2}{2}},$$

and α are positive constants that will be selected below.

Lemma 5.4. *There exists $c > 0$ such that $u_\varepsilon(x,t) \le cv(x,t)$ for all $|x| \ge 1, t \ge 0$.*

Proof. Notice that v satisfies $-\Delta v + c(x,t)v = 0$, $x \in \mathbb{R}^{N-1} \setminus \{0\}$, $t \in \mathbb{R}$ where

$$c(x,t) = -\alpha\frac{(N-2)}{|x|} + \alpha^2 + \frac{(N-2)}{(1+t^2)^2}[(N-1)t^2 - 1].$$

Let us consider $V_\varepsilon = u_\varepsilon / v$. A straightforward calculation yields

$$-\Delta V_\varepsilon + 2\alpha \sum_{i=1}^{N-1} \frac{x_i}{|x|}(V_\varepsilon)_{x_i} + \frac{2(N-2)t}{(1+t^2)} V_t - c(x,t)V_\varepsilon = 0, \quad \Omega_\varepsilon \setminus \{0\}.$$

Consider the set $A_\varepsilon = \{(x,t) \in \mathbb{R}^N_+ : |x| \ge 1, \ t \ge 0\} \cap \Omega_\varepsilon$. We claim that there exists $C > 0$ independent of ε such that

$$\|V_\varepsilon\|_{L^\infty(A_\varepsilon)} \le C. \tag{5.11}$$

Suppose that (5.11) does not hold, that is, there exists $y_\varepsilon = (x_\varepsilon, t_\varepsilon) \in A_\varepsilon$ such that $V_\varepsilon(y_\varepsilon) \to +\infty$. From Lemma 5.3 we conclude that $|y_\varepsilon| \to +\infty$. Let $\eta_\varepsilon = (\eta_1(\varepsilon), \ldots, \eta_{N-1}(\varepsilon), \eta_{t_\varepsilon}) \in \mathbb{R}^N$ be the unit outward normal to $\partial\Omega_\varepsilon$ at $(x_\varepsilon, t_\varepsilon)$. Then, using the Hopf lemma we have $\partial V_\varepsilon / \partial \eta_\varepsilon(y_\varepsilon) > 0$. On the other hand, for ε sufficiently small,

$$\frac{\partial V_\varepsilon}{\partial \eta_\varepsilon}(y_\varepsilon) = \frac{1}{v}\left[\langle \nabla u_\varepsilon, \eta_\varepsilon\rangle - \frac{u_\varepsilon}{v}\langle \nabla v, \eta_\varepsilon\rangle\right] \le 0 \tag{5.12}$$

which is impossible. Thus it remains to prove (5.12). Notice that

$$\begin{aligned} \frac{\partial V_\varepsilon}{\partial \eta_\varepsilon}(y_\varepsilon) =& V_\varepsilon \left[u_\varepsilon^{p-1} - 1 + \alpha \sum_{i=1}^{N-1} \frac{x_i}{|x|}\eta_i(\varepsilon) + \frac{t_\varepsilon(N-2)}{1+t_\varepsilon^2}\eta_{t_\varepsilon}\right] \\ \le& V_\varepsilon \left[u_\varepsilon^{p-1} - 1 + \alpha + \frac{t_\varepsilon(N-2)}{1+t_\varepsilon^2}\eta_{t_\varepsilon}\right]. \end{aligned}$$

Since Ω is strictly convex $t_\varepsilon \to \infty$ as $\varepsilon \to 0$. Thus, taking ε sufficiently small, we can choose $\alpha > 0$ such that $u_\varepsilon^{p-1} - 1 + \alpha + \frac{t_\varepsilon(N-2)}{1+t_\varepsilon^2}\eta_{t_\varepsilon} \le -\frac{1}{2}$ and this completes the proof. □

6. Proof of Theorem 1.2

In this section we will complete the proof of Theorem 1.2, for that it will be crucial in our argument the following lower bound estimate on the minimax level $C_p(\Omega_\varepsilon)$ (see (6.1) below).

For any sequence $\varepsilon_k \to 0$, let $u_k = u_{\varepsilon_k}$ be the solution of (Q_{ε_k}) given in Proposition 2.3. Let us choose $z_k := z_{\varepsilon_k} \in \partial\Omega_{\varepsilon_k}$, such that $u_k(z_k) = \max_{z\in\overline{\Omega}_{\varepsilon_k}} u_k(z)$ we recall that since u_k is harmonic in Ω_{ε_k}, the maximum of u_k in $\overline{\Omega}_{\varepsilon_k}$ must be on $\partial\Omega_{\varepsilon_k}$. With this notation we have the following estimate.

Proposition 6.1. *There exists k_0 such that for $k \geq k_0$, it holds*

$$C_p(\Omega_{\varepsilon_k}) \geq C_p(\mathbb{R}^N_+) - \varepsilon_k \gamma \mathcal{H}(\varepsilon_k z_k) + o(\varepsilon_k). \tag{6.1}$$

Finalizing the proof of Theorem 1.2. Combining Propositions 4.1 and 6.1, we obtain (1.8). From (1.7), (1.8) and Taylor's theorem we obtain (1.9). Moreover, since $\gamma > 0$ we conclude that $H(\overline{z}) \geq \mathcal{H}(z)$, for all $z \in \partial\Omega$. Therefore, $\mathcal{H}(\varepsilon z_\varepsilon) \to \max_{z\in\partial\Omega} \mathcal{H}(z)$, which completes the proof of Theorem 1.2. □

6.1. Proof of Proposition 6.1

Up to a subsequence, we may assume that there exists $\overline{z} \in \partial\Omega$ such that $y_k = \varepsilon_k z_k \to \overline{z}$. Define $u_k(y) = u_{\varepsilon_k}(y + y_k)$, $y \in \Omega_{\varepsilon_k} - y_k$. Thus, after applying suitable rotation and translation, we may assume that $\overline{z} = 0$ and $\Omega \subset \mathbb{R}^N_+$ can be described in a fixed neighborhood $\mathcal{U}$ of $\overline{z}$ as $\{(x,t) : t > G_k(x)\}$ with G_k smooth $G_k(0) = 0$ and $\nabla G_k(0) = 0$. We can take G_k such that G_k converges in C^2_{loc}-topology to G the corresponding parametrization of $\partial\Omega$ at $\overline{z}$. We define $\Omega_k = \Omega_{\varepsilon_k}$ and $\mathcal{U}_k = {\varepsilon_k}^{-1}\mathcal{U}$, and we set

$$\mathcal{V}_k := \{(x,t) \in \mathbb{R}^N : |\varepsilon_k x| \leq r_0 \quad \text{and} \quad 0 \leq \varepsilon_k t \leq t_k\} \subset \mathcal{U}_k$$

where $t_k = \min_{|x|=r_0} G_k(x) > 0$. Since $C_p(\Omega_k) = I_{\Omega_k}(u_k) \geq I_{\Omega_k}(su_k)$ for all $s > 0$, using the decay of u_k, we get $C_p(\Omega_k) \geq I_{\mathcal{V}_k\cap\Omega_k}(su_k) + o(\varepsilon_k)$ for all $s > 0$ where $\Gamma_k = \mathcal{V}_k \cap \partial\Omega_k$ and

$$I_{\mathcal{V}_k}(u_k) = \frac{1}{2}\int_{\mathcal{V}_k\cap\Omega_k} |\nabla u_k|^2 \, dz + \frac{1}{2}\int_{\Gamma_k} |u_k|^2 \, d\sigma - \frac{1}{p}\int_{\Gamma_k} |u_k|^p \, d\sigma.$$

Now we extend u_k to $\mathcal{V}_k$ by defining $\overline{u}_k(x,t)$ in the following way:

$$\overline{u}_k(x,t) := \begin{cases} u_k(x,t) & \text{if} \quad \varepsilon_k t \geq G_k(\varepsilon_k x), \\ u_k(x, \varepsilon_k^{-1} G_k(\varepsilon_k x)) & \\ \quad +(G_k(\varepsilon_k x) - \varepsilon_k t) & \text{if} \quad \varepsilon_k t < G_k(\varepsilon_k x). \\ \qquad \times \left[u_k^p(x, G_k(\varepsilon_k x)) - u_k(x, G_k(\varepsilon_k x))\right] & \end{cases}$$

Using again the decay of u_k we have $C_p(\Omega_k) \geq I_{\mathcal{V}_k}(s\overline{u}_k) - I_{\mathcal{V}_k\setminus(\Omega_k\cap\mathcal{V}_k)}(s\overline{u}_k) + o(k)$. Passing to a subsequence, we may assume that $u_k \to w$ in H^1, where w is a least-energy solution of (P_∞). Now, let $s_k > 0$ be such that $I_{\mathcal{V}_k}(s_k\overline{u}_k) =$

$\sup_{s>0} I_{\mathcal{V}_k}(s\overline{u}_k)$. One can see that $s_k \to 1$ and $I_{\mathcal{V}_k}(s_k\overline{u}_k) \geq C_p(\mathbb{R}^N_+) + o(k)$. From these facts, it follows that

$$C_p(\Omega_k) \geq C_p(\mathbb{R}^N_+) - R_1(k) + R_2(k) + o(\epsilon_k), \tag{6.2}$$

where

$$R_1(k) := \frac{1}{2}\int_{\mathcal{V}_k \setminus (\Omega_k \cap \mathcal{V}_k)} |s_k \nabla \overline{u}_k|^2 \, dz,$$

$$R_2(k) := \frac{1}{2}\int_{\Gamma_k} |s_k u_k|^2 \, dx - \frac{1}{p}\int_{\overline{\mathcal{V}}_k \cap \mathbb{R}^{N-1}} (s_k\overline{u}_k)^p \, dx.$$

Thus we can proceed as in the proof of Proposition 4.1, to obtain the following estimates:

$$2R_1(k) = -\epsilon_k \int_{\mathbb{R}^{N-1}} |\nabla w(x,0)|^2 g(x)\, dx + o(\epsilon_k),$$

$$R_2(k) = \epsilon_k \int_{\mathbb{R}^{N-1}} w_t^2(x,0) g(x)\, dx + o(\epsilon_k),$$

which together with (6.2) implies that estimate (6.1) holds. Thus, we obtain

$$C_p(\Omega_k) \geq C_p(\mathbb{R}^N_+) - \varepsilon_k \int_{\mathbb{R}^{N-1}} \left[\frac{|\nabla w(x,0)|^2}{2} - w_t^2(x,0)\right] g(x)\, \mathrm{d}x + o(\varepsilon_k).$$

This, together with Lemma 8.1 yields $C_p(\Omega_k) \geq C_p(\mathbb{R}^N_+) - \varepsilon_k \gamma \mathcal{H}(\varepsilon_k z_{\varepsilon_k}) + o(\varepsilon_k)$, and this completes the proof. □

7. Proof of Theorem 1.3

In this section we prove the nonexistence of nonconstant positive solution of (P_ε), for ε sufficiently large. Define $\overline{u} = \frac{1}{|\partial\Omega|}\int_{\partial\Omega} u \, \mathrm{d}\sigma$ = average of u over $\partial\Omega$. Using standard argument one can prove the following Poincaré inequality.

Lemma 7.1. *There exists a constant C, depending only on N such that*

$$\|u - \overline{u}\|_{L^2(\partial\Omega)} \leq C\|\nabla u\|_{L^2(\Omega)}, \ \forall u \in H^1(\Omega).$$

Finalizing the proof of Theorem 1.3. We decompose u as $u = \overline{u} + \varphi$, where

$$\overline{u} = \frac{1}{|\partial\Omega|}\int_{\partial\Omega} u \, \mathrm{d}\sigma \text{ and } \int_{\partial\Omega} \varphi \, \mathrm{d}\sigma = 0.$$

Observing that $u^{p-1} - \overline{u}^{p-1} = (p-1)\left(\int_0^1 (\overline{u} + t\varphi)^{p-2}\, dt\right)\varphi$, and using φ as test function in (2.1), we obtain

$$\varepsilon \int_\Omega |\nabla\varphi|^2 \, \mathrm{d}z + \int_{\partial\Omega} \varphi^2 \, \mathrm{d}\sigma = (p-1)\int_{\partial\Omega}\left(\int_0^1 (\overline{u} + t\varphi)^{p-2} dt\right)\varphi^2 \, \mathrm{d}\sigma,$$

which together with Proposition 5.1 implies that

$$\varepsilon \int_\Omega |\nabla\varphi|^2 \, \mathrm{d}z \leq \varepsilon \int_\Omega |\nabla\varphi|^2 \, \mathrm{d}z + \int_{\partial\Omega} \varphi^2 \, \mathrm{d}\sigma \leq C\int_{\partial\Omega} \varphi^2 \, \mathrm{d}\sigma.$$

Using Lemma 7.1 we obtain that φ must be constant for ε sufficiently large. Thus $0 = \int_{\partial\Omega} \varphi \, d\sigma = |\partial\Omega|\varphi$. Therefore $\varphi \equiv 0$ and the proof of Theorem 1.3 is complete. □

8. Appendix (Mean curvature)

Given a positive solution w of (P_∞), as in [4] we define its *restricted energy density* as

$$E(w, y) = \left(\frac{|\nabla w|^2}{2} - w_t^2\right)(y, 0), \quad y \in \mathbb{R}^{N-1}.$$

Then we define the *generalized curvature* at $z = (x, t) \in \partial\Omega$, with $x \in \mathbb{R}^{N-1}$ and $t \in \mathbb{R}$ fixed as the following number:

$$H(z) = \max_{w \in \mathcal{S}} \int_{\mathbb{R}^{N-1}} \langle D^2G(x)y, y\rangle \, E(w, y) \, dy,$$

where $\langle , \rangle$ denotes the usual inner product in $\mathbb{R}^{N-1}$, and $D^2G(x)$ denotes the Hessian matrix of G at x, and $\mathcal{S}$ denotes the set of positive solutions of (P_∞). In [1], the authors have proved there exists a positive constant C such that for all $w \in \mathcal{S}$ we have $\|w\|_{L^\infty(\Omega)} \leq C$ and $w(y, 0)$ is radially symmetric in the variable y. Thus H does not depend on the particular choice of G, but only of z.

In order to relate the *generalized curvature* to the *mean curvature* of $\partial\Omega$ at z for sake of completeness we recall here a few important features about the mean curvature (cf. Trudinger [8]). The eigenvalues of $D^2G(x)$, $\lambda_1, \ldots, \lambda_{N-1}$ are called the *principal curvatures* of $\partial\Omega$ at z and the corresponding eigenvectors are called the *principal directions* of $\partial\Omega$ at z. Furthermore, the *mean curvature* of $\partial\Omega$ at $z = (x, t)$ is given by

$$\mathcal{H}(z) = \frac{1}{N-1} \sum_{i=1}^{N-1} \lambda_i = \frac{-1}{N-1} \operatorname{div}\left(\frac{(-\nabla G(x), 1)}{\sqrt{1 + |\nabla G(x)|^2}}\right) = \frac{1}{N-1} \Delta G(x),$$

whenever $\nabla G(x) = 0$. On the other hand, by a rotation of coordinates we may assume that the $x_1, \ldots, x_{N-1}$ axes lie along principal directions corresponding to $\lambda_1, \ldots, \lambda_{N-1}$ at z. So, the Hessian matrix can be described as

$$D^2G(x) = \begin{bmatrix} \lambda_1 & 0 & \cdots & 0 \\ 0 & \lambda_2 & \cdots & 0 \\ \vdots & \vdots & \ddots & \vdots \\ 0 & 0 & \cdots & \lambda_{N-1} \end{bmatrix}.$$

Thus, at $z = (0, 0)$ we have

$$\int_{\mathbb{R}^{N-1}} \langle D^2G(0)y, y\rangle \, E(w, y) \, dy = \sum_{i=1}^{N-1} \int_{\mathbb{R}^{N-1}} \lambda_i y_i^2 E(w, y) \, dy.$$

By the definition of the *mass moment of inertia* we have that the *moment of inertia* about the y_i-axis, $i = 1, \ldots, N-1$, respectively the *polar moment of inertia* are given by

$$I_{y_i} = \int_{\mathbb{R}^{N-1}} y_i^2 E(w,y)\,\mathrm{d}y, \quad I_0 = \sum_{i=1}^{N-1} I_{y_i} = \sum_{i=1}^{N-1} \int_{\mathbb{R}^{N-1}} y_i^2 E(w,y)\,\mathrm{d}y,$$

respectively. Now, using the fact of $E(w,y)$ is a symmetric function, we conclude that $I_{y_1} = \cdots = I_{y_{N-1}}$, which implies that $I_0 = (N-1)I_{y_1}$. With this notation we have

Lemma 8.1. *For $w \in \mathcal{S}$ it holds,*

$$\int_{\mathbb{R}^{N-1}} \langle D^2G(0)y, y\rangle\, E(w,y)\,\mathrm{d}y = \mathcal{H}(0)\omega_{N-2} \int_0^{+\infty} E(w,r)r^N\,\mathrm{d}r.$$

Proof. Notice that

$$\sum_{i=1}^{N-1} \lambda_i I_{y_i} = I_{y_1} \sum_{i=1}^{N-1} \lambda_i = \left(\frac{1}{N-1}\sum_{i=1}^{N-1} \lambda_i\right) I_0 = \mathcal{H}(0) \int_{\mathbb{R}^{N-1}} |y|^2 E(w,y)\,\mathrm{d}y,$$

which implies the desired result. □

Lemma 8.2. *The constant γ defined in* (4.9) *is positive.*

Proof. Here we proceed as in [5]. Taking $\varphi(x,t) = |x|^2 w_t e^{\lambda t w_t^+}$ as a test function in (P_∞) we get

$$\begin{aligned} 0 = \int_{\mathbb{R}^N_+} \Bigg[& 2x.\nabla_x w w_t + |x|^2 \left(\frac{|\nabla w|^2}{2}\right)_t \\ & + \lambda |x|^2 w_t (\nabla_x w \nabla w_t^+ + w_t^+(t w_{tt}^+ + w_t^+))\Bigg] e^{\lambda t w_t^+}\,\mathrm{d}z \\ + \int_{\mathbb{R}^{N-1}} & |x|^2 w_t^2\,\mathrm{d}x, \end{aligned}$$

where we are using the notation $w_t^+ = (w_t)^+$ and $w_{tt}^+ = (w_t^+)_t$ in the weak sense. Integrating by partes we can estimate $\gamma = \int_{\mathbb{R}^N_+ \cap \{w_t < 0\}} [2w_t x \cdot \nabla_x w]\,\mathrm{d}z + o(\lambda)$ as $\lambda \to -\infty$. Taking into account that $x \cdot \nabla_x = r w_r < 0$ we get $\gamma > 0$, which implies the desired result. □

References

[1] E. Abreu, J.M. do Ó, E. Medeiros, *Properties of positive harmonic functions on the half-space with a nonlinear boundary condition*, J. Differential Equations **248** (2010), 617–637.

[2] Adimurthi, S.L. Yadava, *Some remarks on Sobolev type inequalities*, Calc. Var. Partial Differential Equations **2** (1994), 427–442.

[3] J.F. Bonder, J.D. Rossi, *Asymptotic behavior of the best Sobolev trace constant in expanding and contracting domains*, Commun. Pure Appl. Anal., **1** (2002), 359–378.

[4] M. del Pino, P. Felmer, *Spike-layered of singularly perturbed elliptic problems in a degenerate setting*, Indiana Univ. Math. J., **48** (1999), 883–898.

[5] M. del Pino, C. Flores, *Asymptotic behavior of best constants and extremals for trace embeddings in expanding domains*, Comm. Partial Differential Equations, **26** (2001), 2189–2210.

[6] W.Y. Ding, W.M. Ni, *On the existence of positive entire solutions of a semilinear elliptic equation*, Arch. Rational Mech. Anal., **91** (1986), 283–308.

[7] J.F. Escobar, *Sharp constant in a Sobolev trace inequality*, Indiana Univ. Math. J., **37** (1988), 687–698.

[8] Gilbarg, D., Trudinger, N.S., *Elliptic Partial Differential Equations of Second Order* Berlin-Heidelberg-New York-Tokyo, Springer-Verlag 1983.

[9] B. Hu, *Nonexistence of a positive solution of the Laplace equation with a nonlinear boundary condition*, Differential Integral Equations, **7** (1994), 301–313.

[10] G.M. Lieberman, *Boundary regularity for solutions of degenerate elliptic equations*, Nonlinear Anal., **12** (1988), 1203–1219.

[11] W.M. Ni, I. Takagi, *On the shape of least-energy solution to a semilinear Neumann problem*, Comm. Pure Appl. Math. **41** (1991) 819–851.

[12] D. Pierotti, S. Terracini, S., *On a Neumann problem with critical exponent and critical nonlinearity on the boundary*, Comm. Partial Differential Equations, **20** (1995), 1155–1187.

[13] M. Struwe, M., Variational methods. *Applications to nonlinear partial differential equations and Hamiltonian systems.* Third edition. Springer-Verlag, Berlin 2000.

Emerson Abreu
Departamento de Matemática
Universidade Federal de Minas Gerais
CP702
30161-970 Belo Horizonte, MG, Brazil

João Marcos do Ó and Everaldo Medeiros
Departamento de Matemática
Universidade Federal da Paraíba
58051-900 João Pessoa, PB, Brazil

e-mail: `jmbo@pq.cnpq.br`
`everaldo@mat.ufpb.br`

Progress in Nonlinear Differential Equations
and Their Applications, Vol. 85, 23–38

Multiplicity of Positive Solutions for an Obstacle Problem in $\mathbb{R}$

Claudianor O. Alves and Francisco Julio S.A. Corrêa

Dedicated to Bernhard Ruf on the occasion of his 60th birthday

Abstract. In this paper we establish the existence of two positive solutions for the obstacle problem

$$\int_{\mathbb{R}} [u'(v-u)' + (1+\lambda V(x))u(v-u)] \geq \int_{\mathbb{R}} f(u)(v-u), \forall v \in \mathbb{K}$$

where f is a continuous function verifying some technical conditions and $\mathbb{K}$ is the convex set given by

$$\mathbb{K} = \left\{v \in H^1(\mathbb{R}); v \geq \varphi\right\},$$

with $\varphi \in H^1(\mathbb{R})$ having nontrivial positive part with compact support in $\mathbb{R}$.

Mathematics Subject Classification (2010). 34B18, 35A15, 46E39.

Keywords. Obstacle problem, variational methods, positive solutions.

1. Introduction

In this paper we will be concerned with the question of existence of positive solutions of a kind of obstacle problem. This class of problems has been largely studied due both its mathematical interest and its physical applications. For example, it appears in mechanics, engineering, mathematical programming and optimization, among other things. See, for instance, the classical books Kinderlehrer & Stampacchia [12], Rodrigues [18] and Troianiello [24] and the references therein.

The typical obstacle problem is as follows: Let Ω be a domain in $\mathbb{R}^N$. Given functions $g : \mathbb{R} \to \mathbb{R}$ and $\varphi : \Omega \to \mathbb{R}$, finding $u \in H_0^1(\Omega)$ satisfying

$$\int_{\Omega} \nabla u \cdot \nabla(v-u) \geq \int_{\Omega} g(u)(v-u) \tag{P}$$

The first author was partially supported by CNPq/Brazil 620150/2008-4 and 303080/2009-4. The second author was Partially supported by CNPq/Brazil 300561/2010-5.

for all function v in the convex set

$$\mathbb{K} := \{v \in H_0^1(\Omega); v(x) \geq \varphi(x) \ \ a.e. \ \Omega\} \tag{1.1}$$

where φ is called the obstacle function.

Related to this kind of problem, the reader may consult Jianfu ([10], [11]), where the author uses variational methods, Le [13] in which is used subsolution-supersolution techniques, Chang [4] where it is considered an obstacle problem related to discontinuous nonlinearities and Rodrigues [19] who considers combination of the obstacle problem with nonlocal equations in a class of free boundary problems. For more recent references we may cite Matzeu & Servadei [16], in which the authors adapt for inequalities the iterative technique contained in de Figueiredo, Girardi & Matzeu [6] for elliptic equations, Matzeu & Servadei [17] where the stability of solutions obtained in [16] are analized. Other results may be found in Servadei & Valdinoci [22], Mancini & Musina [15], Servadei ([21], [20]), Magrone, Mugnai & Servadei [14].

These works and the references therein show clearly the mathematical importance and the wide variety of practical situations in which obstacle problems may be found and applied.

Here we are interested in the unidimensional counterpart of problem (P). More precisely, we consider the problem

$$\int_{\mathbb{R}} [u'(v-u)' + (1+\lambda V(x))u(v-u)] \geq \int_{\mathbb{R}} f(u)(v-u), \forall v \in \mathbb{K}, \tag{P_λ}$$

where u is a nonnegative function belonging to the convex set $\mathbb{K}$ given by

$$\mathbb{K} := \{v \in H^1(\mathbb{R}); v \geq \varphi\}, \tag{$\mathbb{K}$}$$

where $\varphi \in H^1(\mathbb{R})$ is assumed to have nontrivial positive part, that is, $\varphi_+ = \max\{\varphi\,, 0\} \not\equiv 0$. Moreover, $\lambda > 0$ is a parameter and $f : \mathbb{R} \to \mathbb{R}$ is a nondecreasing continuous function verifying the following assumptions:

$$\frac{f(t)}{t} \to 0 \text{ as } |t| \to 0 \tag{f_1}$$

and the Ambrosetti & Rabinowitz Condition, that is, there is $\theta > 2$ such that

$$0 < \theta F(t) \leq f(t)t \ \ \forall t \in \mathbb{R} \setminus \{0\} \tag{f_2}$$

where $F(t) = \int_0^t f(s)ds$. We assume that $V : \mathbb{R} \to \mathbb{R}$ is a nonnegative continuous function such that

$$\mathcal{O} := int \ \left((V^{-1}(\{0\}))\right) \neq \emptyset$$

is a bounded open set of $\mathbb{R}$ containing the support of φ_+, that is, Supp $(\varphi_+) \subset \mathcal{O}$. Here, Supp$(\varphi_+)$ denotes the support of φ_+ and

$$V^{-1}(\{0\}) = \{x \in \mathbb{R}; V(x) = 0\}.$$

The present paper was motivated by recent works involving the following class of problems

$$\begin{cases} -\Delta u + (1+\lambda V(x))u = f(u) & \text{in } \mathbb{R}^N \\ u(x) > 0 & \text{in } \mathbb{R}^N \end{cases}$$

where λ is a positive parameter, $V : \mathbb{R}^N \to \mathbb{R}$ is a nonnegative function and f is a continuous function satisfying some technical conditions. The reader may find more details in the papers of Alves [1], Bartsch & Wang [3], Clapp & Ding [5], Ding & Tanaka [7] and their references. Here, we adapt some approaches found in these references to study the obstacle problem (P_λ).

Our main result is the following

Theorem 1.1. *Suppose (f_1)–(f_2) hold, then there are $r, \lambda^* > 0$, such that if $\|\varphi_+\|_{H^1(\mathbb{R})} < \frac{r}{2}$, problem (P_λ) has two positive solutions for all $\lambda > \lambda^*$.*

One of the main difficulties to prove Theorem 1.1 is related to the fact that the energy functional associated with the problem (P_λ) does not satisfy in general the well-known Palais–Smale condition, once that we are working in whole $\mathbb{R}$. To overcome this difficulty, we adapt some ideas found in del Pino & Felmer [8], modifying the function f outside the set $\mathcal{O}$, in such way that the energy functional of the modified obstacle problem satisfies the Palais–Smale condition. Using variational methods, we prove the existence of two solutions for the modified obstacle problem. After that, it is proved that under the hypotheses of Theorem 1.1, the solutions found are solutions of the original obstacle problem.

The structure of this paper is as follows: In Section 2 we introduce the modified obstacle problem, in Section 3 we establish the existence of a first solution for the modified obstacle problem by minimization, in Section 4 we show the existence of a second solution for the modified obstacle problem by the Mountain Pass Theorem and in Section 5 we prove Theorem 1.1.

2. The modified obstacle problem

From this time onwards, since we intend to find positive solution, we will assume, without loss of generality, that

$$f(t) = 0 \ \forall t \leq 0.$$

To prove the existence of positive solutions for (P_λ), we will work with a modified obstacle problem, following some ideas found in del Pino & Felmer [8]. To this end, we consider the function $h : \mathbb{R} \to \mathbb{R}$ as follows:

$$h(t) = \begin{cases} f(t) & \text{if } t \leq a, \\ \frac{1}{k}t & \text{if } t \geq a, \end{cases}$$

where $k > \max\left\{\frac{\theta}{\theta-2}, 2\right\}$ and $a > 0$ satisfy $\frac{f(a)}{a} = \frac{1}{k}$. We now set

$$g(x,t) = \chi(x)f(t) + (1-\chi(x))h(t),$$

where $\Omega \subset \mathbb{R}$ is a bounded open set containing $\mathcal{O}$ and χ is the characteristic function of the set Ω, that is,

$$\chi(x) = \begin{cases} 1, & x \in \Omega \\ 0, & x \in \Omega^c. \end{cases}$$

Using the function g, we will show the existence of two positive solutions for the obstacle problem

$$(P_A) \qquad \int_{\mathbb{R}} [u'(v-u)' + (1+\lambda V(x))u(v-u)] \geq \int_{\mathbb{R}} g(x,u)(v-u), \ \forall v \in \mathbb{K}.$$

Remark 2.1. *If u is a solution of (P_A) verifying*

$$u(x) \leq a, \ \forall x \in \Omega^c,$$

then u is a solution of the original obstacle problem. Indeed, if $x \in \Omega$, we have $\chi(x) = 1$ and so

$$g(x, u(x)) = f(u(x)).$$

If $x \notin \Omega$ $(x \in \Omega^c)$, then $\chi(x) = 0$ and so

$$g(x, u(x)) = h(u(x)) = f(u(x)),$$

because $h(u(x)) = f(u(x))$ since $0 \leq u(x) \leq a$ in Ω^c.

Let $E_\lambda \subset H^1(\mathbb{R})$ be the subspace

$$E_\lambda = \left\{ u \in H^1(\mathbb{R}); \int_{\mathbb{R}} V(x)u^2 < \infty \right\}$$

endowed with the norm

$$\|u\|_\lambda = \left(\int_{\mathbb{R}} [|u'|^2 + (1+\lambda V(x))|u|^2] \right)^{\frac{1}{2}}.$$

Hereafter, we denote by $\| \ \|$ the usual norm in $H^1(\mathbb{R})$.

Since we approach our problem by means of variational method, we consider the energy functional associated with the obstacle problem (P_A), $I_\lambda : E_\lambda \to \mathbb{R}$, given by

$$I_\lambda(u) = \frac{1}{2}\|u\|_\lambda^2 - \int_{\mathbb{R}} G(x,u) + \Psi(u),$$

where

$$G(x,t) = \int_0^t g(x,s)ds$$

and $\Psi : E \to (-\infty, \infty]$ is the indicatrix function of the set $\mathbb{K}$, i.e.,

$$\Psi(u) = 0, \ \forall u \in \mathbb{K} \ \text{and} \ \Psi(u) = +\infty, \ \forall u \in \mathbb{K}^c. \tag{2.1}$$

Proposition 2.1. *The functional I_λ satisfies the* (PS) *condition.*

Proof. Let $d \in \mathbb{R}$ and $(u_n) \subset E_\lambda$ be a $(\mathrm{PS})_d$ sequence for I_λ. Then, there is $(z_n) \subset E'_\lambda$ with $z_n \to 0$ such that

$$I_\lambda(u_n) \to d \text{ and } I'_\lambda(u_n)(v - u_n) \geq \langle z_n, v - u_n \rangle \ \forall n \in \mathbb{N} \text{ and } v \in \mathbb{K},$$

that is,

$$\int_{\mathbb{R}} u'_n(v - u_n)' + (1 + \lambda V(x))u_n(v - u_n) - \int_{\mathbb{R}} g(x, u_n)(v - u_n) \geq \langle z_n, v - u_n \rangle,$$

for all $v \in \mathbb{K}$.

Claim 2.1. *(u_n) is a bounded sequence in E_λ.*

We deal separately with the sequences (u_{n+}) and (u_{n-}), where $u_{n-} = \max\{-u_n, 0\}$. Since $u_n = u_{n+} - u_{n-}$, it is enough to show that (u_{n+}) and (u_{n-}) are bounded in E_λ. To show the boundedness of (u_{n-}), we consider the test function $v = u_n + u_{n-} \in \mathbb{K}$. So,

$$\int_{\mathbb{R}} (u'_n(u_{n-})' + (1 + \lambda V(x))u_n u_{n-}) - \int_{\mathbb{R}} g(x, u_n)u_{n-} \geq \langle z_n, u_{n-} \rangle.$$

Because $\displaystyle\int_{\mathbb{R}} (1 + \lambda V(x))u_{n+}u_{n-} = \int_{\mathbb{R}} g(x, u_n)u_n^- = 0$, we obtain

$$-\|u_{n-}\|_\lambda^2 \geq \langle z_n, u_{n-} \rangle,$$

which leads to

$$\|u_{n-}\|_\lambda^2 \leq \|z_n\| \|u_{n-}\|_\lambda.$$

As $z_n \to 0$ in E'_λ, we conclude that $u_{n-} \to 0$ in E_λ, and thus, (u_{n-}) is bounded in E_λ.

With respect to (u_{n+}), fixing the test function $v = u_n + u_{n+} \in \mathbb{K}$, we derive that

$$\|u_{n+}\|_\lambda^2 - \int_{\mathbb{R}} g(x, u_n)u_n^+ \geq \langle z_n, u_n^+ \rangle, \tag{2.2}$$

leading to

$$-\int_\Omega f(u_n)u_n \geq -\|u_{n+}\|_\lambda^2 + \langle z_n, u_n^+ \rangle. \tag{2.3}$$

On the other hand, we know that

$$d = \frac{1}{2}\|u_n\|_\lambda^2 - \int_\Omega F(u_n) - \int_{\Omega^c} G(x, u_n) + o_n(1).$$

Using the definition of g, it is easy to prove that

$$2G(x, t) \leq g(x, t)t \leq \frac{1}{k}(1 + \lambda V(x))|t|^2 \ \forall x \in \Omega^c \text{ and } t \in \mathbb{R}. \tag{2.4}$$

Thereby, from (f_2) and (2.4)

$$d \geq \frac{1}{2}\|u_{n+}\|_\lambda^2 - \frac{1}{\theta}\int_\Omega f(u_n)u_n - \frac{1}{2k}\int_{\Omega^c} (1 + \lambda V(x))|u_n|^2 + o_n(1). \tag{2.5}$$

Combining (2.3) and (2.5),

$$d \geq \left[\left(\frac{1}{2}-\frac{1}{\theta}\right)-\frac{1}{2k}\right]\|u_{n+}\|_{\lambda}^{2}-\|z_n\|\|u_{n+}\|+o_n(1).$$

Since $k > \frac{\theta}{\theta-2}$ and $z_n \to 0$ in E'_{λ}, the last inequality implies that (u_{n+}) is bounded in E_{λ}. Therefore, (u_n) is bounded in E_{λ}.

Now, we will show that (u_n) has a subsequence that converges strongly in E_{λ}. Since (u_{n-}) converges to 0 in E_{λ}, without loss of generality, we will assume that $u_n \geq 0$ for all $n \in \mathbb{N}$. We begin by fixing $R > 0$ so large in order that $\Omega \subset \left(-\frac{R}{2}, \frac{R}{2}\right)$ and a function $\eta \in C^1(\mathbb{R}, \mathbb{R})$ satisfying

- $0 \leq \eta(t) \leq 1, \forall t \in \mathbb{R}$;
- $\eta(t) = 0, |t| \leq \frac{R}{2}$;
- $\eta(t) = 1, |t| \geq R$;
- $|\eta'(t)| \leq \frac{C}{R}, \forall t \in \mathbb{R}$.

Claim 2.2. *Given $\delta > 0$, there is $R > 0$ such that*

$$\int_{|x|\geq R}(|u_n'|^2+|u_n|^2) < \delta.$$

Assuming that this claim is true, we continue with our proof. Considering the test function $v = u_n - \eta(u_n - \varphi_+) = u_n - \eta u_n \in \mathbb{K}$, it follows that

$$\int_{\mathbb{R}}[u_n'(\eta u_n)' + (1+\lambda V(x))u_n(\eta u_n)] \leq \int_{\mathbb{R}} g(x,u_n)(\eta u_n) + o_n(1)$$

or, equivalently,

$$\int_{\mathbb{R}}\eta|u_n'|^2 + \int_{\mathbb{R}}u_n'\eta' u_n + \int_{\mathbb{R}}(1+\lambda V(x))\eta|u_n|^2 \leq \int_{|x|\geq\frac{R}{2}} g(x,u_n)\eta u_n + o_n(1)$$

implying that

$$\begin{aligned}\int_{|x|\geq R}|u_n'|^2 + \int_{|t|\leq R}u_n'\eta' u_n + \int_{|x|\geq\frac{R}{2}}(1+\lambda V(x))\eta|u_n|^2 \\ \leq \int_{|x|\geq\frac{R}{2}}\frac{1}{k}(1+\lambda V(x))\eta|u_n|^2 + o_n(1).\end{aligned}$$

Because $k > 2$, it follows that

$$\begin{aligned}\int_{|x|\geq R}|u_n'|^2 + \int_{|t|\leq R}u_n'\eta' u_n + \int_{|x|\geq\frac{R}{2}}(1+\lambda V(x))\eta|u_n|^2 \\ \leq \int_{|x|\geq\frac{R}{2}}\left(\frac{1+\lambda V(x)}{2}\right)|u_n|^2 + o_n(1)\end{aligned}$$

and so,

$$\int_{|x|\geq R}|u_n'|^2 + \frac{1}{2}\int_{|x|\geq\frac{R}{2}}(1+\lambda V(x))\eta|u_n|^2 \leq \int_{|x|\leq R}|u_n'||\eta'||u_n| \leq \frac{C}{R} + o_n(1).$$

Thereby,

$$\int_{|x|\geq R} |u_n'|^2 + \int_{|x|\geq R} (1+\lambda V(x))|u_n|^2 \leq \frac{C}{R} + o_n(1),$$

showing that

$$\limsup_{n\to+\infty} \int_{|x|\geq R} (|u_n'|^2 + |u_n|^2) \leq \frac{C}{R}.$$

Now, we choose $R > 0$ so large in order

$$\limsup_{n\to+\infty} \int_{|x|\geq R} (|u_n'|^2 + |u_n|^2) < \delta,$$

proving the Claim 2.2.

Recalling that for each $R > 0$, the Sobolev embedding

$$H^1(\mathbb{R}) \hookrightarrow C([-R, R])$$

is compact, we have that

$$u_n \to u \text{ in } C([-R, R]).$$

This limit, combined with the Claim 2.2, asserts that

$$\int_{\mathbb{R}} g(x, u_n)u_n \to \int_{\mathbb{R}} g(x, u)u \tag{2.6}$$

and

$$\int_{\mathbb{R}} g(x, u_n)v \to \int_{\mathbb{R}} g(x, u)v \;\; \forall v \in \mathbb{K}, \tag{2.7}$$

where $u \in \mathbb{K}$ is the weak limit of (u_n) in E_λ.

Since (u_n) is a bounded Palais–Smale sequence for I_λ, we have

$$\int_{\mathbb{R}} u_n'(v-u_n)' + (1+\lambda V(x))u_n(v-u_n) \geq \int_{\mathbb{R}} g(x, u_n)(v-u_n) + o_n(1) \;\; \forall v \in \mathbb{K} \tag{2.8}$$

or equivalently

$$\int_{\mathbb{R}} [u_n'v' + (1+\lambda V(x))u_n v] \geq \int_{\mathbb{R}} [|u_n'|^2 + (1+\lambda V(x))|u_n|^2] + \int_{\mathbb{R}} g(x, u_n)(v-u_n) + o_n(1)$$

for all $v \in \mathbb{K}$. Taking the inferior limits on both sides of the above inequality and using (2.6) and (2.7), we get

$$\int_{\mathbb{R}} [u'v' + (1+\lambda V(x))uv] \geq \int_{\mathbb{R}} [|u'|^2 + (1+\lambda V(x))|u|^2] + \int_{\mathbb{R}} g(x, u)(v-u) + o_n(1)$$

that is,

$$\int_{\mathbb{R}} [u'(v-u)' + (1+\lambda V(x))u(v-u)] \geq \int_{\mathbb{R}} g(x, u)(v-u), \;\; \forall v \in \mathbb{K}$$

from where it follows that u is a critical point of I_λ.

Using u as a test function in (2.8) and the limit (2.7), it follows that

$$\limsup_{n\to+\infty} \|u_n\|_\lambda^2 \leq \|u\|_\lambda^2.$$

Since E_λ is a Hilbert space, the last inequality leads to $u_n \to u$ in E_λ, finishing the proof of proposition. □

3. First solution for (P_A)

The first positive solution of (P_A) will be obtained via Ekeland's Variational Principle [9]. In this section, we denote by B_r and $\mathbb{K}_r$ the following sets

$$B_r = \{u \in E_\lambda;\ \|u\|_\lambda < r\} \quad \text{and} \quad \mathbb{K}_r = \mathbb{K} \cap \overline{B}_r.$$

Theorem 3.1. *There is $r > 0$, such that if $\|\varphi_+\|_{H^1(\mathbb{R})} < \frac{1}{2}\sqrt{r}$, the variational problem*

$$m = \inf\{I_\lambda(u) : u \in \mathbb{K}_r\} \tag{3.1}$$

has a solution for all $\lambda > 0$. Moreover, this solution is a positive solution of (P_A).

Proof. First of all, we observe that

$$\int_{\mathbb{R}} G(x, u(x)) = \int_\Omega F(u) + \int_{\Omega^c} G(x, u(x)).$$

From (f_1), if $\|u\|_\lambda = r$ and r is small enough, we have that

$$\int_\Omega F(u) \leq \frac{1}{4}\int_\Omega |u|^2 \leq \frac{1}{4}\|u\|_\lambda^2.$$

Hence

$$\int_{\mathbb{R}} G(x, u(x)) \leq \frac{1}{4}\|u\|_\lambda^2 + \int_{\Omega^c} G(x, u(x)),$$

and so, by (2.4),

$$\int_{\mathbb{R}} G(x, u(x)) \leq \frac{1}{4}\|u\|_\lambda^2 + \frac{1}{2k}\int_{\Omega^c} (1 + \lambda V(x))|u|^2.$$

Thereby,

$$I_\lambda(u) \geq \frac{1}{4}\|u\|_\lambda^2 - \frac{1}{2k}\int_{\Omega^c} (1 + \lambda V(x))|u|^2 + \Psi(u) \tag{3.2}$$

from where it follows that

$$I_\lambda(u) \geq \left(\frac{1}{4} - \frac{1}{2k}\right)\|u\|_\lambda^2 + \Psi(u), \forall u \in E_\lambda. \tag{3.3}$$

Since $k > 2$,

$$I(u) \geq \frac{1}{8}\|u\|_\lambda^2, \forall u \in \mathbb{K}_r. \tag{3.4}$$

From the above study, we have that m is well defined and $m \in [0, +\infty)$. Therefore, there is $(u_n) \subset \mathbb{K}_r$ such that

$$I_\lambda(u_n) \to m.$$

Once that (u_n) is bounded, because $(u_n) \subset \overline{B}_r(0)$, we can assume, without loss of generality, that

$$u_n \rightharpoonup u \text{ in } E_\lambda \quad \text{and} \quad u_n(x) \to u(x) \text{ a.e. in } \mathbb{R}.$$

By Ekeland's Variational Principle, we also assume that

$$m \le I_\lambda(u_n) \le m + \frac{1}{n^2} \ \forall n \in \mathbb{N}$$

and

$$I_\lambda(u) \ge I_\lambda(u_n) - \frac{1}{n}\|u - u_n\|_\lambda \ \forall u \in \mathbb{K}_r.$$

Observing that $\varphi_+ \in \mathbb{K}_r$, by (3.4),

$$\frac{1}{8}\|u_n\|_\lambda^2 \le I_\lambda(u_n) \le m + \frac{1}{n^2} \le I_\lambda(\varphi_+) + \frac{1}{n^2} \le \frac{1}{2}\|\varphi_+\|^2 + \frac{1}{n^2}$$

leading to

$$\limsup_{n\to+\infty} \|u_n\|_\lambda^2 \le 4\,\|\varphi_+\|^2 < r.$$

Thus, there is $n_0 \in \mathbb{N}$ such that

$$\|u_n\|_\lambda^2 < r \ \ \forall n \ge n_0.$$

Now, repeating the same arguments found in [11], we have that (u_n) is a $(\mathrm{PS})_m$ sequence for I_λ, that is,

$$I_\lambda(u_n) \to m \text{ and } I'_\lambda(u_n)(v - u_n) \ge \langle z_n, v - u_n\rangle \ \ \forall v \in \mathbb{K} \tag{3.5}$$

with $z_n \to 0$ in E'_λ. Using Proposition 2.1, there are a subsequence of (u_n), still denoted by (u_n), and u in E_λ such that

$$u_n \to u \text{ in } E_\lambda. \tag{3.6}$$

From this, $u \in \mathbb{K}_r$ and $I_\lambda(u) = m$, showing that u is a solution for (3.1). Now, combining (3.5) and (3.6), it follows that

$$\int_{\mathbb{R}} [u'(v-u)' + (1 + \lambda V(x))u(v-u)] \ge \int_{\mathbb{R}} g(x,u)(v-u) \ \ \forall v \in \mathbb{K}. \tag{3.7}$$

Using the test function $v = u + u_- \in \mathbb{K}$, a direct computation implies that $u_- = 0$, consequently u is nonnegative. The positivity of u is obtained by applying the maximum principle. □

4. Second solution for (P_A)

In this section, we will apply the Mountain Pass Theorem due to Szulkin [23] to get a second positive solution for problem (P_A). Here, we denote by u_λ the solution obtained in Theorem 3.1.

Lemma 4.1. *The energy functional I_λ verifies the geometry of the Mountain Pass Theorem.*

Proof. Note that, by Theorem 3.1,

$$I_\lambda(u) \geq I_\lambda(u_\lambda) \ \forall u \in \mathbb{K}_r.$$

Since $\Psi(u) = +\infty$ for all $u \in \mathbb{K}_r^c$, it follows that

$$I_\lambda(u) \geq I_\lambda(u_\lambda) \ \forall u \in \overline{B}_r. \tag{4.1}$$

Moreover, if $\rho = \frac{1}{8}r^2$, (3.4) gives

$$I_\lambda(u) \geq \rho > 0, \qquad \text{for all} \quad u \in \partial\overline{B}_r.$$

On the other hand, since $\|\varphi_+\|^2 < \frac{1}{4}r^2$, we have that $\varphi_+ \in \mathbb{K}_r$, and so,

$$I_\lambda(u_\lambda) \leq I_\lambda(\varphi_+) \leq \frac{1}{2}\|\varphi_+\|^2 < \rho, \tag{4.2}$$

from where it follows that

$$\inf_{u\in\partial B_r} I_\lambda(u) > I_\lambda(u_\lambda). \tag{4.3}$$

We now observe that, for $t \geq 1$, $t\varphi_+ \in \mathbb{K}$. Then, $\Psi(t\varphi_+) = 0$ and

$$I_\lambda(t\varphi_+) = \frac{t^2}{2}\int_{\mathbb{R}}(|\varphi_+'|^2 + |\varphi_+|^2) - \int_{\mathbb{R}} F(t\varphi_+).$$

By (f_2), there are $A, B > 0$ such that

$$F(s) \geq As^\theta - B \ \forall s \geq 0.$$

Consequently,

$$I_\lambda(t\varphi_+) \leq \frac{t^2}{2}\int_{\mathbb{R}}(|\varphi_+'|^2 + |\varphi_+|^2) - t^\theta A\int_D(\varphi_+)^\theta + B|D|,$$

where D is a mensurable set with finite measure verifying $D \cap \operatorname{Supp}(\varphi_+) \neq \emptyset$. From this,

$$I_\lambda(t\varphi_+) \to -\infty \ \text{ as } \ t \to +\infty,$$

and thus, setting $e = t\varphi_+$ for t large enough, we derive that

$$\|e\| > r \ \text{ and } \ I_\lambda(e) < I_\lambda(u_\lambda). \tag{4.4}$$

From (4.1)–(4.4), we deduce that I_λ satisfies the mountain pass geometry, see [23, Theorem 3.2]. □

Theorem 4.1. *Under the assumptions of Theorem* 3.1, *Problem* (P_A) *has a positive solution at the mountain pass level for all* $\lambda > 0$, *that is, there is* $w_\lambda \in \mathbb{K}$ *verifying*

$$I_\lambda(w_\lambda) = c_\lambda \ \text{ and } \ I_\lambda'(w_\lambda)(v - w_\lambda) \geq 0 \ \forall v \in \mathbb{K},$$

where c_λ *is the mountain pass level of* I_λ.

Proof. Combining Lemma 4.1 and Proposition 2.1 with the Mountain Pass Theorem, we have that the mountain pass level c_λ associated with I_λ is a critical value, hence there is $w_\lambda \in \mathbb{K}$ such that

$$I_\lambda(w_\lambda) = c_\lambda \text{ and } I'_\lambda(w_\lambda)(v - w_\lambda) \geq 0 \ \forall v \in \mathbb{K}.$$

Using the test function $v = w_\lambda + w_{\lambda-} \in \mathbb{K}$, a direct computation implies that $w_{\lambda-} = 0$, consequently w_λ is nonnegative. The positivity of w_λ is obtained by applying maximum principle. □

Corollary 4.1. *Under the assumptions of Theorem* 3.1, *problem* (P_A) *has two positive solutions for all* $\lambda > 0$.

Proof. From the previous study, we have two solutions denoted by u_λ and w_λ, where u_λ was obtained by minimization and w_λ by Mountain Pass Theorem. Moreover, by (4.2),

$$m = I_\lambda(u_\lambda) < \rho \quad \text{and} \quad I_\lambda(w_\lambda) = c_\lambda \geq \rho.$$

Thus,

$$I_\lambda(u_\lambda) < I_\lambda(w_\lambda),$$

from where it follows that u_λ and w_λ are different. Hence, problem (P_A) has two positive solutions. □

5. Proof of Theorem 1.1

In what follows, our main goal is to show that there is $\lambda^* > 0$ such that if $\lambda \geq \lambda^*$, the solutions u_λ and w_λ obtained in Corollary 4.1 satisfy

$$w_\lambda(x), u_\lambda(x) \leq a, \ \forall x \in \Omega^c. \tag{5.1}$$

From this, by using Remark 2.1, we will conclude that w_λ and u_λ are positive solutions of (P_λ) if $\lambda \geq \lambda^*$.

Hereafter, $\lambda_n \to +\infty$, $u_n = u_{\lambda_n}$ and $w_n = w_{\lambda_n}$. From Theorem 3.1, we know that $u_n \in \mathbb{K}_r$ for all $n \in \mathbb{N}$, thus (u_n) is bounded in $H^1(\mathbb{R})$. Next, we will show that (w_n) is also bounded in $H^1(\mathbb{R})$.

Lemma 5.1. *The sequence* (w_n) *is bounded in* $H^1(\mathbb{R})$. *More precisely, there is* $M > 0$ *such that*

$$\|w_n\|_{\lambda_n} \leq M \ \forall n \in \mathbb{N}.$$

Proof. Since w_n is a solution of (P_{λ_n}), it follows that

$$\int_{\mathbb{R}} [w'_n(v - w_n)' + (1 + \lambda_n V(x)) w_n (v - w_n)] \geq \int_{\mathbb{R}} g(x, w_n)(v - w_n), \ \forall v \in \mathbb{K}. \tag{5.2}$$

Repeating the same arguments used in the proof of Proposition 2.1, we derive that

$$I_{\lambda_n}(w_n) \geq \left[\left(\frac{1}{2} - \frac{1}{\theta}\right) - \frac{1}{2k}\right] \|w_n\|^2_{\lambda_n} \ \forall n \in \mathbb{N}. \tag{5.3}$$

Now, considering the path $\gamma(t) = tt^*\varphi_+$ for $t \in [0,1]$ and t^* large enough and setting

$$\Sigma = \max_{t\in[0,1]} J(\gamma(t)) > 0,$$

where

$$J(u) = \frac{1}{2}\int_\Omega [|u'|^2 + |u|^2] - \int_\Omega F(u),$$

it follows that

$$I_{\lambda_n}(w_n) \leq \max_{t\in[0,1]} I_{\lambda_n}(\gamma(t)) = \max_{t\in[0,1]} J(\gamma(t)) = \Sigma \ \forall n \in \mathbb{N},$$

because $I_{\lambda_n}(\gamma(t)) = J(\gamma(t))$ for all $n \in \mathbb{N}$ and $t \in [0,1]$.

This combined with (5.3) implies that $(\|w_n\|_{\lambda_n})$ is bounded in $\mathbb{R}$. □

Lemma 5.2. *There are subsequences of (u_n) and (w_n), still denoted by itself, which are strongly convergent in $H^1(\mathbb{R})$.*

Proof. In what follows, we will prove the lemma only for (u_n), because the same arguments can be applied to (w_n). Following the same arguments used in the proof of Proposition 2.1, for each $\delta > 0$, there is $R > 0$ such that

$$\limsup_{n\to+\infty} \int_{|x|\geq R} [|u_n'|^2 + |u_n|^2] < \delta.$$

The above limit yields

$$\int_{\mathbb{R}} g(x,u_n)u_n \to \int_{\mathbb{R}} g(x,u)u \tag{5.4}$$

and

$$\int_{\mathbb{R}} g(x,u_n)v \to \int_{\mathbb{R}} g(x,u)v \ \forall v \in \mathbb{K}, \tag{5.5}$$

where $u \in \mathbb{K}$ is the weak limit of (u_n) in $H^1(\mathbb{R})$.

Claim 5.1. *The weak limit u is null in $\overline{\mathcal{O}^c}$, that is,*

$$u(t) = 0 \ \forall t \in \overline{\mathcal{O}^c}.$$

Hence, $u \in H_0^1(\mathcal{O})$.

In fact, for each $m \in \mathbb{N}$, we define

$$\Delta_m = \left\{t \in \mathbb{R};\, V(t) > \frac{1}{m}\right\}.$$

It is immediate to see that

$$P = \{t \in \mathbb{R};\, V(t) > 0\} = \bigcup_{m=1}^{\infty} \Delta_m.$$

Note that

$$\int_{\Delta_m} |u_n|^2 \leq \frac{m}{\lambda_n}\|u_n\|_{\lambda_n}^2 \leq \frac{m}{\lambda_n}r^2 \ \forall n, m \in \mathbb{N}$$

where r is the constant given in Theorem 3.1. The last inequality, together with Fatou's Lemma, lead to

$$\int_{\Delta_m} |u|^2 = 0 \ \forall m \in \mathbb{N}.$$

Thereby, $u = 0$ a.e in Δ_m for all $m \in \mathbb{N}$, implying that $u = 0$ a.e. in P. Now, the claim follows using the continuity of u.

Using $v = u$ as a test function in (3.7),

$$\int_{\mathbb{R}} |u_n'|^2 + \int_{\mathbb{R}} (1 + \lambda_n V)|u_n|^2 \leq \int_{\mathbb{R}} (1 + \lambda_n V)u_n u + \int_{\mathbb{R}} u_n' u' - \int_{\mathbb{R}} g(x, u_n)(u - u_n). \tag{5.6}$$

Once that $V(t) \geq 0$ and $u = 0$ in $\overline{\Omega^c}$,

$$\int_{\mathbb{R}} |u_n'|^2 + \int_{\mathbb{R}} |u_n|^2 \leq \int_{\mathbb{R}} u_n' u' + \int_{\mathbb{R}} u_n u - \int_{\mathbb{R}} g(x, u_n)(u - u_n).$$

Taking the limit of $n \to +\infty$ and using (5.4)–(5.6),

$$\limsup_{n \to +\infty} \int_{\mathbb{R}} [|u_n'|^2 + |u_n|^2] \leq \int_{\mathbb{R}} [|u'|^2 + |u|^2].$$

Since $H^1(\mathbb{R})$ is a Hilbert space and $u_n \rightharpoonup u$ in $H^1(\mathbb{R})$, the above limit implies that $u_n \to u$ in $H^1(\mathbb{R})$. □

As a consequence of the lemmas proved in this section, we have the following results

Corollary 5.1. *The sequences* (u_n) *and* (w_n) *satisfy*

$$\lambda_n \int_{\mathbb{R}} V(x)|u_n|^2 \to 0 \ \ as \ n \to +\infty \tag{5.7}$$

and

$$\lambda_n \int_{\mathbb{R}} V(x)|w_n|^2 \to 0 \ \ as \ n \to +\infty, \tag{5.8}$$

for some subsequence. Moreover, the weak limits u *and* w *of* (u_n) *and* (w_n) *respectively, belong to* $H_0^1(\mathcal{O})$ *and they are positive solutions of the obstacle problem*

$$\int_{\mathcal{O}} [\psi'(v - \psi)' + \psi(v - \psi)] \geq \int_{\mathcal{O}} f(\psi)(v - \psi) \ \forall v \in \widehat{\mathbb{K}} \tag{$P_{\mathcal{O}}$}$$

where

$$\widehat{\mathbb{K}} := \left\{ v \in H_0^1(\mathcal{O}); v(x) \geq \varphi(x) \ \ a.e. \ \mathcal{O} \right\}.$$

Proof. From now on, we will prove the lemma only for the sequence (u_n), because the same arguments can be applied to (w_n). Repeating the same type of arguments explored in the proof of Claim 5.1, we get again an equality like (5.6), that is,

$$\int_{\mathbb{R}} |u_n'|^2 + \int_{\mathbb{R}} (1 + \lambda_n V)|u_n|^2 \leq \int_{\mathbb{R}} (1 + \lambda_n V)u_n u + \int_{\mathbb{R}} u_n' u' - \int_{\mathbb{R}} g(x, u_n)(u - u_n).$$

Using the fact that $V(t)u(t) = 0$ for all $t \in \mathbb{R}$, it follows that

$$\int_{\mathbb{R}} |u_n'|^2 + \int_{\mathbb{R}} (1 + \lambda_n V)|u_n|^2 \leq \int_{\mathbb{R}} u_n' u' + \int_{\mathbb{R}} u_n u - \int_{\mathbb{R}} g(x, u_n)(u - u_n). \quad (5.9)$$

From Theorem 5.2, $u_n \to u$ in $H^1(\mathbb{R})$ for some subsequence. Hence,

$$\liminf_{n \to +\infty} \int_{\mathbb{R}} (|u_n'|^2 + |u_n|^2) = \int_{\mathbb{R}} (|u'|^2 + |u|^2),$$

$$\lim_{n \to +\infty} \int_{\mathbb{R}} (u_n' u' + u_n u) = \int_{\mathbb{R}} (|u'|^2 + |u|^2),$$

and

$$\lim_{n \to +\infty} \int_{\mathbb{R}} g(x, u_n)(u - u_n) = 0.$$

The above limits combined with (5.9) yield

$$\lambda_n \int_{\mathbb{R}} V|u_n|^2 \to 0.$$

To prove that $(P_{\mathcal{O}})$ holds, we begin recalling that for all $v \in \mathbb{K}$,

$$\int_{\mathbb{R}} [u_n'(v - u_n)' + (1 + \lambda_n V(x))u_n(v - u_n)] \geq \int_{\mathbb{R}} g(x, u_n)(v - u_n).$$

Choosing $v \in \widehat{\mathbb{K}}$, we get

$$\int_{\mathbb{R}} [u_n'(v - u_n)' + u_n(v - u_n) - \lambda_n V(x)|u_n|^2] \geq \int_{\mathbb{R}} g(x, u_n)(v - u_n).$$

Taking the limit of $n \to \infty$ and using Lemma 5.2 and (5.7), we derive that

$$\int_{\mathcal{O}} [u'(v - u)' + u(v - u)] \geq \int_{\mathcal{O}} f(u)(v - u) \ \ \forall v \in \widehat{\mathbb{K}},$$

finishing the proof. □

Corollary 5.2. *The sequences (u_n) and (w_n) satisfy the following limits*

$$\|w_n\|_{L^\infty(\overline{\mathcal{O}}^c)}, \|u_n\|_{L^\infty(\overline{\mathcal{O}}^c)} \to 0 \ \ as \ \ n \to +\infty.$$

Proof. These limits are an immediate consequence of the continuous embedding $H^1(\overline{\Omega}^c) \hookrightarrow L^\infty(\overline{\mathcal{O}}^c)$ together with the limits $u_n \to u$ and $w_n \to w$ in $H^1(\mathbb{R})$ and of the fact that $u = w = 0$ in $\mathcal{O}^c$. □

Proof of Theorem 1.1. The study made in this section allows us to prove that (5.1) holds for λ large enough. We will show only (5.1) to (u_n), because the argument is the same for (w_n). Arguing by contradiction, we assume that there is $\lambda_n \to +\infty$ such that

$$\|u_n\|_{L^\infty(\Omega^c)} > a \quad \forall n \in \mathbb{N}. \quad (5.10)$$

From Lemma 5.2, there is a subsequence of (u_n), still denoted by itself, and $u \in H_0^1(\mathcal{O})$ such that

$$u_n \to u \ \text{ in } \ H^1(\mathbb{R}).$$

By Corollary 5.2, the below limit holds

$$\|u_n\|_{L^\infty(\overline{\mathcal{O}}^c)} \to 0 \text{ as } n \to +\infty,$$

which implies that there is $n_0 \in \mathbb{N}$ such that

$$\|u_n\|_{L^\infty(\Omega^c)} < \frac{a}{2} \quad \forall n \geq n_0,$$

obtaining a contradiction with (5.10). This way, it follows that there is $\lambda^* > 0$ such that the solution u_λ satisfies

$$u_\lambda(x) \leq a \ \forall x \in \Omega^c \text{ and } \lambda \geq \lambda^*.$$

Now, by Remark 2.1, we can conclude that u_λ is a positive solution for (P_λ) for $\lambda \geq \lambda^*$. □

Acknowledgement

Part of this work was done while the second author was visiting the Faculdade de Matemática of the Universidade Federal do Pará (Brazil). In particular, he would like to express his deep gratitude to Prof. Giovany Figueiredo (UFPA) for his kind hospitality. The authors would like to thank to the referee for his/her suggestions to improve this paper.

References

[1] C.O. Alves, *Existence of multi-bump solutions for a class of quasilinear problems.* Adv. Nonlinear Stud., 6 (2006), 491–509.

[2] T. Bartsch & Z.Q. Wang, *Existence and multiplicity results for some superlinear elliptic problems in* $\mathbb{R}^N$, Comm. Partial Differential Equations 20 (1995), 1725–1741.

[3] T. Bartsch & Z.Q. Wang, *Multiple positive solutions for a nonlinear Schrödinger equation*, Z. Angew Math. Phys. 51 (2000).

[4] K.C. Chang, *The Obstacle Problem and Partial Differential Equations with Discontinuous Nonlinearities*, Comm. Pure Appl. Math., Vol. XXXIII, 117–146 (1980).

[5] M. Clapp & Y.H. Ding, *Positive solutions of a Schrodinger equations with critical nonlinearity*, Z. Angew. Math. Phys, 55 (2004), 592–605.

[6] D.G. de Figueiredo, M. Girardi & M. Matzeu, *Semilinear elliptic equations with dependence on the gradient via mountain pass techniques*, Differential Integral Equations, 17 (1-2) (2004), 199–210.

[7] Y.H. Ding & K. Tanaka, *Multiplicity of positive solutions of a nonlinear Schrödinger equation*, Manuscripta Math., 112 (2003).

[8] M. del Pino & P.L. Felmer, *Local mountain passes for semilinear elliptic problems in unbounded domains*, Calc. Var. Partial Differential Equations, 4 (1996), 121–137.

[9] I. Ekeland, *Nonconvex minimization problems*, Bull. Amer. Math. Soc. 1, 1979, 443–474.

[10] J. Yang, *Positive solutions of an obstacle problem,* Ann. Fac. Sci. Toulouse Math., 6ème Série, Tome 4, N. 2, 339–366 (1995).

$N-m \geq 2$, $m \geq 1$, $\widetilde{\Omega} \subset \mathbb{R}^m$ open smooth bounded. We restrict to the case where $g(x,u)$ behaves as a superlinear but subcritical power u^{p-1} at infinity with $2 < p < 2^*$, where $2^* = \frac{2N}{N-2}$.

In bounded domains, many authors (see, e.g., Rabinowitz [27] and references within) have looked for solutions of elliptic equations with zero Dirichlet boundary conditions by variational methods.

In particular, if $f = 0$, Ambrosetti and Rabinowitz have proved the existence of infinitely many solutions to $(\mathcal{P}_0)$ by means of the Symmetric Mountain Pass Theorem (see [2]) exploiting the symmetry of the problem.

Always in the bounded case, if $f \not\equiv 0$, some perturbative techniques have been employed to find multiplicity results for values of p not much larger than 2 (see [8, 9, 28, 30, 31]). Later on, when the symmetry of the problem is broken also by the presence of non-homogeneous boundary conditions (see, e.g., [15, 18, 19]) more restrictive assumptions for the exponent p have been found. Candela, Palmieri and Salvatore in [17] have improved the previous results for problems with broken symmetry by exploiting further radial symmetry assumption. In particular, in the case of homogeneous boundary data, they state the existence of infinitely many radial solutions in a ball for any p such that $2 < p < 2^*$.

On the other hand, if $\Omega = \mathbb{R}^N$, $g = g(u)$ and $f \equiv 0$, in [11, 12, 13] thanks to the radial symmetry of the problem Berestycki and Lions overcome the lack of compactness and find infinitely many radial solutions of $(\mathcal{P}_0)$ if g satisfies the so-called double-power growth condition, namely $g(u)$ behaves as a subcritical power u^{p-1} at infinity and a supercritical power u^{q-1} near the origin, where $2 < p < 2^* < q$. This condition have allowed later Benci and Fortunato in [10] to introduce as a natural framework the Orlicz spaces $L^p + L^q$ which have been used also in more general cases by other authors such as Badiale, Pisani and Rolando in [7].

Recently, if $f \not\equiv 0$, in [4] the authors have obtained, by Bolle's perturbation method and variational techniques, multiplicity results of radial solutions in $\mathbb{R}^N$ when $g(x,u) = |u|^{p-2}u$ for any p such that $2 < p < 2^*$. These results without any difficulty can be proved for more general nonlinearities $g(x,u)$, thus extending to $\mathbb{R}^N$ the results obtained in [17] for problems with zero Dirichlet boundary conditions.

On the contrary, if Ω is an unbounded cylinder, to our knowledge, few existence and multiplicity results have been proved until now only for the unperturbed problem $(\mathcal{P}_0)$. Indeed, in [3] Azzollini and Pomponio have obtained existence and multiplicity results to $(\mathcal{P}_0)$, by exploiting compact imbeddings of cylindrical symmetric spaces in the Orlicz spaces $L^p + L^q$, in the three-dimensional autonomous case. Indeed, they consider $\Omega = \widetilde{\Omega} \times \mathbb{R}^2$, $\widetilde{\Omega}$ a bounded interval of $\mathbb{R}$, and $g \in C(\mathbb{R},\mathbb{R})$ and its primitive function $G(s) = \int_0^s g(\sigma)d\sigma$ verifying the following assumptions:

(g_1') there exists $\mu > 2$ such that for all $s \in \mathbb{R}$: $\mu G(s) \leq g(s)s$;
(g_2') for all $s \in \mathbb{R}$: $|g(s)| \leq c \min(|s|^{q-1}, |s|^{p-1})$;
(g_3') for all $s \in \mathbb{R}$: $G(s) \geq c' \min(|s|^q, |s|^p)$;

with $2 < p < 6 < q$ and $c, c' > 0$, thus proving an existence result and, if in addition g is odd, a multiplicity result.

Moreover, in [22] Fan and Zhao found multiple solutions for the p-Laplacian problem on unbounded cylinders in $\mathbb{R}^N$, $N \geq 3$; their results apply in particular to problem $(\mathcal{P}_0)$ (see [22, Remark 2]) but in their paper there aren't the details of the proof.

Anyway, up to now, no existence and multiplicity results have been stated for problem $(\mathcal{P}_f)$ with $f \not\equiv 0$ on Ω unbounded cylinder.

So, as concerns as the unperturbed case, the aim of this paper is to prove that, by using some compactness imbeddings concerning "partially" spherically symmetric Sobolev spaces and by means of Mountain Pass Theorem and its symmetric version, the results in [3] can be extended to dimensions $N \geq 3$ and to more general non autonomous nonlinearities g (see Remark 1.2) without using the Orlicz spaces $L^p + L^q$.

More in general, in this paper, we establish existence and multiplicity results to $(\mathcal{P}_f)$ on unbounded cylinders also in the perturbed case $f \not\equiv 0$.

For the unperturbed case, we have the following result.

Theorem 1.1. *Suppose that $g \in C(\Omega \times \mathbb{R}, \mathbb{R})$ verifies*

(g_0) $g(\widetilde{x}, y_1, s) = g(\widetilde{x}, y_2, s)$ *for every* $s \in \mathbb{R}$, $\widetilde{x} \in \widetilde{\Omega}$ *and* $y_1, y_2 \in \mathbb{R}^{N-m}$, $|y_1| = |y_2|$, *i.e.,* $g(\widetilde{x}, \cdot, s)$ *is spherically symmetric on* $\mathbb{R}^{N-m}$;

(g_1) *there exists* $\mu > 2$ *such that*

$$0 < \mu G(x, s) \leq g(x, s)s \quad \text{for all } x \in \Omega \text{ and } s \in \mathbb{R} \setminus \{0\},$$

where $G(x, s) = \int_0^s g(x, \sigma)\, d\sigma$;

(g_2) $\displaystyle\lim_{s \to 0} \frac{g(x, s)}{s} = 0$ *uniformly with respect to* $x \in \Omega$.

(g_3) *there exist* $2 < p < 2^*$, $a_0, a_1 \geq 0$ *such that*

$$|g(x, s)| \leq a_0 |s|^{p-1} + a_1 \quad \text{for all } x \in \Omega \text{ and } s \in \mathbb{R}. \tag{1.1}$$

Then, problem $(\mathcal{P}_0)$ has at least one nontrivial weak solution. Moreover, if in addition

(g_4) $g(x, s)$ *is odd with respect to* s

holds, then $(\mathcal{P}_0)$ has infinitely many weak solutions.

Remark 1.2. Let us point out that Theorem 1.1 extends the results in [3] to dimensions $N \geq 3$ and to more general non autonomous nonlinearities $g(x, u)$. In fact, in the case $g(x, u) = g(u)$, conditions (g_1') and (g_3') obviously imply (g_1) while condition (g_2') implies (g_2). Indeed, if $|s| \leq 1$, by (g_2') and $p < q$ it is $|g(s)| \leq c\, |s|^{q-1}$. Thus for all $s \neq 0$, $|s| \leq 1$ we have

$$0 \leq \left| \frac{g(s)}{s} \right| \leq c|s|^{q-2}$$

and (g_2) follows since $q > 2$.

Furthermore (g_2') implies also (g_3). Indeed, by (g_2') and $p < q$ it is $|g(s)| \leq c\,|s|^{q-1} \leq c\,|s|^{p-1}$ when $|s| \leq 1$ while $|g(s)| \leq c\,|s|^{p-1}$ when $|s| \geq 1$ so that (g_3) is satisfied with $a_0 = c$ and $a_1 = 0$.

Remark 1.3. It is easy to prove (see [5]) that by (g_1) and (g_2), for any $s_0 > 0$ small enough, there exists a function $\gamma_{s_0} \in L^\infty(\Omega)$, $\gamma_{s_0}(x) > 0$ for every $x \in \Omega$, such that

$$G(x,s) \geq \gamma_{s_0}(x)|s|^\mu \quad \text{for all } x \in \Omega \text{ and } s \in \mathbb{R} \text{ s.t. } |s| \geq s_0. \tag{1.2}$$

Indeed, from (g_1), fixed $s_0 > 0$, we have that

$$G(x,s) \geq \frac{G(x,s_0)}{|s_0|^\mu}|s|^\mu \quad \text{for all } x \in \Omega \text{ and } s \in \mathbb{R} \text{ s.t. } |s| \geq s_0.$$

Setting $\gamma_{s_0}(x) = \frac{G(x,s_0)}{|s_0|^\mu}$ for every $x \in \Omega$, it follows that $\gamma_{s_0}(x) > 0$. Now, we prove that $\gamma_{s_0} \in L^\infty(\Omega)$ for s_0 sufficiently small. Indeed, from (g_2) and l'Hôpital's rule, we have that

$$\lim_{s\to 0} \frac{G(x,s)}{s^2} = 0, \quad \text{uniformly with respect to } x, \tag{1.3}$$

hence, fixing $\varepsilon = 1$, there exists $\delta_1 > 0$ such that

$$G(x,s) \leq |s|^2 \quad \text{for every } x \in \Omega \text{ and } s \in \mathbb{R} \text{ s.t. } |s| < \delta_1.$$

Chosen $s_0 < \delta_1$, it follows that $G(x,s_0) \leq \delta_1^2$ for every $x \in \Omega$, so $\operatorname{supess}_{\Omega} \gamma_{s_0}(x)$ is finite and $\gamma_{s_0} \in L^\infty(\Omega)$.

As concerns as the perturbed case, we have the following result.

Theorem 1.4. *Let $N - m \geq 3$ and $g \in C(\Omega \times \mathbb{R}, \mathbb{R})$ satisfying (g_0)–(g_4) and*

(g_5) there exists $\gamma_0 > 0$ such that

$$G(x,s) \geq \gamma_0|s|^\mu, \quad \text{for every } x \in \Omega \text{ and } s \in \mathbb{R}.$$

Taken any function $f \in L^{\frac{\mu}{\mu-1}}(\Omega)$ such that

(f_0) $f(\widetilde{x}, y_1) = f(\widetilde{x}, y_2)$ for every $\widetilde{x} \in \widetilde{\Omega}$ and $y_1, y_2 \in \mathbb{R}^{N-m}$, $|y_1| = |y_2|$, i.e., $f(\widetilde{x}, \cdot)$ is spherically symmetric on $\mathbb{R}^{N-m}$,

problem $(\mathcal{P}_f)$ has infinitely many weak solutions for all p verifying

$$2 < p < \overline{p}_{N,m} \tag{1.4}$$

where $\overline{p}_{N,m} = 2 + \frac{4}{(N-m)(m+1)-2}$.

Remark 1.5. Since $\overline{p}_{N,m} \leq 2 + \frac{2}{N-2}$, the result obtained above in the cylindrical case extends in some sense the one obtained for problem $(\mathcal{P}_f)$ when Ω is bounded (see [9, 31]). Nevertheless, it doesn't cover the whole interval $(2, 2^*)$ as in the radial case (see [17] for the bounded case and [4] for the unbounded case).

Remark 1.6. In the perturbed case, we need the further assumption (g_5) which is not guaranteed by (g_1) and (g_2) which imply only condition (1.2), as pointed out in Remark 1.3.

The paper is organized as follows: in Section 2, we introduce the variational formulation of the problems and suitable "partially" spherically symmetric Sobolev spaces. We overcome the lack of compactness of the problems by recalling some compact Sobolev imbeddings for such spaces. Section 3 is devoted to the proof of Theorem 1.1. In Section 4, we recall Bolle's perturbation method and multiplicity results for perturbed problems. In Section 5, we present some preliminary lemmas and in Section 6, we prove Theorem 1.4. At last, in the Appendix 7, we improve a classical estimate of the number $N_-(-\Delta+V)$, useful in the proof of Theorem 1.4.

Notation

- If $x, y \in \mathbb{R}^N$, $x \cdot y$ denotes the Euclidean product in $\mathbb{R}^N$;
- If X is a Banach space, $\|\cdot\|_X$ denotes its norm;
- $L^t(\Omega)$, with $1 \le t \le +\infty$, denotes the Lebesgue space with the usual norm $|\cdot|_{L^t(\Omega)}$;
- for all $t > 1$, t' is its conjugate exponent, i.e., $\frac{1}{t} + \frac{1}{t'} = 1$;
- a_i, c_i, C_i denote suitable positive constants.

2. Variational framework

In order to prove that problem $(\mathcal{P}_f)$ and consequently $(\mathcal{P}_0)$ has a variational structure, let us consider the space $W_0^{1,2}(\Omega)$ endowed with the norm

$$||u|| = \left(\int_\Omega |\nabla u|^2\, dx\right)^{\frac{1}{2}}. \tag{2.1}$$

Let us point out that the norm in $W^{1,2}(\Omega)$ given by

$$\left(\int_\Omega |\nabla u|^2\, dx + \int_\Omega |u|^2\, dx\right)^{\frac{1}{2}}$$

is equivalent to the norm $\|\cdot\|$. Indeed, as Ω is a subset of $\mathbb{R}^N$ which lies between two hyperplanes, by Poincaré inequality a constant $k > 0$ exists such that

$$\int_\Omega |\nabla u|^2\, dx + \int_\Omega |u|^2\, dx \le k \int_\Omega |\nabla u|^2\, dx$$

(see [1, p. 159]). Hence, $\|\cdot\|$ is equivalent to the classical norm in $W_0^{1,2}(\Omega)$. For this reason, even if we have a problem with "zero mass" we don't use the space $\mathcal{D}^{1,2}(\Omega)$ but we study $(\mathcal{P}_f)$ in $W_0^{1,2}(\Omega)$ as in the "positive mass case", thus simplifying the argument in [3]. From now on, we denote by E the space $W_0^{1,2}(\Omega)$ endowed with the norm given by (2.1) and by $(E', \|\cdot\|_{E'})$ its dual space. Then, by the Sobolev imbedding theorems (see [16, Corollary 9.14]) it follows

$$E \hookrightarrow L^t(\Omega) \qquad \text{if} \quad 2 \le t \le 2^*. \tag{2.2}$$

Let us point out that, since Ω is unbounded, the previous imbeddings are not compact.

Now, it is possible to state the following variational principle.

Proposition 2.1. *Let $g \in C(\Omega \times \mathbb{R}, \mathbb{R})$ satisfying (g_2) and (g_3) and $f \in L^{\mu'}(\Omega)$. Then, the weak solutions of $(\mathcal{P}_f)$ (resp. $(\mathcal{P}_0)$) are the critical points of the energy functional defined on E by*

$$J_1(u) = \frac{1}{2}\int_\Omega |\nabla u|^2\, dx - \int_\Omega G(x,u)\, dx - \int_\Omega fu\, dx$$
$$\left(\text{resp. } J_0(u) = \frac{1}{2}\int_\Omega |\nabla u|^2\, dx - \int_\Omega G(x,u)\, dx.\right)$$

More precisely, $J_1 \in C^1(E)$ (resp. J_0) and its differential $J_1' : E \to E'$ (resp. J_0') is defined as

$$J_1'(u)[\zeta] = \int_\Omega [\nabla u \cdot \nabla \zeta - g(x,u)\zeta - f\zeta]\, dx \tag{2.3}$$
$$\left(\text{resp. } J_0'(u)[\zeta] = \int_\Omega [\nabla u \cdot \nabla \zeta - g(x,u)\zeta]\, dx\right)$$

for all $u, \zeta \in E$.

Proof. It is sufficient to prove the above proposition for the functional J_1, namely that the functional

$$J_1(u) = \frac{1}{2}\|u\|^2 - \int_\Omega G(x,u)\, dx - \int_\Omega fu\, dx, \qquad u \in E,$$

is well defined and its Fréchet differential given in (2.3) is a continuous operator from E to E'. We study separately the maps

$$\varphi_0(u) = \frac{1}{2}\|u\|^2, \qquad \varphi_1(u) = \int_\Omega G(x,u)dx, \qquad \varphi_2(u) = \int_\Omega fudx.$$

It is standard to prove that the map φ_0 and φ_2 are of class $C^1(E)$ with differentials

$$\varphi_0'(u)[\zeta] = \int_\Omega \nabla u \cdot \nabla \zeta\, dx \quad \text{and} \quad \varphi_2'(u)[\zeta] = \int_\Omega f\zeta\, dx \quad \text{for all } u, \zeta \in E.$$

Now, we have to prove that also φ_1 is well defined and C^1 in E and

$$\varphi_1'(u)[\zeta] = \int_\Omega g(x,u)\zeta dx \qquad \text{for all } u, \zeta \in E. \tag{2.4}$$

Let us point out that, from (g_2) and (g_3), it follows that fixing any $\varepsilon > 0$ a constant $c_\varepsilon > 0$ exists such that

$$|g(x,s)| \ \le\ \varepsilon|s| + c_\varepsilon|s|^{p-1} \tag{2.5}$$

for all $x \in \Omega$ and $s \in \mathbb{R}$. Indeed, by (g_2), for any $\varepsilon > 0$ there exists $\delta_\varepsilon > 0$ such that for every $|s| < \delta_\varepsilon$ it is $|g(x,s)| \le \varepsilon|s|$ while by (g_3) if $|s| \ge 1$ it is $|g(x,s)| \le a_0|s|^{p-1} + a_1 \le (a_0 + a_1)|s|^{p-1}$ for all $x \in \Omega$. If $\delta_\varepsilon \le |s| \le 1$, $|g(x,s)| \le$

$M_\varepsilon|s|^{p-1}$ with $M_\varepsilon = \max_{\delta_\varepsilon \le |s| \le 1}\left|\frac{g(x,s)}{s^{p-1}}\right|$. Then, (2.5) easily follows with $c_\varepsilon = \max\{M_\varepsilon, a_0 + a_1\}$. Now, by integrating (2.5), it is

$$|G(x,s)| \le \varepsilon|s|^2 + c_\varepsilon|s|^p \tag{2.6}$$

for all $x \in \Omega$ and $s \in \mathbb{R}$. Thus, by (2.2) with $t = 2$ and $t = p$, it follows that $\varphi_1 \in C^1(E,\mathbb{R})$ and its Fréchet differential is as in (2.4) (see, e.g., [32, Theorem 1.22]). □

At this point, in order to overcome the lack of compactness of the problem, we introduce a subspace E_G of E involving spherical symmetry such that E_G is a "natural constraint" for J_1 (resp. J_0) and E_G is compactly imbedded in $L^t(\Omega)$ for suitable t's.

Definition 2.2. Let G be a subgroup of $O(N)$ defined by $G = id_{\mathbb{R}^m} \times O(N-m)$. The action of G on the space $\mathcal{F}(\Omega,\mathbb{R})$ of the real functions defined on Ω is such that, for all $g = id_{\mathbb{R}^m} \times g_1 \in G$,

$$gu(\widetilde{x},y) := u(\widetilde{x}, g_1^{-1}y), \quad \text{for every } (\widetilde{x},y) \in \widetilde{\Omega} \times \mathbb{R}^{N-m}.$$

The subspace of the fixed points of G is defined by

$$\mathrm{Fix}(G) = \{u \in \mathcal{F}(\Omega,\mathbb{R}) : \ gu = u \quad \text{for all } g \in G\}.$$

Let $E_G = E \cap \mathrm{Fix}(G)$. In other words, a function $u \in E$ belongs to E_G if and only if $u(\widetilde{x},\cdot)$ is spherically symmetric on $\mathbb{R}^{N-m}$. Obviously, the action of G on E_G is isometric, that is,

$$\|gu\| = \|u\|, \quad \text{for all } g \in G.$$

Clearly, $E_G \hookrightarrow E$. We denote by $(E'_G, \|\cdot\|_{E'_G})$ the dual space of E_G. Moreover, by (g_0) and (f_0) (resp. by (g_0)) the functional J_1 (resp. J_0) is invariant under the action of the group G, i.e., $J_1 \circ g = J_1$ (resp. $J_0 \circ g = J_0$) for all $g \in G$. Hence, by the Principle of Symmetric Criticality by Palais in [26], any critical point of $J_1|_{E_G}$ (resp. $J_0|_{E_G}$) is a "free" critical point of J_1 (resp. J_0). Therefore, from now on, we look for critical points of J_1 (resp. J_0) constrained to E_G and, for simplicity, we still denote $J_1|_{E_G}$ (resp. $J_0|_{E_G}$) by J_1 (resp. J_0).

In the following, we denote by $W_G^{1,2}(\Omega) = W^{1,2}(\Omega) \cap \mathrm{Fix}(G)$. Let us recall that the imbeddings of $W_G^{1,2}(\Omega)$ in $L^t(\Omega)$ spaces are compact as proved by P.L. Lions in [25, Lemma III.2] (see also [6] for related results).

Since $E_G = W_0^{1,2}(\Omega) \cap \mathrm{Fix}(G) \hookrightarrow W_G^{1,2}(\Omega)$, it follows that

$$E_G \hookrightarrow\hookrightarrow L^t(\Omega) \quad \text{for } 2 < t < 2^*. \tag{2.7}$$

At this point, it is possible to prove the compactness of the Fréchet differential of the functional φ_1 introduced in Proposition 2.1.

Proposition 2.3. *Under the same assumptions in Proposition* 2.1, *it follows that* $d\varphi_1$ *is compact from* E_G *in* E'_G.

Proof. Let us point out that there exists $t_0 > 1$ such that

$$E \hookrightarrow L^{t_0(p-1)}(\Omega) \quad \text{and} \quad L^{t_0}(\Omega) \hookrightarrow E', \text{ i.e., } E \hookrightarrow L^{\frac{t_0}{t_0-1}}(\Omega). \tag{2.8}$$

Indeed, the system

$$\begin{cases} 2 < t_0(p-1) < \ 2^*, \\ 2 < \dfrac{t_0}{t_0-1} \quad < \ 2^*, \end{cases} \qquad \text{i.e.,} \qquad \begin{cases} \dfrac{2}{p-1} < t_0 < \dfrac{2^*}{p-1}, \\ \dfrac{2^*}{2^*-1} < t_0 < 2, \end{cases}$$

is solvable because, as $2 < p < 2^*$,

$$\max\left\{\frac{2}{p-1}, \frac{2^*}{2^*-1}\right\} < \min\left\{\frac{2^*}{p-1}, 2\right\}.$$

Since $2 < t_0(p-1) < 2^*$, the conclusion follows by (2.7), (g_3) and the fact that, if $\|u_n - u\|_{L^{t_0(p-1)}(\Omega)} \to 0$ as $n \to +\infty$, then $\|g(\cdot, u_n) - g(\cdot, u)\|_{L^{t_0}(\Omega)} \to 0$ as $n \to +\infty$ (see [32, Lemma 1.20]). □

Remark 2.4. We cannot apply directly the last part of Theorem 1.22 in [32] as the compact imbedding (2.7) does not hold for $t = 2$.

3. Unperturbed case

Our aim is to find weak solutions of problem $(\mathcal{P}_0)$ by applying the Mountain Pass Theorem (see [2, Theorem 2.1]) and its symmetric version (see [2, Corollary 2.9]) to the functional J_0. In order to do this, we first recall the following Palais–Smale condition, briefly (PS).

Definition 3.1. The functional J_0 satisfies the (PS) condition if any sequence $(u_n)_n \subset E_G$ such that

$$(J_0(u_n))_n \text{ is bounded} \tag{3.1}$$

and

$$dJ_0(u_n) \to 0 \quad \text{as } n \to +\infty, \tag{3.2}$$

converges in E_G, up to subsequences.

Proposition 3.2. *Let $g \in C(\Omega \times \mathbb{R}, \mathbb{R})$ satisfying (g_0), (g_1), (g_2) and (g_3). Then, the functional J_0 satisfies the (PS) condition.*

Proof. Let $(u_n)_n$ be a sequence verifying (3.1) and (3.2), then by (g_1) and (2.3) it follows

$$\begin{aligned} c_1 + \varepsilon_n \|u_n\| &\geq \mu J_0(u_n) - J_0'(u_n)[u_n] \\ &= \left(\frac{\mu}{2} - 1\right) \|u_n\|^2 + \int_\Omega (g(x, u_n)u_n - \mu G(x, u_n))\, dx \\ &\geq \left(\frac{\mu}{2} - 1\right) \|u_n\|^2 \end{aligned} \tag{3.3}$$

where $\varepsilon_n \to 0$ as $n \to +\infty$ and c_1 is a suitable positive constant. Hence, by (3.3), $(u_n)_n$ is bounded in E_G. So, $u \in E_G$ exists such that, up to subsequences, $u_n \rightharpoonup u$ weakly in E_G and, from Proposition 2.3,

$$\varphi_1'(u_n) \to \varphi_1'(u) \text{ in } E_G'. \tag{3.4}$$

Thus, (3.2) and (3.4) imply

$$(\varphi_0'(u_n) - \varphi_0'(u))[u_n - u] \to 0 \quad \text{if } n \to +\infty,$$

i.e.,

$$\int_\Omega |\nabla(u_n - u)|^2 dx \to 0. \tag{3.5}$$

and the conclusion follows. □

Proof of Theorem 1.1. By Proposition 2.1 and by (g_0), the functional $J_0 \in C^1(E_G)$. Proposition 3.2 implies that the functional J_0 satisfies the (PS) condition in E_G. Obviously, $J_0(0) = 0$; we claim that there exist α, $\varrho > 0$ such that

$$J_0(u) \;\geq\; \alpha \quad \text{if } u \in E_G,\ \|u\| = \varrho. \tag{3.6}$$

In fact, fixing any $\varepsilon > 0$, from (2.6) and (2.2) two positive constants c_2 and c_3 exist such that

$$J_0(u) \geq \frac{1}{2}\ (1 - \varepsilon c_2)\ \|u\|^2 - c_3\|u\|^p \quad \text{for all } u \in E_G.$$

Thus, since $p > 2$, taking $\|u\| = \varrho$ with ε and ϱ small enough, (3.6) holds for a suitable $\alpha > 0$.

Now, fix $u \in E_G$ with $u \neq 0$ and $s_0 > 0$ with $|s_0| \leq 1$. Denote $\Omega_{u,s_0} = \{x \in \Omega : |u(x)| \geq s_0\}$. By (g_1), (1.2) and $\mu > 2$, we have that

$$\begin{aligned}
J_0(u) &\leq \frac{1}{2}\|u\|^2 - \int_{\Omega_{u,s_0}} \gamma_{s_0}(x)|u|^\mu\, dx \\
&= \frac{1}{2}\|u\|^2 - \int_\Omega \gamma_{s_0}(x)|u|^\mu\, dx + \int_{\Omega\setminus\Omega_{u,s_0}} \gamma_{s_0}(x)|u|^\mu\, dx \\
&\leq \frac{1}{2}\|u\|^2 - \int_\Omega \gamma_{s_0}(x)|u|^\mu\, dx + \|\gamma_{s_0}\|_{L^\infty(\Omega)} \int_{\Omega\setminus\Omega_{u,s_0}} |u|^\mu\, dx \\
&\leq \frac{1}{2}\|u\|^2 - \int_\Omega \gamma_{s_0}(x)|u|^\mu\, dx + \|\gamma_{s_0}\|_{L^\infty(\Omega)} \int_\Omega |u|^2\, dx.
\end{aligned}$$

Then, by Sobolev imbeddings (2.2) a positive constant c_4 exists such that

$$J_0(u) \leq c_4\|u\|^2 - \int_\Omega \gamma_{s_0}(x)|u|^\mu\, dx. \tag{3.7}$$

By (3.7), it follows that $J_0(tu) \to -\infty$ as $t \to +\infty$. Whence, the classical Mountain Pass Theorem applies (see [2, Theorem 2.1]) and a non zero critical point of J_0 in E_G, hence a non trivial weak solution of system $(\mathcal{P}_0)$ exists. Furthermore, if also condition (g_4) holds, the functional J_0 is even. Let V be a finite-dimensional subspace of E_G. Now, it is easy to prove that the term $\left(\int_\Omega \gamma_{s_0}(x)|u|^\mu\, dx\right)^{\frac{1}{\mu}}$ is a norm in E_G, hence by (3.7) and the equivalence of all norms in V, there exists a positive constant $R = R(V)$ such that

$$J_0(u) \le 0 \quad \text{if } u \in V,\ \|u\| \ge R. \tag{3.8}$$

So the symmetric version of Mountain Pass Theorem applies (see [2, Corollary 2.9]) and I has an unbounded sequence of critical levels. □

4. Bolle's perturbation arguments

From now on, we will study the perturbation problem $(\mathcal{P}_f)$ by applying the method introduced by Bolle in [14] for dealing with problems with broken symmetry. First, we recall the main abstract theorem as stated in [15]. Remark that the theorem holds for C^2 functionals, but here we assume they are C^1 according to [20]. The idea is to consider a continuous path of functionals starting from a symmetric functional J_0 and to prove a preservation result for min-max critical levels in order to get critical points also for the end-point functional J_1 (which is the "true" functional associated to the non-symmetric problem).

Let H be a Hilbert space equipped with the norm $\|\cdot\|_H$. Assume that $H = H_- \oplus H_+$, where $\dim(H_-) < \infty$, and let $(e_k)_k$ be an orthonormal basis of H_+. Consider

$$H_0 = H_-, \quad H_{k+1} = H_k \oplus \mathbb{R}e_{k+1} \quad \text{if } k \in \mathbb{N},$$

so $(H_k)_k$ is an increasing sequence of finite-dimensional subspaces of H.

Let $J : [0,1] \times H \to \mathbb{R}$ be a C^1-functional and, taken $\theta \in [0,1]$, let us set $J_\theta = J(\theta,\cdot) : H \to \mathbb{R}$ and $J'_\theta(u) = \partial J(\theta,u)/\partial u$. Furthermore, let

$$\Gamma = \{\gamma \in C(H,H) : \gamma \text{ odd and there exists } L > 0 \text{ s.t. } \gamma(u) = u \text{ if } \|u\|_H \ge L\},$$

$$c_k = \inf_{\gamma\in\Gamma} \sup_{u\in H_k} J_0(\gamma(u)).$$

Assume that:

(A_1) J satisfies the following weaker form of the classical Palais–Smale condition: any $((\theta_n,u_n))_n \subset [0,1] \times H$ such that

$$(J(\theta_n,u_n))_n \text{ is bounded}, \qquad \lim_{n\to\infty} J'_{\theta_n}(u_n) = 0 \tag{4.1}$$

converges up to a subsequence;

(A_2) for any $b > 0$ there exists $\overline{C}_b > 0$ such that if $(\theta,u) \in [0,1] \times H$ then

$$|J_\theta(u)| \le b \Longrightarrow \left|\frac{\partial J}{\partial \theta}(\theta,u)\right| \le \overline{C}_b(\|J'_\theta(u)\|_{H'} + 1)(\|u\|_H + 1);$$

(A_3) there exist two continuous maps $\eta_1, \eta_2 : [0,1] \times \mathbb{R} \to \mathbb{R}$, Lipschitz continuous with respect to the second variable, such that $\eta_1(\theta,\cdot) \le \eta_2(\theta,\cdot)$ and if $(\theta,u) \in$

$[0,1] \times H$ then

$$J'_\theta(u) = 0 \Longrightarrow \eta_1(\theta, J_\theta(u)) \le \frac{\partial J}{\partial \theta}(\theta, u) \le \eta_2(\theta, J_\theta(u));$$

(A_4) J_0 is even and for each finite-dimensional subspace V of H it results

$$\lim_{\substack{u\in V \\ \|u\|_H \to +\infty}} \sup_{\theta\in[0,1]} J(\theta, u) = -\infty.$$

For $i \in \{1,2\}$, let $\Psi_i : [0,1] \times \mathbb{R} \to \mathbb{R}$ be the flow associated to η_i, i.e., the solution of problem

$$\begin{cases} \dfrac{\partial \Psi_i}{\partial \theta}(\theta, s) = \eta_i(\theta, \Psi_i(\theta, s)), \\ \Psi_i(0, s) = s. \end{cases}$$

Note that $\Psi_i(\theta, \cdot)$ is continuous, non-decreasing on $\mathbb{R}$ and $\Psi_1(\theta, \cdot) \le \Psi_2(\theta, \cdot)$. Set

$$\overline{\eta}_1(s) = \sup_{\theta\in[0,1]} |\eta_1(\theta, s)|, \quad \overline{\eta}_2(s) = \sup_{\theta\in[0,1]} |\eta_2(\theta, s)|.$$

In this framework, the following abstract result can be proved (see [14, Theorem 3] and [15, Theorem 2.2]).

Theorem 4.1. *There exists $\widetilde{C} \in \mathbb{R}$ such that if $k \in \mathbb{N}$ then*

(i) *either J_1 has a critical level $\widetilde{c}_k$ with $\Psi_2(1, c_k) < \Psi_1(1, c_{k+1}) \le \widetilde{c}_k$,*

(ii) *or $c_{k+1} - c_k \le \widetilde{C}(\overline{\eta}_1(c_{k+1}) + \overline{\eta}_2(c_k) + 1)$.*

Remark 4.2. If $\eta_2 \ge 0$ in $[0,1] \times \mathbb{R}$, the function $\Psi_2(\cdot, s)$ is non-decreasing on $[0,1]$. Hence, $c_k \le \widetilde{c}_k$ for every c_k satisfying case (i).

5. Preliminary lemmas

Now, in order to find multiple critical points of the non-even functional J_1 associated to $(\mathcal{P}_f)$ (see Proposition 2.1), we want to apply Bolle's perturbation method. Thus, consider the family of functionals $J : [0,1] \times E_G \to \mathbb{R}$ defined as

$$\begin{aligned} J(\theta, u) &= \frac{1}{2} \int_\Omega |\nabla u|^2 \, dx - \int_\Omega G(x, u) \, dx - \theta \int_\Omega f u \, dx \\ &= J(0, u) - \theta \int_\Omega f u \, dx. \end{aligned} \tag{5.1}$$

Clearly, $J(0, \cdot) = J_0$ is an even functional while $J(1, \cdot) = J_1$. By Proposition 2.1 we have that J is a C^1-functional with

$$\frac{\partial J}{\partial \theta}(\theta, \zeta) = -\int_\Omega f \zeta \, dx,$$

$$J'_\theta(u)[\zeta] = \frac{\partial J}{\partial u}(\theta, u)[\zeta] = \int_\Omega \nabla u \cdot \nabla \zeta \, dx - \int_\Omega g(x, u) \zeta \, dx - \theta \int_\Omega f \zeta \, dx$$

for every $\theta \in [0,1]$ and $u, \zeta \in E_G$.

The following technical lemmas state that the functional J in (5.1) satisfies assumptions (A_1)–(A_4) introduced in the previous section.

Lemma 5.1. *Let $g \in C(\Omega\times\mathbb{R},\mathbb{R})$ satisfying (g_0), (g_1), (g_2) and (g_3) and $f \in L^{\mu'}(\Omega)$. Then, if $(\theta_n, u_n)_n \subset [0,1]\times E_G$ is a sequence verifying (4.1), then it converges up to a subsequence.*

Proof. Let $(\theta_n, u_n)_n \subset [0,1]\times E_G$ be a sequence satisfying (4.1), hence

$$\frac{1}{2}\int_\Omega |\nabla u_n|^2\, dx - \int_\Omega G(x,u_n)\, dx - \theta_n \int_\Omega f u_n\, dx \le a_1$$

and

$$|J'_{\theta_n}(u_n)[u_n]| = \left|\int_\Omega |\nabla u_n|^2\, dx - \int_\Omega g(x,u_n)u_n\, dx - \theta_n \int_\Omega f u_n\, dx\right| \le \varepsilon_n \|u_n\|$$

where $\varepsilon_n \to 0$ as $n \to \infty$. As by (g_1)

$$\begin{aligned} a_2 + \varepsilon_n\|u_n\| &\ge \mu J_{\theta_n}(u_n) - J'_{\theta_n}(u_n)[u_n] \ge \left(\frac{\mu}{2}-1\right)\|u_n\|^2 - (\mu-1)\,\theta_n \int_\Omega f u_n\, dx \\ &\ge a_3\|u_n\|^2 - a_4\|u_n\|, \end{aligned}$$

it follows that $(u_n)_n$ is bounded in E_G; hence, it converges weakly in E_G up to a subsequence. Thus, the proof follows easily by (2.7) and standard arguments. □

Lemma 5.2. *Let $g \in C(\Omega\times\mathbb{R},\mathbb{R})$ satisfying (g_0), (g_2) and (g_3) and $f \in L^{\mu'}(\Omega)$. Then, for any $b > 0$ there exists $\overline{C}_b > 0$ such that if $(\theta, u) \in [0,1]\times E_G$ it is*

$$|J_\theta(u)| \le b \Longrightarrow \left|\frac{\partial J}{\partial \theta}(\theta,u)\right| \le \overline{C}_b(\|J'_\theta(u)\|_{E'_G} + 1)(\|u\| + 1).$$

Proof. The expression of $\frac{\partial J}{\partial \theta}(\theta,u)$, the Hölder inequality and Sobolev embeddings imply

$$\left|\frac{\partial J}{\partial \theta}(\theta,u)\right| \le \|f\|_{L^{\mu'}(\Omega)}\|u\|_{L^\mu(\Omega)} \le a_5\|u\| \quad \text{for all } (\theta,u)\in[0,1]\times E_G,$$

so the conclusion follows. □

Since we want to determine the "control" functions $\eta_i(\theta, s)$ in (A_3), we prove the following

Lemma 5.3. *Let $g \in C(\Omega\times\mathbb{R},\mathbb{R})$ satisfying (g_0), (g_1), (g_2), (g_3) and (g_5) and $f \in L^{\mu'}(\Omega)$. Then, there exists a constant $\overline{C} > 0$ such that*

$$(\theta,u)\in[0,1]\times E_G,\quad J'_\theta(u) = 0 \Longrightarrow \left|\frac{\partial J}{\partial \theta}(\theta,u)\right| \le \overline{C}(J^2_\theta(u)+1)^{\frac{1}{2\mu}}.$$

Proof. Let $(\theta,u)\in[0,1]\times E_G$ such that $J'_\theta(u)=0$. By (g_1), (g_5) and $f\in L^{\mu'}(\Omega)$, we have that

$$\begin{aligned} J_\theta(u)=J_\theta(u)-\frac{1}{2}J'_\theta(u)[u] &= \int_\Omega(\frac{1}{2}g(x,u)u-G(x,u)\,dx-\frac{\theta}{2}\int_\Omega fu\,dx \\ &\geq \Big(\frac{\mu}{2}-1\Big)\int_\Omega G(x,u)\,dx-\frac{\theta}{2}\int_\Omega fu\,dx \\ &\geq a_6\|u\|^\mu_{L^\mu(\Omega)}-\|f\|^{\mu'}_{L^{\mu'}(\Omega)}\|u\|_{L(\Omega)} \\ &\geq a_7(\|u\|^\mu_{L^\mu(\Omega)}-1). \end{aligned}$$

Then, by the Hölder inequality

$$\left|\frac{\partial J}{\partial\theta}(\theta,u)\right|\leq\|f\|_{L^{\mu'}(\Omega)}\|u\|_{L^\mu(\Omega)}\leq\overline{C}(J^2_\theta(u)+1)^{\frac{1}{2\mu}}$$

and (A_3) holds with $\eta_2(\theta,s)=-\eta_1(\theta,s)=\overline{C}(1+s^2)^{\frac{1}{2\mu}}$. □

Lemma 5.4. *Let $g\in C(\Omega\times\mathbb{R},\mathbb{R})$ satisfying (g_0), (g_2), (g_3) and (g_5) and $f\in L^{\mu'}(\Omega)$. Then, for each finite-dimensional subspace V of E_G it results*

$$\lim_{\substack{u\in V\\ \|u\|\to+\infty}}\ \sup_{\theta\in[0,1]} J_\theta(u)=-\infty.$$

Proof. Since by (g_5) and the Hölder inequality

$$J(\theta,u)\leq\frac{1}{2}\|u\|^2-\gamma_0\|u\|^\mu+\|f\|_{L^{\mu'}(\Omega)}\|u\|_{L^\mu(\Omega)},$$

then the conclusion follows by $\mu>2$ and the equivalence of all norms in a finite-dimensional space. □

Remark 5.5. Let us point out that, arguing as in Section 3, the proof of Lemma 5.4 holds again by using (g_1) and (g_2) instead of (g_5). On the contrary, assumption (g_5) needs in the proof of Lemma 5.3.

6. Growth estimates and proof of Theorem 1.4

In order to apply Theorem 4.1, we consider a sequence of finite-dimensional subspaces of E_G as follows. Let us consider a Schauder basis $(u_k)_k$ in E_G. Then, we can define

$$H_k=\operatorname{span}\{u_1,\ldots,u_k\},\quad H^\perp_{k-1}=\overline{\operatorname{span}\{u_k,u_{k+1},\ldots\}},$$

and a corresponding sequence of min-max levels as

$$c_k=\inf_{\gamma\in\Gamma}\sup_{u\in H_k}J_0(\gamma(u)),\quad \text{for each } k\geq1, \tag{6.1}$$

where Γ is as in Section 4 with $H=E_G$. By the lemmas in the previous Section the path of functionals $(J_\theta)_{\theta\in[0,1]}$ satisfies assumptions (A_1)–(A_4) of Theorem 4.1 with

$$-\eta_1(\theta,s)=\eta_2(\theta,s)=\overline{C}(1+s^2)^{\frac{1}{2\mu}}. \tag{6.2}$$

So, Theorem 4.1 applies. In order to state the existence of infinitely many solutions of problem $(\mathcal{P}_f)$, by Remark 4.2, we have to prove that alternative (i) occurs for all k large enough. If, by contradiction, we assume that alternative (ii) occurs for k large enough, by the form of $\eta_i(\theta, s)$ in (6.2), it follows that

$$c_{k+1} - c_k \leq \widetilde{C}(c_{k+1}^{\frac{1}{\mu}} + c_k^{\frac{1}{\mu}} + 1).$$

Therefore, arguing as in [8, Lemma 5.3], a constant C_1 and an integer k_0 exist such that

$$c_k \leq C_1 k^{\frac{\mu}{\mu-1}} \quad \text{for all } k \geq k_0. \tag{6.3}$$

In order to have a contradiction, we need a suitable lower estimate of the growth of $c_k's$.

Following an idea of Tanaka in [31], let us point out that, by (2.6) for $\varepsilon = \frac{1}{4}$, a positive constant $C > 0$ exists such that

$$J_0(u) \geq \frac{1}{4}\|u\|^2 - C\|u\|^p_{L^p(\Omega)}$$

so it is

$$c_k \geq b_k \tag{6.4}$$

where

$$b_k = \inf_{\gamma\in\Gamma} \sup_{u\in H_k} K(\gamma(u)), \quad K(u) = \frac{1}{4}\|u\|^2 - C\|u\|^p_{L^p(\Omega)}.$$

Then, by (2.7), we can apply Theorem B in [31] so that, for all $k \in \mathbb{N}$ there exists a critical point $u_k \in E_G$ of K such that

$$K(u_k) \leq b_k \tag{6.5}$$

and its large Morse index is greater or equal than k, i.e., the operator

$$K''(u_k) = -\frac{1}{2}\Delta - Cp(p-1)|u_k|^{p-2} \quad \text{or equivalently} \quad -\Delta - 2Cp(p-1)|u_k|^{p-2}$$

has at least k non-positive eigenvalues.

Now, we deal with the case $m \geq 2$ so that $m + 1 \geq 3$. The case $m = 1$ can be treated in a simpler way. Thanks to the partially spherically symmetry of u_k for all integer k, we can apply Proposition 7.3 in Appendix 7 with the potential $V = -2Cp(p-1)|u_k|^{p-2}$, so we have

$$\begin{aligned} k &\leq N_-(-\Delta - 2Cp(p-1)|u_k|^{p-2}) \\ &\leq \overline{C}_m \int_{\widetilde{\Omega}\times[0,+\infty)} |u_k(\widetilde{x}, \rho)|^{\frac{(p-2)(m+1)}{2}} \, d\widetilde{x}d\rho. \end{aligned} \tag{6.6}$$

Moreover, since u_k is a critical point of K, by (6.5) it follows that

$$b_k \geq \frac{p-2}{4p}\|u_k\|^2 = \frac{p-2}{2}C\|u_k\|^p_{L^p(\Omega)}. \tag{6.7}$$

Now, our aim is to estimate the last integral in (6.6). Then, arguing as in [17] (see also [4]), by the Hölder inequality, it follows that

$$\int_{\tilde{\Omega}\times[0,1]} |u_k(\tilde{x},\rho)|^{\frac{(p-2)(m+1)}{2}}\, d\tilde{x}d\rho \tag{6.8}$$
$$\leq \left(\int_{\tilde{\Omega}\times[0,1]} \rho^l\, d\tilde{x}d\rho\right)^{\frac{2p-(p-2)(m+1)}{2p}} \left(\int_{\tilde{\Omega}\times[0,1]} \rho^{N-m-1}|u_k(\tilde{x},\rho)|^p\, d\tilde{x}d\rho\right)^{\frac{(p-2)(m+1)}{2p}}$$

with $l = -\frac{(N-m-1)(p-2)(m+1)}{2p-(p-2)(m+1)}$. Let us point out that $N-m \geq 2$ implies that $2p-(p-2)(m+1) > 0$, for any $p \in (2, 2^*)$. Clearly, if

$$p < 2 + \frac{4}{(N-m)(m+1)-2}, \tag{6.9}$$

then $l > -1$ and (6.7) and (6.8) imply

$$\int_{\tilde{\Omega}\times[0,1]} |u_k(\tilde{x},\rho)|^{\frac{(p-2)(m+1)}{2}}\, d\tilde{x}d\rho \leq C_2\|u_k\|_{L^p(\Omega)}^{\frac{(p-2)(m+1)}{2}} \leq C_3 b_k^{\frac{(p-2)(m+1)}{2p}}. \tag{6.10}$$

On the other hand, applying again the Hölder inequality, for suitable $r > 2$, it is

$$\begin{aligned}\int_{\tilde{\Omega}\times[1,+\infty)} |u_k(\tilde{x},\rho)|^{\frac{(p-2)(m+1)}{2}}\, d\tilde{x}d\rho &\leq \left(\int_{\tilde{\Omega}\times[1,+\infty)} \rho^L\, d\tilde{x}d\rho\right)^{\frac{2r-(r-2)(m+1)}{2r}} \\ &\times \left(\int_{\tilde{\Omega}\times[1,+\infty)} \rho^{N-m-1}|u_k(\tilde{x},\rho)|^{\frac{p-2}{r-2}r}\, d\tilde{x}d\rho\right)^{\frac{(r-2)(m+1)}{2r}}\end{aligned} \tag{6.11}$$

with $L < -1$ where $L = -\frac{(N-m-1)(r-2)(m+1)}{2r-(r-2)(m+1)}$. Hence, by (6.11)

$$\begin{aligned}&\int_{\tilde{\Omega}\times[1,+\infty)} |u_k(\tilde{x},\rho)|^{\frac{(p-2)(m+1)}{2}}\, d\tilde{x}d\rho \\ &\leq C_4\left(\int_{\tilde{\Omega}\times[1,+\infty)} \rho^{N-m-1}|u_k(\tilde{x},\rho)|^{\frac{p-2}{r-2}r}\, d\tilde{x}d\rho\right)^{\frac{(r-2)(m+1)}{2r}} \\ &\leq C_4\, \|u_k\|_{L^{\frac{p-2}{r-2}r}(\Omega)}^{\frac{(p-2)(m+1)}{2}}.\end{aligned} \tag{6.12}$$

Let us point out that in the previous estimates we need to choose r such that

$$\begin{cases} 0 < \dfrac{2r-(r-2)(m+1)}{2r} < 1, \\ \dfrac{(N-m-1)(r-2)(m+1)}{2r-(r-2)(m+1)} > 1, \\ 2 \leq \dfrac{p-2}{r-2}r < p. \end{cases}$$

From direct calculations, recalling that $N \geq 4$ and then $2 < p < 4$, this system is solvable if we choose $r > 2$ such that

$$\max\left\{p, \frac{2(N-m)(m+1)}{(N-m)(m+1)-2}\right\} < r < \min\left\{\frac{4}{4-p}, 2+\frac{4}{m-1}\right\} \tag{6.13}$$

and this is possible for all p verifying

$$2+\frac{4}{(N-m)(m+1)} < p < 2^*. \tag{6.14}$$

Clearly, if we assume p satisfying (6.9) and (6.14), condition (6.13) becomes

$$\frac{2(N-m)(m+1)}{(N-m)(m+1)-2} < r < \min\left\{\frac{4}{4-p}, 2+\frac{4}{m-1}\right\}. \tag{6.15}$$

So, if (6.14) holds, by (6.12), (6.7) and the Gagliardo–Nirenberg interpolation inequality (see [16]) it follows that

$$\begin{aligned}\int_{\widetilde{\Omega}\times[1,+\infty)} |u_k(\widetilde{x},\rho)|^{\frac{(p-2)(m+1)}{2}}\, d\widetilde{x}d\rho &\leq C_4\|u_k\|_{L^{\frac{p-2}{r-2}r}(\Omega)}^{\frac{(p-2)(m+1)}{2}}\\ &\leq C_5\left(\|u_k\|_{L^2(\Omega)}^{a}\|u_k\|_{L^p(\Omega)}^{1-a}\right)^{\frac{(p-2)(m+1)}{2}}\\ &\leq C_6\left(b_k^{\frac{a}{2}}b_k^{\frac{1-a}{p}}\right)^{\frac{(p-2)(m+1)}{2}} = C_6\, b_k^{\frac{(r-2)(m+1)}{2r}}\end{aligned} \tag{6.16}$$

where $0 < a < 1$ and $\frac{a}{2}+\frac{(1-a)}{p} = \frac{r-2}{(p-2)r}$. Then, for p as in (6.9) and (6.14), by (6.6), (6.10) and (6.16) for any integer k we have that

$$k \leq C_7\, b_k^{\frac{(p-2)(m+1)}{2p}} + C_8\, b_k^{\frac{(r-2)(m+1)}{2r}};$$

therefore $b_k \to +\infty$ and, since $p < r$, it follows that, for k large,

$$k \leq C_8\, b_k^{\frac{(r-2)(m+1)}{2r}}$$

and consequently by (6.4)

$$c_k \geq C_9\, k^{\frac{2r}{(r-2)(m+1)}} \quad \text{if } k \text{ is large enough.}$$

Since $\frac{2r}{(r-2)(m+1)} \uparrow N-m$ if $r \downarrow \frac{2(N-m)(m+1)}{(N-m)(m+1)-2}$ (see (6.15)), it follows that, if p satisfies (6.9) and (6.14), for any $\delta > 0$ and k large, it is

$$c_k \geq C_\delta\, k^{N-m-\delta}. \tag{6.17}$$

Really, (6.17) holds for any $2 < p < 2+\frac{4}{(N-m)(m+1)-2}$. To this aim, we recall the following result.

Lemma 6.1. *Assume that* $2 < p \leq 2+\frac{4}{(N-m)(m+1)}$. *Then, for some* $\overline{p}$ *such that* $2+\frac{4}{(N-m)(m+1)} < \overline{p} < 2+\frac{4}{(N-m)(m+1)-2}$, *for all* $\varepsilon > 0$ *a positive constant* $A_\varepsilon > 0$ *exists such that*

$$\int_{\mathbb{R}^N} |u|^p\, dx \leq \varepsilon \int_{\mathbb{R}^N} |u|^2\, dx + A_\varepsilon \int_{\mathbb{R}^N} |u|^{\overline{p}}\, dx, \quad \textit{for all } u \in E_G.$$

Proof. Let $s > 1$ and $\alpha > 0$ real numbers that will be fixed later. By applying Young's inequality, we have that

$$|u|^p = (\varepsilon s)^{\frac{1}{s}} |u|^\alpha \frac{1}{(\varepsilon s)^{\frac{1}{s}}} |u|^{p-\alpha} \le \varepsilon |u|^{\alpha s} + \frac{1}{s'(\varepsilon s)^{\frac{s'}{s}}} |u|^{(p-\alpha)s'}.$$

Our aim is to prove the existence of suitable α and s such that

$$\begin{cases} \alpha s = 2, \\ 2 + \frac{4}{(N-m)(m+1)} < (p-\alpha)\frac{s}{s-1} < 2 + \frac{4}{(N-m)(m+1)-2} \end{cases}$$

or, equivalently, if $2 < p < 2 + \frac{4}{(N-m)(m+1)}$,

$$\begin{cases} \alpha = \frac{2}{s}, \\ \frac{4}{2p-(p-2)(N-m)(m+1)} < s < \frac{4}{4-(p-2)(N-m)(m+1)}. \end{cases} \tag{6.18}$$

Since it is easy to see that

$$1 < \frac{4}{2p-(p-2)(N-m)(m+1)} < \frac{4}{4-(p-2)(N-m)(m+1)},$$

taking s as in (6.18), the proof concludes with $\overline{p} = (p-\alpha)\frac{s}{s-1}$, i.e., $\overline{p} = \frac{sp-2}{s-1}$. If $p = 2 + \frac{4}{(N-m)(m+1)}$, the thesis follows from simpler arguments. □

The previous lemma implies that, if $2 < p \le 2 + \frac{4}{(N-m)(m+1)}$, for a suitable $\overline{p}$ such that

$$2 + \frac{4}{(N-m)(m+1)} < \overline{p} < 2 + \frac{4}{(N-m)(m+1)-2},$$

for $\varepsilon > 0$ small enough two positive constants B_ε and C_ε exist such that, for all $u \in E_G$,

$$J_0(u) \ge \overline{K}(u) \quad \text{with} \quad \overline{K}(u) = B_\varepsilon \|u\|^2 - C_\varepsilon \|u\|^{\overline{p}}_{L^{\overline{p}}(\Omega)}.$$

So, denoting by $\overline{b}_k$ the min-max levels of $\overline{K}$ defined by (6.1), it follows that

$$c_k \ge \overline{b}_k. \tag{6.19}$$

On the other hand, by applying to the functional $\overline{K}$ Theorem B in [31], Proposition 7.3 in Appendix 7 and the subsequent arguments, estimate (6.17) holds also for the critical levels $\overline{b}_k$ of $\overline{K}$ and then, from (6.19) for the critical levels c_k of J_0. Hence, we conclude that (6.17) holds for all p between 2 and $2 + \frac{4}{(N-m)(m+1)-2}$. Finally, since

$$\frac{\mu}{\mu-1} < N-m$$

holds since $N - m \ge 3$, (6.17) is in contradiction with (6.3) for k large. So, alternative (i) of Theorem 4.1 and Remark 4.2 occur for all k large enough and the existence of infinitely many weak solutions of problem $(\mathcal{P}_f)$ follows if p is such that $2 < p < \overline{p}_{N,m}$ where $\overline{p}_{N,m} = 2 + \frac{4}{(N-m)(m+1)-2}$.

7. Appendix

In this section, we give a suitable estimate of the number $N_-(-\Delta+V(x))$ of the non-positive eigenvalues of the operator $-\Delta+V(x)$ in E_G. Let us recall that for a general potential V the following proposition holds (see [21, 23, 24, 29, 31]).

Proposition 7.1. *Let $N\geq 2$ and $V:\Omega\to\mathbb{R}$. Then,*

(i) *if $N\geq 3$, there is a constant $\widetilde{C}_N>0$ such that*

$$N_-(-\Delta+V(x))\leq \widetilde{C}_N\|V_-(x)\|^{\frac{N}{2}}_{L^{\frac{N}{2}}(\Omega)};$$

(ii) *if $N=2$, for any $\varepsilon>0$ there is a constant $\widetilde{C}_\varepsilon>0$ such that*

$$N_-(-\Delta+V(x))\leq \widetilde{C}_\varepsilon\|V_-(x)\|^{\varepsilon+1}_{L^{\varepsilon+1}(\Omega)}$$

where $V_-(x)=\min\{0,V(x)\}$.

Now, assume that the potential $V:\Omega\to\mathbb{R}$ is partially spherically symmetric, i.e.,
(V_0) $V(\widetilde{x},y_1)=V(\widetilde{x},y_2)$ for every $\widetilde{x}\in\widetilde{\Omega}$ and $y_1,y_2\in\mathbb{R}^{N-m}$, $|y_1|=|y_2|$.
Setting $\rho=|y|$ with $y\in\mathbb{R}^{N-m}$, we state the following result useful for improving the previous estimates of the number $N_-(-\Delta+V(x))$ when partially spherical symmetry occurs.

Lemma 7.2. *Let $N-m\geq 3$ and $V:\Omega\to\mathbb{R}$ verifying (V_0). Then, λ is an eigenvalue of $-\Delta+V(x)$ in E_G with eigenfunction φ if and only if it is an eigenvalue of*

$$-\Delta_{\widetilde{x}}-\frac{\partial^2}{\partial\rho^2}+\frac{(N-m-1)(N-m-3)}{4\rho^2}+V(\widetilde{x},\rho)$$

in $W^{1,2}_0(\widetilde{\Omega}\times(0,+\infty))$ with eigenfunction $\widetilde{\varphi}=\varphi\rho^{\frac{N-m-1}{2}}$.

Proof. By the partially spherical symmetry of the problem, λ is an eigenvalue of $-\Delta+V(x)$ in E_G with eigenfunction φ if and only if λ is an eigenvalue of

$$-\Delta_{\widetilde{x}}-\frac{\partial^2}{\partial\rho^2}-\frac{N-m-1}{\rho}\frac{\partial}{\partial\rho}+V(\widetilde{x},\rho)$$

in $W^{1,2}_0(\widetilde{\Omega}\times(0,+\infty))$ with eigenfunction φ. This is equivalent to study the equation

$$\Delta_{\widetilde{x}}\varphi+\frac{\partial^2\varphi}{\partial\rho^2}+p_1(\rho)\frac{\partial\varphi}{\partial\rho}+p_2(\widetilde{x},\rho)\varphi=0 \tag{7.1}$$

where $p_1(\rho)=\frac{N-m-1}{\rho}$ and $p_2(\widetilde{x},\rho)=\lambda-V(\widetilde{x},\rho)$. By the classical change of variable $\varphi=\widetilde{\varphi}e^{-\frac{1}{2}\int_1^\rho p_1(s)\,ds}$, (7.1) is equivalent to

$$\Delta_{\widetilde{x}}\widetilde{\varphi}+\frac{\partial^2\widetilde{\varphi}}{\partial\rho^2}+\left(-\frac{1}{4}p_1^2(\rho)-\frac{1}{2}p_1'(\rho)+p_2(\widetilde{x},\rho)\right)\widetilde{\varphi}=0$$

namely

$$\Delta_{\widetilde{x}}\widetilde{\varphi}+\frac{\partial^2\widetilde{\varphi}}{\partial\rho^2}+\left(-\frac{1}{4}\left(\frac{N-m-1}{\rho}\right)^2+\frac{1}{2}\frac{N-m-1}{\rho^2}+(\lambda-V(\widetilde{x},\rho))\right)\widetilde{\varphi}=0.$$

Then, λ is an eigenvalue of

$$-\Delta_{\widetilde{x}} - \frac{\partial^2}{\partial \rho^2} + \frac{(N-m-1)(N-m-3)}{4\rho^2} + V(\widetilde{x}, \rho)$$

in $W_0^{1,2}(\widetilde{\Omega} \times (0, +\infty))$ with eigenfunction $\widetilde{\varphi} = \varphi \rho^{\frac{N-m-1}{2}}$. □

Now, we are ready to prove the following result.

Proposition 7.3. *Let $N - m \geq 3$ and $V : \Omega \to \mathbb{R}$ verifying (V_0).*

(i) *If $m \geq 2$ and $V \in L^{\frac{m+1}{2}}(\Omega)$, there exists a constant $\overline{C}_m > 0$ such that*

$$N_-(-\Delta + V(x)) \leq \overline{C}_m \int_{\widetilde{\Omega} \times [0,+\infty)} |V_-(\widetilde{x}, \rho)|^{\frac{m+1}{2}} \, d\widetilde{x} d\rho.$$

(ii) *If $m = 1$ and $V \in L^{\varepsilon+1}(\Omega)$, for any $\varepsilon > 0$ there is a constant $\overline{C}_\varepsilon > 0$ such that*

$$N_-(-\Delta + V(x)) \leq \overline{C}_\varepsilon \int_{\widetilde{\Omega} \times [0,+\infty)} |V_-(\widetilde{x}, \rho)|^{\varepsilon+1} \, d\widetilde{x} d\rho.$$

Proof. (i) By Lemma 7.2 it is

$$N_-(-\Delta + V(x)) = N_-(-\Delta_{\widetilde{x}} - \frac{\partial^2}{\partial \rho^2} + \frac{(N-m-1)(N-m-3)}{4\rho^2} + V(\widetilde{x}, \rho)).$$

Hence, $N - m \geq 3$ and Proposition 7.1 (i) imply

$$\begin{aligned} &N_-(-\Delta_{\widetilde{x}} - \frac{\partial^2}{\partial \rho^2} + \frac{(N-m-1)(N-m-3)}{4\rho^2} + V(\widetilde{x}, \rho)) \\ &\leq N_-(-\Delta_{\widetilde{x}} - \frac{\partial^2}{\partial \rho^2} + V(\widetilde{x}, \rho)) \leq \overline{C}_m \int_{\widetilde{\Omega} \times [0,+\infty)} |V_-(\widetilde{x}, \rho)|^{\frac{m+1}{2}} \, d\widetilde{x} d\rho. \end{aligned}$$

(ii) It is enough to apply Lemma 7.2 and Proposition 7.1 (ii). □

References

[1] R.A. Adams, *Sobolev Spaces*, Academic Press, New York, 1975.

[2] A. Ambrosetti, P.H. Rabinowitz, *Dual variational methods in critical point theory and applications*, J. Funct. Anal. 14 (1973), 349–381.

[3] A. Azzollini, A. Pomponio, *Compactness results and applications to some "zero mass" elliptic problems*, Nonlinear Anal. 69 (2008), 3559–3576.

[4] S. Barile, A. Salvatore, *Radial solutions of semilinear elliptic equations with broken symmetry on unbounded domains*, Discrete and Continuous Dynamical Systems, Supplement 2013, 41–49.

[5] S. Barile, A. Salvatore, *Weighted elliptic systems of Lane–Emden type in unbounded domains*, Mediterr. J. Math. 9 (2012), 409–422.

[6] S. Barile, A. Salvatore, *Existence and multiplicity results for some elliptic systems in unbounded cylinders*, Milan J. Math. 81 (2013), 99–120.

[7] M. Badiale, L. Pisani, S. Rolando, *Sum of weighted spaces and nonlinear elliptic equations*, NoDEA Nonlinear Differential Equations Appl. 18 (2011), 369–405.

Progress in Nonlinear Differential Equations
and Their Applications, Vol. 85, 61–86

Basic Properties of Ultrafunctions

Vieri Benci and Lorenzo Luperi Baglini

Dedicated to Bernard Ruf in occasion of his 60th birthday

Abstract. Ultrafunctions are a particular class of functions defined on a non-Archimedean field $\mathbb{R}^* \supset \mathbb{R}$. They provide generalized solutions to functional equations which do not have any solutions among the real functions or the distributions. In this paper we analyze systematically some basic properties of the spaces of ultrafunctions.

Mathematics Subject Classification (2010). 26E30, 26E35, 46F30.

Keywords. Ultrafunctions, delta function, distributions, non-Archimedean mathematics, non-standard analysis.

1. Introduction

In some recent papers the notion of ultrafunction has been introduced ([1], [2]). Ultrafunctions are a particular class of functions defined on a non-Archimedean field $\mathbb{R}^* \supset \mathbb{R}$. We recall that a non-Archimedean field is an ordered field which contain infinite and infinitesimal numbers.

To any continuous function $f : \mathbb{R}^N \to \mathbb{R}$ we associate in a canonical way an ultrafunction $\widetilde{f} : (\mathbb{R}^*)^N \to \mathbb{R}^*$ which extends f; more exactly, to any functional vector space $V(\Omega) \subseteq L^2(\Omega) \cap \mathcal{C}(\overline{\Omega})$, we associate a space of ultrafunctions $\widetilde{V}(\Omega)$. The ultrafunctions are much more than the functions and among them we can find solutions of functional equations which do not have any solutions among the real functions or the distributions.

A typical example of this situation is analyzed in [2] where a simple Physical model is studied. In this problem there is a material point interacting with a field and, as it usually happens, the energy is infinite. Therefore the need to use infinite numbers arises naturally. Other situations in which infinite and infinitesimal numbers appear in a natural way are studied in [5], in [6] and in section 4.4.

In this paper we analyze systematically some basic properties of the spaces of ultrafunctions $\widetilde{V}(\Omega)$. In particular we will show that:

- to any measurable function f we can associate an unique ultrafunction $\widetilde{f}$ such that $f(x) = \widetilde{f}(x)$ if f is continuous in a neighborhood of x;

- to every distribution T we can associate an ultrafunction $\widetilde{T}(x)$ such that $\forall\varphi \in \mathcal{D}$, $\langle T,\varphi\rangle = \int^* \widetilde{T}(x)\widetilde{\varphi}(x)dx$ where $\int^*$ is a suitable extension of the integral to the ultrafunctions;
- the vector space of ultrafunctions $\widetilde{V}(\Omega)$ is hyperfinite, namely it shares many properties of finite vector spaces (see Sect. 2.4);
- the vector space of ultafunctions $\widetilde{V}(\Omega)$ has an hyperfinite basis $\{\delta_a(x)\}_{a\in\Sigma}$ where δ_a is the "Dirac ultrafunction in a" (see Def. 18) and $\Sigma \subset (\mathbb{R}^*)^N$ is a suitable set;
- any ultrafunction u can be represented as follows:

$$u(x) = \sum_{q\in\Sigma} u(q)\sigma_q(x),$$

  where $\{\sigma_a(x)\}_{a\in\Sigma}$ is the dual basis of $\{\delta_a(x)\}_{a\in\Sigma}$;
- any operator $F : V(\Omega) \to \mathcal{D}'(\Omega)$, can be extended to an operator

$$\widetilde{F} : \widetilde{V}(\Omega) \to \widetilde{V}(\Omega);$$

  the extension of the derivative and the Fourier transform will be analyzed in some detail.

The techniques on which the notion of ultrafunction is based are related to non-Archimedean Mathematics (NAM) and to non-standard analysis (NSA). The first section of this paper is devoted to a relatively elementary presentation of the basic notions of NAM and NSA inspired by [3] and [4]. Some technicalities have been avoided by presenting the matter in an axiomatic way. Of course, it is necessary to prove the consistency of the axioms. This is done in the appendix; however in the appendix we have assumed the reader to be familiar with NSA.

1.1. Notation

Let Ω be a subset of $\mathbb{R}^N$: then

- $\mathcal{C}(\Omega)$ denotes the set of continuous functions defined on $\Omega \subset \mathbb{R}^N$;
- $\mathcal{C}^k(\Omega)$ denotes the set of functions defined on $\Omega \subset \mathbb{R}^N$ which have continuous derivatives up to the order k;
- $\mathcal{D}(\Omega)$ denotes the set of the infinitely differentiable functions with compact support defined on $\Omega \subset \mathbb{R}^N$; $\mathcal{D}'(\Omega)$ denotes the topological dual of $\mathcal{D}(\Omega)$, namely the set of distributions on Ω;
- $H^{1,p}(\Omega)$ is the usual Sobolev space defined as the set of functions in $L^p(\Omega)$ such that $\nabla u \in L^p(\Omega)^N$;
- $H^1(\Omega) = H^{1,2}(\Omega)$;
- if V is a finite-dimensional vector space, V' will denote its dual; if V is a Banach space, V' will denote its (topological) dual;
- $\mathfrak{supp}(f) = \overline{\{x \in \mathbb{R}^N : f(x) \neq 0\}}$;
- $\mathfrak{mon}(x) = \{y \in \mathbb{R}^N : x \sim y\}$;
- $\mathfrak{gal}(x) = \{y \in \mathbb{R}^N : x \sim_f y\}$.

2. Λ-theory

In this section we present the basic notions of non-Archimedean Mathematics and of non-standard analysis following a method inspired by [3] (see also [1] and [2]).

2.1. Non-Archimedean fields

Here, we recall the basic definitions and facts regarding non-Archimedean fields, namely fields that contain infinite and infinitesimal numbers. In the following, $\mathbb{K}$ will denote an ordered field. We recall that such a field contains (a copy of) the rational numbers. Its elements will be called numbers.

Definition 1. Let $\mathbb{K}$ be an ordered field. Let $\xi \in \mathbb{K}$. We say that:

- ξ is infinitesimal if, for all positive $n \in \mathbb{N}$, $|\xi| < \frac{1}{n}$;
- ξ is finite if there exists $n \in \mathbb{N}$ such as $|\xi| < n$;
- ξ is infinite if, for all $n \in \mathbb{N}$, $|\xi| > n$ (equivalently, if ξ is not finite).

Definition 2. An ordered field $\mathbb{K}$ is called non-Archimedean if it contains an infinitesimal $\xi \neq 0$.

It is easily seen that all infinitesimal are finite, that the inverse of an infinite number is a nonzero infinitesimal number, and that the inverse of a nonzero infinitesimal number is infinite.

Definition 3. A superreal field is an ordered field $\mathbb{K}$ that properly extends $\mathbb{R}$.

It is easy to show, due to the completeness of $\mathbb{R}$, that there are nonzero infinitesimal numbers and infinite numbers in any superreal field. Infinitesimal numbers can be used to formalize a new notion of "closeness":

Definition 4. We say that two numbers $\xi, \zeta \in \mathbb{K}$ are infinitely close if $\xi - \zeta$ is infinitesimal. In this case, we write $\xi \sim \zeta$. Moreover, we say that ξ, ζ are finitely close if $\xi - \zeta$ is finite. In this case, we write $\xi \sim_f \zeta$.

Clearly, the relation "$\sim$" of infinite closeness is an equivalence relation. This leads to consider its equivalence classes:

Definition 5. Let $\mathbb{K}$ be a superreal field, and $\xi \in \mathbb{K}$ a number. The monad of ξ is the set of all numbers that are infinitely close to it:

$$\mathfrak{mon}(\xi) = \{\zeta \in \mathbb{K} : \xi \sim \zeta\},$$

and the galaxy of ξ is the set of all numbers that are finitely close to it:

$$\mathfrak{gal}(\xi) = \{\zeta \in \mathbb{K} : \xi \sim_f \zeta \}.$$

By definition, it follows that the set of infinitesimal numbers is $\mathfrak{mon}(0)$ and that the set of finite numbers is $\mathfrak{gal}(0)$. In particular, given any infinitesimal number ξ, ξ is infinitely near to a real number (the number 0). So we could argue that, similarly, the monad of any number $\xi \in \mathbb{K}$ contains (exactly) one real number. This is clearly false if we take ξ infinite, but it is true whenever ξ is finite:

Theorem 6. *If $\mathbb{K}$ is a superreal field, every finite number $\xi \in \mathbb{K}$ is infinitely close to a unique real number $r \sim \xi$, called the* **shadow** *or the* **standard part** *of ξ. The number r is the only real number which determines the section*

$$\{x \in \mathbb{R} : x \leq \zeta\}, \{x \in \mathbb{R} : x > \zeta\}.$$

Theorem 6 is a well-known result in non-Archimedean analysis (see, e.g., [7]). Basically, Theorem 6 shows that the finite part of every non-Archimedean superreal field can be thought of as constructed starting from $\mathbb{R}$ and "surrounding" each real number r with a cloud of numbers that are infinitely close to r.

From now on, given a finite number ξ we will denote its shadow as $sh(\xi)$, and we put $sh(\xi) = +\infty$ ($sh(\xi) = -\infty$) if $\xi \in \mathbb{K}$ is a positive (negative) infinite number.

2.2. The Λ-limit

In this section we will introduce a superreal field $\mathbb{K}$ and we will analyze its main properties by mean of the Λ-theory (see also [1], [2]).

To formalize the Λ-theory we need a "mathematical universe" $\mathbb{U}$, i.e., a set that contains all the usual objects of analysis such as real numbers, real functions and so on. For our applications a good choice of $\mathbb{U}$ is given by the superstructure on $\mathbb{R}$:

$$\mathbb{U} = \bigcup_{n=0}^{\infty} \mathbb{U}_n$$

where $\mathbb{U}_n$ is defined by induction as follows:

$$\mathbb{U}_0 = \mathbb{R};$$
$$\mathbb{U}_{n+1} = \mathbb{U}_n \cup \mathcal{P}(\mathbb{U}_n).$$

Here $\mathcal{P}(E)$ denotes the power set of E. Identifying the couples with the Kuratowski pairs and the functions and the relations with their graphs, it follows that $\mathbb{U}$ contains almost every usual mathematical object. Given the universe $\mathbb{U}$, we denote by $\mathcal{F}$ the family of finite subsets of $\mathbb{U}$. Clearly $(\mathcal{F}, \subset)$ is a directed set. We recall that a directed set is a partially ordered set $(D, \prec)$ such that, $\forall a, b \in D$, $\exists c \in D$ such that

$$a \prec c \text{ and } b \prec c$$

A function $\varphi : D \to E$, defined on a directed set will be called *net* (with values in E). A net φ is the generalization of the notion of sequence and it has been constructed in such a way that the Weierstrass definition of limit makes sense: if φ_λ is a real net, we have that

$$\lim_{\lambda \to \infty} \varphi(\lambda) = L$$

if and only if

$$\forall \varepsilon > 0, \exists \lambda_0 > 0, \text{ such that } \forall \lambda > \lambda_0, \ |\varphi(\lambda) - L| < \varepsilon. \quad (1)$$

The key notion of the Λ-theory is the Λ-limit. Also the Λ-limit is defined for real nets but differs from the limit defined by (1) mainly for the fact that there exists a non-Archimedean field in which every real net admits a limit.

Now, we will present the notion of Λ-limit axiomatically:

Axioms of the $\mathbf{\Lambda}$-limit

- (Λ-1) **Existence Axiom.** *There is a superreal field $\mathbb{K} \supset \mathbb{R}$ such that every net $\varphi : \mathcal{F} \to \mathbb{R}$ has a unique limit $L \in \mathbb{K}$ (called the "Λ-limit" of φ.) The Λ-limit of φ will be denoted as*
$$L = \lim_{\lambda \uparrow \mathbb{U}} \varphi(\lambda).$$
Moreover we assume that every $\xi \in \mathbb{K}$ is the Λ-limit of some real function $\varphi : \mathcal{F} \to \mathbb{R}$.
- (Λ-2) **Real numbers axiom.** *If $\varphi(\lambda)$ is eventually constant, namely $\exists \lambda_0 \in \mathcal{F}, r \in \mathbb{R}$ such that $\forall \lambda \supset \lambda_0$, $\varphi(\lambda) = r$, then*
$$\lim_{\lambda \uparrow \mathbb{U}} \varphi(\lambda) = r.$$
- (Λ-3) **Sum and product Axiom.** *For all $\varphi, \psi : \mathcal{F} \to \mathbb{R}$:*
$$\lim_{\lambda \uparrow \mathbb{U}} \varphi(\lambda) + \lim_{\lambda \uparrow \mathbb{U}} \psi(\lambda) = \lim_{\lambda \uparrow \mathbb{U}} (\varphi(\lambda) + \psi(\lambda));$$
$$\lim_{\lambda \uparrow \mathbb{U}} \varphi(\lambda) \cdot \lim_{\lambda \uparrow \mathbb{U}} \psi(\lambda) = \lim_{\lambda \uparrow \mathbb{U}} (\varphi(\lambda) \cdot \psi(\lambda)).$$

These axioms state in a precise way the properties of the Λ-limit that we discussed before: (Λ-1) states that every net $\varphi : \mathcal{F} \to \mathbb{R}$ has a limit, and that every number $\xi \in \mathbb{K}$ is the limit of some net; (Λ-2) are familiar notions which holds also for the Weierstrass limit (1). The following theorem states that these axioms are consistent.

Theorem 7. *The set of axioms* $\{(\Lambda\text{-}1), (\Lambda\text{-}2), (\Lambda\text{-}3)\}$ *is consistent.*

Theorem 7 will be proved in the Appendix.

Now we want to extend the definition of the Λ-limit to any bounded net of mathematical objects in $\mathbb{U}$ (a net $\varphi : \mathcal{F} \to \mathbb{U}$ is called bounded if there exists n such that $\forall \lambda \in \mathcal{F}, \varphi(\lambda) \in \mathbb{U}_n$). To this aim, consider a net
$$\varphi : \mathcal{F} \to \mathbb{U}_n. \tag{2}$$
We will define $\lim_{\lambda \uparrow \mathbb{U}} \varphi(\lambda)$ by induction on n. For $n = 0$, $\lim_{\lambda \uparrow \mathbb{U}} \varphi(\lambda)$ is defined by the axioms (Λ-1), (Λ-2), (Λ-3); so by induction we may assume that the limit is defined for $n-1$ and we define it for the net (2) as follows:
$$\lim_{\lambda \uparrow \mathbb{U}} \varphi(\lambda) = \left\{ \lim_{\lambda \uparrow \mathbb{U}} \psi(\lambda) \mid \psi : \mathcal{F} \to \mathbb{U}_{n-1} \text{ and } \forall \lambda \in \mathcal{F},\ \psi(\lambda) \in \varphi(\lambda) \right\}.$$

Definition 8. A mathematical entity (number, set, function or relation) which is the Λ-limit of a net is called **internal**.

2.3. Natural extensions of sets and functions

Definition 9. The **natural extension** of a set $E \subset \mathbb{R}$ is given by

$$E^* := \lim_{\lambda \uparrow \mathbb{U}} c_E(\lambda) = \left\{ \lim_{\lambda \uparrow \mathbb{U}} \psi(\lambda) \mid \psi(\lambda) \in E \right\}$$

where $c_E(\lambda)$ is the net identically equal to E.

This definition, combined with axiom (Λ-1), entails that

$$\mathbb{K} = \mathbb{R}^*.$$

Since a function f can be identified with its graph then the natural extension of a function is defined by the above definition. Moreover we have the following result:

Theorem 10. *The* **natural extension** *of a function*

$$f : E \to F \quad \textit{is a function} \quad f^* : E^* \to F^*$$

and for every net $\varphi : \mathcal{F} \cap \mathcal{P}(E) \to E$, *and every function* $f : E \to F$, *we have that*

$$\lim_{\lambda \uparrow \mathbb{U}} f(\varphi(\lambda)) = f^* \left(\lim_{\lambda \uparrow \mathbb{U}} \varphi(\lambda) \right).$$

When dealing with functions, sometimes the "$*$" will be omitted if the domain of the function is clear from the context. For example, if $\eta \in \mathbb{R}^*$ is an infinitesimal, then clearly e^η denotes $\exp^*(\eta)$.

The following theorem is a fundamental tool in using the Λ-limit:

Theorem 11 (Leibniz Principle). *Let* $\mathcal{R}$ *be a relation in* $\mathbb{U}_n$ *for some* $n \geq 0$ *and let* $\varphi, \psi : \mathcal{F} \to \mathbb{U}_n$. *If*

$$\forall \lambda \in \mathcal{F}, \ \varphi(\lambda) \mathcal{R} \psi(\lambda) \quad \textit{then} \quad \left(\lim_{\lambda \uparrow \mathbb{U}} \varphi(\lambda) \right) \mathcal{R}^* \left(\lim_{\lambda \uparrow \mathbb{U}} \psi(\lambda) \right).$$

When $\mathcal{R}$ is $\in$ or $=$ we will not use the symbol $*$ to denote their extensions, since their meaning is unaltered in universe constructed over $\mathbb{R}^*$.

2.4. Hyperfinite extensions

Definition 12. An internal set is called **hyperfinite** if it is the Λ-limit of a net $\varphi : \mathcal{F} \to \mathcal{F}$.

So the hyperfinite sets are the Λ-limit of finite sets. There importance relies on the fact that, by virtue of Theorem 11 they share many properties of finite set even if, in general, they have a large cardinality.

Definition 13. Given any set $E \in \mathbb{U}$, the hyperfinite extension of E is defined as follows:

$$E^\circ := \lim_{\lambda \uparrow \mathbb{U}} (E \cap \lambda).$$

All the internal finite sets are hyperfinite, but there are hyperfinite sets which are not finite. For example the set

$$\mathbb{R}^{\circ} := \lim_{\lambda \uparrow \mathbb{U}} (\mathbb{R} \cap \lambda)$$

is not finite. The hyperfinite sets are very important since they inherit many properties of finite sets via Leibniz principle. For example, $\mathbb{R}^{\circ}$ has the maximum and the minimum and every internal function

$$f : \mathbb{R}^{\circ} \to \mathbb{R}^*$$

has the maximum and the minimum as well.

Also, it is possible to add the elements of an hyperfinite set of numbers or vectors as follows: let

$$A := \lim_{\lambda \uparrow \mathbb{U}} A_{\lambda}$$

be a hyperfinite set; then the hyperfinite sum is defined in the following way:

$$\sum_{a \in A} a = \lim_{\lambda \uparrow \mathbb{U}} \sum_{a \in A_{\lambda}} a.$$

In particular, if $A_{\lambda} = \{a_1(\lambda), \ldots, a_{\beta(\lambda)}(\lambda)\}$ with $\beta(\lambda) \in \mathbb{N}$, then setting

$$\beta = \lim_{\lambda \uparrow \mathbb{U}} \beta(\lambda) \in \mathbb{N}^*$$

we use the notation

$$\sum_{j=1}^{\beta} a_j = \lim_{\lambda \uparrow \mathbb{U}} \sum_{j=1}^{\beta(\lambda)} a_j(\lambda).$$

2.5. Qualified sets

When we have a net $\varphi : Q \to \mathbb{U}_n$, where $Q \subset \mathcal{F}$, we can define the Λ-limit of φ by posing

$$\lim_{\lambda \in Q} \varphi(\lambda) = \lim_{\lambda \uparrow \mathbb{U}} \widetilde{\varphi}(\lambda)$$

where

$$\widetilde{\varphi}(\lambda) = \begin{cases} \varphi(\lambda) & \text{for } \lambda \in Q; \\ \varnothing & \text{for } \lambda \notin Q. \end{cases}$$

As one can expect, if two nets φ, ψ are equal on a "large" or a "qualified" subset of $\mathcal{F}$ then they share the same Λ-limit. The notion of "qualified" subset of $\mathcal{F}$ can be precisely defined as follows:

Definition 14. We say that a set $Q \subset \mathcal{F}$ is qualified if for every bounded net φ we have that

$$\lim_{\lambda \uparrow \mathbb{U}} \varphi(\lambda) = \lim_{\lambda \in Q} \varphi(\lambda).$$

By the above definition, we have that the Λ-limit of a net φ depends only on the values that φ takes on a qualified set (it is in this sense that we could imagine Q to be "large"). It is easy to see that (nontrivial) qualified sets exist. For example by (Λ-2) we deduce that, for every $\lambda_0 \in \mathcal{F}$, the set

$$Q(\lambda_0) := \{\lambda \in \mathcal{F} \mid \lambda_0 \subseteq \lambda\}$$

is qualified. In this paper, we will use the notion of qualified set via the following theorem:

Theorem 15. *Let $\mathcal{R}$ be a relation in $\mathbb{U}_n$ for some $n \geq 0$ and let $\varphi, \psi : \mathcal{F} \to \mathbb{U}_n$. Then the following statements are equivalent:*

- *there exists a qualified set Q such that*

$$\forall \lambda \in Q,\ \varphi(\lambda)\mathcal{R}\psi(\lambda);$$

- *we have*

$$\left(\lim_{\lambda \uparrow \mathbb{U}} \varphi(\lambda)\right) \mathcal{R}^* \left(\lim_{\lambda \uparrow \mathbb{U}} \psi(\lambda)\right).$$

Proof. It is an immediate consequence of Theorem 11 and the definition of qualified set. □

3. Ultrafunctions

In this section, we will introduce the notion of ultrafunction and we will analyze its first properties.

3.1. Definition of ultrafunctions

By now, we just stated that "ultrafunctions are generalized functions", but we never stated which properties this space of generalized functions satisfy. The question is: which properties would we like a space of generalized functions to have? In what follows, Ω denotes an open set in $\mathbb{R}^N$ and by ultrafunctions we mean ultrafunctions defined on Ω^*.

First of all, since we started by saying that one of the aims of ultrafunctions is to generalize distributions, it is natural to request that every distribution is an ultrafunction (or, more precisely, that the space of distributions $\mathcal{D}(\overline{\Omega})$ can be embedded into the space of ultrafunctions).

Moreover, we would like to have only "good" functions as ultrafunctions. Of course, the notion of "good function" depends on the context. In this context, a function is "good" if it is continuous on $\overline{\Omega}$. So our second request is that the ultrafunctions are in the space $\mathcal{C}(\overline{\Omega})^*$.

Also, we would like to have a scalar product. A natural scalar product in functional analysis is $\langle f, g\rangle = \int f(x)g(x)dx$. To be sure that $\langle \cdot, \cdot \rangle$ is, in fact, a scalar product on the space of ultrafunctions, we require that the space of ultrafunctions is a vector subspace of $\left(L^2(\Omega)\right)^*$.

Finally, our last request is to have "enough compactness" to get existence results in a very large class of problems. In order to satisfy this request, we construct the space of ultrafunctions as a hyperfinite-dimensional vector space. One would argue that this leads to a contradiction with our first request; as we will show, our choice of the setting for generalized functions avoids this problem.

Now, let us formalize correctly our requests: let Ω be a set in $\mathbb{R}^N$, and let $V(\Omega)$ be a (real or complex) vector space such that $\mathcal{D}(\overline{\Omega}) \subseteq V(\Omega) \subseteq L^2(\Omega) \cap \mathcal{C}(\overline{\Omega})$.

Definition 16. Given the function space $V(\Omega)$ we set

$$\widetilde{V}(\Omega) := \lim_{\lambda \uparrow \mathbb{U}} V_\lambda(\Omega) = \mathrm{Span}^*(V(\Omega)^\circ),$$

where

$$V_\lambda(\Omega) = \mathrm{Span}(V(\Omega) \cap \lambda).$$

$\widetilde{V}(\Omega)$ will be called the **space of ultrafunctions** generated by $V(\Omega)$.

So, given any vector space of functions $V(\Omega)$, the space of ultrafunction generated by $V(\Omega)$ is a vector space of hyperfinite dimension that includes $V(\Omega)$, and the ultrafunctions are Λ-limits of functions in V_λ. Hence the ultrafunctions are particular internal functions

$$u : (\mathbb{R}^*)^N \to \mathbb{C}^*.$$

Observe that, by definition, the dimension of $\widetilde{V}(\Omega)$ (that we denote by β) is equal to the internal cardinality of any of its bases, and the following formula holds:

$$\beta = \lim_{\lambda \uparrow \mathbb{U}} \dim(V_\lambda(\Omega)).$$

Since $\widetilde{V}(\Omega) \subset \left[L^2(\mathbb{R})\right]^*$, it can be equipped with the following scalar product

$$(u, v) = \int^* u(x)\overline{v(x)}\, dx,$$

where $\int^*$ is the natural extension of the Lebesgue integral considered as a functional

$$\int : L^1(\Omega) \to \mathbb{C}.$$

Notice that the Euclidean structure of $\widetilde{V}(\Omega)$ is the Λ-limit of the Euclidean structure of every V_λ given by the usual L^2 scalar product. The norm of an ultrafunction will be given by

$$\|u\| = \left(\int^* |u(x)|^2\, dx\right)^{\frac{1}{2}}.$$

Remark 17. Notice that the natural extension f^* of a function f is an ultrafunction if and only if $f \in V(\Omega)$.

3. *if two ultrafunctions u and v coincide on a set of independent points then they are equal;*
4. *if Σ is a set of independent points and $a, b \in \Sigma$ then $\sigma_a(b) = \delta_{ab}$;*
5. *for any $q \in \Omega^*$, $\sigma_q(x)$ is well defined.*

Proof. 1. It is an immediate consequence of the definition of dual basis.

2. Since $\{\delta_q(x)\}_{q\in\Sigma}$ is the dual basis of $\{\sigma_q(x)\}_{q\in\Sigma}$ we have that

$$u(x) = \sum_{q\in\Sigma} \left(\int \delta_q(\xi)u(\xi)d\xi \right) \sigma_q(x) = \sum_{q\in\Sigma} u(q)\sigma_q(x).$$

3. It follows directly from 2.

4. If follows directly by equation (4)

5. Given any point $q \in \Omega^*$ clearly there is a Delta-basis $\{\delta_a(x)\}_{a\in\Sigma}$ with $q \in \Sigma$. Then $\sigma_q(x)$ can be defined by mean of the basis $\{\delta_a(x)\}_{a\in\Sigma}$. We have to prove that, given another Delta-basis $\{\delta_a(x)\}_{a\in\Sigma'}$ with $q \in \Sigma'$, the corresponding $\sigma'_q(x)$ is equal to $\sigma_q(x)$. Using (2), with $u(x) = \sigma'_q(x)$, we have that

$$\sigma'_q(x) = \sum_{a\in\Sigma} \sigma'_q(a)\sigma_a(x).$$

Then, by (4), it follows that $\sigma'_q(x) = \sigma_q(x)$. □

Let Σ be a set of independent points, and let $L_\Sigma : \widetilde{V}(\Omega) \to \widetilde{V}(\Omega)$ be the linear operator such that

$$L_\Sigma \sigma_a(x) = \delta_a(x)$$

for every $a \in \Sigma$.

Proposition 23. L_Σ is selfadjoint, positive and

$$\int^* L_\Sigma u(x)v(x)dx = \sum_{a\in\Sigma} u(a)v(a).$$

Proof. Since $u(x) = \sum_{a\in\Sigma} u(a)\sigma_a(x)$ and $v(x) = \sum_{a\in\Sigma} v(a)\sigma_a(x)$, then

$$\int^* L_\Sigma u(x)v(x)dx = \int^* L_\Sigma \left(\sum_{a\in\Sigma} u(a)\sigma_a(x) \right) \left(\sum_{b\in\Sigma} v(b)\sigma_b(x) \right) dx$$
$$= \sum_{a\in\Sigma}\sum_{b\in\Sigma} u(a)v(b) \int^* \delta_a(x)\sigma_b(x)dx = \sum_{a\in\Sigma} u(a)v(a).$$

Hence, clearly, L_Σ is selfadjoint and positive. □

From now on, we consider the set Σ fixed once for all and we simply denote the operator L_Σ by L. Since L is a positive selfadjoint operator, $A = L^{1/2}$ is a well-defined positive selfadjoint operator. For every $a \in \Sigma$ we set

$$\theta_a(x) = A\sigma_a(x).$$

Theorem 24. *The following properties hold:*

1. $\{\theta_a(x)\}_{a\in\Sigma}$ *is an orthonormal basis;*
2. *for every* $a, b \in \Sigma$, $\theta_a(b) = \theta_b(a)$*;*
3. *for every ultrafunction* u *we have*
$$u(x) = \sum_{a\in\Sigma} u(a)\sigma_a(x) = \sum_{a\in\Sigma} \underline{u}(a)\theta_a(x) = \sum_{a\in\Sigma} \underline{\underline{u}}(a)\delta_a(x),$$
where we have set, for every $a \in \Sigma$,
$$\underline{u}(a) := (A^{-1}u)(a) = \int^* \theta_a(\xi)u(\xi)d\xi;$$
$$\underline{\underline{u}}(a) = (A^{-1}\underline{u})(a) = (L^{-1}u)(a) = \int^* \sigma_a(\xi)u(\xi)d\xi;$$
4. *for every ultrafunctions* u, v *we have*
$$\int^* u(x)v(x)dx = \sum_{a\in\Sigma} \underline{u}(a)\underline{v}(a) = \sum_{a\in\Sigma} \underline{\underline{u}}(a)v(a);$$
5. *for every ultrafunction* u *we have*
$$\int^* u(x)dx = \sum_{a\in\Sigma} \underline{\underline{u}}(a).$$

Proof. 1) $\{\theta_a(x)\}_{a\in\Sigma}$ is a basis since it is the image of the basis $\{\sigma_a(x)\}_{a\in\Sigma}$ respect to the invertible linear application L. It is orthonormal: for every $a, b \in \Sigma$ we have
$$\int^* \theta_a(x)\theta_b(x)dx = \int^* A\sigma_a(x)A\sigma_b(x)dx = \int^* L\sigma_a(x)\sigma_b(x) = \sigma_b(a) = \delta_{ab}.$$

2) We have
$$\theta_a(b) = \int^* \theta_a(x)\delta_b(x)dx = \int^* \theta_a(x)A\theta_b(x)dx$$
$$= \int^* A\theta_a(x)\theta_b(x)dx = \int^* \delta_a(x)\theta_b(x)dx = \theta_b(a).$$

3) The equality
$$u(x) = \sum_{a\in\Sigma} u(a)\sigma_a(x)$$
has been proved in Theorem 22, (5); the equality
$$u(x) = \sum_{a\in\Sigma} \underline{u}(a)\theta_a(x),$$
where $\underline{u}(a) = \int^* \theta_a(\xi)u(\xi)d\xi$, follows since $\{\theta_a(x)\}_{a\in\Sigma}$ is an orthonormal basis. And
$$(A^{-1}u)(a) = \int^* \delta_a(\xi)A^{-1}u(\xi)d\xi = \int^* A^{-1}\delta_a(\xi)u(\xi)d\xi = \int^* \theta_a(\xi)u(\xi)d\xi$$
since A (and, so, A^{-1}) is selfadjoint.

The equality

$$u(x) = \sum_{a \in \Sigma} \underline{\underline{u}}(a)\delta_a(x),$$

where $\underline{\underline{u}}(a) = \int^* \sigma_a(\xi)u(\xi)d\xi$, follows by point (1) in Theorem 22. And

$$\underline{\underline{u}}(a) = \int^* \sigma_a(\xi)u(\xi)d\xi = \int^* L^{-1}\delta_a(\xi)u(\xi)d\xi = \int^* \delta_a(\xi)L^{-1}u(\xi)d\xi = (L^{-1}u)(a).$$

4) We have that $\int^* u(x)v(x)dx = \sum_{a\in\Sigma} \underline{u}(a)\underline{v}(a)$ since $\{\theta_a(x)\}_{a\in\Sigma}$ is an orthonormal basis:

$$\begin{aligned}\int^* u(x)v(x)dx &= \int^* \left(\sum_{a\in\Sigma} \underline{u}(a)\theta_a(x)\right)\left(\sum_{b\in\Sigma} \underline{v}(b)\theta_b(x)dx\right) \\ &= \sum_{a\in\Sigma}\sum_{b\in\Sigma} \underline{u}(a)\underline{v}(b)\int^* \theta_a(x)\theta_b(x)dx = \sum_{a\in\Sigma} \underline{u}(a)\underline{v}(a);\end{aligned}$$

the equality $\int^* u(x)v(x)dx = \sum_{a\in\Sigma} \underline{\underline{u}}(a)v(a)$ follows by expressing $u(x)$ in the Delta-basis and $v(x)$ in the Sigma-basis:

$$\begin{aligned}\int^* u(x)v(x)dx &= \int^* \left(\sum_{a\in\Sigma} \underline{\underline{u}}(a)\delta_a(x)\right)\left(\sum_{b\in\Sigma} v(b)\sigma_b(x)\right)dx \\ &= \sum_{a\in\Sigma}\sum_{b\in\Sigma} v(b)\underline{\underline{u}}(a)\int^* \delta_a(x)\sigma_b(x)dx = \sum_{a\in\Sigma} \underline{\underline{u}}(a)v(a).\end{aligned}$$

5) This follows by expressing $u(x)$ in the Delta-basis:

$$\int^* u(x)dx = \int^* \sum_{a\in\Sigma} \underline{\underline{u}}(a)\delta_a(x)dx = \sum_{a\in\Sigma} \underline{\underline{u}}(a)\int^* \delta_a(x)dx = \sum_{a\in\Sigma} \underline{\underline{u}}(a). \qquad \square$$

3.3. Canonical extension of a function

Let $V'(\Omega)$ denote the dual of $V(\Omega)$ and let $\mathfrak{M}$ denote the set of measurable functions in $\mathbb{R}^N$. If $T \in V'(\Omega)$ and if there is a function $f \in \mathfrak{M}$ such that

$$\forall v \in V(\Omega),\ \langle T, v\rangle = \int f(x)v(x)dx$$

then T and f will be identified, and with some abuse of notation we shall write $T = f \in V'(\Omega) \cap \mathfrak{M}$. With this identification, $V'(\Omega) \cap \mathfrak{M} \subset L^2$.

Definition 25. If $T \in [V'(\Omega)]^*$, there exists a unique ultrafunction $\widetilde{T}(x)$ such that

$$\forall v \in \widetilde{V}(\Omega),\ \langle T, v\rangle = \int^* \widetilde{T}(x)v(x)dx.$$

In particular, if $u \in [V'(\Omega) \cap \mathfrak{M}]^*$, $\widetilde{u}$ will denote the unique ultrafunction such that

$$\forall v \in \widetilde{V}(\Omega),\ \int^* u(x)v(x)dx = \int^* \widetilde{u}(x)v(x)dx.$$

Notice that $V'(\Omega) \cap \mathfrak{M}$ is a space of distributions which contains the delta measures, so to every Delta distribution δ_q is associated an ultrafunction which, by definition, is the Delta ultrafunction centered in q, as expected. Moreover, the definition is well posed: in fact, if $T \in [V'(\Omega)]^*$, then the function $f_T : \widetilde{V}(\Omega) \to \mathbb{R}^*$ such that for every ultrafunction u

$$f_T(u) = \langle T, u \rangle$$

is a linear functional on $\widetilde{V}(\Omega)$. So $f_T \in \left(\widetilde{V}(\Omega)\right)'$ and by Riesz's Representation Theorem it follows that there is one (and only one) element $\widetilde{T}(x)$ in $\widetilde{V}(\Omega)$ such that for every $u \in \widetilde{V}(\Omega)$ we have

$$f_T(u) = \int^* \widetilde{T}(x)v(x)dx.$$

Since $f_T(u) = \langle T, u \rangle$, we have proved that the definition is well posed.

Definition 26. If $f \in V'(\Omega) \cap \mathfrak{M}$, $\widetilde{(f^*)}$ is called the canonical extension of f. In the following, since f and f^* can be identified, we will write $\widetilde{f}$ instead of $\widetilde{(f^*)}$.

Thus any function

$$f : \mathbb{R}^N \to \mathbb{R}$$

can be extended to the function

$$f^* : (\mathbb{R}^*)^N \to \mathbb{R}^*$$

which is called the natural extension of f and if $f \in V'(\Omega) \cap \mathfrak{M}$, we have also the canonical extension of f given by

$$\widetilde{f} : (\mathbb{R}^*)^N \to \mathbb{R}^*$$

If $f \notin V(\Omega)$, by Remark 17, $\widetilde{f} \neq f^*$, thus $f^* \notin \widetilde{V}(\Omega)$.

Example: if $\Omega = (-1, 1)$, then $|x|^{-1/2} \in V(-1,1)' \cap \mathfrak{M}$; the ultrafunction $\widetilde{|x|^{-1/2}}$ is different from $\left(|x|^{-1/2}\right)^*$ since the latter is not defined for $x = 0$, while

$$\left(\widetilde{|x|^{-1/2}}\right)_{x=0} = \int^* |x|^{-1/2}\delta_0(x)dx.$$

Theorem 27. *If* $T \in [V(\Omega)']^*$, *then*

$$\begin{aligned}\widetilde{T}(x) &= \sum_{q\in\Sigma} \langle T, \delta_q \rangle \, \sigma_q(x) \\ &= \sum_{q\in\Sigma} \langle T, \theta_q \rangle \, \theta_q(x) \\ &= \sum_{q\in\Sigma} \langle T, \sigma_q \rangle \, \delta_q(x).\end{aligned}$$

In particular, if $f \in [V'(\Omega) \cap \mathfrak{M}]^*$

$$\widetilde{f}(x) = \sum_{q\in\Sigma} \left[\int f^*(\xi)\delta_q(\xi)d\xi \right] \sigma_q(x) \tag{6}$$

$$= \sum_{q\in\Sigma} \left[\int f^*(\xi)\theta_q(\xi)d\xi \right] \theta_q(x) \tag{7}$$

$$= \sum_{q\in\Sigma} \left[\int f^*(\xi)\sigma_q(\xi)d\xi \right] \delta_q(x). \tag{8}$$

Proof. It is sufficient to prove that

$$\forall v \in V(\Omega), \quad \int \sum_{q\in\Sigma} \langle T, \delta_q \rangle \, \sigma_q(x)v(x)dx = \langle T, v \rangle .$$

We have that

$$\int \sum_{q\in\Sigma} \langle T, \delta_q \rangle \, \sigma_q(x)v(x)dx = \sum_{q\in\Sigma} \langle T, \delta_q \rangle \int \sigma_q(x)v(x)dx$$

$$= \left\langle T, \sum_{q\in\Sigma} \left(\int \sigma_q(x)v(x)dx \right) \delta_q \right\rangle = \langle T, v \rangle .$$

The other equalities can be proved similarly. □

3.4. Ultrafunctions and distributions

In this section we will show that the space of ultrafunctions is reacher than the space of distribution, in the sense that any distribution can be represented by an ultrafunction and that the converse is not true.

Definition 28. Let $D \subset \widetilde{V}(\Omega)$ be a vector space. We say that two ultrafunctions u and v are D-equivalent if

$$\forall \varphi \in D, \quad \int^* (u(x) - v(x)) \, \varphi(x)dx = 0.$$

We say that two ultrafunctions u and v are distributionally equivalent if they are $\mathcal{D}(\Omega)$-equivalent.

Theorem 29. *Given* $T \in \mathcal{D}'$, *there exists an ultrafunction* u *such that*

$$\forall \varphi \in \mathcal{D}(\Omega), \quad \int^* u(x)\varphi^*(x)dx = \langle T, \varphi \rangle . \tag{9}$$

Proof. Let $\{e_j(x)\}_{j\in J}$ be an orthonormal basis of the hyperfinite space $\widetilde{V}(\Omega) \cap \mathcal{D}(\Omega)^*$ and take

$$u(x) = \sum_{j\in J} \langle T^*, e_j \rangle \; e_j(x).$$

Now take $\varphi \in \mathcal{D}$. Since $\varphi^* \in \widetilde{V}(\Omega) \cap \mathcal{D}(\Omega)^*$, we have that

$$\varphi^*(x) = \sum_{j \in J} \left(\int^* \varphi^*(\xi) e_j(\xi) d\xi \right) e_j(x).$$

Thus

$$\int^* u(x)\varphi^*(x)dx = \int^* \sum_{j \in J} \langle T^*, e_j \rangle \; e_j(x)\varphi^*(x)dx = \sum_{j \in J} \left\langle T^*, e_j \int^* e_j(x)\varphi^*(x)dx \right\rangle$$

$$= \left\langle T^*, \sum_{j \in J} \left(\int^* e_j(x)\varphi^*(x)dx \right) e_j \right\rangle = \langle T^*, \varphi^* \rangle = \langle T, \varphi \rangle . \qquad \square$$

The following proposition shows that the ultrafunction u associated to the distribution T by (9) is not unique:

Proposition 30. Take $T \in \mathcal{D}'(\Omega)$ and let

$$V_T = \{u \in \widetilde{V}(\Omega) : \forall \varphi \in \mathcal{D}(\Omega), \; \int^* u(x)\varphi^*(x)dx = \langle T, \varphi \rangle\},$$

let $u \in V_T$ and let v be any ultrafunction. Then

1. $v \in V_T$ if and only if u and v are $\mathcal{D}(\Omega)$-equivalent;
2. V_T is infinite.

Proof. 1) If $v \in V_T$ then $\forall \varphi \in \mathcal{D}(\Omega)$, $\int^*(u(x) - v(x))\varphi^*(x)dx = \langle T, \varphi \rangle - \langle T, \varphi \rangle = 0$, so u and v are $\mathcal{D}(\Omega)$-equivalent; conversely, if u and v are $\mathcal{D}$-equivalent then $\forall \varphi \in \mathcal{D}(\Omega)$, $\int^* u(x)\varphi^*(x)dx = \int^* v(x)\varphi^*(x)dx$. Since $\int^* u(x)\varphi^*(x)dx = \langle T, \varphi \rangle$ then $v \in V_T$.

2) Let $v \neq 0$ be any ultrafunction in the orthogonal (in $\widetilde{V}(\Omega)$) of $\widetilde{V}(\Omega) \cap \mathcal{D}(\Omega)^*$. Then u and $u + v$ are $\mathcal{D}(\Omega)$-equivalent, since $\int^*(u(x) + v(x))\varphi^*(x)dx = \int^* u(x)\varphi^*(x)dx + \int^* v(x)\varphi^*(x)dx = \int^* u(x)\varphi^*(x)dx + 0$. Since the orthogonal of $\widetilde{V}(\Omega) \cap \mathcal{D}(\Omega)^*$ is infinite, we obtain the thesis. $\square$

Remark 31. There is a natural way to associate a unique ultrafunction to a distribution (see also [1]). In order to do this it is sufficient to split $\widetilde{V}(\Omega)$ in two orthogonal component: $\widetilde{V}(\Omega) \cap \mathcal{D}(\Omega)^*$ and $\left(\widetilde{V}(\Omega) \cap \mathcal{D}(\Omega)^*\right)^{\perp}$. As we have seen in the proof of the above theorem every ultrafunction in V_T can be spitted in two components, $u + v$ where $v \in \left(\widetilde{V}(\Omega) \cap \mathcal{D}(\Omega)^*\right)^{\perp}$ and $u \in \widetilde{V}(\Omega) \cap \mathcal{D}(\Omega)^*$ is univocally determined. Then, we have an injective map

$$i : \mathcal{D}'(\Omega) \to \widetilde{V}(\Omega)$$

given by $i(T) = u$ where $u \in V_T \cap \mathcal{D}(\Omega)^*$.

Remark 32. The space of ultrafunctions is richer than the space of distributions; for example consider the function

$$u(x) := f(x) \min\left(x^{-2}, \alpha\right)$$

where $\alpha > 0$ is an infinite number and $f(x)$ is a function with compact support such that $f(0) = 1$. Since $u \in \left[V'(\Omega) \cap \mathfrak{M}\right]^*$, $\widetilde{u}$ is well defined (see Def. 25). On the other hand, $\widetilde{u}$ does not correspond to any distribution since

$$\int^* \widetilde{u}(x)\varphi^*(x)dx = \int^* f^*(x) \min\left(x^{-2}, \alpha\right) \varphi^*(x)dx$$

is infinite when $\varphi(x) \geq 0$ and $\varphi(0) > 0$. In [1] Section 6, it is presented an elliptic problem which has a solution in the space of ultrafunctions, but no solution in the space of distributions.

4. Operations with ultrafunctions

4.1. Extension of operators

Definition 33. Given the operator $F : V(\Omega) \to \mathcal{D}'(\Omega)$, the map

$$\widetilde{F} : \widetilde{V}(\Omega) \to \widetilde{V}(\Omega)$$

defined by

$$\widetilde{F}(u) = \widetilde{F^*(u)} \tag{10}$$

is called **canonical** extension of F ("$\sim$" is defined by Definition 25).

By the definition of $\widetilde{F}$, we have that

$$\forall v \in \widetilde{V}(\Omega), \int^* \widetilde{F}(u(x))\, v(x)\; dx = \int^* F^*(u(x))\, v(x) dx. \tag{11}$$

Comparing Definition 33 with Theorem 27 we have that

$$\begin{aligned}\widetilde{F}(u(x)) &= \sum_{q\in\Sigma} \langle F^*(u), \delta_q\rangle\, \sigma_q(x)\\ &= \sum_{q\in\Sigma} \langle F^*(u), \theta_q\rangle\, \theta_q(x)\\ &= \sum_{q\in\Sigma} \langle F^*(u), \sigma_q\rangle\, \delta_q(x).\end{aligned}$$

In particular, if $F : V(\Omega) \to V'(\Omega) \cap \mathfrak{M}^*$:

$$\begin{aligned}\widetilde{F}(u(x)) &= \sum_{q\in\Sigma} \left[\int F^*(u(\xi))\, \delta_q(\xi) d\xi\right] \sigma_q(x)\\ &= \sum_{q\in\Sigma} \left[\int F^*(u(\xi))\, \theta_q(\xi) d\xi\right] \theta_q(x)\\ &= \sum_{q\in\Sigma} \left[\int F^*(u(\xi))\, \sigma_q(\xi) d\xi\right] \delta_q(x).\end{aligned} \tag{12}$$

4.2. Derivative

A good generating space to define the derivative of an ultrafunction is the following one:

$$V^1(\Omega) = H^{1,1}(\Omega) \cap \mathcal{C}(\overline{\Omega}) \subseteq L^2(\Omega) \cap \mathcal{C}(\overline{\Omega}).$$

In order to simplify the exposition, we will assume that $\Omega \subseteq \mathbb{R}$. The generalization of the notions exposed in this section when $\Omega \subseteq \mathbb{R}^N$ is immediate.

Let $u \in \widetilde{V^1}(\Omega)$ be a ultrafunction. Since $V^1(\Omega)^* \subset H^1(\Omega)^*$, we have that the derivative $\frac{du}{dx} = \partial u = u'$ is in $L^2(\Omega) \subset \left[\left(V^1(\Omega)\right)' \cap \mathfrak{M}\right]^*$, (here $\left(V^1(\Omega)\right)'$ denotes the topological dual of $V^1(\Omega)$). Then we can apply Definition 33:

Definition 34. We set

$$Du = \widetilde{\partial} u = \widetilde{\partial u}.$$

The operator

$$D : \widetilde{V^1}(\Omega) \to \widetilde{V^1}(\Omega)$$

is called (generalized) derivative of the ultrafunction u.

By (12) we have the following representation of the derivative:

$$\forall u \in \widetilde{V^1}(\Omega),\ Du(x) = \sum_{q \in \Sigma} \left[\int^* u'(\xi)\delta_q(\xi)d\xi\right] \sigma_q(x).$$

If $u' \in \widetilde{V^1}(\Omega) \subset \left[V^1(\Omega)\right]^*$, we have that

$$Du(x) = \sum_{q \in \Sigma} u'(q)\sigma_q(x) = u'(x).$$

In particular, if $u \in H^{2,1}(\Omega) \cap \mathcal{C}^1(\overline{\Omega})$, $Du = u'$ and so D extends the operator $\frac{d}{dx} : H^{2,1}(\Omega) \cap \mathcal{C}^1(\overline{\Omega}) \to V^1(\Omega)$ to the operator $D : \widetilde{V^1}(\Omega) \to \widetilde{V^1}(\Omega)$.

4.3. Fourier transform

In this section we will investigate the extension of the one-dimensional Fourier transform. A good space to work with the Fourier transform is the space

$$V^{\mathfrak{F}}(\mathbb{R}) = H^1(\mathbb{R}) \cap L^2(\mathbb{R},\ |x|^2).$$

It is easy to see that the space $V^{\mathfrak{F}}(\mathbb{R})$ can be characterized as follows:

$$V^{\mathfrak{F}}(\mathbb{R}) = \left\{u \in H^1(\mathbb{R}) : \hat{u} \in H^1(\mathbb{R})\right\}.$$

In fact, if $\hat{u} \in H^1(\mathbb{R})$, then $\int |\nabla u(\xi)|^2 \, d\xi < +\infty$ and hence $\int |u(x)|^2 |x|^2 \, dx < +\infty$.

Then $V^{\mathfrak{F}}(\mathbb{R}) \subset L^2(\mathbb{R}, |x|^2)$, so $V^{\mathfrak{F}}(\mathbb{R}) \subset H^1(\mathbb{R}) \cap L^2(\mathbb{R}, |x|^2)$ which is a Hilbert space equipped with the norm

$$\|u\|^2_{V^{\mathfrak{F}}(\mathbb{R})} = \int |u(x)|^2 \, |x|^2 dx + \int |\hat{u}(\xi)|^2 \, |\xi|^2 d\xi.$$

Moreover

$$\begin{aligned}\int |u(x)|\,dx &= \int |u(x)|\,(1+|x|)\frac{1}{1+|x|}dx \\ &\le \left(\int |u(x)|^2\,(1+|x|)^2 dx\right)^{\frac{1}{2}} \left(\int \frac{1}{(1+|x|)^2}dx\right)^{\frac{1}{2}} \\ &\le \text{const.}\left(\|u\|_{L^2(\mathbb{R})} + \|u\|_{L^2(\mathbb{R},|x|^2)}\right).\end{aligned}$$

Thus, $V^{\mathfrak{F}}(\mathbb{R}) \subset L^1(\mathbb{R})$. Recalling that the functions in $H^1(\mathbb{R})$ are continuous, we have that

$$V^{\mathfrak{F}}(\mathbb{R}) \subset \mathcal{C}(\mathbb{R}) \cap H^1(\mathbb{R}) \cap L^1(\mathbb{R}) \cap L^2(\mathbb{R},|x|^2).$$

We use the following definitions of the Fourier transform: if $u \in \widetilde{V^{\mathfrak{F}}}(\mathbb{R})$, we set

$$\mathfrak{F}(u)(k) = \widehat{u}(k) = \frac{1}{\sqrt{2\pi}}\int^{*} u(x)\ e^{-ikx}\ dx; \tag{13}$$

$$\mathfrak{F}^{-1}(u)(x) = \frac{1}{\sqrt{2\pi}}\int^{*} \widehat{u}(k)\ e^{ikx}\ dx. \tag{14}$$

Now, in order to deal with the Fourier transform in an easier way, we need a new axiom whose consistency is easy to be verified (see Appendix):

Axiom 35 (FTA, Fourier Transform Axiom). *If $u \in \widetilde{V^{\mathfrak{F}}}(\mathbb{R})$ then $\mathfrak{F}^{*}(u) \in \widetilde{V^{\mathfrak{F}}}(\mathbb{R})$ and $\bar{u} \in \widetilde{V^{\mathfrak{F}}}(\mathbb{R})$ (here $\bar{u}$ is the complex conjugate of u).*

It is immediate to see that, by this axiom, for every ultrafunction, u we have

$$\mathfrak{F}^{*}(u) = \widetilde{\mathfrak{F}}(u)$$

and hence, since there is no risk of ambiguity, we will simply write $\mathfrak{F}(u)$.

It is well known that in the theory of tempered distributions we have that:

$$\mathfrak{F}(\delta_a) = \frac{e^{-iak}}{\sqrt{2\pi}};$$

$$\mathfrak{F}\left(\frac{e^{iax}}{\sqrt{2\pi}}\right) = \delta_a.$$

In the theory of ultrafunctions an analogous result holds:

Proposition 36. We have that:

1. $\mathfrak{F}\left(\widetilde{\frac{e^{iax}}{\sqrt{2\pi}}}\right) = \delta_a(k)$;
2. $\mathfrak{F}\left(\delta_a(x)\right) = \widetilde{\frac{e^{-iak}}{\sqrt{2\pi}}}$;
3. $\frac{1}{2\pi}\int^{*} \widetilde{e^{-iax}}\ \widetilde{e^{ikx}}\ dx = \delta_a(k)$.

Proof. 1. For every $v \in V^{\mathfrak{F}}$,

$$\int^* \mathfrak{F}\left(\frac{\widetilde{e^{iax}}}{\sqrt{2\pi}}\right) v(k)dk = \int^* \left(\frac{1}{2\pi}\int^* \widetilde{e^{-iak}}\ \widetilde{e^{ixk}}dx\right) v(k)dk$$
$$= \frac{1}{2\pi}\int^* \int^* \widetilde{e^{-iak}}\ \widetilde{e^{ixk}}v(k)dkdx$$
$$= \frac{1}{\sqrt{2\pi}}\int^* \widetilde{e^{-iak}}\mathfrak{F}^{-1}(v(k))dx = v(a).$$

Hence, 1 holds.

2. We have

$$\mathfrak{F}\left(\delta_a(x)\right) = \int^* \delta_a(x)e^{-ikx}dx = \int^* \delta_a(x)\widetilde{e^{-ikx}}dx = \widetilde{e^{-ika}}.$$

3. We have

$$\frac{1}{2\pi}\int^* \widetilde{e^{iax}}\ \widetilde{e^{-ikx}}\ dx = \frac{1}{2\pi}\int^* \widetilde{e^{iax}}\ e^{-ikx}dx = \mathfrak{F}\left(\frac{\widetilde{e^{iax}}}{\sqrt{2\pi}}\right) = \delta_a(k).$$ □

By our definitions we have that:

$$\widetilde{e^{ikx}} = \sum_{q\in\Sigma}\left[\int^* e^{ik\xi}\delta_q(\xi)d\xi\right]\sigma_q(x);$$
$$\widetilde{e^{ixk}} = \sum_{q\in\Sigma}\left[\int^* e^{ix\xi}\delta_q(\xi)d\xi\right]\sigma_q(k).$$

Therefore it is not evident whether $\widetilde{e^{ikx}} = \widetilde{e^{ixk}}$ or not. The following corollary answers this question.

Corollary 37. *We have that:*

$$\widetilde{e^{ikx}} = \widetilde{e^{ixk}}.$$

Proof. By the previous proposition, we have that

$$\widetilde{e^{-ikx}} = \sqrt{2\pi}\mathfrak{F}\left(\delta_k(x)\right) = \int^* \delta_k(x)e^{-ixk}dk = \int^* \delta_x(k)e^{-ixk}dx = \widetilde{e^{-ixk}}.$$

Replacing x with $-x$ we get the result. □

Since $\mathfrak{F}: V^{\mathfrak{F}}(\mathbb{R}) \to V^{\mathfrak{F}}(\mathbb{R})$ is an isomorphism, it follows that, for any Delta-basis $\{\delta_a\}_{a\in\Sigma}$, the set

$$\left\{\frac{\widetilde{e^{iax}}}{\sqrt{2\pi}}\right\}_{a\in\Sigma} = \left\{\mathfrak{F}\left(\delta_{-a}\right)\right\}_{a\in\Sigma}$$

is a basis and we get the following result:

Theorem 38. *If $u \in V^{\mathfrak{F}}(\mathbb{R})$, then*

$$u(x) = \frac{1}{\sqrt{2\pi}} \sum_{k \in \Sigma} \widehat{\underline{\underline{u}}}(k) \widetilde{e^{ikx}}.$$

where we have set (see Theorem 24)

$$\widehat{\underline{\underline{u}}}(k) = \int^{*} \widehat{u}(\xi) \sigma_k(\xi) d\xi.$$

Proof. Since $\left\{ \frac{\widetilde{e^{ikx}}}{\sqrt{2\pi}} \right\}_{k \in \Sigma}$ is a basis, any $u \in V^{\mathfrak{F}}(\mathbb{R})$ has the following representation:

$$u(x) = \frac{1}{\sqrt{2\pi}} \sum_{k \in \Sigma} u_k \widetilde{e^{ikx}}.$$

Let us compute the u_k's: we have

$$\int \delta_k(x) \overline{\sigma_b(x)} dx = \int \delta_k(x) \sigma_b(x) dx = \delta_{kb} \quad \text{and so} \quad \int \widehat{\delta_k}(x) \overline{\widehat{\sigma_b}}(x) dx = \delta_{kb}$$

and by Proposition 36,

$$\int \frac{\widetilde{e^{-ikx}}}{\sqrt{2\pi}} \overline{\widehat{\sigma_b}}(x) dx = \delta_{kb}.$$

Hence $\{\widehat{\sigma_k}(x)\}_{k \in \Sigma}$ is the dual basis of $\left\{ \frac{\widetilde{e^{-ikx}}}{\sqrt{2\pi}} \right\}_{k \in \Sigma}$,namely $\{\widehat{\sigma_k}(-x)\}_{k \in \Sigma}$ is the dual basis of $\left\{ \frac{\widetilde{e^{ikx}}}{\sqrt{2\pi}} \right\}_{k \in \Sigma}$. Hence, since $\widehat{\widehat{v}}(x) = v(-x)$, we have:

$$\begin{aligned} u_k &= \int u(\xi) \overline{\widehat{\sigma_k}(-\xi)} d\xi = \int u(\xi) \overline{\widehat{\widehat{\sigma_k}}(-\xi)} d\xi \\ &= \int \widehat{u}(\xi) \overline{\sigma_k(\xi)} d\xi = \int \widehat{u}(\xi) \sigma_k(\xi) d\xi = \widehat{\underline{\underline{u}}}(k). \end{aligned}$$

□

4.4. A trivial example of generalized solution

For the applications of ultrafunctions theory we refer to [1] and [2]. Also we are working with ultrafunctions in more sophisticated environments such as Morse theory where it seems that we can get interesting results. In this section we will give a (relatively) trivial result with the only purpose to make the reader to understand how ultrafunctions provide generalized solutions to "classical problems".

Let us consider a classical problem of calculus of variations: minimize the functional

$$J(u) = \int F(x, u, \nabla u) dx \tag{15}$$

in the function space

$$\mathcal{C}^1_g(\Omega) = \left\{ u \in \mathcal{C}^1(\Omega) \cap \mathcal{C}(\overline{\Omega}) \mid \forall x \in \partial\Omega,\ u(x) = g(x) \right\}, \quad g \in \mathcal{C}(\partial\Omega).$$

It is well known that in general this problem has no solution even when F is coercive and hence the infimum exists. However, if F is convex and g is sufficiently smooth it is possible to find a minimizer in some Sobolev space (or in some "Sobolev type" space such as Orlicz spaces).

If F is not convex it is not possible to find a minimizer, not even among the generalized functions of "Sobolev" type as the following example shows:

$$\text{minimize} \quad J_1(u) = \int_0^1 \left[\left(|\nabla u|^2 - 1 \right)^2 + |u|^2 \right] dx \quad \text{in} \quad \mathcal{C}_0^1(0,1). \tag{16}$$

It is not difficult to realize that J_1 has a minimizing sequence u_n which converges uniformly to 0 and such that $J_1(u_n) \to 0$, but $J_1(u) > 0$ for any $u \in \mathcal{C}_0^1(0,1)$ (and also for any $u \in H_0^1(0,1)$).

On the contrary, it is possible to show that these problems have minimizers in spaces of ultrafunctions; a natural space to work in is

$$\widetilde{W_g^1}(\Omega) = \left\{ u \in \widetilde{W^1}(\Omega) \mid \forall x \in \partial\Omega^*,\ u(x) = g(x) \right\}$$

where

$$W^1(\Omega) = \mathcal{C}^1(\Omega) \cap \mathcal{C}(\overline{\Omega}).$$

So our problem becomes

$$\min_{u \in \widetilde{W_g^1}(\Omega)} J^*(u).$$

Theorem 39. *Assume that F is continuous and that*

$$F(x,u,\xi) \geq a(\xi) - M \tag{17}$$

where $a(\xi) \to +\infty$ as $\xi \to +\infty$ and M is a constant. Then

$$\min_{u \in \widetilde{W_g^1}(\Omega)} J^*(u)$$

exists

Proof. Set

$$W_\lambda = \left\{ u \in W^1(\Omega) \mid \forall x \in \partial\Omega^*,\ u(x) = g(x) \right\} \cap \mathrm{Span}(\lambda).$$

If λ is sufficiently large, then $W_\lambda \neq \varnothing$. By (17), $J|_{W_\lambda}$ is coercive and, since it is continuous, it has a minimizer; namely

$$\exists u_\lambda \in W_\lambda,\ \forall v \in W_\lambda,\ J(u_\lambda) \geq J(v).$$

Now set $\bar{u} = \lim_{\lambda \uparrow \mathbb{U}} u_\lambda$ and apply Theorem 15 where

$$u \mathcal{R} W := \forall v \in W,\ J(u) \geq J(v)$$

Then, since $\widetilde{W_g^1}(\Omega) = \lim_{\lambda \uparrow \mathbb{U}} W_\lambda$, the following relation holds:

$$\forall v \in \widetilde{W_g^1}(\Omega),\ J^*(\bar{u}) \geq J^*(v).$$ □

Example: By the above theorem, the functional J_1^* (where J_1 is defined by (16)) has a minimizer $\bar{u}$ in $\widetilde{W_0^1}(0,1)$. It is not difficult to show that $\forall x \in (0,1)^*$, $\bar{u}(x) \sim 0$ and that $J_1(\bar{u})$ is a positive infinitesimal.

5. Appendix

In this section we prove that the axiomatic construction of ultrafunctions is consistent. We assume that the reader knows the key concepts in non-standard analysis (see, e.g., [7]).

The following result has already been proved in [1]. Here we give an alternative proof of this result based on non-standard analysis:

Theorem 40. *The set of axioms* $\{(\Lambda\text{-}1), (\Lambda\text{-}2), (\Lambda\text{-}3)\}$ *is consistent.*

Proof. Let $\mathbb{U}$, $\mathbb{V}$ be mathematical universes and let $\langle \mathbb{U}, \mathbb{V}, \star \rangle$ be a nonstandard extension of $\mathbb{U}$ that is $|\mathbb{U}|^+$-saturated. We denote by $\mathcal{F}$ the set of finite subsets of $\mathbb{U}$ and, for every $\lambda \in \mathcal{F}$, we pose

$$F_\lambda = \{S \subset \mathbb{V} |\ S \text{ is hyperfinite and } \lambda^\star \subset S\}.$$

By saturation $\bigcap_{\lambda \in \mathcal{F}} F_\lambda \neq \emptyset$. We take $\Lambda \in \bigcap_{\lambda \in \mathcal{F}} F_\lambda$.

For any given net $\varphi : \mathcal{F} \to \mathbb{U}$ we define its Λ-limit as

$$\lim_{\lambda \uparrow \mathbb{U}} \varphi(\lambda) = \varphi^\star(\Lambda)$$

and we pose

$$\mathbb{K} = \lim_{\lambda \uparrow \mathbb{U}} \mathbb{R} = \left\{ \lim_{\lambda \uparrow \mathbb{U}} \varphi(\lambda) \mid \varphi : \mathcal{F} \to \mathbb{R} \right\}.$$

With these choices the Λ-limit satisfies the axioms $(\Lambda\text{-}1)$, $(\Lambda\text{-}2)$, $(\Lambda\text{-}3)$: the only nontrivial fact is $(\Lambda\text{-}2)$. Let φ be an eventually constant net, and let $\lambda_0 \in \mathcal{F}, r \in \mathbb{R}$ be such that $\forall \lambda \in \{\eta \in \mathcal{F} \mid \lambda_0 \subset \eta\}$

$$\varphi(\lambda) = r.$$

By transfer it follows that $\forall \lambda \in \{\eta \in \mathcal{F} \mid \lambda_0 \subset \eta\}^\star = \{\eta \in \mathcal{F}^\star \mid \lambda_0^\star \subset \eta\}$ we have:

$$\varphi^\star(\lambda) = r^\star.$$

But $r = r^\star$ and $\lambda_0^\star \subset \Lambda$ by construction. So, since $\Lambda \in \mathcal{F}^\star$, $\varphi^\star(\Lambda) = r$. □

Having defined the Λ-limit, from now on we use the symbol $*$ to denote the extensions of objects in $\mathbb{U}$ in the sense of Λ-limit (not to be confused with the extensions obtained by applying the star map $\star$: e.g., the field $\mathbb{K} = \mathbb{R}^*$ is a subfield of $\mathbb{R}^\star$).

We observe that, given a set S in $\mathbb{U}$, its hyperfinite extension (in the sense of the Λ-limit) is

$$S^\circ = \lim_{\lambda \uparrow \mathbb{U}} (S \cap \lambda) = S^\star \cap \Lambda$$

and we use this observation to prove that, given a set of functions $V(\Omega)$, by posing

$$\widetilde{V}(\Omega) = \mathrm{Span}(V(\Omega)^\circ) = \mathrm{Span}(V(\Omega)^\star \cap \Lambda)$$

we obtain the set of ultrafunctions generated by $V(\Omega)$.

The only nontrivial fact to prove is that, for every function $f \in V(\Omega)$, its natural extension f^* is an ultrafunction. First of all, we observe that, by definition,

$f^* = f^\star$. Also, since $f \in V(\Omega)$, by transfer it follows that $f^\star \in V(\Omega)^\star$. And, by our choice of Λ, we also have that $f^\star \in \Lambda$ since, by construction, $\{f\}^\star = \{f^\star\} \subset \Lambda$.

It remains to prove the coherence of the axioms (Λ-1), (Λ-2), (Λ-3) combined with the Fourier Transform Axiom (FTA).

Theorem 41. *The set of axioms* $\{(\Lambda\text{-}1), (\Lambda\text{-}2), (\Lambda\text{-}3), \text{FTA}\}$ *is consistent.*

Proof. The basic idea is to chose an hyperfinite set $\Lambda \in \bigcap_{\lambda\in\mathcal{F}} F_\lambda$,where F_λ is defined in Theorem 40 (which automatically ensures the satisfaction of (Λ-1), (Λ-2), (Λ-3)), with one more particular property that will ensure the satisfaction of FTA.

We start by considering a generic hyperfinite set $\Lambda' \in \bigcap_{\lambda\in\mathcal{F}} F_\lambda$ and we let

$$B' = \{e_i(x) | i \in I\}$$

be any hyperfinite basis for $\text{Span}(V^{\mathfrak{F}}(\mathbb{R})^\star \cap \Lambda')$. Now we pose

$$B = \{\mathfrak{F}^j(e_i(x)) : 0 \leq j \leq 3, i \in I\} \cup \{\overline{\mathfrak{F}^j(e_i(x))} : 0 \leq j \leq 3, i \in I\},$$

where $\mathfrak{F}$ denotes the Fourier transform. Since $\mathfrak{F}^4 = id$, we have that B is closed by Fourier transform and complex conjugate. We now pose

$$\Lambda = \Lambda' \cup B$$

and it is immediate to prove that, with this choice, FTA is ensured, because B is a set of generators for $\widetilde{V^{\mathfrak{F}}}(\mathbb{R})$ closed by Fourier transform and complex conjugate. □

References

[1] Benci V., *Ultrafunctions and generalized solutions*, Advanced Nonlinear Studies, vol. 13,(2013), pp. 461–486, ISSN: 1536-1365, arXiv:1206.2257.

[2] Benci V., Luperi Baglini L., *A model problem for ultrafunctions*, to appear, arXiv:1212.1370.

[3] Benci, V., Di Nasso, M. *Alpha-theory: an elementary axiomatic for nonstandard analysis*, Expositiones Mathematicae **21** (2003) pp. 355–386.

[4] Benci V., *An algebraic approach to nonstandard analysis*, in: Calculus of Variations and Partial differential equations, (G. Buttazzo, et al., eds.), Springer, Berlin, (1999), 285–326.

[5] Benci V., Galatolo S., Ghimenti M., *An elementary approach to Stochastic Differential Equations using the infinitesimals*, in Contemporary Mathematics, **530**, Ultrafilters across Mathematics, American Mathematical Society, (2010), 1–22.

[6] Benci V., Horsten H., Wenmackers S., *Non-Archimedean probability*, Milan J. Math., (2012), arXiv:1106.1524

[7] Keisler H.J., *Foundations of Infinitesimal Calculus*, Prindle, Weber & Schmidt, Boston, (1976). [This book is now freely downloadable at: http://www.math.wisc.edu/~keisler/foundations.html]

perturbations of the p-Laplacian, and usually described as conditions of *multiplicity near resonance*.

Because of the radial symmetry, we look for radial solutions of the problem. So, letting $r = |x|$ and $v(x) = u(r)$, we reduce it to the one-dimensional Neumann problem

$$-\left(r^{N-1}\frac{u'}{\sqrt{1-u'^2}}\right)' + \lambda r^{N-1}|u|^{m-2}u = r^{N-1}[g(u) - h(r)] \text{ in } (\rho, R),$$
$$u'(\rho) = 0 = u'(R).$$

The results and proofs are given in Section 4. The approach is variational and based upon Szulkin's critical point theory [29] for smooth perturbations of some convex functionals in a Banach space.

2. Multiplicity near resonance for semilinear elliptic problems

Let $\Omega \subset \mathbb{R}^N$ be a bounded domain, $g : \mathbb{R} \to \mathbb{R}$ be continuous and bounded, $h \in L^2(\Omega)$, $\lambda_1 > 0$ be the principal eigenvalue of $-\Delta$ with Dirichlet boundary conditions on Ω, and let φ_1 be the corresponding positive principal eigenfunction normalized by

$$\int_\Omega \varphi_1 = 1.$$

Let us consider the semilinear Dirichlet problem

$$-\Delta u - \lambda_1 u = g(u) - h(x) \text{ in } \Omega, \quad u = 0 \text{ on } \partial\Omega. \tag{1}$$

If we assume in addition that

$$\lim_{u\to-\infty} g(u) := g(-\infty) \text{ and } \lim_{u\to+\infty} g(u) := g(+\infty) \tag{2}$$

exist and that, for all $u \in \mathbb{R}$, either

$$g(-\infty) \le g(u) \le g(+\infty), \tag{3}$$

or

$$g(+\infty) \le g(u) \le g(-\infty), \tag{4}$$

then, multiplying both members of (1) by φ_1, integrating by parts, it is easy to see that *if conditions* (2) *together with* (3) *hold, a necessary condition for the existence of a solution to* (1) *is that*

$$g(-\infty) \le \int_\Omega h\varphi_1 \le g(+\infty) \tag{5}$$

and *if conditions* (2) *together with* (4) *hold, a necessary condition for the existence of a solution to* (1) *is that*

$$g(+\infty) \le \int_\Omega h\varphi_1 \le g(-\infty). \tag{6}$$

When $g \equiv 0$, conditions (5) and (6) reduce to the usual orthogonality condition

$$\int_\Omega h\varphi_1 = 0 \tag{7}$$

upon h to avoid *resonance*.

A famous result of 1970 by Landesman and Lazer [15] implies that *when condition* (2) *holds, the slightly strengthened condition* (5)

$$g(-\infty) < \int_\Omega h\varphi_1 < g(+\infty), \tag{8}$$

or the slightly strengthened condition (6)

$$g(+\infty) < \int_\Omega h\varphi_1 < g(-\infty) \tag{9}$$

is sufficient for the existence of a solution to problem (1). In other words, the presence of a bounded nonlinearity g having a *gap* between its limiting values at $-\infty$ and $+\infty$ *increases* the range of the linear operator $-\Delta - \lambda_1 I$ with Dirichlet conditions on Ω from the co-dimensional one vector subspace of $L^2(\Omega)$

$$\widetilde{L}^2(\Omega) := \left\{h \in L^2(\Omega) : \int_\Omega h\varphi_1 = 0\right\}$$

to the open strip of $L^2(\Omega)$

$$(g(-\infty), g(+\infty)) \oplus \widetilde{L}^2(\Omega) \text{ or } (g(+\infty), g(-\infty)) \oplus \widetilde{L}^2(\Omega).$$

This result was proved by a clever and technically involved used of Schauder's fixed point theorem. It has inspired a very large number of refinements, extensions and of generalizations, and much more transparent proofs have been given using Leray–Schauder's degree. In 1976, Ahmad, Lazer and Paul [1] have shown, using variational techniques, that *condition* (8) *could be replaced by the more general one*

$$\lim_{|c|\to\infty} \left[G(c\varphi_1) - c\int_\Omega h\varphi_1\right] = +\infty, \tag{10}$$

and *condition* (9) *by the more general one*

$$\lim_{|c|\to\infty} \left[G(c\varphi_1) - c\int_\Omega h\varphi_1\right] = -\infty, \tag{11}$$

where G is the indefinite integral of g defined by

$$G(u) = \int_0^u g(s)\,ds. \tag{12}$$

In 1988, Schmitt and one of the authors [23] have considered the corresponding parameter dependent problem

$$-\Delta u - \lambda_1 u + \lambda u = g(u) - h(x) \text{ in } \Omega, \quad u = 0 \text{ on } \partial\Omega, \tag{13}$$

and have shown as special case of a more general abstract result that *if the Landesman–Lazer condition* (8) *holds, there exists $\lambda_0 > 0$ such that* (13) *has at* *least one solution for* $\lambda \in (-\lambda_0, 0]$ *and at least three solutions for* $\lambda \in (0, \lambda_0)$, and *if the Landesman–Lazer condition* (9) *holds, there exists* $\lambda_0 > 0$ *such that* (13) *has at least three solutions for* $\lambda \in (-\lambda_0, 0)$ *and at least one solution for* $\lambda \in [0, \lambda_0)$. The idea of the proof consists in using the Leray–Schauder degree to prove the existence of at least one solution for $|\lambda|$ sufficiently small (and in particular for $\lambda = 0$, which is the Landesman–Lazer case), and obtaining the two other ones using bifurcation from infinity at the eigenvalue λ_0 based upon Krasnosel'skii's results [14].

When $h \in L^\infty(\Omega)$ *and satisfies the orthogonality condition* (7), de Figueiredo and Ni [12] have shown in 1979 that *the Landesman–Lazer conditions* (8) *and* (9) *can be respectively replaced by the sign condition*

$$g(u)u > 0, \quad \forall\, u \neq 0. \tag{14}$$

or the sign condition

$$g(u)u < 0, \quad \forall\, u \neq 0. \tag{15}$$

In 1989, Schmitt and one of the authors [24] have shown, by a similar combination of Leray–Schauder degree and bifurcation from infinity, that *if condition* (14) *holds, there exists* $\lambda_0 > 0$ *such that* (13) *has at least one solution for* $\lambda \in (-\lambda_0, 0]$ *and at least three solutions for* $\lambda \in (0, \lambda_0)$, and *if condition* (15) *holds, there exists* $\lambda_0 > 0$ *such that* (13) *has at least three solutions for* $\lambda \in (-\lambda_0, 0)$ *and at least one solution for* $\lambda \in [0, \lambda_0)$.

If $\partial_\nu u$ denotes the normal derivative of u, similar results hold for the *Neumann boundary value problem*

$$-\Delta u + \lambda u = g(u) - h(x) \ \text{ in } \Omega, \ \ \partial_\nu u = 0 \ \text{ on } \partial\Omega, \tag{16}$$

around the principal eigenvalue $\lambda_1 = 0$ with normalized constant principal eigenfunction $\varphi_1 \equiv |\Omega|^{-1}$. In the statements, *assumptions* (8), (9), (10), (11), (7) *have to be respectively replaced by*

$$g(-\infty) < |\Omega|^{-1} \int_\Omega h < g(+\infty), \tag{17}$$

$$g(+\infty) < |\Omega|^{-1} \int_\Omega h < g(-\infty) \tag{18}$$

$$\lim_{|c|\to\infty} \left[G(c) - c \int_\Omega h \right] = +\infty, \tag{19}$$

$$\lim_{|c|\to\infty} \left[G(c) - c \int_\Omega h \right] = -\infty, \tag{20}$$

$$\int_\Omega h = 0. \tag{21}$$

A first contribution in this direction was already given in 1973 in [20].

Results of this type are called *multiplicity results near resonance*. They have been generalized or applied to more general equations, using similar *topological* techniques in [2, 3, 10, 11, 16, 22]. A *variational* approach to study multiplicity results near resonance was first introduced by Ma, Ramos and Sanchez in [28, 18] for semilinear and quasilinear Dirichlet problems involving the p-Laplacian. See also [19, 17, 25, 13, 27] for a similar variational treatment of various semilinear or quasilinear equations, systems or inequalities with Dirichlet conditions, and [26] for perturbations of the p-Laplacian with Neumann boundary conditions.

3. Quasilinear problems involving the mean extrinsic curvature operator

In the Euclidian space $\mathbb{R}^N$, given an open bounded domain $\Omega \subset \mathbb{R}^N$, the graph of a function $u \in C^1(\Omega, \mathbb{R})$ can be seen as a hypersurface of $\mathbb{R}^{N+1}$. The corresponding *mean curvature operator* is defined by

$$\mathcal{C}(u) := \nabla \cdot \frac{\nabla u}{\sqrt{1+|\nabla u|^2}},$$

and plays an important role in the study of minimal surfaces (zero mean curvature), or more generally in the study of surfaces with prescribed mean curvature.

In a flat Minkowski space $\mathbb{L}^{N+1} = \{(x,t) : x \in \mathbb{R}^N, t \in \mathbb{R}\}$, with metric $\sum_{j=1}^{N}(dx_j)^2 - (dt)^2$, given a bounded domain $\Omega \subset \{(x,t) \in \mathbb{L}^{N+1} : t = 0\} \simeq \mathbb{R}^N$, the graph of a function $u \in C^1(\Omega, \mathbb{R})$ can be seen as a space-like hypersurface of $\mathbb{L}^{N+1}$. The associated *mean extrinsic curvature operator* defined by

$$\mathcal{M}(u) := \nabla \cdot \frac{\nabla u}{\sqrt{1-|\nabla u|^2}},$$

plays an important role in various questions of geometry and relativity [4].

In recent papers [5, 6, 7, 8], the authors have obtained various existence and multiplicity theorems for the radial solutions of quasilinear elliptic equations of the form

$$-\mathcal{M}(v) = f(|x|, v, \partial_\nu v) \text{ in } A(\rho, R), \tag{22}$$

on an annulus or a ball $A(\rho, R)$, where $f : [\rho, R] \times \mathbb{R} \times \mathbb{R} \to \mathbb{R}$ is continuous. The boundary conditions are either Dirichlet ones

$$v = 0 \text{ on } \partial A(\rho, R), \tag{23}$$

or Neumann ones

$$\partial_\nu v = 0 \text{ on } \partial A(\rho, R). \tag{24}$$

Viewing the radial symmetry, letting $r = |x|$, they searched for radial solutions of the form $v(x) = u(r)$, which reduces (22) to the ordinary differential equation

$$-\left(r^{N-1}\frac{u'}{\sqrt{1-u'^2}}\right)' = r^{N-1}f(r,u,u') \tag{25}$$

the Dirichlet boundary conditions (23) to

$$u(\rho) = 0 = u(R) \text{ if } \rho > 0, \quad u'(\rho) = 0 = u(R) \text{ if } \rho = 0 \tag{26}$$

and the Neumann boundary conditions (24) to

$$u'(\rho) = 0 = u'(R). \tag{27}$$

Using a suitable reduction of (25)–(26) to a fixed point problem and Schauder's fixed point theorem, the authors have proved in [5] that *the Dirichlet problem* (25)–(26) *has at least one solution for every continuous* $f : [\rho, R] \times \mathbb{R} \times \mathbb{R} \to \mathbb{R}$. In particular *the Dirichlet problem*

$$\begin{gathered} -r^{1-N}\left(r^{N-1}\frac{u'}{\sqrt{1-u'^2}}\right)' + \lambda k(u) = h(r) \text{ in } (\rho, R), \\ u(\rho) = 0 = u(R) \text{ if } \rho > 0, \ u'(\rho) = 0 = u(R) \text{ if } \rho = 0 \end{gathered} \tag{28}$$

has at least one solution for all $\lambda \in \mathbb{R}$, *and all continuous* $k : \mathbb{R} \to \mathbb{R}$ *and continuous* $h : [\rho, R] \to \mathbb{R}$. By analogy with the Dirichlet problem for the classical Laplacian, one can say that no "eigenvalues" exist for the radial Dirichlet problem associated to the differential operator $\mathcal{M}$. Consequently, the Landesman–Lazer problem and the associated multiplicity result near resonance is meaningless for the radial solutions of (22)–(23).

The situation is different for the Neumann problem. It is immediately seen, by integrating both members over $A(\rho, R)$ and integrating by parts that *a necessary condition for the existence of a radial solution to the problem*

$$-\nabla \cdot \left(\frac{\nabla v}{\sqrt{1-|\nabla v|^2}}\right) = h(|x|) \ \text{in } A(\rho, R), \ \partial_\nu v = 0 \ \text{on } \partial A(\rho, R) \tag{29}$$

is that

$$\int_{A(\rho,R)} h(|x|)\, dx = 0 \ \text{ or equivalently } \int_\rho^R h(r) r^{N-1}\, dr = 0. \tag{30}$$

This condition is also sufficient for radial solutions, because it can easily be shown that, if condition (30) holds, the Neumann problem (29) for radial solutions

$$\begin{gathered} -r^{1-N}\left(r^{N-1}\frac{u'}{\sqrt{1-u'^2}}\right)' = h(r) \ \text{in } (\rho, R), \\ u'(\rho) = 0 = u'(R) \end{gathered}$$

has the one-dimensional linear manifold of solutions

$$u(r) = c + \int_\rho^r \frac{H(s)}{\sqrt{1+|H(s)|^2}}\, ds, \ \text{ where } H(r) = r^{1-N}\int_\rho^r h(s) s^{N-1}\, ds, \ \ c \in \mathbb{R}.$$

On the other hand, the authors have proved in [6], using a suitable fixed point reduction of the equivalent ordinary differential problem and Leray–Schauder degree, that, *for any continuous* $k : \mathbb{R} \to \mathbb{R}$ *such that either*

$$\limsup_{u \to -\infty} k(u) = -\infty, \quad \liminf_{u \to +\infty} k(u) = +\infty$$

or

$$\liminf_{u \to -\infty} k(u) = +\infty, \quad \limsup_{u \to +\infty} k(u) = -\infty,$$

the Neumann problem

$$-r^{1-N}\left(r^{N-1}\frac{u'}{\sqrt{1-u'^2}}\right)' + \lambda k(u) = h(r) \quad in \quad (\rho, R),$$
$$u'(\rho) = 0 = u'(R)$$

has a solution for any $\lambda \neq 0$ *and any continuous* $h : [\rho, R] \to \mathbb{R}$. By analogy with the Neumann problem for the classical Laplacian, we can say that zero is the unique "eigenvalue" for the radial Neumann problem associated to the differential operator $\mathcal{M}$.

Consequently, the Landesman–Lazer problem and the associated multiplicity result near resonance for the radial solutions of the Neumann problem (22)–(24), or equivalently for the solutions of the Neumann problem (25)–(27) are only meaningful near this zero "eigenvalue". On the other hand, as far as we know, no bifurcation from infinity results are known for nonlinear perturbations of the operator $-\mathcal{M}$, so that variational methods seem to be the way for trying to extend the multiplicity results near resonance to the radial solutions of some classes of Neumann problems of the form

$$\begin{aligned} -\mathcal{M}(v) + \lambda k(v) &= g(v) - h(|x|) \quad \text{in} \quad A(\rho, R), \\ \partial_\nu v &= 0 \quad \text{on} \quad \partial A(\rho, R), \end{aligned} \tag{31}$$

for a suitable choice of k and Landesman–Lazer or Ahmad–Lazer–Paul type assumptions upon g and h.

In the case of the classical Laplacian, $k(u) = u$, and the first condition upon g is its sublinearity with respect to u. Here, we shall take for k a mapping of the type $k(u) = |u|^{m-2}u$ for some $m > 1$ and a perturbation term g which is of lower order at infinity. Namely, we will consider Neumann problems of the form

$$\begin{aligned} -\nabla \cdot \left(\frac{\nabla v}{\sqrt{1-|\nabla v|^2}}\right) + \lambda |v|^{m-2} v &= g(v) - h(|x|) \quad \text{in} \quad A(\rho, R), \\ \partial_\nu v = 0 \quad &\text{on} \quad \partial A(\rho, R), \end{aligned} \tag{32}$$

has the structure required by Szulkin's critical point theory [29], that we now recall briefly.

Let $(X, \|\cdot\|)$ be a real Banach space and I be a functional of the type

$$I = \mathcal{F} + \psi,$$

where $\psi : X \to (-\infty, +\infty]$ is proper, convex, lower semicontinuous (in short, l.s.c.) and $\mathcal{F} \in C^1(X; \mathbb{R})$. According to Szulkin [29], $u \in X$ is said to be *a critical point* of I if it satisfies the inequality

$$\langle \mathcal{F}'(u), v - u \rangle + \psi(v) - \psi(u) \geq 0, \quad \forall\, v \in X.$$

A number $c \in \mathbb{R}$ such that $I^{-1}(c)$ contains a critical point is called *a critical value* of I. The functional I is said *to satisfy the Palais–Smale* (in short, (PS)) *condition* if every sequence $\{u_n\} \subset X$ for which $I(u_n) \to c \in \mathbb{R}$ and

$$\langle \mathcal{F}'(u_n), v - u_n \rangle + \psi(v) - \psi(u_n) \geq -\varepsilon_n \|v - u_n\|, \quad \forall v \in X,$$

where $\varepsilon_n \to 0$, (called (PS)*-sequence*), possesses a convergent subsequence. The following result is part of [29, Corollary 3.3].

Lemma 1. *Suppose that $I = \mathcal{F} + \psi$ satisfies the (PS)-condition. If I has two local minima, then it has at least three critical points.*

5. The multiplicity result

The search of solutions of problem (33) is reduced to finding critical points of the energy functional I_λ defined in (39) by the following proposition, which is proved in [7, Proposition 1].

Proposition 1. *If $u \in C$ is a critical point of I_λ, then u is a solution of* (33).

We now state and prove the multiplicity result for problem (33). The proof is based upon two preliminary lemmas, the first one is proved in [7, Lemma 4].

Lemma 2. *Let $s \geq 1$ be a real number. Then*

$$|u(r)|^s \geq |\overline{u}|^s - s(R - \rho)|\overline{u}|^{s-1}, \ \forall\, u \in K, \ \forall\, r \in [\rho, R]. \tag{40}$$

The second lemma is inspired from [18, 28].

Lemma 3. *Assume that conditions (H_R), (H_G) and (H_L) hold. Then there exists $\lambda_+ > 0$ such that, for any $0 < \lambda < \lambda_+$, problem* (33) *has at least one solution $u_\lambda > 0$, which minimize I_λ on $C^+ = \{v \in C : v \geq 0\}$. Moreover, u_λ is a local minimum for I_λ.*

Proof. First, notice that from (38), we obtain

$$\|\widetilde{u}\|_\infty \leq R - \rho \ \forall\, u \in K. \tag{41}$$

This implies that

$$\overline{u} - (R - \rho) \leq u(r) \leq \overline{u} + (R - \rho) \quad \text{for all} \quad u \in K, \tag{42}$$

On the other hand, the authors have proved in [6], using a suitable fixed point reduction of the equivalent ordinary differential problem and Leray–Schauder degree, that, *for any continuous* $k : \mathbb{R} \to \mathbb{R}$ *such that either*

$$\limsup_{u \to -\infty} k(u) = -\infty, \quad \liminf_{u \to +\infty} k(u) = +\infty$$

or

$$\liminf_{u \to -\infty} k(u) = +\infty, \quad \limsup_{u \to +\infty} k(u) = -\infty,$$

the Neumann problem

$$-r^{1-N}\left(r^{N-1}\frac{u'}{\sqrt{1-u'^2}}\right)' + \lambda k(u) = h(r) \quad in \quad (\rho, R),$$
$$u'(\rho) = 0 = u'(R)$$

has a solution for any $\lambda \neq 0$ *and any continuous* $h : [\rho, R] \to \mathbb{R}$. By analogy with the Neumann problem for the classical Laplacian, we can say that zero is the unique "eigenvalue" for the radial Neumann problem associated to the differential operator $\mathcal{M}$.

Consequently, the Landesman–Lazer problem and the associated multiplicity result near resonance for the radial solutions of the Neumann problem (22)–(24), or equivalently for the solutions of the Neumann problem (25)–(27) are only meaningful near this zero "eigenvalue". On the other hand, as far as we know, no bifurcation from infinity results are known for nonlinear perturbations of the operator $-\mathcal{M}$, so that variational methods seem to be the way for trying to extend the multiplicity results near resonance to the radial solutions of some classes of Neumann problems of the form

$$\begin{aligned} -\mathcal{M}(v) + \lambda k(v) &= g(v) - h(|x|) \quad \text{in} \quad A(\rho, R), \\ \partial_\nu v &= 0 \quad \text{on} \quad \partial A(\rho, R), \end{aligned} \tag{31}$$

for a suitable choice of k and Landesman–Lazer or Ahmad–Lazer–Paul type assumptions upon g and h.

In the case of the classical Laplacian, $k(u) = u$, and the first condition upon g is its sublinearity with respect to u. Here, we shall take for k a mapping of the type $k(u) = |u|^{m-2}u$ for some $m > 1$ and a perturbation term g which is of lower order at infinity. Namely, we will consider Neumann problems of the form

$$\begin{aligned} -\nabla \cdot \left(\frac{\nabla v}{\sqrt{1-|\nabla v|^2}}\right) + \lambda |v|^{m-2} v &= g(v) - h(|x|) \quad \text{in} \quad A(\rho, R), \\ \partial_\nu v &= 0 \quad \text{on} \quad \partial A(\rho, R), \end{aligned} \tag{32}$$

Consequently,

$$\inf_{C^+} I_\lambda \to -\infty \quad \text{as} \quad \lambda \to 0,$$

which, together with (42) imply (46), as claimed. □

Theorem 1. *Assume that conditions* (H_R), (H_G) *and* (H_L) *hold. Then there exists some* $\lambda_0 > 0$ *such that, for any* $\lambda \in (0, \lambda_0)$, *problem* (33) *has at least three solutions.*

Proof. From Lemma 3, it follows that there exists $\lambda_+ > 0$ such that I_λ has a local minimum at some $u_{\lambda,1} > 0$ for any $0 < \lambda < \lambda_+$. Using exactly the same strategy, we can find $\lambda_- > 0$ such that I_λ has a local minimum at some $u_{\lambda,2} < 0$ for any $0 < \lambda < \lambda_-$. Taking $\lambda_0 = \min\{\lambda_-, \lambda_+\}$ it follows that I_λ has two local minima for any $\lambda \in (0, \lambda_0)$. On the other hand, from the proof of Lemma 3, it is easy to see that I_λ is coercive on C, implying that I_λ satisfies the (PS) condition for any $\lambda > 0$. Hence, from Lemma 1, we infer that I_λ has at least three critical points for all $\lambda \in (0, \lambda_0)$, which are solutions of (33) by Proposition 1. □

Corollary 1. *Under the assumptions of Theorem* 1, *there exists* $\lambda_0 > 0$ *such that problem* (32) *has at least three radial solutions for any* $\lambda \in (0, \lambda_0)$.

The following examples are easy consequences of Theorem 1.

Example 1. For any $m \geq 2$, any $\sigma \in (1, m)$, and any $h \in C$, there exists $\lambda_0 > 0$ such that the Neumann problem

$$\begin{aligned} -\mathcal{M}(v) + \lambda |v|^{m-2} v &= |v|^{\sigma-2} v - h(|x|) \ \text{ in } \ B(\rho, R), \\ \partial_\nu v &= 0 \ \text{ on } \ \partial B(\rho, R) \end{aligned} \tag{49}$$

has at least three radial solutions when $\lambda \in (0, \lambda_0)$.

Remark 1. For any $m \geq 2$, any $\sigma \in (1, m)$, and any $h \in C$, problem (49) has at least one solution for all $\lambda \in \mathbb{R}$. *This is a consequence of* [6, Theorem 3.1].

Example 2. For any $m \geq 2$, and any $h \in C$ such that $-1 < \overline{h} < 1$, there exists $\lambda_0 > 0$ such that the Neumann problem

$$\begin{aligned} -\mathcal{M}(v) + \lambda |v|^{m-2} v &= \frac{v}{\sqrt{1+v^2}} - h(|x|) \ \text{ in } \ B(\rho, R), \\ \partial_\nu v &= 0 \ \text{ on } \ \partial B(\rho, R) \end{aligned} \tag{50}$$

has at least three radial solutions when $\lambda \in (0, \lambda_0)$.

Remark 2. For any $m > 1$, and any $h \in C$, problem (50) has at least one solution for all $\lambda \in \mathbb{R} \setminus \{0\}$. If $\lambda = 0$, problem (50) has at least one solution for any $h \in C$ such that $-1 < \overline{h} < 1$. *This is a consequence of* [6, Theorem 3.1].

Example 3. For any $m \geq 2$, and any $h \in \widetilde{C}$, there exists $\lambda_0 > 0$ such that the Neumann problem

$$\begin{aligned} -\mathcal{M}(v) + \lambda |v|^{m-2}v &= \frac{v}{1+v^2} + h(|x|) \text{ in } B(\rho, R) \\ \partial_\nu v &= 0 \text{ on } \partial B(\rho, R) \end{aligned} \tag{51}$$

has at least three radial solutions when $\lambda \in (0, \lambda_0)$.

Remark 3. For any $m > 1$, and any $h \in C$, problem (51) has at least one solution for all $\lambda \in \mathbb{R} \setminus \{0\}$. If $\lambda = 0$, problem (51) has at least one solution for any $h \in \widetilde{C}$. *This is a consequence of* [6, Theorem 3.1].

Acknowledgement

Support of C. Bereanu and P. Jebelean from the grant PN-II-RU-TE-2011-3-0157 (CNCS-Romania) is gratefully acknowledged.

References

[1] S. Ahmad, A.C. Lazer and J.L. Paul, Elementary critical point theory and perturbations of elliptic boundary value problems at resonance, Indiana Univ. Math. J. 25 (1976), 933–944.

[2] D. Arcoya and J.L. Gámez, Bifurcation theory and related problems: anti-maximum principle and resonance, Comm. Partial Differential Equations 26 (2001), 1879–1911.

[3] M. Badiale and D. Lupo, Some remarks on a multiplicity result by Mawhin and Schmitt, Bull. Cl. Sci. Acad. R. Belgique (5) 75 (1989), 210–224.

[4] R. Bartnik and L. Simon, Spacelike hypersurfaces with prescribed boundary values and mean curvature, Commun. Math. Phys. 87 (1982), 131–152.

[5] C. Bereanu, P. Jebelean, and J. Mawhin, Radial solutions for some nonlinear problems involving mean curvature operators in Euclidian and Minkowski spaces, Proc. Amer. Math. Soc. 137 (2009), 161–169.

[6] C. Bereanu, P. Jebelean, and J. Mawhin, Radial solutions for Neumann problems involving mean curvature operators in Euclidean and Minkowski spaces, Math. Nachr. 283 (2010), 379–391.

[7] C. Bereanu, P. Jebelean, and J. Mawhin, Variational methods for nonlinear perturbations of singular ϕ-Laplacians, Rend. Lincei Mat. Appl. 22 (2011), 89–111.

[8] C. Bereanu, P. Jebelean, and J. Mawhin, Radial solutions for Neumann problems involving mean extrinsic curvature and periodic nonlinearities, Calculus of Variations Partial Differential Equations 46 (2013), 113–122.

[9] C. Bereanu, P. Jebelean, and J. Mawhin, Multiple solutions for Neumann and periodic problems with singular ϕ-Laplacian, J. Functional Anal. 261 (2011), 3226–3246.

[10] R. Chiappinelli and D.G. de Figueiredo, Bifurcation from infinity and multiple solutions for an elliptic system, Differential Integral Equations 6 (1993), 757–771.

[11] R. Chiappinelli, J. Mawhin and R. Nugari, Bifurcation from infinity and multiple solutions for some Dirichlet problems with unbounded nonlinearities, Nonlinear Anal. 18 (1992), 1099–1112.

[12] D.G. de Figueiredo, and Wei Ming Ni, Perturbations of second order linear elliptic problems by nonlinearities without Landesman–Lazer condition, Nonlinear Anal. 3 (1979), 629–634.

[13] P. De Nápoli and M.C. Mariani, Three solutions for quasilinear equations in $\mathbb{R}^N$, *in* Proceedings USA-Chile Workshop on Nonlinear Analysis (Viña del Mar-Valparaiso, 2000), Southwest Texas State University, Texas, 2001, 131–140.

[14] M.A. Krasnosel'skii, Topological Methods in Nonlinear Integral Equations, Pergamon, Oxford, 1963.

[15] E. Landesman and A.C. Lazer, Nonlinear perturbations of linear elliptic boundary value problems at resonance, Indiana Univ. Math. J. 19 (1970), 609–623.

[16] D. Lupo and M. Ramos, Some multiplicity results for two-point boundary value problems near resonance, Rend. Sem. Mat. Univ. Politec. Torino 48 (1990),125–135 (1992).

[17] T.F. Ma and M.L. Pelicer, Perturbations near resonance for the p-Laplacian in $\mathbb{R}^N$, Abstract Appl. Anal. 7:6 (2002), 323–334.

[18] T.F. Ma, M. Ramos and L. Sanchez, Multiple solutions for a class of nonlinear boundary value problems near resonance: a variational approach, Nonlinear Analysis 30 (1997), 3301–3311.

[19] T.F. Ma and L. Sanchez, Three solutions of a quasilinear elliptic problem near resonance, Math. Slovaca 47 (1997), 451–457.

[20] J. Mawhin, Problèmes aux limites du type de Neumann pour certaines équations différentielles ou aux dérivées partielles non linéaires, in *Equations différentielles et fonctionnelles non linéaires* (Bruxelles-Louvain-la-Neuve 1973), P. Janssens, J. Mawhin and N. Rouche eds., Hermann, Paris, 1973, 123–134.

[21] J. Mawhin, Bifurcation from infinity and nonlinear boundary value problems, in Ordinary and Partial Differential Equations (Dundee 1988), Sleeman and Jarvis eds., Longman, Harlow, 1989, vol. 2, 119–129.

[22] J. Mawhin and W. Omana, Bounded nonlinear perturbations of singular boundary value problems at resonance (with W. Omana), Bull. Cl. Sci. Acad. R. Belgique (6) 2 (1991) 343–356.

[23] J. Mawhin and K. Schmitt, Landesman–Lazer type problems at an eigenvalue of odd multiplicity, Results in Math. 14 (1988), 138–146.

[24] J. Mawhin and K. Schmitt, Nonlinear eigenvalue problems with the parameter near resonance, Ann. Polon. Math. 60 (1990), 241–248.

[25] D. Motreanu, V.V. Motreanu, and N.S. Papageorgiou, Positive solutions and multiple solutions at non-resonance, resonance and near resonance for hemivariational inequalities with p-Laplacian, Trans. Amer. Math. Soc. 360 (2008), 2527–2545.

[26] D. Motreanu, V.V. Motreanu, and N.S. Papageorgiou, Nonlinear Neumann problems near resonance, Indiana Univ. Math. J. 58 (2009), 1257–1279.

[27] Z.Q. Ou and C.M. Tang, Existence and multiplicity results for some elliptic systems at resonance, Nonlinear Anal. 71 (2009), 2660–2666.

[28] M. Ramos and L. Sanchez, A variational approach to multiplicity in elliptic problems near resonance, Proc. Roy. Soc. Edinburgh Sect. A 127 (1997), 385–394.

[29] A. Szulkin, Minimax principles for lower semicontinuous functions and applications to nonlinear boundary value problems, Ann. Inst. H. Poincaré Anal. Non Linéaire 3 (1986), 77–109.

Cristian Bereanu
Institute of Mathematics "Simion Stoilow"
Romanian Academy
21, Calea Griviţei
RO-010702-Bucharest, Sector 1, Romania
e-mail: cristian.bereanu@imar.ro

Petru Jebelean
Department of Mathematics
West University of Timişoara
4, Blvd. V. Pârvan
RO-300223-Timişoara, Romania
e-mail: jebelean@math.uvt.ro

Jean Mawhin
Institut de Recherche en Mathématique et Physique
Université Catholique de Louvain
2, Chemin du Cyclotron
B-1348 Louvain-la-Neuve, Belgique
e-mail: jean.mawhin@uclouvain.be

certain sense, this problem is complementary to our study of an Implicit Function Theorem for such variational problems, see [9], namely characterizing when it fails.

Equivariant bifurcation for a 1-parameter family of gradient-like operators invariant under the action of a Lie group on a Banach space was pioneered by the work of J. Smoller and A. Wasserman [21]. They found sufficient conditions for the existence of a bifurcation instant in a 1-parameter family of zeros of such a path of operators, when these zeros are fixed points of the action. The sufficient condition is stated in terms of the induced isotropy representations on the negative eigenspace of the linearized operators. This result was then successfully used to obtain bifurcation of radial solutions to semilinear elliptic PDEs in a disk with homogeneous linear boundary conditions, among other similar applications.

Nevertheless, in applications to geometric variational problems, it is too restrictive to assume that the starting 1-parameter family of solutions is formed only by *fixed points* of the action. Typically, variational problems involving maps with values in Riemannian manifolds are invariant under the isometry group of the target manifold, which acts by left-composition. It often is a natural situation that the given family of critical points is only invariant under a smaller group of isometries, i.e., the orbits of such points may not consist of single points, although they may also have nontrivial isotropy. It is also important to observe that, in many cases, the action of the symmetry group is not everywhere differentiable. For instance, the (left-composition) action of the isometry group of $\overline{M}$ on the space of $\mathcal{C}^k$ *unparameterized embeddings* of a compact manifold M into $\overline{M}$ is only continuous, and differentiable only at $\mathcal{C}^\infty$ embeddings, see [2]. This is the action one has to consider when studying the CMC variational problem. Finally, it is also common to have only a *local* action of a symmetry group (which is also the case in the CMC variational problem).

In the present paper, we take into account all of the above observations and extend the classic equivariant bifurcation result of J. Smoller and A. Wasserman [21] to this more general situation. Let us describe with more details our main abstract bifurcation results, Theorems 4.3 and 4.5. Assume $\mathcal{M}$ is a Banach manifold endowed with a connection and G is a compact Lie group acting[1] continuously by affine diffeomorphisms on $\mathcal{M}$. Let $\mathfrak{f}_\lambda\colon \mathcal{M} \to \mathbb{R}$ be a family of smooth G-invariant functionals, parameterized by $\lambda \in [a,b]$, and $\lambda \mapsto x_\lambda$ be a curve of critical points in $\mathcal{M}$, i.e., $\mathrm{d}\mathfrak{f}_\lambda(x_\lambda) = 0$, for all λ. Under the appropriate Fredholmness assumptions on the second derivative of $\mathfrak{f}_\lambda$ at x_λ, we prove that if the following conditions are satisfied, there exists equivariant bifurcation at some $\lambda_* \in \,]a,b[\,$:

- *Constant isotropy*: the isotropy group H of x_λ is a *nice group*[2] (in the sense of [21]) and independent of λ;

[1]To simplify our discussion, we suppose here that the action of G is globally defined, although the results described in the sequel also hold for the more general case of *local actions*.

[2]e.g., this is satisfied if H is a closed subgroup of G with less than 5 connected components, see Example 4.2.

- *Equivariant nondegeneracy*: the kernel of the second derivatives $\mathrm{d}^2\mathfrak{f}_a(x_a)$ and $\mathrm{d}^2\mathfrak{f}_b(x_b)$ coincides with the tangent space to the G-orbit of x_a and of x_b, respectively;
- *Jump of negative isotropy representation*: the linear representations of H on the "negative eigenspaces" of $\mathrm{d}^2 f_a(x_a)$ and $\mathrm{d}^2 f_b(x_b)$ are not equivalent.

In other words, the above three conditions imply that there exists a sequence $(x_n)_n$ in $\mathcal{M}$ and a sequence $(\lambda_n)_n$ in $[a, b]$, with $x_n \to x_{\lambda_*}$ and $\lambda_n \to \lambda_*$ as $n \to \infty$, such that for all n, $\mathrm{d}\mathfrak{f}_{\lambda_n}(x_n) = 0$ and the orbit $G \cdot x_n$ is disjoint from the orbit $G \cdot x_{\lambda_n}$, see also Definition 4.1. A particular case of the third condition above is when there is a change of the Morse index (the sum of dimensions of the negative eigenspaces) from x_a to x_b. Clearly, having the same dimension is a necessary condition for two representations to be equivalent, so a jump of the Morse index also determines existence of equivariant bifurcation.

The key idea for the proof of the above results is the construction of slices for group actions, and the reduction of the variational problem to a given slice (where a nonlinear formulation of the classic result of J. Smoller and A. Wasserman [21] can be applied). Although slices for continuous group actions exist in a general topological setting (see [11]), when using variational calculus, one needs a stronger notion of slice (with some differentiability properties). Typically, differentiable slices are constructed applying the exponential map to the normal space of an orbit. This does not work in the general case of actions on Banach manifolds, that may not admit a (complete) invariant inner product. Our central observation is that, in a Banach manifold setting, a similar construction can be performed using the exponential of any *invariant connection*, which exists naturally in many interesting situations. This exponential is then applied to some invariant closed complement of the tangent space to a differentiable group orbit. Invariant closed complements always exist in the case of strongly continuous group actions on Banach spaces (see Lemma 3.2). Thus, the core of the present paper consists in a description of the main properties of connections on infinite-dimensional Banach manifolds (or Banach vector bundles), and the construction of smooth slices for continuous affine (local) actions.

As an example of application of this theory, we obtain bifurcation results for families of CMC hypersurfaces, see Theorems 5.4 and 5.8. Those are then applied to concrete families of Clifford tori in round and Berger spheres, and of rotationally symmetric surfaces in $\mathbb{R}^3$. In those cases, a few recent bifurcation results by the second named author and others are reobtained, see [3, 17, 19]. Other bifurcation results obtained by the first and second named authors for geometric variational problems with symmetries with a similar framework can be found in [6, 7, 8].

Some natural questions arise regarding further generalizations, e.g., when one considers the case in which the isotropy group of x_λ depends on the parameter λ, described in Example 5.10. This is a topic of current research by the authors, as well as the study of other geometric applications, e.g., to the variational problem of constant anisotropic mean curvature hypersurfaces.

The paper is organized as follows. Section 2 contains general facts about connections on (infinite-dimensional) Banach vector bundles. Special emphasis is given to Banach bundles of sections of finite-dimensional vector bundles, endowed with an affine connection, over differentiable manifolds. This is the case of main interest in applications. In this context, the two main results (Proposition 2.5 and Corollary 2.7) are that the map of right-composition with diffeomorphisms of the base manifolds is affine, as well as the map of left-composition with an affine map. Section 3 deals with the question of existence of slices for group actions on infinite-dimensional Banach manifolds. The main result of this section, Theorem 3.4, gives the existence of a slice through a point x for affine actions, under the assumption of compactness of the isotropy of x. The case of local group actions is also discussed, see Subsection 3.1. Section 4 contains the main equivariant bifurcation results (Theorems 4.3 and 4.5), which generalize [21, Thm. 2.1] and [21, Thm. 3.3] respectively. Section 5 contains a geometric application of the two abstract bifurcation results in the context of CMC embeddings, which was the original motivation for the development of the theory. Concrete examples of bifurcation of CMC embeddings recently discovered are briefly presented in the end of this section. Finally, Appendix A describes the basic framework for the used nonlinear formulation of the results of J. Smoller and A. Wasserman [21].

2. Connections on infinite-dimensional manifolds

We start by studying the notion of connection on a Banach vector bundle. Given a connection on a finite-dimensional vector bundle $\pi^E\colon E \to M$ and a smooth manifold D (possibly with boundary), we describe the construction of a naturally associated connection on the bundle $\pi^{\mathcal{E}}: \mathcal{E} \to \mathcal{M}$, where $\mathcal{E} = \mathcal{C}^k(D, E)$, $\mathcal{M} = \mathcal{C}^k(D, M)$ and $\pi^{\mathcal{E}}$ is the left composition with π^E. This is characterized as the unique connection for which the evaluation maps $\mathrm{ev}_p\colon \mathcal{E} \to E$ are affine. We show the invariance of this connection by the right action of the diffeomorphism group of D. When $E = TM$ and the connection on TM is the Levi–Civita connection of some semi-Riemannian metric tensor g on M, then the associated connection on $\mathcal{C}^k(D, M)$ is also invariant by the left action of the isometry group of g. A classic reference on these topics is [13].

2.1. Banach vector bundles

Let $\mathcal{M}$ be a smooth Banach manifold, and $\pi^{\mathcal{E}}: \mathcal{E} \to \mathcal{M}$ be a smooth Banach vector bundle on $\mathcal{M}$. This means that $\mathcal{E} = \bigcup_{x\in\mathcal{M}} \mathcal{E}_x$, with $\mathcal{E}_x = \pi^{-1}(x)$, is a collection of vector spaces, and that it is given an *atlas of compatible trivializations of* $\mathcal{E}$. Given a Banach space $\mathcal{E}_0$, write

$$\mathrm{Fr}_{\mathcal{E}_0}(\mathcal{E}) := \bigcup_{x\in\mathcal{M}} \mathrm{Iso}(\mathcal{E}_0, \mathcal{E}_x),$$

where $\mathrm{Iso}(\mathcal{E}_0, \mathcal{E}_x)$ is the set of Banach space isomorphisms (bi-Lipschitz linear isomorphisms) from $\mathcal{E}_0$ to $\mathcal{E}_x$. For a vector bundle $\pi^{\mathcal{E}}: \mathcal{E} \to \mathcal{M}$ with fibers of finite dimension n, we will write $\mathrm{Fr}(\mathcal{E})$ for $\mathrm{Fr}_{\mathbb{R}^n}(\mathcal{E})$. A local trivialization of $\pi^{\mathcal{E}}: \mathcal{E} \to \mathcal{M}$ with domain $\mathcal{U} \subset \mathcal{M}$ and typical fiber $\mathcal{E}_0$ is a local section $s: \mathcal{U} \to \mathrm{Fr}_{\mathcal{E}_0}(\mathcal{E})$. Two local trivializations s_i, with domain $\mathcal{U}_i \subset \mathcal{M}$ and typical fibers $\mathcal{E}_i$, $i = 1, 2$, are compatible if the transition map $s_2^{-1} s_1: \mathcal{U}_1 \cap \mathcal{U}_2 \to \mathrm{Iso}(\mathcal{E}_1, \mathcal{E}_2)$ is smooth. A collection $(\mathcal{U}_i, s_i, \mathcal{E}_i)_{i \in I}$ of local trivializations of $\mathcal{E}$ is an atlas if the domains $\mathcal{U}_i$ cover $\mathcal{M}$. For details on the structure of such Banach vector bundles, see [20].

2.2. Connections on Banach vector bundles

A *connection* on the Banach vector bundle $\pi^{\mathcal{E}}: \mathcal{E} \to \mathcal{M}$ is a smooth map $P^{\mathcal{E}}: T\mathcal{E} \to \mathcal{E}$ such that:

(a) for all $x \in \mathcal{M}$ and $e \in \mathcal{E}_x$, the restriction $P^{\mathcal{E}}_e = P^{\mathcal{E}}|_{T_e\mathcal{E}}$ is a linear map with values in $\mathcal{E}_x$;

(b) for any local trivialization $s: \mathcal{U} \to \mathrm{Fr}_{\mathcal{E}_0}(\mathcal{E})$, there exists a smooth map

$$\mathcal{U} \ni x \longmapsto \Gamma_x \in \mathrm{Bil}(T_x\mathcal{M} \times \mathcal{E}_x, \mathcal{E}_x)$$

such that, denoting by $\widetilde{s}: \mathcal{E}|_{\mathcal{U}} \to \mathcal{E}_0$ the map $\widetilde{s}(e) = s\big(\pi(e)\big)^{-1}(e)$, the following identity holds for all $x \in \mathcal{U}$, $e \in \mathcal{E}_x$ and $\eta \in T_e\mathcal{E}$:

$$P^{\mathcal{E}}(\eta) = s(x)\big(\mathrm{d}\widetilde{s}_e(\eta)\big) + \Gamma_x\big(\mathrm{d}\pi^{\mathcal{E}}_e(\eta), e\big).$$

A standard argument shows that it suffices to have property (b) satisfied only for the set of local trivializations of an atlas.

A connection $\mathcal{P}^{\mathcal{E}}$ defines a distribution $\mathrm{Hor}(P^{\mathcal{E}})$ on the total space $\mathcal{E}$, called the *horizontal distribution*, given by $\mathrm{Hor}(P^{\mathcal{E}})_e = \mathrm{Ker}(P^{\mathcal{E}}_e)$. A vector $v \in T_e\mathcal{E}$ will be called *horizontal* if it belongs to $\mathrm{Hor}(P^{\mathcal{E}})_e$.

2.3. Connections and exponential maps on Banach manifolds

By a *manifold with connection*, we mean a Banach manifold $\mathcal{M}$ with a connection on its tangent bundle $\pi: T\mathcal{M} \to \mathcal{M}$. If P is a connection on $T\mathcal{M}$, one has a vector field $X(P)$ on $T\mathcal{M}$, called *geodesic field*, defined by the following: for $x \in \mathcal{M}$ and $v \in T_x\mathcal{M}$, $X(P)_v$ is the unique horizontal vector in $T_v(T\mathcal{M})$ that projects onto $v \in T_x\mathcal{M}$ (by the differential $\mathrm{d}\pi_v$). A curve $\gamma: I \to \mathcal{M}$ is a *P-geodesic* if it is the projection of an integral curve $\Gamma: I \to T\mathcal{M}$ of $X(P)$. If $\mathcal{M}$ is a manifold with connection P, then one has an exponential map

$$\exp^P: \mathrm{Dom}(\exp^P) \subset T\mathcal{M} \longrightarrow \mathcal{M},$$

defined on an open subset $\mathrm{Dom}(\exp^P) \subset T\mathcal{M}$ containing the zero section, with properties totally analogous to the exponential map of a connection on a finite-dimensional manifold. In particular, for all $x \in \mathcal{M}$, $\mathcal{A}_x = \mathrm{Dom}(\exp^P) \cap T_x\mathcal{M}$ is a star-shaped open neighborhood of 0 in $T_x\mathcal{M}$, and, denoting by $\exp^P_x$ the restriction of $\exp^P$ to $\mathcal{A}_x$, one has $\mathrm{d}\exp^P_x(0) = \mathrm{Id}$. In particular, $\exp^P_x$ is a diffeomorphism from an open neighborhood of 0 in $T_x\mathcal{M}$ onto an open neighborhood of x in $\mathcal{M}$.

2.4. Banach bundles of sections of a finite-dimensional vector bundle

We now describe an important example of the abstract setting of Subsection 2.2. Consider a vector bundle $\pi^E : E \to M$ over a finite-dimensional differentiable manifold M, and let $P^E : TE \to E$ be a connection in E. Let D be a compact differentiable manifold (possibly with boundary), and for some $k \geq 2$, set $\mathcal{M} = \mathcal{C}^k(D, M)$ and $\mathcal{E} = \mathcal{C}^k(D, E)$. There exists a natural map $\pi^{\mathcal{E}} : \mathcal{E} \to \mathcal{M}$, namely the map $(\pi^E)_*$ of left-composition with π^E. The sets $\mathcal{M}$ and $\mathcal{E}$ admit a natural structure of Banach manifold, making $\pi^{\mathcal{E}} : \mathcal{E} \to \mathcal{M}$ an infinite-dimensional Banach vector bundle. More precisely, for $f \in \mathcal{M}$, the fiber $\mathcal{E}_f$ is the Banach space of $\mathcal{C}^k$ sections of the pull-back bundle $f^*(E)$, also called sections of E *along* f,

$$\mathcal{E}_f = \big\{ F \in \mathcal{C}^k(D, E) : F(x) \in E_{f(x)} \text{ for all } x \in D \big\}.$$

There is also a natural identification of the tangent bundle of $\mathcal{E}$ as

$$T\mathcal{E} \cong \mathcal{C}^k(D, TE),$$

and if f is in $\mathcal{M}$ (which can be thought of as the zero section of $\mathcal{E}$), there is a canonical splitting $T_f\mathcal{E} \cong T_f\mathcal{M} \oplus \mathcal{E}_f$ in horizontal and vertical parts respectively. The horizontal subspace of $T_f\mathcal{E}$ in this particular case is $T_f\mathcal{M} \cong \mathcal{C}^k(D, TM)$. To have a notion of horizontal subspace at the tangent space to $\mathcal{E}$ at points outside the zero section, we need a connection on this Banach vector bundle.

A connection $P^{\mathcal{E}} : T\mathcal{E} \to \mathcal{E}$ can be defined as being the map $(P^E)_*$ of left-composition with P^E. Let us show that this satisfies the axioms (a) and (b) described in Subsection 2.2.

First, given $f \in \mathcal{M}$ and $F \in \mathcal{E}_f$, we have that the restriction $P^{\mathcal{E}}_F$ of $P^{\mathcal{E}}$ to $T_F\mathcal{E}$ maps a section η of $F^*(TE)$ (i.e., an element of $T_F\mathcal{E}$) to the section $P^E \circ \eta$ of f^*E (i.e., an element of $\mathcal{E}_f$), see the commutative diagram below. This map $P^{\mathcal{E}}_F$ is clearly linear, since it is given by left-composition with P^E, proving that axiom (a) holds.

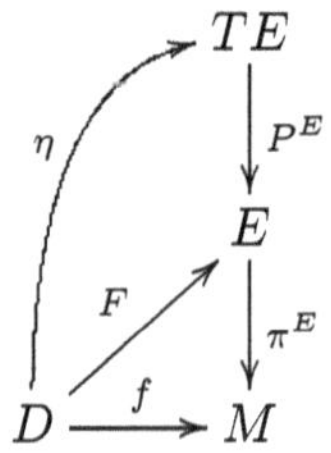

Second, we observe that an atlas of trivializations of $\pi^{\mathcal{E}} : \mathcal{E} \to \mathcal{M}$ can be constructed using smooth maps $s : \mathrm{Dom}(s) \subset D \times M \to \mathrm{Fr}(E)$ such that $\pi^E \circ s(p, x) = x$ for all $(p, x) \in \mathrm{Dom}(s)$, and such that $s(p, \cdot)$ is a local trivialization of $\pi^E : E \to M$. Once such a map s is given, a trivialization $\mathfrak{s} : \mathrm{Dom}(\mathfrak{s}) \subset \mathcal{C}^k(D, M) \to \mathrm{Fr}_{\mathcal{E}_0}\big(\mathcal{C}^k(D, E)\big)$, with $\mathcal{E}_0 = \mathcal{C}^k(D, \mathbb{R}^n)$, is defined by setting, for all

$x \in \mathrm{Dom}(\mathfrak{s}) = \{x \in \mathcal{C}^k(D, M) : \mathrm{Gr}(x) \subset \mathrm{Dom}(s)\}$,[3]

$$\begin{aligned} \mathfrak{s}(x)\colon \mathcal{C}^k(D, \mathbb{R}^n) &\longrightarrow \mathcal{C}^k(D, E), \\ \mathfrak{s}(x)(v)_p &= s\big(p, x(p)\big)v(p), \end{aligned}$$

for all $v \in \mathcal{C}^k(D, \mathbb{R}^n)$ and $p \in D$. Given $p \in D$, $F \in \mathcal{E}$ and $\eta \in T_e\mathcal{E}$, then:

$$\begin{aligned} P_F^{\mathcal{E}}(\eta)(p) &= P_{F(p)}^{E}\big(\eta(p)\big) \\ &= s\big(p, x(p)\big)\big[\mathrm{d}\widetilde{s}_p\big(F(p)\big)\eta(p)\big] + \Gamma_{x(p)}^{P^E}\big(\mathrm{d}\pi_{F(p)}^{E}\big(\eta(p), F(p)\big)\big), \end{aligned}$$

which says that the Christoffel symbol $\widetilde{\Gamma}$ of $P^{\mathcal{E}}$ associated to the trivialization $\mathfrak{s}$ is given by:

$$\begin{aligned} \widetilde{\Gamma}_x\colon \mathcal{C}^k(D, TM; x) \times \mathcal{C}^k(D, E; x) &\longrightarrow \mathcal{C}^k(D, E; x) \\ \widetilde{\Gamma}_x(v, e)(p) &= \Gamma_{x(p)}^{P}\big(v(p), e(p)\big). \end{aligned} \tag{2.1}$$

Here, $\mathcal{C}^k(D, TM; x)$ and $\mathcal{C}^k(D, E; x)$ respectively denote the spaces of $\mathcal{C}^k$ sections of the pull-back bundles $x^*(TM)$ and $x^*(E)$.

An interesting particular case of the above construction is when $E = TM$ is the tangent bundle of M. Recall that D is a smooth manifold (possibly with boundary), M is a manifold whose tangent bundle TM has a connection P^{TM} and the Banach vector bundle $\mathcal{E} = \mathcal{C}^k(D, TM)$ is the tangent bundle of the Banach manifold $\mathcal{M} = \mathcal{C}^k(D, M)$, under the identification

$$\mathcal{E} = \mathcal{C}^k(D, TM) \cong T\mathcal{C}^k(D, M) = T\mathcal{M}.$$

Endowing $T\mathcal{M}$ with the naturally induced connection $P^{T\mathcal{M}}$ described above, the $P^{T\mathcal{M}}$-geodesics in $\mathcal{M}$ are smooth curves $s \mapsto x_s \in \mathcal{C}^k(D, M)$ such that, for all $p \in D$, the curve $s \mapsto x_s(p) \in M$ is a P^{TM}-geodesic in M. This is a manifestation of the fact we will see next that the induced connection $P^{T\mathcal{M}}$ is characterized by every evaluation map $\mathrm{ev}_p\colon \mathcal{M} \to M$ being *affine*, see Proposition 2.4.

2.5. Affine maps

Let us now consider two Banach vector bundles $\pi^{\mathcal{E}} : \mathcal{E} \to \mathcal{M}$ and $\pi^{\mathcal{E}'} : \mathcal{E}' \to \mathcal{M}'$ endowed with connections $P^{\mathcal{E}}$ and $P^{\mathcal{E}'}$ respectively. Let $f\colon \mathcal{M} \to \mathcal{M}'$ be a smooth map and $T\colon \mathcal{E} \to \mathcal{E}'$ a smooth Banach bundle morphism for which the following diagram commutes.

$$\begin{array}{ccc} \mathcal{E} & \xrightarrow{\;T\;} & \mathcal{E}' \\ {\scriptstyle \pi^{\mathcal{E}}}\big\downarrow & & \big\downarrow{\scriptstyle \pi^{\mathcal{E}'}} \\ \mathcal{M} & \xrightarrow[\;f\;]{} & \mathcal{M}'. \end{array}$$

[3] $\mathrm{Gr}(x)$ denotes the graph of $x \in \mathcal{C}^k(D, E)$.

When $\mathcal{M}$ is endowed with a connection, we will say that the group action is *affine* if G acts by affine diffeomorphisms of $\mathcal{M}$.

Define the auxiliary maps, with $g \in G$, $y \in \mathcal{M}$,

$$\begin{aligned} \beta_x \colon G &\longrightarrow \mathcal{M} & \qquad \phi_g \colon \mathcal{M} &\longrightarrow \mathcal{M} \\ g &\longmapsto g \cdot x & y &\longmapsto g \cdot y. \end{aligned} \tag{3.1}$$

From assumption (b), ϕ_g is a diffeomorphism for each $g \in G$. Assumption (c) means that β_x is differentiable. In particular, the G-orbit of x is a submanifold of $\mathcal{M}$, whose tangent space at x is given by the image of $\mathrm{d}\beta_x(1) \colon \mathfrak{g} \to T_x\mathcal{M}$, where $\mathfrak{g}$ is the Lie algebra of G. Let us denote by G_x the isotropy (or stabilizer) of x, which is the closed subgroup of G given by $G_x = \{g \in G : \phi_g(x) = x\}$.

Definition 3.1. A *slice* for the action of G on $\mathcal{M}$ at x is a smooth submanifold $\mathcal{S} \subset \mathcal{M}$ containing x, such that

1. the tangent space $T_x\mathcal{S} \subset T_x\mathcal{M}$ is a closed complement to $\mathrm{Im}(\mathrm{d}\beta_x(1))$, i.e., $T_x\mathcal{M} = \mathrm{Im}(\mathrm{d}\beta_x(1)) \oplus T_x\mathcal{S}$;
2. $G \cdot \mathcal{S}$ is a neighborhood of the orbit $G \cdot x$, i.e., the orbit of every $y \in \mathcal{M}$ sufficiently close to x must intersect $\mathcal{S}$;
3. $\mathcal{S}$ is invariant under the isotropy group G_x.

We will prove the existence of slices for affine actions of compact Lie groups. Towards this goal, we need an auxiliary result on linear actions of compact groups.

Lemma 3.2. *Let G be a compact Hausdorff topological group with a* strongly continuous[4] *linear representation on a Banach space $\mathcal{X}$. Then:*

(a) *if $S \subset \mathcal{X}$ is a closed G-invariant complemented subspace of $\mathcal{X}$, then S admits a G-invariant closed complement;*

(b) *the origin of $\mathcal{X}$ has a fundamental system of G-invariant neighborhoods.*

Proof. First, observe that by the uniform boundedness principle, the linear operators on $\mathcal{X}$ associated to the action of elements $g \in G$ have norm bounded by a constant which is independent of g. By a simple argument, it follows that the action defines a continuous function $G \times \mathcal{X} \to \mathcal{X}$. For part (a), let $P \colon \mathcal{X} \to \mathcal{X}$ be a projector (i.e., bounded linear idempotent) with image S. Define $\widetilde{P} \colon \mathcal{X} \to \mathcal{X}$ as the Bochner integral $\widetilde{P}(x) = \int_G gPg^{-1}x \,\mathrm{d}g$, where $\mathrm{d}g$ is the Haar measure of G. It is easy to see that $\widetilde{P}$ is a well-defined bounded linear operator on $\mathcal{X}$, with image contained in S. Furthermore, it fixes the elements of S, and commutes with the G-action. It follows that $\widetilde{P}$ is also a projector with image S, and its kernel is the desired G-invariant closed complement to S.

As to part (b), let V be an arbitrary neighborhood of the origin of $\mathcal{X}$. The inverse image of V by the action $G \times \mathcal{X} \to \mathcal{X}$ is an open subset Z of the product $G \times \mathcal{X}$ that contains $G \times \{0\}$. Since G is compact, there exists a neighborhood U of 0 in $\mathcal{X}$ such that $G \times U$ is contained in Z, i.e., $g \cdot x \in V$ for all $g \in G$, $x \in U$. The union $\bigcup_{g \in G} gU$ is a G-invariant open neighborhood of 0 in $\mathcal{X}$, contained in V. □

---

[4]i.e., the maps $G \ni g \mapsto g \cdot x \in \mathcal{X}$ are continuous for all $x \in \mathcal{X}$.

We also observe the following interesting fact.

Lemma 3.3. *Let $\rho\colon G \times \mathcal{X} \to \mathcal{X}$ be a continuous action of a compact group G on a topological space $\mathcal{X}$. Assume $x_0 \in \mathcal{X}$ is a fixed point of G. Then x_0 admits a fundamental system of G-invariant (open) neighborhoods.*

Proof. Let V be an arbitrary neighborhood of x_0, and set $W = \bigcap_{g\in G} gV$. Clearly W is G-invariant, and $W \subset V$. Let us show that W is a neighborhood of x_0. The set $\rho^{-1}(V)$ is an open subset of $G \times \mathcal{X}$ that contains $G \times \{x_0\}$, and by the compactness of G, it also contains the product $G \times U$, where U is some open neighborhood of x_0. Thus, $U \subset W$, and W is a neighborhood of x_0. The interior of W is also G-invariant. □

We can now prove our result on the existence of slices.

Theorem 3.4. *In the above situation, assume that $\mathcal{M}$ is endowed with a connection which is G-invariant (i.e., each diffeomorhism ϕ_g is affine), and that G_x is compact. Then there exists a slice $\mathcal{S}$ through x.*

Proof. Consider the isotropy representation of G_x on $T_x\mathcal{M}$, given by $g \mapsto \mathrm{d}\phi_g(x)$. The finite-dimensional subspace $\mathrm{Im}\big(\mathrm{d}\beta_x(1)\big)$ is clearly invariant under this linear action. By part (a) of Lemma 3.2, there exists a closed G_x-invariant complement S of $\mathrm{Im}\big(\mathrm{d}\beta_x(1)\big)$. Denote by $\exp_x$ the exponential map of the G-invariant connection at x, and let $U_0 \subset T_x\mathcal{M}$ be an open neighborhood of 0 on which $\exp_x$ is a diffeomorphism. By part (b) of Lemma 3.2, there exists an open neighborhood $\widetilde{U}_0 \subset U_0$ of 0 such that $\widetilde{U}_0 \cap S$ is G_x-invariant. Set

$$\mathcal{S} := \exp_x(\widetilde{U}_0 \cap S).$$

We claim that $\mathcal{S}$ is a slice for the action of G at x. Property (1) of slices is clearly satisfied, since $\mathrm{d}\exp_x(0) = \mathrm{Id}$, and S is a closed complement to $\mathrm{Im}\big(\mathrm{d}\beta_x(1)\big)$. Property (2) would follow immediately from the transversality condition (1) under the hypothesis of differentiability of the group action, which we do not assume. A slightly more involved topological argument based on degree theory is required for the continuous case, and this is discussed separately in Proposition 3.5, which is to be applied with $A = N = \mathcal{M}$, $M = G$, $Q = \mathcal{S}$, χ being the action, $a_0 = x$, and $m_0 = 1$. For property (3), observe that since the connection is G-invariant, then $\phi_g \circ \exp_x = \exp_{\phi_g(x)} \circ \mathrm{d}\phi_g(x)$, for all $g \in G$. Thus, given $v \in \widetilde{U}_0 \cap S$ and $g \in G_x$, $\phi_g\big(\exp_x(v)\big) = \exp_x\big(\mathrm{d}\phi_g(x)v\big) \in \mathcal{S}$, because $\widetilde{U}_0 \cap S$ is G_x-invariant, i.e., $\mathcal{S}$ is G_x-invariant. □

Proposition 3.5. *Let M be a finite-dimensional manifold, N a (possibly infinite-dimensional) Banach manifold, $Q \subset N$ a Banach submanifold, and A a topological space. Assume that $\chi\colon A \times M \to N$ is a continuous function such that there exists $a_0 \in A$ and $m_0 \in M$ with:*

(a) $\chi(a_0, m_0) \in Q$; (b) $\chi(a_0, \cdot)\colon M \to N$ *is of class* C^1;
(c) $\partial_2\chi(a_0, m_0)\big(T_{m_0}M\big) + T_{\chi(a_0,m_0)}Q = T_{\chi(a_0,m_0)}N$.

Then, for $a \in A$ near a_0, $\chi(a, M) \cap Q \neq \emptyset$.

Proof. Let $f\colon U \subset \mathbb{R}^d \to \mathbb{R}^d$ be a $\mathcal{C}^1$ function, defined on an open neighborhood U of 0, such that $f(0) = 0$ and $\mathrm{d}f(0)$ an isomorphism. The induced map $\widetilde{f}\colon \mathbb{S}^{d-1} \to \mathbb{S}^{d-1}$ is defined by $\widetilde{f}(x) = \|f(rx)\|^{-1} f(rx)$, where $r > 0$ is such that 0 is the unique zero of f in the closed ball $\overline{B(0,r)}$ of $\mathbb{R}^d$. This induced map must have topological degree equal to ± 1.

If A is any topological space, $f\colon A \times U \to \mathbb{R}^d$ is continuous, and $a_0 \in A$ is such that $f(a_0, \cdot)$ is of class $\mathcal{C}^1$, $f(a_0, 0) = 0$ and $\partial_2 f(a_0, 0)$ is an isomorphism, for a near a_0, and $r > 0$ sufficiently small, $0 \in f\big(a, \overline{B(0,r)}\big)$. This follows from the continuity of the topological degree. The same conclusion holds for a function $f\colon A \times U \to \mathbb{R}^d$, where now U is an open neighborhood of 0 in $\mathbb{R}^s$, with $s \geq d$, under the assumption that $f(a_0, \cdot)$ is of class $\mathcal{C}^1$, $f(a_0, 0) = 0$, and $\partial_2 f(a_0, 0)$ be surjective. Namely, it suffices to apply the argument above to the function obtained by restricting f to a d-dimensional subspace where $\partial_2 f(a_0, 0)$ is an isomorphism.

To finish the proof, use local coordinates adapted to Q in N, and assume that M, Q and N are Banach spaces, with $N = Q \oplus \mathbb{R}^d$, $d \leq s = \dim(M)$ is the codimension of Q, and $m_0 = 0$. In this situation, the conclusion is obtained applying the argument above to the function $f\colon A \times M \to \mathbb{R}^d$ given by $f(a, m) = \pi\big(\chi(a, m)\big)$, where $\pi\colon N \to \mathbb{R}^d$ is the projection relative to the decomposition $N = Q \oplus \mathbb{R}^d$. Clearly, $f(a, m) = 0$ if and only if $\chi(a, m) \in Q$. Assumption (a) gives $f(a_0, 0) = 0$, and assumption (c) implies that $\partial_2 f(a_0, 0)$ is surjective. □

3.1. Local actions

The existence of slices proved in Theorem 3.4 holds in the more general case of *local* group actions. Let us briefly recall the definition and a few basic facts about local actions.

Let G be a Lie group and $\mathcal{M}$ a topological manifold. By a *local action* of G on $\mathcal{M}$, we mean a continuous map $\rho\colon \mathrm{Dom}(\rho) \subset G \times \mathcal{M} \to \mathcal{M}$, defined on an open subset $\mathrm{Dom}(\rho) \subset G \times \mathcal{M}$ containing $\{1\} \times \mathcal{M}$ satisfying:

(a) $\rho(1, x) = x$ for all $x \in \mathcal{M}$;
(b) if $(g_2, x) \in \mathrm{Dom}(\rho)$ and $\big(g_1, \rho(g_2, x)\big) \in \mathrm{Dom}(\rho)$, then $(g_1 g_2, x) \in \mathrm{Dom}(\rho)$, and $\rho\big(g_1, \rho(g_2, x)\big) = \rho(g_1 g_2, x)$.

Usual group actions can be obtained as the particular case in which the domain $\mathrm{Dom}(\rho)$ coincides with the entire $G \times \mathcal{M}$. Local actions can be *restricted*, in the following sense. If $\mathcal{N} \subset \mathcal{M}$ is a submanifold, then one has a local action $\widetilde{\rho}$ of G on $\mathcal{N}$ by setting $\mathrm{Dom}(\widetilde{\rho}) = \big\{(g, x) \in (G \times \mathcal{N}) \cap \mathrm{Dom}(\rho) : \rho(g, x) \in \mathcal{N}\big\}$, and $\widetilde{\rho} = \rho|_{\mathrm{Dom}(\widetilde{\rho})}$. In fact, the most natural occurrence[5] of local actions is when one has a (global) action of a group G on a topological manifold $\mathcal{X}$, and $\mathcal{M}$ is an open (not necessarily G-invariant) subset of $\mathcal{X}$. The restriction of the action of G to $\mathcal{M}$ in the above sense is a local action of G on $\mathcal{M}$.

[5] In fact, local actions of groups are always restrictions of global actions. In the literature, these are known as *enveloping actions* of the local action, see [1].

Assumption (b) implies that for all $x \in M$, denoting by

$$G_x = \{g \in G : (g, x) \in \mathrm{Dom}(\rho),\ \rho(g, x) = x\}$$

the isotropy of x, then G_x is a closed subgroup of G.

Given a local action ρ of G on $\mathcal{M}$, for $g \in G$, let ρ_g denote the map $\rho(g, \cdot)$, defined on a (possibly empty) open set $\mathrm{Dom}(\rho_g) = \mathrm{Dom}(\rho) \cap \{g\} \times \mathcal{M}$. The following follow easily from the definition.

Lemma 3.6. *Let* $\rho\colon \mathrm{Dom}(\rho) \subset G \times \mathcal{M} \to \mathcal{M}$ *be a local action of* G *on* M. *Then*

(a) *for all* $g \in G$, *the map* $\rho_g\colon \rho_g^{-1}(\mathrm{Dom}(\rho_{g^{-1}})) \to \rho_{g^{-1}}^{-1}(\mathrm{Dom}(\rho_g))$ *is a homeomorphism;*

(b) *the set* $\{(g, x) \in G \times \mathcal{M} : x \in \rho_g^{-1}(\mathrm{Dom}(\rho_{g^{-1}}))\}$ *is an open subset that contains* $\{1\} \times \mathcal{M}$*; in particular:*

(c) *for all* $x \in \mathcal{M}$, *there exists an open neighborhood* U_x *of* 1 *in* G *such that for all* $g \in U_x$, $x \in \rho_g^{-1}(\mathrm{Dom}(\rho_{g^{-1}}))$.

In view of (c) above, one can define a map $\beta_x\colon \mathrm{Dom}(\beta_x) \subset G \to \mathcal{M}$ on a neighborhood $\mathrm{Dom}(\beta_x)$ of 1 in G, by $\beta_x(g) = \rho(g, x)$, compare with (3.1). In particular, if $x \in \mathcal{M}$ is such that the map β_x is differentiable (at 1), then one has a well-defined linear map $\mathrm{d}\beta_x(1)\colon \mathfrak{g} \to T_x\mathcal{M}$. A subset $C \subset \mathcal{M}$ will be called G-invariant if, given $x \in C$, then $\rho(g, x) \in C$ for all $g \in \mathrm{Dom}(\beta_x)$.

In view of the above, the definition of slice for local actions is totally analogous to Definition 3.1. Furthermore, the statement and proof of Theorem 3.4 carry over *verbatim* to the case of local affine actions.

4. Equivariant bifurcation

Let us define what we intend by equivariant bifurcation. To simplify our discussion, we restrict to the case when there is a globally defined action (opposed to a local action). Consider the same setup (a), (b) and (c) of Section 3. Let $[a, b] \ni \lambda \mapsto \mathfrak{f}_\lambda$ be a continuous path of C^k-functionals $\mathfrak{f}_\lambda\colon \mathcal{M} \to \mathbb{R}$, $k \geq 2$, which are G-invariant, i.e., $\mathfrak{f}_\lambda(g \cdot y) = \mathfrak{f}_\lambda(y)$ for all $y \in \mathcal{M}$, $g \in G$ and $\lambda \in [a, b]$. We are interested in studying bifurcation of solutions to the equation $\mathrm{d}\mathfrak{f}_\lambda(x) = 0$.

Definition 4.1. Given $\lambda_0 \in [a, b]$, we say that *equivariant bifurcation* of critical points of the family $(\mathfrak{f}_\lambda)_{\lambda \in [a,b]}$ occurs at $(x_{\lambda_0}, \lambda_0)$ if there is a sequence $(x_n, \lambda_n) \in \mathcal{M} \times [a, b]$ such that

1. $\lim\limits_{n\to\infty} (x_n, \lambda_n) = (x_{\lambda_0}, \lambda_0)$;
2. $\mathrm{d}\mathfrak{f}_{\lambda_n}(x_n) = 0$, for all n;
3. $x_n \notin G \cdot x_{\lambda_n}$, for all n.

We now discuss our central result, which is a sufficient condition for equivariant bifurcation in the above sense. It will be obtained by combining the slice theory developed in the previous section with a nonlinear formulation of a celebrated bifurcation result of J. Smoller and A. Wasserman [21]. In order to deal with the important general case of functionals defined on *Banach* manifolds (rather than *Hilbert* manifolds), we will need an appropriate framework described by a set of assumptions on an auxiliary Hilbert/Fredholm structure of the problem.

Let B_2 and B_0 be Banach spaces and H be a Hilbert space with inner product $\langle\cdot,\cdot\rangle$. To keep things in perspective, in our geometric applications to a finite-dimensional manifold M, we will set $B_2 = \mathcal{C}^{2,\alpha}(M)$, $B_0 = \mathcal{C}^{0,\alpha}(M)$ and $H = L^2(M)$. Assume that $\mathcal{M}$ is modelled on B_2 and is endowed with an affine G-invariant connection. Let $[a,b] \ni \lambda \mapsto x_\lambda \in \mathcal{M}$ be a continuous path, such that for all λ, x_λ is a critical point of $\mathfrak{f}_\lambda$, which actually implies that the entire orbit $G\cdot x_\lambda$ consists of critical points of $\mathfrak{f}_\lambda$. Also, assume that a sufficiently small open set $U \subset \mathcal{M}$ containing all x_λ admits continuous embeddings $U \subset B_0 \subset H$, such that the following are satisfied. First, the local G-action on U extends continuously to a local G-action on B_0 and on H. Second, there exists a continuous path $\lambda \mapsto \partial\mathfrak{f}_\lambda$ of G-equivariant $\mathcal{C}^{k-1}$-maps $\partial\mathfrak{f}_\lambda \colon U \to B_0$ satisfying

$$\mathrm{d}\mathfrak{f}_\lambda(y)\xi = \langle \partial\mathfrak{f}_\lambda(y), \xi\rangle, \tag{4.1}$$

for all $y \in U$, $\xi \in T_yU \cong B_2$ and λ. In particular, we have

$$\partial\mathfrak{f}_\lambda(x_\lambda) = 0, \qquad \text{for all } \lambda \in [a,b].$$

The map $\partial\mathfrak{f}_\lambda$ plays the role of the *gradient* of $\mathfrak{f}_\lambda$, which does not exist in the usual sense due to the lack of a complete inner product on B_2.

For all $\lambda \in [a,b]$, let G_λ be the isotropy of x_λ, which is a closed subgroup of G. Given $\varepsilon > 0$, set

$$N_\lambda(\varepsilon) := \operatorname{span}\big\{v \in B_2 : \mathrm{d}(\partial\mathfrak{f}_\lambda)_{x_\lambda}(v) = \mu v, \text{ for some } \mu \le \varepsilon\big\}. \tag{4.2}$$

We define the *generalized negative eigenspace*[6] of $\mathrm{d}(\partial\mathfrak{f}_\lambda)_{x_\lambda}$ to be

$$N_\lambda := N_\lambda(0). \tag{4.3}$$

Before stating the main result of this section, we briefly recall yet another notion used by J. Smoller and A. Wasserman [21, p. 73]. A group G is said to be *nice* if, given unitary representations π_1 and π_2 of G on Hilbert spaces V_1 and V_2 respectively, such that $B_1(V_1)/S_1(V_1)$ and $B_1(V_2)/S_1(V_2)$ have the same (equivariant) homotopy type as G-spaces, then π_1 and π_2 are equivalent. Here, B_1 and S_1 denote respectively the unit ball and the unit sphere in the specified Hilbert space, and the quotient $B_1(V_i)/S_1(V_i)$ is meant in the topological sense.[7]

Example 4.2. Any compact connected Lie group G is nice. More generally, any compact Lie group with less than 5 connected components is nice. Denoting by G^0 the identity connected component of G, then G is nice if the discrete part G/G^0 is either the product of a finite number of copies of $\mathbb{Z}_2$ (e.g., the case $G = \mathrm{O}(n)$); or the product of a finite number of copies of $\mathbb{Z}_3$; or if $G/G^0 = \mathbb{Z}_4$, see [18].

We are now ready to state and prove our main result.

---

[6] In the terminology of J. Smoller and A. Wasserman [21], this is the *eigenspace* of $\mathrm{d}(\partial\mathfrak{f}_\lambda)_{x_\lambda}$.

[7] i.e., it denotes the unit ball of V_i with its boundary contracted to one point.

Theorem 4.3. *In the above setup, assume that*

(a) *there exists* $\varepsilon > 0$ *such that* $\dim(N_\lambda(\varepsilon)) < +\infty$, *for all* $\lambda \in [a, b]$;
(b) *for all* λ, G_λ *is a fixed compact nice subgroup* G_0 *of* G;
(c) $\mathrm{Ker}(\mathrm{d}(\partial\mathfrak{f}_a)_{x_a}) = T_{x_a}(G \cdot x_a)$ *and* $\mathrm{Ker}(\mathrm{d}(\partial\mathfrak{f}_b)_{x_b}) = T_{x_b}(G \cdot x_b)$;
(d) $\dim(N_a) \neq \dim(N_b)$.

Then, equivariant bifurcation of the family $(x_\lambda)_\lambda$ *of critical points of* $(\mathfrak{f}_\lambda)_\lambda$ *occurs at some* $(x_{\lambda_0}, \lambda_0)$, *with* $\lambda_0 \in\,]a, b[$.

Proof. Under the above hypotheses, Theorem 3.4 ensures the existence of a slice $\mathcal{S}$ invariant under the action of G_0 by diffeomorphisms. We have a family $T_\lambda = \mathrm{d}\mathfrak{f}_\lambda$ of G_0-equivariant sections of $T\mathcal{S}$. Note that, since $\mathfrak{f}_\lambda$ is constant along the orbits and by the transversality property (1) of the slice, we have that $\mathcal{S}$ is a *natural constraint.* In other words, the constrained critical points of the restriction $\mathfrak{f}_\lambda\big|_{\mathcal{S}}$ of $\mathfrak{f}_\lambda$ to $\mathcal{S}$ actually satisfy $\mathrm{d}\mathfrak{f}_\lambda(x_\lambda) = 0$. Assumption (c) means that x_a and x_b are (equivariantly) nondegenerate critical points. The result then follows from [21, Thm 2.1], in its nonlinear formulation explained in Appendix A. □

Remark 4.4. Assumption (a) in Theorem 4.3 is satisfied, for instance, when $\lambda \mapsto \mathrm{d}\,\partial\mathfrak{f}_\lambda(x_\lambda)\colon B_2 \to B_0$ is a continuous path of Fredholm operators that are *essentially positive.* By definition, this means that $\mathrm{d}\partial\,\mathfrak{f}_\lambda(x_\lambda)$ are Fredholm operators of the form $P_\lambda + K_\lambda$, where $P_\lambda\colon B_2 \to B_0$ is a symmetric isomorphism (relatively to the inner product of H) and satisfies $\langle P_\lambda x, x\rangle > 0$ for all $x \in B_2 \setminus \{0\}$; and $K_\lambda\colon B_2 \to B_0$ is a compact symmetric operator (also relatively to to the inner product of H). In this situation, the space $N_\lambda(\varepsilon)$ is the direct sum of the eigenspaces of the compact operator $P_\lambda^{-1}K_\lambda$ (which is symmetric with respect to the inner product defined by P_λ, hence diagonalizable) corresponding to eigenvalues less than or equal to $\varepsilon - 1 < 0$. Assuming $\varepsilon < 1$, the operator $P_\lambda^{-1}K_\lambda$ has only a finite number of such eigenvalues, and each of them has finite multiplicity. By continuity, one can give a uniform estimate on the dimension of $N_\lambda(\varepsilon)$, for $\lambda \in [a, b]$.

Assumption (d) in Theorem 4.3 means that there is a *jump* of the Morse index of x_λ, as λ goes from a to b. We now present a subtler criterion for equivariant bifurcation, where this assumption is weakened. Recall the *isotropy representation* π_λ of G_λ on $T_{x_\lambda}\mathcal{M}$ is the linear representation defined by $\pi_\lambda(g) = \mathrm{d}\phi_g(x_\lambda)$. Since $\partial\mathfrak{f}_\lambda$ is equivariant, it is easy to see that $N_\lambda(\varepsilon)$ is invariant under π_λ, for all $\varepsilon > 0$. Define the *negative isotropy representation* π_λ^- to be the restriction

$$\pi_\lambda^- := \pi_\lambda|_{N_\lambda}\colon N_\lambda \longrightarrow N_\lambda. \tag{4.4}$$

Observe that $\dim N_\lambda$ is the Morse index of x_λ.

Theorem 4.5. *Replace the assumption* (d) *of Theorem* 4.3 *with*

(d′) *the negative isotropy representations* π_a^- *and* π_b^- *are not equivalent.*[8]

---

[8]Two representations $\pi_i : H \to \mathrm{GL}(V_i)$, $i = 1, 2$, of the group H on the vector spaces V_1 and V_2 respectively, are *equivalent* if there exists a H-equivariant isomorphism $T : V_1 \to V_2$, i.e., an isomorphism satisfying $T(\pi_1(h)v) = \pi_2(h)(T(v))$ for all $h \in H$ and all $v \in V_1$. In particular, $\dim V_1 = \dim V_2$.

Then, the same conclusion holds, i.e., equivariant bifurcation of $(x_\lambda)_{\lambda\in[a,b]}$ *occurs at some* $(x_{\lambda_0},\lambda_0)$*, with* $\lambda_0 \in \,]a,b[$.

Proof. The same proof of Theorem 4.3 applies, using [21, Thm 3.3], in its nonlinear formulation (explained in Appendix A), to obtain the conclusion. □

Remark 4.6. All the results stated above carry over *verbatim* to the case of local affine actions, using the same standard procedures mentioned before.

5. Geometric applications on CMC hypersurfaces

In this section, we apply our abstract equivariant bifurcation results (Theorems 4.3 and 4.5) to the geometric variational problem of *constant mean curvature (CMC)* embeddings. Bifurcation phenomena for 1-parameter families of CMC embeddings have been studied in the last years by several authors, see, e.g., [3, 6, 17, 19]. We will state and prove general bifurcation results for CMC embeddings (Theorems 5.4 and 5.8) and discuss how some explicit bifurcation examples can be reobtained from these general results.

5.1. Variational setup

The problem of finding constant mean curvature H embeddings of a compact m-dimensional manifold M into a complete Riemannian manifold $(\overline{M},\overline{g})$ with $\dim(\overline{M}) = m+1$ is equivalent to finding critical points of the area functional with a fixed volume constraint, where H is the Lagrange multiplier (which will play the role of the parameter λ). More precisely, assume for simplicity that M and $\overline{M}$ are oriented, and consider the 1-parameter family of functionals $(\mathfrak{f}_H)_H$ given by

$$\mathfrak{f}_H(x) = \mathrm{Area}(x) + mH\,\mathrm{Vol}(x), \tag{5.1}$$

where $x\colon M \to \overline{M}$ is an embedding, $\mathrm{Area}(x) = \int_M \mathrm{vol}_{x^*(\overline{g})}$ is the volume of $x(M) \subset \overline{M}$, $\mathrm{vol}_{x^*(\overline{g})}$ is the volume form of the pull-back metric $x^*(\overline{g})$ and $\mathrm{Vol}(x)$ is the *volume enclosed*[9] by $x(M)$. Then $x\colon M \to \overline{M}$ is a critical point of $\mathfrak{f}_H$ if and only if it is an embedding of constant mean curvature H, see [4, 5]. As we will see later, a convenient regularity assumption is that $\mathfrak{f}_H$ acts on the space of Hölder $C^{2,\alpha}$ embeddings.

More precisely, assuming that the embedding x is transversely oriented (i.e., the normal bundle to x is oriented), we may parameterize embeddings close to x by functions on M using the normal exponential map. An embedding $x_f\colon M \to \overline{M}$ that is $C^{2,\alpha}$-close to x can be written as

$$x_f(p) = \exp^{\perp}_{x(p)}\big(f(p)\,N_x(p)\big), \quad p \in M, \tag{5.2}$$

where $\exp^{\perp}$ is the normal exponential map of $x(M) \subset \overline{M}$ and N_x is a unit normal vector field along x. We thus identify x_f with the function $f \in C^{2,\alpha}(M)$, which

[9]This notion will be clarified by the end of this subsection. For now, one may assume for simplicity that $x(M) = \partial\Omega$ is the boundary of an open bounded region $\Omega \subset \overline{M}$, and then $\mathrm{Vol}(x) = \int_\Omega \mathrm{vol}_{\overline{g}}$ is the volume of this enclosed region.

is close to zero. This also gives an identification of the tangent space at x to the space of $\mathcal{C}^{2,\alpha}$ embeddings (which is formed by normal vector fields along x) with the Banach space $\mathcal{C}^{2,\alpha}(M)$. With this identification, the first variation formula for (5.1) is given by

$$\mathrm{d}\mathfrak{f}_H(x)(f) = \int_M m\big(H - \mathcal{H}(x)\big) f \,\mathrm{vol}_{x^*(\overline{g})}, \quad f \in \mathcal{C}^{2,\alpha}(M), \tag{5.3}$$

where $\mathcal{H}(x)$ is the mean curvature function of the embedding x. From (5.3), it follows that x is a critical point of $\mathfrak{f}_H$ if and only if $\mathcal{H}(x) = H$ (i.e., x has constant mean curvature H), as we claimed above.

We will also need to consider the second variation of (5.1) at a critical point x, which under the above identifications, is the symmetric bilinear form on $\mathcal{C}^{2,\alpha}(M)$ given by

$$\mathrm{d}^2\mathfrak{f}_H(x)(f_1, f_2) = -\int_M J_x(f_1) f_2 \,\mathrm{vol}_{x^*(\overline{g})}, \quad f_1, f_2 \in \mathcal{C}^{2,\alpha}(M), \tag{5.4}$$

where J_x is the Jacobi operator

$$J_x = \Delta_x + \|A_x\|^2 + m\,\mathrm{Ric}_{\overline{M}}(N_x), \tag{5.5}$$

where Δ_x is the Laplacian of the pull-back metric $x^*(g)$ on M, $\|A_x\|$ is the norm of the second fundamental form of x, $\mathrm{Ric}_{\overline{M}}$ is the (normalized) Ricci curvature of the ambient space $(\overline{M}, \overline{g})$ and N_x is a unit normal field along x. Functions f in the kernel of J_x are called *Jacobi fields* along x. The number of negative eigenvalues of J_x (counted with multiplicity) is the *Morse index* of x, that we denote $\mathrm{i}_{\mathrm{Morse}}(x)$.

The ambient isometry group $G = \mathrm{Iso}(\overline{M}, \overline{g})$ acts on the space of embeddings, and composing a CMC embedding with an element of G trivially gives rise to a new CMC embedding. Recall that from the Myers–Steenrod Theorem, G is a Lie group, and is compact if $\overline{M}$ is compact (see [10]). In addition, since $(\overline{M}, \overline{g})$ is complete, the Lie algebra of G is identified with the space of Killing vector fields of $(\overline{M}, \overline{g})$. We are interested in *G-equivariant* bifurcation of CMC embeddings, i.e., getting new embeddings that are not merely obtained by composing a pre-existing one with an isometry of the ambient manifold. Another way in which one could trivially obtain a new CMC embedding is by reparameterizing it, i.e., composing on the right with a diffeomorphism of M. Two CMC embeddings $x_i\colon M \to \overline{M}$, $i = 1, 2$, are said to be *isometrically congruent* if there exists a diffeomorphism ϕ of M and an isometry ψ of $(\overline{M}, \overline{g})$ such that $x_2 = \psi \circ x_1 \circ \phi$.

Infinitesimally, the action of G provides some trivial Jacobi fields along any critical point. Namely, if K is a Killing vector field of $(\overline{M}, \overline{g})$, then $f = \overline{g}(K, N_x)$ is a Jacobi field along x. Denote by Jac_x the (finite-dimensional) vector space of Jacobi fields along x, and by Jac_x^K the subspace of Jac_x spanned by the functions $\overline{g}(K, N_x)$, where K is a Killing vector field of $(\overline{M}, \overline{g})$. The CMC embedding x will be called *nondegenerate* if $\mathrm{Jac}_x^K = \mathrm{Jac}_x$, i.e., if every Jacobi field along x arises from a Killing field of the ambient space.

It is natural to expect that, with the above equivariant notion of nondegeneracy, an equivariant implicit function theorem should hold. Indeed, the following is proved in [9, Prop 4.1].

Theorem 5.1. *Let $x\colon M \to \overline{M}$ be a nondegenerate CMC embedding, with mean curvature equal to H_0. Then, there exists $\varepsilon > 0$ and a smooth map*

$$]H_0 - \varepsilon, H_0 + \varepsilon[\ \ni H \longmapsto x_H \in \mathcal{C}^{2,\alpha}(M, \overline{M}),$$

such that for all H, $x_H\colon M \to \overline{M}$ is a CMC embedding of mean curvature H and

(a) *$x_{H_0} = x$;*

(b) *if $y\colon M \to \overline{M}$ is a CMC embedding sufficiently close to x in the $\mathcal{C}^{2,\alpha}$-topology, then there exists $H \in\ ]H_0 - \varepsilon, H_0 + \varepsilon[$ such that y is isometrically congruent to x_H.*

5.2. A few technicalities

Let us now deal with some technicalities we omitted in the above explanation of the variational setup for the CMC problem.

First, we need to give a more precise definition of the space where (5.1) is defined. For reasons[10] that will later be clear, it is convenient to consider the space of embeddings $x\colon M \to \overline{M}$ endowed with a $\mathcal{C}^{2,\alpha}$-topology. More precisely, consider the set $\mathrm{Emb}^{2,\alpha}(M, \overline{M})$ of embeddings of class $\mathcal{C}^{2,\alpha}$ of M into $\overline{M}$. This is an open subset of the Banach manifold $\mathcal{C}^{2,\alpha}(M, \overline{M})$, and hence inherits a natural differential structure, becoming a Banach manifold. Since we want to identify embeddings obtained by reparameterizations of a given embedding, we have to consider the action of the group $\mathrm{Diff}(M)$ of diffeomorphisms of M by right-composition on $\mathrm{Emb}^{2,\alpha}(M, \overline{M})$. We denote the orbit space of this action by

$$\mathcal{E}(M, \overline{M}) = \mathrm{Emb}^{2,\alpha}(M, \overline{M})/\mathrm{Diff}(M), \tag{5.6}$$

and its elements are called *unparameterized embeddings.* This set has the structure of an infinite-dimensional *topological* manifold modelled on the Banach space $\mathcal{C}^{2,\alpha}(M)$. Its geometry and local differential structure are studied in detail in [2]. Given $x \in \mathcal{C}^{2,\alpha}(M, \overline{M})$, we denote by $[x]$ its class in $\mathcal{E}(M, \overline{M})$. Henceforth, we are assuming for simplicity that $x(M)$ is transversely oriented in $\overline{M}$.

If we take $x \in \mathrm{Emb}^{2,\alpha}(M, \overline{M})$ in the dense subset of *smooth* embeddings, there exists an open neighborhood of $[x]$ in $\mathcal{E}(M, \overline{M})$ and a bijection from this neighborhood to a neighborhood of the origin of $\mathcal{C}^{2,\alpha}(M)$, whose image is identified (using the normal exponential map) with $\mathcal{C}^{2,\alpha}$ embeddings equivalent to x under the action of $\mathrm{Diff}(M)$. As x runs in the set of smooth embeddings, those maps form an atlas for $\mathcal{E}(M, \overline{M})$ whose charts are continuously compatible. Moreover, if a smooth functional defined in $\mathrm{Emb}^{2,\alpha}(M, \overline{M})$ is invariant under $\mathrm{Diff}(M)$, then

[10] This choice has to do with the nature of the second variation of f_H, which we will want to be a Fredholm operator under the appropriate identification.

the induced functional in $\mathcal{E}(M,\overline{M})$ is smooth[11] in every local chart. Using these charts, we also have an identification

$$T_{[x]}\mathcal{E}(M,\overline{M}) \cong C^{2,\alpha}(M) \tag{5.7}$$

of this tangent space with the Banach space of (real-valued) $C^{2,\alpha}$ functions on M. For more details on this standard construction, see [2].

Second, note that, if $x\colon M \to \overline{M}$ is an embedding, unless $x(M) \subset \overline{M}$ is the boundary of a bounded open set of $\overline{M}$, then the *enclosed volume* $\mathrm{Vol}(x)$ is not well defined. Moreover, it is not clear that such quantity should be invariant under the action of G. To overcome these problems, we recall the notion of *invariant volume functionals* for embeddings of M into $\overline{M}$ developed in [9, Appendix B].

Definition 5.2. Let $\mathcal{U} \subset C^{2,\alpha}(M,\overline{M})$ be an open subset of embeddings $x\colon M \to \overline{M}$. An *invariant volume functional* on $\mathcal{U}$ is a real-valued function $\mathrm{Vol}\colon \mathcal{U} \to \mathbb{R}$ satisfying:

(a) $\mathrm{Vol}(x) = \int_M x^*(\eta)$, where η is an m-form defined on an open subset $U \subset \overline{M}$ such that $\mathrm{d}\eta = \mathrm{vol}_{\overline{g}}$ is the volume form of $\overline{g}$ in U;

(b) given $x \in \mathcal{U}$, for all $\phi \in \mathrm{Iso}(\overline{M},\overline{g})$ sufficiently close to the identity, $\mathrm{Vol}(\phi\circ x) = \mathrm{Vol}(x)$.

If M has boundary, the invariance property (b) is required to hold only for isometries ϕ near the identity that preserve $x(\partial M)$, i.e., $\phi\big(x(\partial M)\big) = x(\partial M)$. An embedding will be called *regular* if it is contained in some open set $\mathcal{U}$ of $C^{2,\alpha}(M,\overline{M})$ which is the domain of some invariant volume functional.

Example 5.3. If $x(M)$ is the boundary of a bounded open subset of $\overline{M}$, then x is regular. If $\overline{M}$ is non-compact, and $\mathrm{Iso}(\overline{M},\overline{g})$ is compact, then every embedding into $\overline{M}$ is regular. If $x\colon M \to \overline{M}$ has image contained in some open subset $U \subset \overline{M}$ whose mth de Rham cohomology vanishes, then x is regular. In particular, if $\overline{M} = \mathbb{R}^{m+1}$ or $\overline{M} = \mathbb{S}^{m+1}$, then every embedding into $\overline{M}$ is regular, see [9, Ex 5].

Third, when considering an invariant volume functional as above (defined in a neighborhood of a given embedding), the left-composition action of $\mathrm{Iso}(\overline{M},\overline{g})$ has to be restricted to this domain, giving rise to a local action. As remarked above, standard techniques apply to have the necessary results also in the case of local actions.

With the above considerations on the (topological) manifold $\mathcal{E}(M,\overline{M})$ of unparameterized embeddings of class $C^{2,\alpha}$ and the local existence of an invariant volume functional, we may study the CMC variational problem in this precise global analytical setup. The functional (5.1) is then well defined and smooth in a neighborhood of a smooth unparameterized regular embedding, and formulas (5.3) and (5.4) hold with the appropriate identifications above mentioned.

---

[11] As a map from a neighborhood of the origin in $C^{2,\alpha}(M)$ to $\mathbb{R}$.

5.3. Equivariant bifurcation using Morse index

We will now use our abstract equivariant bifurcation result to obtain a bifurcation result for CMC embeddings when there is a jump of the Morse index. Let us recall some basic terminology. Assume that $[a,b] \ni r \mapsto x_r \in \mathcal{C}^{2,\alpha}(M,\overline{M})$ is a continuous family of CMC embeddings of M into $\overline{M}$ (which already implies that $x_r\colon M \to \overline{M}$ is smooth[12]) and let H_r denote the value of the mean curvature of x_r. An instant $r_* \in \,]a,b[$ is a *bifurcation instant* for the family $(x_r)_{r\in[a,b]}$ if there exists a sequence r_n in $[a,b]$ tending to r_* as $n\to\infty$ and a sequence x_n of CMC embeddings of M into $\overline{M}$, with the mean curvature of x_n equal to H_{r_n}, such that x_n tends to x_{r_*} in $\mathcal{C}^{2,\alpha}(M,\overline{M})$ as $n\to\infty$, and, for every n, x_n is *not* isometrically congruent to x_{r_n}.

Given a CMC embedding $x\colon M\to\overline{M}$, let G_x denote the closed subgroup of $\mathrm{Iso}(\overline{M},\overline{g})$ consisting of isometries ψ that leave $x(M)$ invariant, i.e., such that $\psi(x(M))\subset x(M)$. In other words, G_x is the *isotropy* of x under the action of G. Since M is compact and the action of G is proper, G_x is compact. The Lie algebra $\mathfrak{g}_x$ of G_x is identified with the space of Killing vector fields of $(\overline{M},\overline{g})$ that are everywhere tangent to $x(M)$. The codimension of G_x in G is equal to the dimension of Jac_x^K.

Theorem 5.4. *Let $[a,b]\ni r\mapsto x_r\in\mathcal{C}^{2,\alpha}(M,\overline{M})$ be a $\mathcal{C}^1$-map, where $x_r\colon M\to\overline{M}$ is a regular CMC embedding for all r, having mean curvature equal to H_r. Let $r_*\in\,]a,b[$ be an instant where*

(a) *the derivative H'_{r_*} of the map $[a,b]\ni r\mapsto H_r\in\mathbb{R}$ at r_* is nonzero;*

(b) *for $\varepsilon>0$ small enough:*

  (b1) *$x_{r_*-\varepsilon}$ and $x_{r_*+\varepsilon}$ are nondegenerate;*

  (b2) *the identity connected component G_r^0 of the isotropy G_{x_r} does not depend on r, for $r\in[r_*-\varepsilon,r_*+\varepsilon]$;*

  (b3) *$\mathrm{i}_{\mathrm{Morse}}(x_{r_*-\varepsilon})\neq\mathrm{i}_{\mathrm{Morse}}(x_{r_*+\varepsilon})$.*

Then, r_* is a bifurcation instant for the family $(x_r)_r$.

Proof. We first verify that the CMC variational problem satisfies the required conditions and then use Theorem 4.3 to obtain the conclusion. In the notation of Section 4, we have $B_2=\mathcal{C}^{2,\alpha}(M)$, $B_0=\mathcal{C}^{0,\alpha}(M)$ and $H=L^2(M,\nu)$, where ν is an arbitrarily fixed volume form (or density) on M. It will be convenient to choose ν to be the volume form of the pull-back metric $x_{r_*}^*(\overline{g})$.

Let $\overline{\nabla}$ be the Levi–Civita connection of $(\overline{M},\overline{g})$. Using this connection, one can define an associated natural connection on $\mathrm{Emb}^{2,\alpha}(M,\overline{M})$, as in Example 2.3. This connection is defined on the entire manifold $\mathcal{C}^{2,\alpha}(M,\overline{M})$, and is characterized by the fact that the evaluation maps $\mathrm{ev}_p\colon\mathcal{C}^k(M,\overline{M})\to\overline{M}$, $p\in M$, are affine (Proposition 2.4).[13]

[12] It is well known that CMC hypersurfaces are the solution to a quasilinear elliptic PDE, hence smoothness follows from usual elliptic regularity theory.

[13] An explicit description of the horizontal distribution of this connection is given as follows. The tangent bundle of $\mathcal{C}^{2,\alpha}(M,\overline{M})$ can be naturally identified with $\mathcal{C}^{2,\alpha}(M,T\overline{M})$; an element of

Let G be the identity connected component of $\mathrm{Iso}(\overline{M},\overline{g})$, which is a (finite-dimensional) Lie group, and consider the smooth action of G by left-composition on $\mathcal{C}^{2,\alpha}(M,\overline{M})$. Clearly, $\mathrm{Emb}^{2,\alpha}(M,\overline{M})$ is invariant by left-compositions with diffeomorphisms of $\overline{M}$, so we have an induced action of G on $\mathrm{Emb}^{2,\alpha}(M,\overline{M})$. Since isometries preserve the Levi–Civita connection, the actions of G on both $\mathcal{C}^{2,\alpha}(M,\overline{M})$ and $\mathrm{Emb}^{2,\alpha}(M,\overline{M})$ are by affine diffeomorphisms, see Proposition 2.6. We observe furthermore that the left-action of G on $\mathrm{Emb}^{2,\alpha}(M,\overline{M})$ commutes with the right-action of the diffeomorphism group $\mathrm{Diff}(M)$. This implies that one can define a left-action of G on the quotient space $\mathcal{E}(M,\overline{M})$. Finally, we recall from Proposition 2.5 that the right-action of $\mathrm{Diff}(M)$ on $\mathrm{Emb}^{2,\alpha}(M,\overline{M})$ is by affine diffeomorphisms, so that one has an induced connection on $\mathcal{E}(M,\overline{M})$ which is preserved by the action of G.

Let $x\colon M\to\overline{M}$ be a $\mathcal{C}^{2,\alpha}$ embedding. Since the action of G on $\overline{M}$ is proper, and M is compact, then the isotropy group G_x is a compact subgroup of G. We recall from [2] that there exists a natural (topological) atlas of continuously compatible charts of $\mathcal{E}(M,\overline{M})$ such that, in these charts, the (local) action of G is differentiable at the class $[x]$ of every *smooth* embedding $x\colon M\to\overline{M}$. In particular, if x has constant mean curvature, then the action of G on $\mathcal{E}(M,\overline{M})$ is differentiable at $[x]$. Moreover, by Lemma 3.3, $[x]$ admits arbitrarily small neighborhoods in $\mathcal{E}(M,\overline{M})$ that are invariant by G_x. With this, we are in the variational framework described in Axioms (a), (b) and (c) of Section 3.

By assumption (a), there exists a $\mathcal{C}^1$ function $H\mapsto r(H)$, defined in a neighborhood of H_{r_*}, with the property that the (constant) mean curvature of $x_{r(H)}$ is equal to H, for all H in this neighborhood. Thus, we may assume that the CMC embeddings x_r, for r close to r_*, are parameterized by their mean curvature H_r instead of r, and we write x_{H_r}. Consider an invariant volume functional Vol defined in a neighborhood $\mathcal{U}\subset\mathcal{C}^{2,\alpha}(M,\overline{M})$ of $x_{H_{r_*}}$. For H near H_{r_*} consider the one-parameter family of functional $\mathfrak{f}_H\colon\mathcal{U}\to\mathbb{R}$ given by (5.1). The group G acts by affine diffeomorphisms on the manifold $\mathcal{C}^{2,\alpha}(M,\overline{M})$ by left-composition; in particular, we have a local action on the open subset $\mathcal{U}$. Since both Area and Vol are invariant under composition on the right with isometries of $(\overline{M},\overline{g})$, then $\mathfrak{f}_H$ is invariant under the local action of G. Moreover, Area and Vol are invariant under right-composition with diffeomorphisms of M, so $\mathfrak{f}_H$ gives a well-defined smooth functional on the quotient space $\mathcal{E}(M,\overline{M})$, as discussed before in Subsection 5.2. With the appropriate identifications, the first variation formula for this functional is given by (5.3), which means that the map $\mathfrak{d}\mathfrak{f}_H(x)\colon U\subset B_2\to B_0$ defined in (4.1) is

$$\mathfrak{d}\mathfrak{f}_H(x)=m\big(H-\mathcal{H}(x)\big)\psi_x, \tag{5.8}$$

$\mathcal{C}^{2,\alpha}(M,T\overline{M})$ is a map of class $\mathcal{C}^{2,\alpha}$ from M to $T\overline{M}$, which is a vector field of class $\mathcal{C}^{2,\alpha}$ in $\overline{M}$ along some function $f\colon M\to\overline{M}$ of class $\mathcal{C}^{2,\alpha}$. The tangent space to $\mathcal{C}^{2,\alpha}(M,T\overline{M})$ at the point X is the space of vector fields of class $\mathcal{C}^{2,\alpha}$ in $T\overline{M}$ along X, i.e., maps $\eta\colon M\to T(T\overline{M})$ of class $\mathcal{C}^{2,\alpha}$ such that $\eta(p)$ is a tangent vector to $T\overline{M}$ at the point $X(p)$, for all $p\in M$. The vertical subspace is given by those η's such that $\eta(p)$ is vertical, for all $x\in M$. The horizontal subspace is the space of maps η such that $\eta(p)$ is horizontal relatively to the connection $\overline{\nabla}$ of $\overline{M}$ for all $p\in M$.

where $\psi_x\colon M \to \mathbb{R}^+$ is the unique map satisfying $\psi_x \operatorname{vol}_{(x_{H_{r_*}})^*(\overline{g})} = \operatorname{vol}_{x^*(\overline{g})}$, in particular, $\psi_{x_{H_{r_*}}} \equiv 1$.

As mentioned above, if $[x] \in \mathcal{E}(M, \overline{M})$ is a critical point of $\mathfrak{f}_H$, then the second variation of $\mathfrak{f}_H$ at $[x]$ is identified with the quadratic form (5.4) on $T_{[x]}\mathcal{E}(M,\overline{M}) \cong \mathcal{C}^{2,\alpha}(M)$. The differential $\mathrm{d}(\partial\mathfrak{f}_H)(x_{r_*})\colon B_2 \to B_0$ is the linearization of the mean curvature function, which is precisely the negative Jacobi operator $-J_{x_{H_{r_*}}}$. This is an essentially positive Fredholm operator from $\mathcal{C}^{2,\alpha}(M)$ to $\mathcal{C}^{0,\alpha}(M)$, see [25, 26].[14] Thus, assumption (a) of Theorem 4.3 is satisfied, see Remark 4.4. Assumptions (b1), (b2) and (b3) respectively imply the hypotheses (b),[15] (c) and (d) of Theorem 4.3. The claimed result then follows immediately from Theorem 4.3. □

Remark 5.5. Theorem 5.4 uses the assumption that the mean curvature function $r \mapsto H_r$ has non-vanishing derivative at the bifurcation instant r_*. Such assumption is used in the proof in order to parameterize the trivial branch of CMC immersions through the value of their mean curvature. A natural question is if this assumption is necessary. The following simple examples show that it is indeed necessary, i.e., bifurcation may not occur otherwise.

Example 5.6. Consider the two-variable function $f(x,y) = 4y^3+6xy^2+3xy-3x^2y$ on the plane. We can regard it as a family of functions of y, parameterized by x. For each fixed x, we look at the critical points of the function $y \mapsto f(x,y)$, i.e., we look for the zeros of the partial derivative $\frac{\partial f}{\partial y} = 12y^2 + 12xy - 3x + 3x^2$. Near $(0,0)$, the points (x,y) that solve $\frac{\partial f}{\partial y} = 0$ form a smooth curve[16] contained in the half-plane $x \geq 0$, tangent to the y axis at $(0,0)$. Notice that the Implicit Function Theorem cannot be used in this situation, as $\frac{\partial^2 f}{\partial y^2}(0,0) = 0$. Observe also that the function x is not locally injective on the points of the curve near $(0,0)$, since for each $x > 0$ there are exactly two solutions of $12y^2 + 12xy - 3x + 3x^2 = 0$, one with $y > 0$ and another with $y < 0$. At all points (x,y) on this curve with $y > 0$, the second derivative $\frac{\partial^2 f}{\partial y^2} = 24y + 12x$ is positive, while it is negative at all points (x,y) on the curve with $y < 0$. Thus, there is a *jump of the Morse* index at the point $(0,0)$, but there is *no bifurcation.*

Example 5.7. An explicit counterexample to CMC bifurcation can be given when assumption (a) of Theorem 5.4 is not satisfied. Consider the family $[-1,1] \ni r \mapsto x_r$, where x_r is the embedding into $\mathbb{R}^3$ of the spherical cap above the xy-plane of the round sphere centered at $(0,0,r)$ of radius $\sqrt{1+r^2}$. These spherical caps have the same boundary, which is the circle C of radius 1 in the xy-plane centered at the origin, see Figure 1. Both principal curvatures of x_r are equal to $\frac{1}{\sqrt{1+r^2}}$, hence also its mean curvature is $H_r = \frac{1}{\sqrt{1+r^2}}$. Notice that H_r attains its maximum

[14] Indeed, observe that $-\Delta_{x_{H_r}}$ is a positive isomorphism from $\mathcal{C}^{2,\alpha}(M)$ to $\mathcal{C}^{0,\alpha}(M)$.

[15] The group G_r^0 is *nice* in the sense of [21] because it is connected.

[16] By explicit calculation, the curve is the graph of the function $x = \frac{1}{2}\left(1 - 4y - \sqrt{1-8y}\right)$.

FIGURE 1. Family of spherical caps with the same boundary C, the unit circle in the xy-plane.

$H_0 = 1$ at the half-sphere, hence assumption (a) of Theorem 5.4 is *not* satisfied when $r_* = 0$.

All other assumptions (b1), (b2) and (b3) are satisfied. Namely, the only degeneracy instant[17] of $(x_r)_r$ is precisely $r_* = 0$. A jump of the Morse index can be obtained applying an adequate version of the Morse Index Theorem to $(x_r)_r$. In fact, $\mathrm{i}_{\mathrm{Morse}}(x_r)$ can be written as the sum of degeneracy instants $s \in [-1, r]$ (counted with multiplicity), and hence is a non-decreasing function of r that jumps as r crosses $r_* = 0$.

Finally, bifurcation *does not happen* at $r_* = 0$. Since $(x_r)_r$ are embedded in the half-space $z > 0$ of $\mathbb{R}^3$ and meet the plane $z = 0$ transversely, along the circle C, any bifurcating branch would satisfy the same properties for a short time. From the a maximum principle type argument (the Alexander reflection method), any such CMC surfaces must be spherical caps, see [12].

[17]Note that when $r = 0$, there exists a Jacobi field $f = \langle K, N_{x_0} \rangle$, where $K = \frac{\partial}{\partial z}$ and N_{x_0} is the unit normal field along x_0. This Jacobi field is in Jac_{x_0} but not in $\mathrm{Jac}^K_{x_0}$, since K is not tangent to the half-sphere (but only to its boundary). Hence, x_0 is a *degenerate* CMC embedding.

5.4. Equivariant bifurcation using representations

It is possible to use representation theory to prove a slight generalization of Theorem 5.4, that gives a subtler criterion for equivariant bifurcation, without necessarily having a jump of the Morse index. As we will see in Subsection 5.6, this finer result is efficient in geometric applications where the direct computation of the Morse index is not feasible.

As mentioned above, given a transversely oriented CMC embedding $x\colon M \to \overline{M}$, we identify the tangent space $T_{[x]}\mathcal{E}(M,\overline{M})$ (i.e., the space of normal vector fields along x) with the Banach space $\mathcal{C}^{2,\alpha}(M)$. Under this identification, we may consider the isotropy representation at x, induced by the left-composition action of $\mathrm{Iso}(\overline{M},\overline{g})$, as the representation $\pi\colon G_x \to \mathrm{GL}\big(T_{[x]}\mathcal{E}(M,\overline{M})\big)$ that maps $\psi \in G_x$ to the operator of left-composition with $\mathrm{d}\psi$, i.e.,

$$\begin{aligned} \pi\colon G_x \times T_{[x]}\mathcal{E}(M,\overline{M}) &\longrightarrow T_{[x]}\mathcal{E}(M,\overline{M}) \\ (\psi, f) &\longmapsto \mathrm{d}\psi \circ f \end{aligned}$$

In more elementary terms, $\pi(\psi)$ acts as follows on a normal vector field $f \in \mathcal{C}^{2,\alpha}(M)$ along x. Consider the variation of x induced by f, $x_s = \exp^{\perp}(sfN_x)$, $s \in\,]-\varepsilon,\varepsilon[$. Then $\pi(\psi)f$ is the normal vector field $\frac{\mathrm{d}}{\mathrm{d}s}\psi\circ x_s\big|_{s=0}$ along x.

If $f\colon M \to \mathbb{R}$ is an eigenfunction of the Jacobi operator J_x, see (5.5), then $\pi(\psi)f = \mathrm{d}\psi \circ f$ is another eigenfunction with the same eigenvalue, for all $\psi \in \mathrm{Iso}(\overline{M},\overline{g})$. This means that the isotropy representation π of G_x restricts to a representation of G_x on each eigenspace of the Jacobi operator. More precisely, if λ is in the spectrum $\sigma(J_x)$ of J_x and E_x^λ is the corresponding eigenspace, we let $\pi_x^\lambda\colon G_x \to \mathrm{GL}(E_x^\lambda)$ be the representation defined as the restriction of π to E_x^λ, i.e.,

$$\pi_x^\lambda(\psi)f = \mathrm{d}\psi \circ f, \quad \psi \in G_x,\ f \in E_x^\lambda.$$

Let us define the *negative isotropy representation* π_x^- as the direct sum of all the representations π_x^λ, as λ varies in the set of negative eigenvalues of J_x, i.e.,

$$\pi_x^- = \bigoplus_{\substack{\lambda\in\sigma(J_x)\\ \lambda<0}} \pi_x^\lambda,$$

which is a representation of G_x on $E_x^- = \bigoplus_{\substack{\lambda\in\sigma(J_x)\\ \lambda<0}} E_x^\lambda$, compare to (4.4).

Observe that $\mathrm{i}_{\mathrm{Morse}}(x) = \dim(E_x^-)$. In Theorem 5.4, assumption (b3) can be hence written as $\dim(E^-_{x_{r_*-\varepsilon}}) \neq \dim(E^-_{x_{r_*+\varepsilon}})$. Recall that two representations $\pi_i\colon H \to \mathrm{GL}(V_i)$, $i = 1,2$, of H are *equivalent* if there exists an H-equivariant isomorphism from V_1 to V_2, which in particular implies $\dim V_1 = \dim V_2$. With this notion, we can weaken (b3) and still obtain bifurcation.

Theorem 5.8. *Replace the assumption* (b3) *of Theorem* 5.4 *with*

(b3$'$) *the representations* $\pi^-_{x_{r_*-\varepsilon}}$ *and* $\pi^-_{x_{r_*-\varepsilon}}$ *of* G_r^0 *are not equivalent.*

Then, the same conclusion holds, i.e., r_* *is a bifurcation instant for the family* $(x_r)_r$.

Proof. The same proof of Theorem 5.4 applies, where instead of Theorem 4.3, we use Theorem 4.5 to obtain the conclusion. □

Remark 5.9. Theorems 5.4 and 5.8 hold *verbatim* in the case of CMC embeddings of compact manifolds M with boundary. In this case, a fixed boundary condition is necessary, namely, assume that the embeddings x_r satisfy $x_r(\partial M) = \Sigma$, with Σ a fixed codimension 2 submanifold of $\overline{M}$. In this situation, the notion of nondegeneracy requires a suitable modification. Given a CMC embedding $x\colon M \to \overline{M}$ satisfying $x(\partial M) = \Sigma$, the space Jac_x is the set of Jacobi fields along x that vanish on ∂M, and the space Jac_x^K is defined to be the vector space spanned by all functions of the form $\overline{g}(K, N_x)$, where K is a Killing vector field in $(\overline{M}, \overline{g})$ that is tangent to $x(M)$ along $x(\partial M)$. Then, x is said to be *nondegenerate* if $\mathrm{Jac}_x^K = \mathrm{Jac}_x$.

Remark 5.10. As we saw in Example 5.6, assumption (a) cannot be omitted in Theorems 5.4 and 5.8. However, it seems reasonable that assumption (b2) may be weakened. Consider the more general case in which the identity connected component G_r^0 of the isotropy of x_r is a *continuous family of Lie groups*. This means that the set $\bigcup_{r\in[a,b]} G_r^0$ has the structure of a *topological groupoid* over the base $[a, b]$, with source and range map given by the projection onto $[a, b]$. A notion of continuity (in fact, smoothness) for families of Lie groups is given in [23], and a CMC version of an equivariant implicit function theorem in the case of varying isotropies is discussed in [24], where the authors prove the existence of non-embedded CMC tori in spheres and hyperbolic spaces. Evidently, the validity of an equivariant bifurcation result in the case of varying isotropies would employ a theory of existence of slices for groupoid affine actions, along the lines of the results discussed in Section 3. This is a topic of current research by the authors, see [10].

5.5. Clifford tori in round and Berger spheres

Let us discuss some bifurcation results for CMC hypersurfaces by the second named author and others that can be reobtained as an application of Theorem 5.4.

The family $x_r\colon \mathbb{S}^n \times \mathbb{S}^m \to \mathbb{S}^{n+m+1}$ of CMC Clifford tori in the round sphere, defined by

$$x_r(x, y) = \left(r\,x, \sqrt{1 - r^2}\,y\right), \quad r \in \,]0, 1[\,, \tag{5.9}$$

is studied in [3]. The central result gives the existence of two sequences r_n and r'_n, with $\lim_{n\to\infty} r_n = 0$ and $\lim_{n\to\infty} r'_n = 1$, of degeneracy instants for the embeddings x_r, with bifurcation at each such instant. In the case $n = m = 1$, an explicit description of the bifurcating branches is given in [14]; such branches are formed by rotationally symmetric embeddings of $\mathbb{S}^1 \times \mathbb{S}^1$ that are analogous to the classical *unduloids* in $\mathbb{R}^3$. The connected component of the identity of the isotropy of every Clifford torus x_r is the group $\mathrm{SO}(n + 1) \times \mathrm{SO}(m + 1)$, diagonally embedded into the isometry group $\mathrm{SO}(n + m + 2)$ of the round sphere $\mathbb{S}^{n+m+1}$. The Jacobi operator of Clifford tori has a simple form, due to the fact that the Ricci curvature of the ambient and also the norm of the second fundamental form are constant functions. Moreover, the induced metric is the standard product metric of $\mathbb{S}^n \times \mathbb{S}^m$. Nondegeneracy and jumps of the Morse index are computed explicitly in this situation using the eigenfunctions of the Laplacian on $\mathbb{S}^n \times \mathbb{S}^m$.

A similar analysis is carried out in [19], in the case of embeddings $x_{r,\tau}\colon \mathbb{S}^1 \times \mathbb{S}^1 \to \mathbb{S}^3_\tau$, as in (5.9), into the three-dimensional Berger sphere $\mathbb{S}^3_\tau$, with $r \in \,]0,1[$ and $\tau > 0$. In analogy with the standard round case (which corresponds to $\tau = 1$), also these embeddings are called *Clifford tori*, and they have constant mean curvature. For $\tau \neq 1$, the identity connected component of the isometry group of $\mathbb{S}^3_\tau$ is SU(2), and the isotropy of every Clifford torus $x_{r,\tau}$ is $\mathbb{S}^1 \times \mathbb{S}^1$, diagonally embedded in SU(2). The space of Killing–Jacobi fields along $x_{r,\tau}$ has dimension 3 when $\tau \neq 1$, while the dimension is 4 in the round case $\tau = 1$. The induced metric on the torus is flat, but not equal to the product metric. A spectral analysis of the Jacobi operator, which is the sum of a multiple of the identity and the Laplacian of a flat (but not product) metric on the torus, is carried out in [19], leading to the following bifurcation result.

Theorem 5.11. *For every $\tau > 0$, there exists a countable set $\mathcal{A}_\tau \subset \,]0,1[$ with the following properties:*

(a) $\inf \mathcal{A}_\tau = 0$ *and* $\sup \mathcal{A}_\tau = 1$*;*

(b) *for all $r_* \in \mathcal{A}_\tau$, the family $r \mapsto x_{r,\tau}$* bifurcates *at $r = r_*$*

Furthermore, for every $r \in \,]0,1[$ there exists a countable set $\mathcal{B}_r \subset \,]0,1[\bigcup]1,+\infty[$ with the following properties:

(c) $\sup \mathcal{B}_r = +\infty$*;*

(d) *given $r \in \,]0,1[$, for all $\tau_* \in \mathcal{B}_r$ the family $\tau \mapsto x_{r,\tau}$ bifurcates at $\tau = \tau_*$.*

The above, as well as the bifurcation statement in the case of the round sphere, can be proved as an application of Theorem 5.4.

5.6. Rotationally symmetric surfaces in $\mathbb{R}^3$

Both results discussed above of bifurcation for the families of Clifford tori in round and Berger spheres can be obtained as an application of Theorem 5.4, given that there is a jump of the Morse index at every degeneracy instant. However, an explicit computation of the Morse index is not feasible in many situations, whereas the weaker assumption of Theorem 5.8 on the *jump* of the isotropy representation may actually be an easier task. An example of this situation is provided by rotationally symmetric CMC surfaces in $\mathbb{R}^3$. This problem is studied in detail in [17].

For convenience of notation, write $\mathbb{S}^1 = [0,2\pi]/\{0,2\pi\}$. Let us consider the case of a family of fixed boundary CMC rotationally symmetric surfaces $\mathrm{x}_r\colon [0,L_r] \times \mathbb{S}^1 \to \mathbb{R}^3$, $r \in I \subset \mathbb{R}$, whose boundary in the union of two co-axial circles lying in parallel planes of type $z = const.$, see Figure 2. Assuming that the rotation axis is the line $x = y = 0$, then $\mathrm{x}_r(s,\theta)$ can be parameterized by

$$x(s) = x_r(s)\cos\theta, \quad y(s) = x_r(s)\sin\theta, \quad z(s) = z_r(s),$$

for some smooth functions $x_r > 0$ and z_r, where $s \in [0,L_r]$ is the arc-length parameter of the plane curve $\gamma_r(s) = \big(x_r(s), z_r(s)\big)$, and $\theta \in \mathbb{S}^1$. A direct computation gives that the Laplacian of the induced Riemannian metric on the cylinder

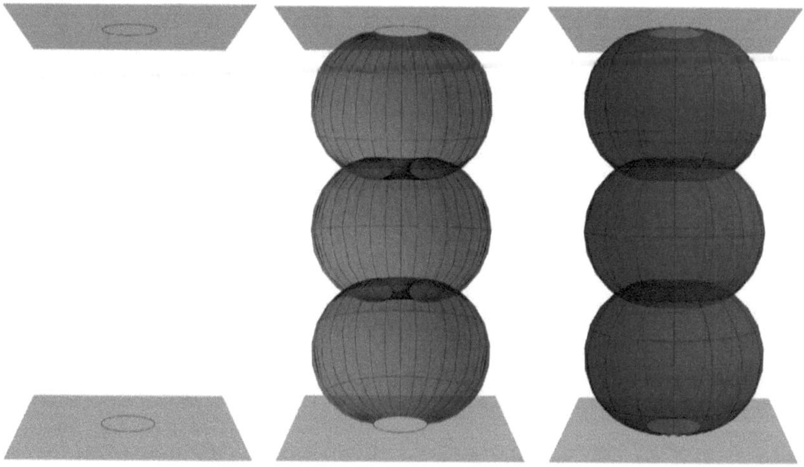

FIGURE 2. The boundary conditions considered for rotationally symmetric CMC surfaces in $\mathbb{R}^3$, and an example of such a surface, a *nodoid* (viewed in half in the second picture and full in the third picture).

$M = [0, L_r] \times \mathbb{S}^1$ is

$$\Delta_r = \frac{1}{x_r}\frac{\partial}{\partial s}\left(x_r \frac{\partial}{\partial s}\right) + \frac{1}{x_r^2}\frac{\partial^2}{\partial \theta^2},$$

and the square norm of the second fundamental form is

$$\|A_{\mathrm{x}_r}\|^2 = (\dot{x}_r \ddot{z}_r - \ddot{x}_r \dot{z}_r)^2 + \frac{\dot{z}_r^2}{x_r^2},$$

where dot represents derivative with respect to s. The eigenvalue problem for the Jacobi equation reads

$$J_r(F) = -\Delta_r F - \|A_{\mathrm{x}_r}\|^2 F = \lambda F, \quad F(0,\theta) = F(L_r, \theta) \equiv 0.$$

Separation of variables $F(s,\theta) = S(s)T(\theta)$ yields the following pair of boundary value problems for ODE's:

$$\begin{aligned} T'' + \kappa T &= 0, & T(0) = T(2\pi),\ T'(0) = T'(2\pi),\\ -(x_r S')' + \left(\tfrac{\kappa}{x_r} - x_r \|A_{\mathrm{x}_r}\|^2\right) S &= \lambda\, x_r S, & S(0) = S(L_r) = 0. \end{aligned}$$

The first problem has nontrivial solutions when $\kappa = n^2$, $n \in \mathbb{Z}$, $n \geq 0$, with corresponding eigenfunctions $\cos n\theta$ and $\sin n\theta$; substituting $\kappa = n^2$ in the second problem we get:

$$\begin{cases} -(x_r S')' + \left(\dfrac{n^2}{x_r} - x_r \|A_{\mathrm{x}_r}\|^2\right) S = \lambda x_r S, \\ S(0) = S(L_r) = 0. \end{cases} \tag{5.10}$$

Every (nontrivial) solution $S_{r,n}$ of the Sturm–Liouville system (5.10) produces two (nontrivial) eigenfunctions of the Jacobi operator along the CMC surface x_r, given by $S_{r,n} \cos n\theta$ and $S_{r,n} \cos n\theta$. The classical Sturm–Liouville theory gives the existence of an unbounded sequence of eigenvalues of (5.10), and the corresponding eigenfunctions are smooth and form a Hilbert basis of $L^2([0, L_r])$. Every solution of the above system with $n > 0$ produces eigenfunctions that are *not* rotationally symmetric. The rotationally symmetric solutions correspond to $n = 0$, in which case the Sturm–Liouville equation reads:

$$\begin{cases} -(x_r S')' - x_r \|A_{\mathrm{x}_r}\|^2 S = \lambda x_r S, \\ S(0) = S(L_r) = 0. \end{cases} \tag{5.11}$$

We will say that r is a *conjugate instant* for the Sturm–Liouville problem if (5.10), has a nontrivial solution with $\lambda = 0$. Evidently, if r is a conjugate instant, then the CMC embedding x_r is degenerate. In this setting, Theorem 5.8 can be applied to obtain the following bifurcation result.

Theorem 5.12. *Consider the family $r \mapsto \mathrm{x}_r$ of rotationally symmetric CMC surfaces in $\mathbb{R}^3$ having fixed boundary described above. For every fixed $n > 0$, let r_n be the first conjugate instant of the Sturm–Liouville equation* (5.10). *Assume that r_n is an isolated degeneracy instant for the family x_r, and that the derivative of the mean curvature function H'_{r_n} is not zero. Then, r_n is a bifurcation instant for the family of CMC surfaces $(\mathrm{x}_r)_r$. Moreover, if r_n is not a conjugate instant also for the Sturm–Liouville problem* (5.11), *then* break of symmetry *occurs at the bifurcating branch, i.e., the bifurcating branch consists of fixed boundary CMC surfaces that are* not *rotationally symmetric.*

Proof. Theorem 5.8 applies here in the following setup. The (identity connected component of the) isotropy of x_r is[18] the group of rotations around the z axis. Assumptions (a), (b1) and (b2) of Theorem 5.4 hold at the instant r_n under our hypotheses. Assumption (b3′) of Theorem 5.8 holds at the first instant at which (5.10) admits a nontrivial solution. Namely, for $r < r_n$, the negative isotropy representation $\pi^-_{\mathrm{x}_r}$ of the group of rotations has no vector whose isotropy is isomorphic to $\mathbb{Z}_n$. On the other hand, for $r > r_n$, with $r - r_n$ sufficiently small, the two Jacobi fields determined by the nontrivial solution of (5.10) belong to the negative eigenspace of J_r, and they have isotropy isomorphic to $\mathbb{Z}_n$. This implies that for $\varepsilon > 0$ small enough, the representations $\pi^-_{\mathrm{x}_{r_n-\varepsilon}}$ and $\pi^-_{\mathrm{x}_{r_n+\varepsilon}}$ are not equivalent. Thus, from Theorem 5.8, bifurcation occurs at r_n. As to the break of symmetry, we observe that if r_n is not a conjugate instant for (5.11), then the symmetrized CMC variational problem is nondegenerate at r_n, and bifurcation by rotationally symmetric CMC embeddings cannot occur. □

We observe that, under the assumptions of Theorem 5.12, jump of the Morse index may *not* occur at r_n. An example where the above result applies is provided by families of fixed boundary *nodoids*, see [17] and Figure 2.

[18]Namely, the subgroup of isometries of $\mathbb{R}^3$ that preserve two co-axial circles lying in parallel planes is generated by the group of rotation around the axis, and, if the two circles have same radius, by the reflection about the plane equidistant to the two parallel planes.

Appendix A. Nonlinear formulation of the bifurcation result

We consider a variational setup similar to that of Section 3, namely $\mathcal{M}$ is a smooth Banach manifold, G is a Lie group acting continuously by diffeomorphisms on $\mathcal{M}$ (recall the auxiliary maps (3.1)), and we also have

(a) $\mathcal{E} \to \mathcal{M}$ is a Banach vector bundle over $\mathcal{M}$;
(b) $[a,b] \ni \lambda \mapsto T_\lambda \in \Gamma(\mathcal{E})$ is a continuous path of G-equivariant sections;
(c) the action of G on $\mathcal{M}$ *lifts* to an action of G on $\mathcal{E}$, which is linear on the fibers;
(d) $[a,b] \ni \lambda \mapsto x_\lambda \in \mathcal{M}$ is a continuous path such that $T_\lambda(x_\lambda) = 0$, for all λ.

Analogously to Definition 4.1, an instant $\lambda_* \in [a,b]$ is an *equivariant bifurcation instant for the family* $(T_\lambda, x_\lambda)_{\lambda\in[a,b]}$ if there is a sequence $(x_n, \lambda_n) \in \mathcal{M} \times [a,b]$ satisfying (1), (3) and $T_{\lambda_n}(x_n) = 0$, for all n, which corresponds to (2). In order to give an existence result for an equivariant bifurcation instant, let us consider the following auxiliary[19] structure

(e) $\mathrm{i}\colon T\mathcal{M} \to \mathcal{E}$ is a G-equivariant continuous inclusion (i.e., an injective morphism of vector bundles) with dense image;
(f) $\langle\cdot,\cdot\rangle$ is a G-invariant continuous (but not necessarily complete) positive-definite inner product in the fibers of $\mathcal{E}$;
(g) $\mathrm{j}_\lambda\colon \mathcal{E}_{x_\lambda} \to T_{x_\lambda}\mathcal{M}^*$ is the map $\mathrm{j}_\lambda(e)v = \langle e, \mathrm{i}(v)\rangle$, and the composition $\mathrm{j}_\lambda \circ (\mathrm{d}^{\mathrm{ver}}T_\lambda)(x_\lambda)\colon T_{x_\lambda}\mathcal{M} \to T_{x_\lambda}\mathcal{M}^*$ is symmetric for all λ.

For all $\lambda \in [a,b]$ and all $\eta \geq 0$, set

$$N_{\lambda,\eta} := \operatorname{span}\big\{v \in T_{x_\lambda}\mathcal{M} : \mathrm{d}^{\mathrm{ver}}T_\lambda(x_\lambda)v = \mu\,\mathrm{i}(v),\ \mu \leq \eta\big\},$$

and $N_\lambda := N_{\lambda,0}$, compare with (4.2) and (4.3). If $G_\lambda \subset G$ is the isotropy of x_λ, we have the isotropy representation $G_\lambda \ni g \mapsto \mathrm{d}\phi_g(x_\lambda) \in \mathrm{GL}(T_{x_\lambda}\mathcal{M})$. For all $\eta \geq 0$, the space $N_{\lambda,\eta}$ is invariant by this action. Denote by $\pi_\lambda^-\colon G_\lambda \to \mathrm{GL}(N_\lambda)$ the restriction of such representation, which is called the *negative isotropy representation* of G_λ (compare with (4.4)).

We can now state the nonlinear formulation of the celebrated result of J. Smoller and A. Wasserman [21, Thm. 3.3], whose proof follows its linear version, using the above auxiliary structure.

Proposition A.1. *In the above setup, assume that*

(a) *there exists* $\varepsilon > 0$ *such that* $\dim(N_{\lambda,\varepsilon}) < +\infty$, *for all* $\lambda \in [a,b]$;
(b) *for all* λ, $G_\lambda = G$;
(c) $\mathrm{d}^{\mathrm{ver}}T_a(x_a)\colon T_{x_a}\mathcal{M} \to \mathcal{E}_{x_a}$ *and* $\mathrm{d}^{\mathrm{ver}}T_b(x_b)\colon T_{x_b}\mathcal{M} \to \mathcal{E}_{x_b}$ *are isomorphisms;*
(d) *the negative isotropy representations* π_a^- *and* π_b^- *are not equivalent.*

Then, there is an equivariant bifurcation instant in $]a,b[$ *for the family* $(T_\lambda, x_\lambda)_\lambda$.

[19]Compare with the structure employed in [9, Section 3].

References

[1] F. Abadie, *Enveloping actions and Takai duality for partial actions*, J. Funct. Anal. 197 (2003), no. 1, 14–67.

[2] L.J. Alías and P. Piccione, *On the manifold structure of the set of unparameterized embeddings with low regularity*, Bull. Braz. Math. Soc., New Series 42 (2) (2011), 171–183.

[3] L.J. Alías and P. Piccione, *Bifurcation of constant mean curvature tori in Euclidean spheres*, J. Geom. Anal. 23 (2013), no. 2, 677–708.

[4] J.L. Barbosa and M. do Carmo, *Stability of hypersurfaces with constant mean curvature*, Math. Z. 185 (1984), 339–353.

[5] J.L. Barbosa, M. do Carmo and J. Eschenburg, *Stability of hypersurfaces with constant mean curvature in Riemannian manifolds*, Math. Z. 197 (1988), 123–138.

[6] R.G. Bettiol and P. Piccione, *Delaunay type hypersurfaces in cohomogeneity one manifolds*, arXiv:1306.6043.

[7] R.G. Bettiol and P. Piccione, *Multiplicity of solutions to the Yamabe problem on collapsing Riemannian submersions.* Pacific J. Math. 266 (2013), no. 1, 1–21.

[8] R.G. Bettiol and P. Piccione, *Bifurcation and local rigidity of homogeneous solutions to the Yamabe problem on spheres*, Calc. Var. PDEs 47 (2013), no. 3-4, 789–807.

[9] R.G. Bettiol, P. Piccione and G. Siciliano, *On the equivariant implicit function theorem with low regularity and applications to geometric variational problems*, arXiv:1009.5721, to appear in Proc. Edinb. Math. Soc.

[10] R.G. Bettiol, P. Piccione and G. Siciliano, *Deforming solutions of geometric variational problems with varying symmetry groups*, to appear.

[11] G.E. Bredon, *Introduction to compact transformation groups*, Pure and Applied Mathematics, Vol. 46, Academic Press, New York–London, 1972.

[12] R. Earp, F. Brito, W.H. Meeks III and H. Rosenberg, *Structure theorems for constant mean curvature surfaces bounded by a planar curve*, Indiana Univ. Math. J. 40 (1991), no. 1, 333–343.

[13] H.I. Elĭasson, *Geometry of manifolds of maps*, J. Differential Geometry 1 (1967), 169–194.

[14] R. Hynd, S.-H. Park and J. McCuan, *Symmetric surfaces of constant mean curvature in* $\mathbb{S}^3$, Pacific J. Math. 241, No. 1 (2009), 63–115.

[15] J. Jost, X. Li-Jost and X.W. Peng, *Bifurcation of minimal surfaces in Riemannian manifolds*, Trans. AMS 347, No. 1 (1995), 51–62.

[16] S. Kobayashi, *Transformation groups in differential geometry*, Reprint of the 1972 edition. Classics in Mathematics. Springer-Verlag, Berlin, 1995.

[17] M. Koiso, B. Palmer and P. Piccione, *Bifurcation and symmetry breaking of nodoids with fixed boundary*, in preparation.

[18] C.N. Lee and A. Wasserman, *On the groups* JO(G), Mem. Amer. Math. Soc. 2 (1975), issue 1, no. 159.

[19] L.L. de Lima, J.H. Lira and P. Piccione, *Bifurcation of Clifford tori in Berger 3-spheres*, to appear in Q. J. Math.

[20] P. Piccione and D. Tausk, *The theory of connections and G-structures. Applications to affine and isometric immersions*, XIV Escola de Geometria Diferencial. [XIV School of Differential Geometry], Instituto de Matemática Pura e Aplicada (IMPA), Rio de Janeiro, 2006.

[21] J. Smoller and A. Wasserman, *Bifurcation and symmetry-breaking*, Invent. Math. 100 (1990), 63–95.

[22] J. Smoller and A. Wasserman, *Symmetry, degeneracy, and universality in semilinear elliptic equations*, J. Funct. Anal. 89 (1990), no. 2, 364–409.

[23] H. Tasaki, M. Umehara and K. Yamada, *Deformations of symmetric spaces of compact type to their noncompact duals*, Japan J. Math. (N.S.) 17 (1991), no. 2, 383–399.

[24] M. Umehara and K. Yamada, *A deformation of tori with constant mean curvature in $\mathbb{R}^3$ to those in other space forms*, Trans. AMS 330, No. 2 (1992), 845–857.

[25] B. White, *The space of m-dimensional surfaces that are stationary for a parametric elliptic functional*, Indiana Univ. Math. J. 36 (1987), 567–602.

[26] B. White, *The space of minimal submanifolds for varying Riemannian metrics*, Indiana Univ. Math. J. 40 (1991), 161–200.

Renato G. Bettiol
University of Notre Dame
Department of Mathematics
255 Hurley Building
Notre Dame, IN, 46556-4618, USA
e-mail: rbettiol@nd.edu

Paolo Piccione and Gaetano Siciliano
Universidade de São Paulo
Departamento de Matemática
Rua do Matão, 1010
São Paulo, SP, 05508-090, Brazil
e-mail: piccione@ime.usp.br
sicilian@ime.usp.br

Progress in Nonlinear Differential Equations
and Their Applications, Vol. 85, 135–143

$W_0^{1,1}$ Solutions in Some Borderline Cases of Elliptic Equations with Degenerate Coercivity

Lucio Boccardo and Gisella Croce

Abstract. We study a degenerate elliptic equation, proving existence results of distributional solutions in some borderline cases.

Mathematics Subject Classification (2010). 35J61, 35J70, 35J75.

Keywords. Elliptic equations; $W^{1,1}$ solutions; degenerate equations.

Bernardo, come vide li occhi miei ...
(Dante, Paradiso XXXI)

1. Introduction

The Sobolev space $W_0^{1,2}(\Omega)$ is the natural functional framework (see [10], [12]) to find weak solutions of nonlinear elliptic problems of the following type

$$\begin{cases} -\mathrm{div}\left(\dfrac{a(x)\nabla u}{(1+|u|)^\theta}\right) = f, & \text{in } \Omega; \\ u = 0, & \text{on } \partial\Omega, \end{cases} \tag{1}$$

where the function f belongs to the dual space of $W_0^{1,2}(\Omega)$, Ω is a bounded, open subset of $\mathbb{R}^N$, with $N > 2$, θ is a real number such that

$$0 \le \theta \le 1\,, \tag{2}$$

and $a : \Omega \to \mathbb{R}$ is a measurable function satisfying the following conditions:

$$\alpha \le a(x) \le \beta\,, \tag{3}$$

This paper contains developments of the results presented by the first author at IX WNDE (João Pessoa, 18.9.2012).

for almost every $x \in \Omega$, where α and β are positive constants. The main difficulty to use the general results of [10], [12] is the fact that

$$A(v) = -\mathrm{div}\left(\frac{a(x)\nabla v}{(1+|v|)^\theta}\right),$$

is not coercive. Papers [7], [4] and [3] deal with the existence and summability of solutions to problem (1) if $f \in L^m(\Omega)$, for some $m \geq 1$.

Despite the lack of coercivity of the differential operator $A(v)$ appearing in problem (1), in the papers [7], [4] and [1], the authors prove the following existence results of solutions of problem (1), under assumption (3):

A) a weak solution $u \in W_0^{1,2}(\Omega) \cap L^\infty(\Omega)$, if $m > \frac{N}{2}$ and (2) holds true;

B) a weak solution $u \in W_0^{1,2}(\Omega) \cap L^{m^{**}(1-\theta)}(\Omega)$, where $m^{**} = (m^*)^* = \frac{mN}{N-2m}$, if

$$0 < \theta < 1, \quad \frac{2N}{N+2-\theta(N-2)} \leq m < \frac{N}{2};$$

C) a distributional solution u in $W_0^{1,q}(\Omega)$, $q = \dfrac{Nm(1-\theta)}{N-m(1+\theta)} < 2$, if

$$\frac{1}{N-1} \leq \theta < 1, \quad \frac{N}{N+1-\theta(N-1)} < m < \frac{2N}{N+2-\theta(N-2)}.$$

D) an entropy solution $u \in M^{m^{**}(1-\theta)}$, with $|\nabla u| \in M^q(\Omega)$, for $1 \leq m \leq \max\left\{1, \frac{N}{N+1-\theta(N-1)}\right\}$.

The borderline case $\theta = 1$ was studied in [3], proving the existence of a solution $u \in W_0^{1,2}(\Omega) \cap L^p(\Omega)$ for every $p < \infty$. The case where the source is $\frac{A}{|x|^2}$ was analyzed too.

About the different notions of solutions mentioned above, we recall that the notion of entropy solution was introduced in [2]. Let

$$T_k(s) = \begin{cases} s & \text{if } |s| \leq k, \\ k\frac{s}{|s|} & \text{if } |s| > k. \end{cases} \tag{4}$$

Then u is an entropy solution to problem (1) if $T_k(u) \in W_0^{1,2}(\Omega)$ for every $k > 0$ and

$$\int_\Omega \frac{a(x)\nabla u}{(1+|u|)^\theta} \cdot \nabla T_k(u-\varphi) \leq \int_\Omega f\, T_k(u-\varphi), \qquad \forall \varphi \in W_0^{1,2}(\Omega) \cap L^\infty(\Omega).$$

Moreover, we say that u is a distributional solution of (1) if

$$\int_\Omega \frac{a(x)\,\nabla u}{(1+|u|)^\theta} \cdot \nabla\varphi = \int_\Omega f\,\varphi, \qquad \forall \varphi \in C_0^\infty(\Omega). \tag{5}$$

The figure on top of the next page can help to summarize the previous results, where the name of a given region corresponds to the results that we have just cited.

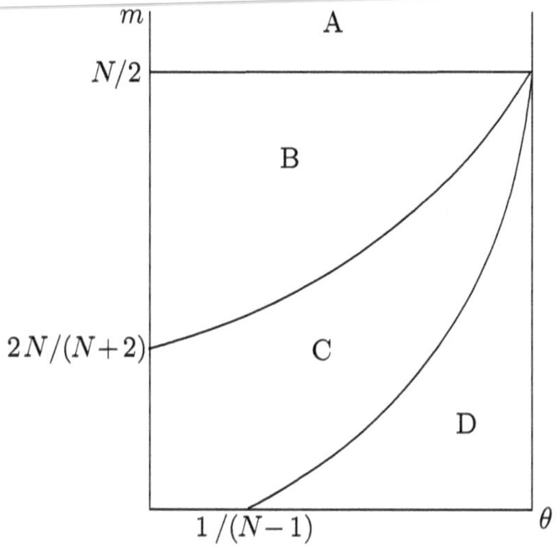

Our results are the following.

Theorem 1.1. *Let f be a function in $L^m(\Omega)$. Assume (3) and*

$$m = \frac{N}{N+1-\theta(N-1)}, \quad \frac{1}{N-1} < \theta < 1 \tag{6}$$

Then there exists a $W_0^{1,1}(\Omega)$ distributional solution to problem (1).

Observe that this case corresponds to the curve which separates the regions C and D of the figure. Note that $m > 1$ if and only if $\theta > \frac{1}{N-1}$.

Also in the following result, we will prove the existence of a $W_0^{1,1}(\Omega)$ solution.

Theorem 1.2. *Let f be a function in $L^m(\Omega)$. Assume (3), $f \log(1+|f|) \in L^1(\Omega)$ and $\theta = \frac{1}{N-1}$. Then there exists a $W_0^{1,1}(\Omega)$ distributional solution of (1).*

We end our introduction just mentioning that a uniqueness result of solutions to problem (1) can be found in [13].

Moreover, in [4, 5, 11, 6] it was showed that the presence of a lower-order term has a regularizing effect on the existence and regularity of the solutions.

To prove our results, we will work by approximation, using the following sequence of problems:

$$\begin{cases} -\text{div}\left(\dfrac{a(x)\nabla u_n}{(1+|u_n|)^\theta}\right) = T_n(f), & \text{in } \Omega; \\ u_n = 0, & \text{on } \partial\Omega. \end{cases} \tag{7}$$

The existence of weak solutions $u_n \in W_0^{1,2}(\Omega) \cap L^\infty(\Omega)$ to problem (7) is due to [7].

2. $W_0^{1,1}(\Omega)$ solutions

In the sequel C will denote a constant depending on $\alpha, N, \text{meas}(\Omega), \theta$ and the $L^m(\Omega)$ norm of the source f.

We are going to prove Theorem 1.1, that is, the existence of a solution to problem (1) in the case where $m = \frac{N}{N+1-\theta(N-1)}$ and $\frac{1}{N-1} < \theta < 1$. Note that $m > 1$ if and only if $\theta > \frac{1}{N-1}$.

Proof of Theorem 1.1. We consider $T_k(u_n)$ as a test function in (7): then

$$\int_\Omega |\nabla T_k(u_n)|^2 \leq \frac{(1+k)^\theta \, \|f\|_{L^1(\Omega)}}{\alpha} \tag{8}$$

by assumption (3) on a.

Choosing $[(1+|u_n|)^p - 1]\text{sign}(u_n)$, for $p = \theta - \frac{1}{N-1}$, as a test function in (7) we have, by Hölder's inequality on the right-hand side and assumption (3) on the left one

$$\begin{aligned} \alpha\, p \int_\Omega \left\{ \frac{|\nabla u_n|}{(1+|u_n|)^{\frac{N}{2(N-1)}}} \right\}^2 &\leq \int_\Omega |f| [(1+|u_n|)^p - 1] \\ &\leq \|f\|_{L^m(\Omega)} \left[\int_\Omega [(1+|u_n|)^p - 1]^{m'} \right]^{\frac{1}{m'}}. \end{aligned} \tag{9}$$

The Sobolev embedding used on the left-hand side implies

$$\left[\int_\Omega \left\{ (1+|u_n|)^{\frac{N-2}{2(N-1)}} - 1 \right\}^{\frac{2N}{N-2}} \right]^{\frac{2}{2^*}} \leq C \left[\int_\Omega [(1+|u_n|)^p - 1]^{m'} \right]^{\frac{1}{m'}}.$$

We observe that $\frac{N-2}{2(N-1)} \frac{2N}{N-2} = pm'$; moreover $\frac{2}{2^*} > \frac{1}{m'}$, since $m < \frac{N}{2}$. Therefore the above inequality implies that

$$\int_\Omega |u_n|^{\frac{N}{N-1}} \leq C. \tag{10}$$

One deduces that

$$\int_\Omega \frac{|\nabla u_n|^2}{(1+|u_n|)^{\frac{N}{N-1}}} \leq C \tag{11}$$

from (10) and (9). Let $v_n = \frac{2(N-1)}{N-2}[(1+|u_n|)^{\frac{N-2}{2(N-1)}} - 1]\text{sign}(u_n)$. Estimate (11) is equivalent to say that $\{v_n\}$ is a bounded sequence in $W_0^{1,2}(\Omega)$; therefore, up to a subsequence, there exists $v \in W_0^{1,2}(\Omega)$ such that $v_n \rightharpoonup v$ weakly in $W_0^{1,2}(\Omega)$ and a.e. in Ω. If we define the function $u = \left(\left[\frac{N-2}{2(N-1)} |v| + 1 \right]^{\frac{2(N-1)}{N-2}} - 1 \right) \text{sign}(v)$, the weak convergence of $\nabla v_n \rightharpoonup \nabla v$ means that

$$\frac{\nabla u_n}{(1+|u_n|)^{\frac{N}{2(N-1)}}} \rightharpoonup \frac{\nabla u}{(1+|u|)^{\frac{N}{2(N-1)}}} \quad \text{weakly in } L^2(\Omega)\,. \tag{12}$$

Moreover, the Sobolev embedding for v_n implies that $u_n \to u$ in $L^s(\Omega)$, for every $1 \le s < \frac{N}{N-1}$.

Hölder's inequality with exponent 2 applied to

$$\int_\Omega |\nabla u_n| = \int_\Omega \frac{|\nabla u_n|}{(1+|u_n|)^{\frac{N}{2N-2}}}(1+|u_n|)^{\frac{N}{2N-2}}$$

gives

$$\int_\Omega |\nabla u_n| \le C, \tag{13}$$

due to (10) and (11). We are now going to estimate $\int_{\{k\le|u_n|\}} |\nabla u_n|$. By using $[(1+|u_n|)^p - (1+k)^p]^+ \mathrm{sign}(u_n)$ as a test function in (7), we have

$$\int_{\{k\le|u_n|\}} \frac{|\nabla u_n|^2}{(1+|u_n|)^{\frac{N}{N-1}}} \le C\Big[\int_{\{k\le|u_n|\}} |f|^m\Big]^{\frac{1}{m}} \Big[\int_\Omega (1+|u_n|)^{\frac{N}{N-1}}\Big]^{\frac{1}{m'}},$$

which implies, by (10),

$$\int_{\{k\le|u_n|\}} \frac{|\nabla u_n|^2}{(1+|u_n|)^{\frac{N}{N-1}}} \le C\Big[\int_{\{k\le|u_n|\}} |f|^m\Big]^{\frac{1}{m}}. \tag{14}$$

Hölder's inequality, estimates (10) and (14) on

$$\int_{\{k\le|u_n|\}} |\nabla u_n| = \int_{\{k\le|u_n|\}} \frac{|\nabla u_n|}{(1+|u_n|)^{\frac{N}{2N-2}}}(1+|u_n|)^{\frac{N}{2N-2}} \le C\Big[\int_{\{k\le|u_n|\}} |f|^m\Big]^{\frac{1}{2m}},$$

give

$$\int_{\{k\le|u_n|\}} |\nabla u_n| = \int_{\{k\le|u_n|\}} \frac{|\nabla u_n|}{(1+|u_n|)^{\frac{N}{2N-2}}}(1+|u_n|)^{\frac{N}{2N-2}} \le C\Big[\int_{\{k\le|u_n|\}} |f|^m\Big]^{\frac{1}{2m}}. \tag{15}$$

Thus, for every measurable subset E, due to (8) and (15), we have

$$\begin{aligned}\int_E \Big|\frac{\partial u_n}{\partial x_i}\Big| &\le \int_E |\nabla u_n| \le \int_E |\nabla T_k(u_n)| + \int_{\{k\le|u_n|\}} |\nabla u_n| \\ &\le \mathrm{meas}(E)^{\frac{1}{2}} \left[\frac{(1+k)^\theta \, \|f\|_{L^1(\Omega)}}{\alpha}\right]^{\frac{1}{2}} + C\Big[\int_{\{k\le|u_n|\}} |f|^m\Big]^{\frac{1}{2m}}.\end{aligned} \tag{16}$$

Now we are going to prove that u_n weakly converges to u in $W_0^{1,1}(\Omega)$ following [5]. Estimates (16) and (10) imply that the sequence $\{\frac{\partial u_n}{\partial x_i}\}$ is equiintegrable. By the Dunford–Pettis theorem, and up to subsequences, there exists Y_i in $L^1(\Omega)$ such that $\frac{\partial u_n}{\partial x_i}$ weakly converges to Y_i in $L^1(\Omega)$. Since $\frac{\partial u_n}{\partial x_i}$ is the distributional partial derivative of u_n, we have, for every n in $\mathbb{N}$,

$$\int_\Omega \frac{\partial u_n}{\partial x_i}\varphi = -\int_\Omega u_n \frac{\partial \varphi}{\partial x_i}, \quad \forall\, \varphi \in C_0^\infty(\Omega).$$

We now pass to the limit in the above identities, using that $\partial_i u_n$ weakly converges to Y_i in $L^1(\Omega)$, and that u_n strongly converges to u in $L^1(\Omega)$: we obtain

$$\int_\Omega Y_i\,\varphi = -\int_\Omega u \frac{\partial \varphi}{\partial x_i}, \quad \forall\, \varphi \in C_0^\infty(\Omega).$$

This implies that $Y_i = \frac{\partial u}{\partial x_i}$, and this result is true for every i. Since Y_i belongs to $L^1(\Omega)$ for every i, u belongs to $W_0^{1,1}(\Omega)$.

We are now going to pass to the limit in problems (7). For the limit of the left-hand side, it is sufficient to observe that $\frac{\nabla u_n}{(1+|u_n|)^{\frac{N}{2(N-1)}}} \rightharpoonup \frac{\nabla u}{(1+|u|)^{\frac{N}{2(N-1)}}}$ weakly in $L^2(\Omega)$ due to (12) and that $|a(x)\nabla\varphi|$ is bounded. □

We prove Theorem 1.2, that is, the existence of a $W_0^{1,1}(\Omega)$ solution in the case where $\theta = \frac{1}{N-1}$ and $f\log(1+|f|) \in L^1(\Omega)$.

Proof of Theorem 1.2. Let $k \geq 0$ and take $[\log(1+|u_n|) - \log(1+k)]^+\operatorname{sign}(u_n)$, as a test function in problems (7). By assumption (3) on a one has

$$\alpha \int_{\{k\leq|u_n|\}} \frac{|\nabla u_n|^2}{(1+|u_n|)^{\theta+1}} \leq \int_{\{k\leq|u_n|\}} |f|\log(1+|u_n|).$$

We now use the following inequality on the left-hand side:

$$a\log(1+b) \leq \frac{a}{\rho}\log\left(1+\frac{a}{\rho}\right) + (1+b)^\rho \tag{17}$$

where a, b are positive real numbers and $0 < \rho < \frac{N-2}{N-1}$. This gives, for any $k \geq 0$

$$\alpha \int_{\{k\leq|u_n|\}} \frac{|\nabla u_n|^2}{(1+|u_n|)^{\theta+1}} \leq \int_{\{k\leq|u_n|\}} \frac{|f|}{\rho}\log\left(1+\frac{|f|}{\rho}\right) + \int_{\{k\leq|u_n|\}} (1+|u_n|)^\rho. \tag{18}$$

In particular, for $k \geq 1$ we have

$$\frac{\alpha}{2^{\theta+1}} \int_{\{k\leq|u_n|\}} \frac{|\nabla u_n|^2}{|u_n|^{\theta+1}} \leq \int_{\{k\leq|u_n|\}} \frac{|f|}{\rho}\log\left(1+\frac{|f|}{\rho}\right) + 2^\rho \int_{\{k\leq|u_n|\}} |u_n|^\rho. \tag{19}$$

Writing the above inequality for $k = 1$ and using the Sobolev inequality on the left-hand side, one has

$$\left[\int_\Omega (|u_n|^{\frac{1-\theta}{2}} - 1)_+^{2^*}\right]^{\frac{2}{2^*}} \le C \int_{\{1\le|u_n|\}} \frac{|f|}{\rho}\log\left(1+\frac{|f|}{\rho}\right) + C\int_{\{1\le|u_n|\}} |u_n|^\rho,$$

which implies that

$$\left[\int_\Omega |u_n|^{\frac{(1-\theta)2^*}{2}}\right]^{\frac{1}{2^*}} \le C + C\sqrt{\int_{\{1\le|u_n|\}} \frac{|f|}{\rho}\log\left(1+\frac{|f|}{\rho}\right)} + C\sqrt{\int_\Omega |u_n|^\rho}\,.$$

By using Hölder's inequality with exponent $\frac{(1-\theta)2^*}{2\rho}$ on the last term of the right-hand side, we get

$$\left[\int_\Omega |u_n|^{\frac{(1-\theta)2^*}{2}}\right]^{\frac{1}{2^*}} \le C + C\sqrt{\int_{\{1\le|u_n|\}} \frac{|f|}{\rho}\log\left(1+\frac{|f|}{\rho}\right)}$$
$$+ C\left[\int_\Omega |u_n|^{\frac{(1-\theta)2^*}{2}}\right]^{\frac{\rho}{(1-\theta)2^*}}.$$

By the choice of ρ, this inequality implies that

$$\int_\Omega |u_n|^{\frac{(1-\theta)2^*}{2}} \le C\,. \tag{20}$$

Inequalities (20) and (18) written for $k = 0$ imply that the sequence $\{v_n\}$, $v_n = \{\frac{2}{1-\theta}[(1+|u_n|)^{\frac{1-\theta}{2}} - 1]\mathrm{sign}(u_n)\}$, is a bounded sequence in $W_0^{1,2}(\Omega)$, as in the proof of Theorem 1.1. Therefore, up to a subsequence there exists $v \in W_0^{1,2}(\Omega)$ such that $v_n \rightharpoonup v$ weakly in $W_0^{1,2}(\Omega)$ and a.e. in Ω. Let $u = \{[\frac{1-\theta}{2}|v| + 1]^{\frac{2}{1-\theta}} - 1\}\mathrm{sign}(v)$; the weak convergence of $\nabla v_n \rightharpoonup \nabla v$ means that

$$\frac{\nabla u_n}{(1+|u_n|)^{\frac{\theta+1}{2}}} \rightharpoonup \frac{\nabla u}{(1+|u|)^{\frac{\theta+1}{2}}} \quad \text{weakly in } L^2(\Omega)\,. \tag{21}$$

Moreover, the Sobolev embedding for v_n implies that $u_n \to u$ in $L^s(\Omega)$, $s < \frac{N}{N-1}$.

By (8) one has

$$\int_\Omega |\nabla u_n| = \int_\Omega |\nabla T_1(u_n)| + \int_{\{1\le|u_n|\}} |\nabla u_n| \le C + \int_{\{1\le|u_n|\}} \frac{|\nabla u_n|}{|u_n|^{\frac{\theta+1}{2}}}|u_n|^{\frac{\theta+1}{2}}\,.$$

Hölder's inequality on the right-hand side, and estimates (19) written with $k = 1$ and (20) imply that the sequence $\{u_n\}$ is bounded in $W_0^{1,1}(\Omega)$.

Moreover, due to (19)

$$\int_{\{k\leq|u_n|\}} |\nabla u_n| = \int_{\{k\leq|u_n|\}} \frac{|\nabla u_n|}{|u_n|^{\frac{\theta+1}{2}}}|u_n|^{\frac{\theta+1}{2}}$$

$$\leq C\sqrt{\int_{\{k\leq|u_n|\}} \frac{|f|}{\rho}\log\left(1+\frac{|f|}{\rho}\right) + \int_{\{k\leq|u_n|\}} |u_n|^{\rho}}\,.$$

For every measurable subset E, the previous inequality and (8) imply

$$\int_E \left|\frac{\partial u_n}{\partial x_i}\right| \leq \int_E |\nabla u_n| \leq \int_E |\nabla T_k(u_n)| + \int_{\{k\leq|u_n|\}} |\nabla u_n|$$

$$\leq C\operatorname{meas}(E)^{\frac{1}{2}}(1+k)^{\frac{\theta}{2}} + C\sqrt{\int_{\{k\leq|u_n|\}} \frac{|f|}{\rho}\log\left(1+\frac{|f|}{\rho}\right) + \int_{\{k\leq|u_n|\}} |u_n|^{\rho}}\,.$$

Since $\rho < \frac{(1-\theta)2^*}{2}$, by using Hölder's inequality on the last term and estimate (20), one has

$$\int_E \left|\frac{\partial u_n}{\partial x_i}\right| \leq C\operatorname{meas}(E)^{\frac{1}{2}}(1+k)^{\frac{\theta}{2}} +$$

$$+ C\sqrt{\int_{\{k\leq|u_n|\}} \frac{|f|}{\rho}\log\left(1+\frac{|f|}{\rho}\right) + \operatorname{meas}(\{|u_n|\geq k\})^{1-\frac{2\rho}{(1-\theta)2^*}}}\,.$$

One can argue as in the proof of Theorem 1.1 to deduce that $u_n \to u$ weakly in $W_0^{1,1}(\Omega)$.

To pass to the limit in problems (7), as in the proof of Theorem 1.1, it is sufficient to observe that $\frac{\nabla u_n}{(1+|u_n|)^{\frac{N}{2(N-1)}}} \rightharpoonup \frac{\nabla u}{(1+|u|)^{\frac{N}{2(N-1)}}}$ weakly in $L^2(\Omega)$, due to (21).

□

References

[1] A. Alvino, L. Boccardo, V. Ferone, L. Orsina, G. Trombetti: Existence results for nonlinear elliptic equations with degenerate coercivity; Ann. Mat. Pura Appl. 182 (2003), 53–79.

[2] P. Benilan, L. Boccardo, T. Gallouët, R. Gariepy, M. Pierre, and J.L. Vazquez: An L^1 theory of existence and uniqueness of nonlinear elliptic equations; Ann. Scuola Norm. Sup. Pisa Cl. Sci. 22 (1995), 240–273.

[3] L. Boccardo: Some elliptic problems with degenerate coercivity; Adv. Nonlinear Stud. 6 (2006), 1–12.

[4] L. Boccardo, H. Brezis: Some remarks on a class of elliptic equations with degenerate coercivity; Boll. Unione Mat. Ital. 6 (2003), 521–530.

[5] L. Boccardo, G. Croce, L. Orsina: Nonlinear degenerate elliptic problems with $W_0^{1,1}(\Omega)$ solutions; Manuscripta Mathematica 137 (2012), 419–439.

[6] L. Boccardo, G. Croce, L. Orsina: A semilinear problem with a $W_0^{1,1}(\Omega)$ solution; Rend. Lincei. Mat. Appl. 23 (2012), 97–103.

[7] L. Boccardo, A. Dall'Aglio, L. Orsina: Existence and regularity results for some elliptic equations with degenerate coercivity; Atti Sem. Mat. Fis. Univ. Modena 46 (1998), 51–81.

[8] L. Boccardo, T. Gallouët: Nonlinear elliptic equations with right-hand side measures; Comm. Partial Differential Equations 17 (1992), 641–655.

[9] L. Boccardo, T. Gallouët: $W_0^{1,1}$ solutions in some borderline cases of Calderòn–Zygmund theory; J. Differential Equations 253 (2012), 2698–2714.

[10] H. Brezis: Equations et inéquations non linéaires dans les espaces vectoriels en dualité; Ann. Inst. Fourier (Grenoble), 18 (1968), 115–175.

[11] G. Croce: The regularizing effects of some lower order terms in an elliptic equation with degenerate coercivity; Rendiconti di Matematica 27 (2007), 299–314.

[12] J. Leray, J.L. Lions: Quelques résultats de Višik sur les problèmes elliptiques semilinéaires par les méthodes de Minty et Browder; Bull. Soc. Math. France, 93 (1965), 97–107.

[13] A. Porretta: Uniqueness and homogenization for a class of non coercive operators in divergence form; Atti Sem. Mat. Fis. Univ. Modena 46 (1998), 915–936.

Lucio Boccardo
Dipartimento di Matematica
"Sapienza" Università di Roma
P.le A. Moro 2
I-00185 Roma, Italy
e-mail: boccardo@mat.uniroma1.it

Gisella Croce
Laboratoire de Mathématiques Appliquées du Havre
Université du Havre
25, rue Philippe Lebon
F-76063 Le Havre, France
e-mail: gisella.croce@univ-lehavre.fr

Progress in Nonlinear Differential Equations
and Their Applications, Vol. 85, 145–161

Remarks on p-Laplacian Problems Depending on the Gradient

H. Bueno and G. Ercole

Abstract. This paper collects and summarizes results of existence of positive solutions for the p-Laplacian problem $-\Delta_p u = \omega(x) f(u, |\nabla u|)$ with Dirichlet boundary condition in a bounded domain $\Omega \subset \mathbb{R}^N$, where ω is a weight function and also for the problem in two positive parameters λ and β:

$$\begin{cases} -\Delta_p u = \lambda h(x,u) + \beta f(x,u,\nabla u) & \text{in } \Omega \\ u = 0 & \text{on } \partial\Omega. \end{cases}$$

Mathematics Subject Classification (2010). 35B09, 35J66, 35J70, 35J92.

Keywords. p-Laplacian, positive solution, sub- and super-solution method, dependence on the gradient.

1. Introduction

Existence of positive solutions for p-Laplacian problems depending on the gradient has been attracting considerable interest among researchers of elliptic PDE's, but no general method to deal with this kind of problem has been established. The dependence on the gradient requests a priori bounds on the solutions and in their derivatives, what brings additional difficulties. Since this problem is not suitable for variational techniques, topological methods (as fixed-point or degree results) and/or blow-up arguments are normally employed to solve it ([3] and references therein).

Maybe because of the belief that the sub- and super-solution method does not handle elliptic problems which are super-linear at the origin, such approach is rare in the literature. One of the main purposes of this paper is to prove that this belief is not true.

Both authors were supported in part by FAPEMIG and CNPq-Brazil.

In this paper, we consider problems in the form

$$\begin{cases} -\Delta_p u = f(x, u, \nabla u) & \text{in } \Omega \\ u = 0 & \text{on } \partial\Omega \end{cases} \tag{1}$$

for various types of continuous nonlinearities f. Special attention is given to the Dirichlet problem

$$\begin{cases} -\Delta_p u = \omega(x) f(u, |\nabla u|) & \text{in } \Omega, \\ u = 0 & \text{on } \partial\Omega \end{cases} \tag{2}$$

where $\Omega \subset \mathbb{R}^N$ $(N > 1)$ is a smooth, bounded domain, $\Delta_p u := \operatorname{div}\left(|\nabla u|^{p-2}\nabla u\right)$ is the p-Laplacian, $1 < p < \infty$, $\omega\colon \overline{\Omega} \to \mathbb{R}$ is a continuous, nonnegative function with isolated zeros (which we will call *weight function*) and the C^1-nonlinearity $f\colon [0,\infty) \times [0,\infty) \to [0,\infty)$ satisfies simple hypotheses.

Adapting methods and techniques developed in [7], where the nonlinearity does not depend on ∇u, we start by obtaining radial, positive solutions u for the class of problem

$$\begin{cases} -\Delta_p u = \omega_\rho(|x - x_0|) f(u, |\nabla u|) & \text{in } B_\rho, \\ u = 0 & \text{on } \partial B_\rho, \end{cases} \tag{3}$$

where B_ρ is the ball with radius ρ centered at x_0 and ω_ρ is a *radial* weight function. The application of the Schauder Fixed Point Theorem yields a radial solution u of (3) for a large class of functions f, including nonlinearities that are super-linear both at the origin and at $+\infty$. (The continuous function ω has isolated zeroes only to simplify the presentation. It is enough that $\omega(x_0) > 0$ for some $x_0 \in \Omega$.)

For this, no asymptotic behavior on f is assumed but, instead, simple local hypotheses on the nonlinearity f. Our hypotheses on the nonlinearity f are not usual in the literature: we assume that f has a *local* behavior satisfying hypotheses of the type

(H1) $0 \le f(u, |v|) \le k_1 M^{p-1}$, if $0 \le u \le M$, $|v| \le \gamma M$,

(H2) $f(u, |v|) \ge k_2 \delta^{p-1}$, if $0 < \delta \le u \le M$, $|v| \le \gamma M$,

where the constants k_1, k_2 and γ are defined later on in this paper and δ, M are arbitrary. These constants depend strongly on the weight function ω and it must be stressed that they can be explicitly calculated in some special cases (for example, if $\omega \equiv 1$; see Example 9). In [4] it was proved that $k_1 < \lambda_1 < k_2$, where λ_1 stands for the first eigenvalue of the p-Laplacian.

Hypotheses (H1) and (H2) are geometrically interpreted in Figure 1. Observe that, since no assumption is made both at the origin or at infinity, such a super-linear behavior is permitted.

Hypotheses of this type will be considered in the scenarios of both the radial problem (3) and the general problem (2).

In the case of the radial problem (3) no further hypotheses are necessary. It will be considered in Section 3.

To apply the sub- and super-solution method to solve problem (2), a condition of the Bernstein–Nagumo type is always assumed; in [2] the nonlinearity f is a

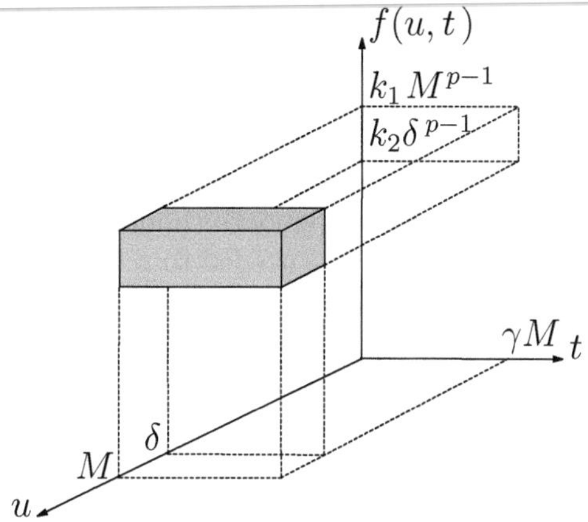

FIGURE 1. For $t = |v|$, the graph of f stays below $k_1 M^{p-1}$ in $[0, M] \times [0, \gamma M]$ and passes through the gray box.

Carathéodory function (i.e., measurable in the x-variable and continuous in the (u, v)-variable) such that

(H3) $f(x, u, v) \leq C(|u|)(1 + |v|^p)$ $(u, v) \in \mathbb{R} \times \mathbb{R}^N$, *a.e.* $x \in \Omega$ for some increasing function $C\colon [0, \infty] \to [0, \infty]$.

This assumption is merely technical and can be chosen as any hypothesis that guarantees the existence of a solution of (1) from an ordered sub- and super-solution pair.

However, the acceptance of the growth condition (H3) poses another problem: what happens in the case $|\nabla u|^b$, if $b > p$? So, we also consider in this paper an example where the exponent b in $|\nabla v|^b$ is greater than p, see Section 6.

Our approach of problem (2) starts by considering the solution of (3) in a subdomain $B_\rho \subset \Omega$. A connection between both problem is achieved by defining the weight ω_ρ in terms of ω by

$$\omega_\rho(s) = \begin{cases} \min\limits_{|x-x_0|=s} \omega(x), & \text{if } 0 < s \leq \rho, \\ \omega(x_0), & \text{if } s = 0. \end{cases} \tag{4}$$

By choosing a ball $B_\rho \subset \Omega$ and such a radialization of the weight function ω, we consider a problem in the radial form (3) in the sub-domain B_ρ, which has a solution u_ρ as a consequence of our study of this problem. The chosen ball B_ρ determines the value of the constants k_2 and γ and the radial solution $u_\rho\colon \overline{B_\rho} \to \mathbb{R}$ produces a sub-solution $\underline{u}$ of problem (2), when we consider the extension $\underline{u}$ of u_ρ defined by $\underline{u}(x) = 0$, if $x \in \Omega \backslash B_\rho$. So, the solution of (3) gives rise to a sub-solution of problem (2).

In order to obtain a super-solution $\overline{u}$ for problem (2), we impose that

$$\frac{\|\nabla\overline{u}\|_\infty}{\|\overline{u}\|_\infty} \leq \gamma, \tag{5}$$

an estimate that is suggested by hypothesis (H1). So, we look for a super-solution of (2) satisfying (5) and defined in a (smooth, bounded) domain $\Omega_2 \supset \Omega$, which determines the value of the constant k_1 needed to solve (2).

In the general setting of the domain Ω_2, the super-solution $\overline{u}$ turns out to be a multiple of the solution ϕ_{Ω_2} of the problem

$$\begin{cases} -\Delta_p\phi_{\Omega_2} = \|\omega\|_\infty \text{ in } \Omega_2, \\ \phi_{\Omega_2} = 0 \quad\quad \text{on } \partial\Omega_2, \end{cases} \tag{6}$$

if ϕ_{Ω_2} satisfies (5). In this setting, the existence of a positive solution for (2) is stated in Section 4.

We give two applications of this result for general nonlinearities in Section 5. In the first application, given in Subsection 5.1, we choose a ball $\Omega_2 = B_R$ such that $\Omega \subset B_R$ and prove that, if R is large enough, it is possible to obtain a super-solution for (2) satisfying (5).

The second application considers the case where Ω_2 is the domain Ω itself. In order to control the quotient (5), we assume Ω to be convex and apply a maximum principle proved in Payne and Philippin [9]. In some cases, if we choose Ω_2 as the convex hull of Ω, the same method produces a better solution than considering $\Omega \subset B_R$ for R large enough.

Inspired by the classical paper of Ambrosetti, Brezis and Cerami [1], in the final Section 6 we consider a problem in two parameters, where $|\nabla u|$ has an exponent higher than p.

2. Preliminaries

Let D be a bounded, smooth domain in $\mathbb{R}^N$, $N > 1$.

We define

$$k_1(D) := \|\phi_D\|_\infty^{-(p-1)} \tag{7}$$

where $\phi_D \in C^{1,\alpha}(\overline{D}) \cap W_0^{1,p}(D)$ is the solution of

$$\begin{cases} -\Delta_p\phi_D = \omega_D \text{ in } D, \\ \phi_D = 0 \quad \text{on } \partial D, \end{cases} \tag{8}$$

where $\omega_D \not\equiv 0$ is any continuous, non-negative function. By the maximum principle, $\phi_D > 0$ in D and $k_1(D)$ is well defined.

Remark 1. By applying the comparison principle in the domains $\Omega_1 \subset \Omega_2$ with $\omega_{\Omega_1} \leq \omega_{\Omega_2}$, it follows immediately that

$$k_1(\Omega_2) = \|\phi_{\Omega_2}\|_\infty^{-(p-1)} \leq \|\phi_{\Omega_1}\|_\infty^{-(p-1)} = k_1(\Omega_1).$$

In the special case $D = B_\rho$, a ball of radius ρ centered at $x_0 \in \Omega$, let us consider the Dirichlet problem

$$\begin{cases} -\Delta_p \phi_\rho = \omega_\rho(|x-x_0|) & \text{in } B_\rho, \\ \phi_\rho = 0 & \text{on } \partial B_\rho, \end{cases} \tag{9}$$

where $\omega_\rho\colon [0,\rho] \to \mathbb{R}$.

It is straightforward to verify that the solution of (9) is given by

$$\phi_\rho(|x-x_0|) = \int_{|x-x_0|}^{\rho} \left(\int_0^\theta K(s,\theta)ds \right)^{\frac{1}{p-1}} d\theta, \ |x-x_0| \le \rho, \tag{10}$$

where

$$K(s,\theta) = \left(\frac{s}{\theta}\right)^{N-1} \omega_\rho(s). \tag{11}$$

The solution ϕ_ρ satisfies $\phi_\rho \in C^2(\overline{B_\rho})$ if $1 < p \le 2$ and $\phi_\rho \in C^{1,\alpha}(\overline{B_\rho})$ if $p > 2$, where $\alpha = 1/(p-1)$. (See [3], Lemma 2 for details.)

We also define another constant that will play an essential role in our technique:

$$\begin{aligned} k_2(B_\rho) &= \left[\int_t^\rho \left(\int_0^t K(s,\theta)ds \right)^{\frac{1}{p-1}} d\theta \right]^{1-p} \\ &= \left[\max_{0 \le r \le \rho} \int_r^\rho \left(\int_0^r K(s,\theta)ds \right)^{\frac{1}{p-1}} d\theta \right]^{1-p}. \end{aligned} \tag{12}$$

Since ω_ρ has isolated zeroes and the function

$$\beta \to \int_\beta^\rho \left(\int_0^\beta K(s,\theta)ds \right)^{\frac{1}{p-1}} d\theta$$

is nonnegative and vanishes both at $\beta = 0$ and at $\beta = \rho$, we have $t > 0$.

We now establish the relation between $k_1(D)$ and $k_2(B_\rho)$, also valid in the case $D = B_\rho$. Its proof follows by applying a comparison principle.

Lemma 2. *Let D be a smooth domain in $\mathbb{R}^N$ ($N > 1$), $B_\rho \subseteq D$ a ball of center x_0 and radius $\rho > 0$ and $k_1(D)$, $k_2(B_\rho)$ the constants defined by (7) and (12), respectively, where ω_ρ is a radial weight function such that $\omega_D \ge \omega_\rho$ in B_ρ.*

Then, $k_1(D) < k_2(B_\rho)$.

3. Radial solutions

In this section we study the radial version of (2), that is

$$\begin{cases} -\Delta_p u = \omega_\rho(|x-x_0|) f(u, |\nabla u|) & \text{in } B_\rho, \\ u = 0 & \text{on } \partial B_\rho, \end{cases} \tag{2}$$

where B_ρ is a ball of radius ρ centered at x_0 and $\omega_\rho\colon [0,\rho] \to \mathbb{R}$.

A solution of (3) will be obtained by applying the Schauder Fixed Point Theorem in the space $C^1(B_\rho)$. So, the hypothesis (H3) is not necessary; we only need f to be continuous and to satisfy hypotheses (H1$_r$) and (H2$_r$) given below.

The radial boundary value problem equivalent to (3) is

$$\begin{cases} \frac{d}{dr}\left(-r^{N-1}\varphi_p(u'(r))\right) = r^{N-1}\omega_\rho(r) f(u, |u'(r)|), & 0 < r < \rho \\ u'(0) = 0, & \\ u(\rho) = 0, & \end{cases} \tag{13}$$

where $\varphi_p(\xi) = |\xi|^{p-2}\xi$.

If $q = p/(p-1)$ and $u > 0$, the function φ_q, inverse of φ_p, is given by

$$\varphi_q(u) = |u|^{q-2}u = u^{q-1} = u^{\frac{p}{p-1}-1} = u^{\frac{1}{p-1}}.$$

It is not difficult to see that $\|\nabla\phi_\rho\|_\infty = \max_{0\le r\le\rho}|\phi'_\rho(r)|$.

To prove the existence of solutions for problem (3), we suppose the existence of δ and M, with $0 < \delta < M$, such that the nonlinearity f satisfies

(H1$_r$) $0 \le f(u, |v|) \le k_1(B_\rho)M^{p-1}$, if $0 \le u \le M$, $|v| \le \gamma_\rho M$;
(H2$_r$) $f(u, |v|) \ge k_2(B_\rho)\delta^{p-1}$, if $\delta \le u \le M$, $|v| \le \gamma_\rho M$,

with $k_1(B_\rho)$ and $k_2(B_\rho)$ defined by (7) and (12), respectively, and γ_ρ defined by

$$\gamma_\rho = \frac{\|\nabla\phi_\rho\|_\infty}{\|\phi_\rho\|_\infty}. \tag{14}$$

Note that

$$\begin{aligned} \frac{\|\nabla\phi_\rho\|_\infty}{\|\phi_\rho\|_\infty} &= k_1(B_\rho)^{1/(p-1)} \max_{0\le r\le\rho} |\phi'_\rho(r)| \\ &= \varphi_q(k_1(B_\rho)) \max_{0\le r\le\rho} \varphi_q\left(\int_0^r K(s,r)ds\right) \\ &= \max_{0\le r\le\rho} \varphi_q\left(k_1(B_\rho)) \int_0^r K(s,r)ds\right). \end{aligned}$$

We remark that $k_1(B_\rho)$, $k_2(B_\rho)$ and γ_ρ depend only on ρ and ω_ρ. The hypothesis (H2$_r$) aims to discard $u \equiv 0$ as a solution of (3), in the case $f(0, |v|) = 0$.

We also define the continuous functions Ψ_δ, Φ_M and Γ_M by

$$\Psi_\delta(r) = \begin{cases} \delta, & \text{if } 0 \le r \le t, \\ \delta \int_r^\rho \varphi_q\left(k_2(B_\rho)\int_0^t K(s,\theta)\,ds\right)d\theta, & \text{if } t < r \le \rho, \end{cases} \tag{15}$$

where t is defined in (12),

$$\Phi_M(r) = M\int_r^\rho \varphi_q\left(k_1(B_\rho)\int_0^\theta K(s,\theta)\,ds\right)d\theta = M\frac{\phi_\rho(r)}{\|\phi_\rho\|_\infty}, \text{ if } 0 < r \le \rho, \tag{16}$$

and

$$\Gamma_M(r) = M\varphi_q\left(k_1(B_\rho)\int_0^r K(s,r)\,ds\right) = M\frac{|\phi'_\rho(r)|}{\|\phi_\rho\|_\infty}, \text{ if } 0 < r \le \rho. \tag{17}$$

It is not difficult to prove[1]

Lemma 3. *We have*

(i) $0 \leq \Phi_M(r) \leq M$;
(ii) $0 \leq \Gamma_M(r) \leq \gamma_\rho M$;
(iii) $0 \leq \Psi_\delta(r) \leq \Phi_M(r)$.

The proof of Theorem 4 follows by applying Schauder's fixed point theorem and Lemma 3.

Theorem 4. *Suppose that the continuous nonlinearity f satisfies* (H1$_r$) *and* (H2$_r$). *Then the problem*

$$\begin{cases} -\Delta_p u = \omega_\rho(|x - x_0|) f(u, |\nabla u|) \text{ in } B_\rho, \\ \quad u = 0 \qquad\qquad\qquad\qquad\quad \text{on } \partial B_\rho, \end{cases} \tag{3}$$

has at least one positive solution $u_\rho(|x - x_0|)$ satisfying

$$\Psi_\delta \leq u_\rho \leq \Phi_M \text{ and } |\nabla u_\rho| \leq \Gamma_M$$

(and so $\delta \leq \|u_\rho\|_\infty \leq M$ and $\|\nabla u_\rho\|_\infty \leq \gamma_\rho M$).

4. Existence of solutions in general domains

In this section we state and prove our main result: the existence of a positive solution for

$$\begin{cases} -\Delta_p u = \omega(x) f(u, |\nabla u|) \text{ in } \Omega, \\ \quad u = 0 \qquad\qquad\qquad \text{on } \partial\Omega. \end{cases} \tag{2}$$

We start by defining the parameters we need to formulate our hypotheses. Let Ω_2 be a bounded, smooth domain such that $\Omega_2 \supset \Omega$ and define

$$k_1(\Omega_2) := \|\phi_{\Omega_2}\|_\infty^{1-p},$$

where ϕ_{Ω_2} is the solution of

$$\begin{cases} -\Delta_p \phi_{\Omega_2} = \|\omega\|_\infty \text{ in } \Omega_2, \\ \quad \phi_{\Omega_2} = 0 \qquad \text{on } \partial\Omega_2. \end{cases} \tag{18}$$

Now, for any ball $B_\rho \subset \Omega$ with center in $x_0 \in \Omega$ and radius $\rho > 0$, let us to denote by ω_ρ the radial function defined by (4). Thus, by using this function we consider $k_2(B_\rho)$ and γ_ρ, defined in accordance to the former definitions (12) and (14), respectively.

At last, we fix $\rho > 0$ such that (see Remark 6, below)

$$\frac{\|\nabla \phi_{\Omega_2}\|_\infty}{\|\phi_{\Omega_2}\|_\infty} \leq \gamma_\rho \tag{19}$$

and then we set the parameters

$$k_1 := k_1(\Omega_2), \quad k_2 := k_2(B_\rho) \quad \text{and} \quad \gamma = \gamma_\rho.$$

[1] For both Lemma 3 and Theorem 4, see [3] for details.

Theorem 5. *Suppose that, for arbitrary δ, M such that $0 < \delta < M$, the nonlinearity f satisfies:*

(H1) $0 \leq f(u, |v|) \leq k_1 M^{p-1}$, *if* $0 \leq u \leq M$, $|v| \leq \gamma M$;
(H2) $f(u, |v|) \geq k_2 \delta^{p-1}$, *if* $\delta \leq u \leq M$, $|v| \leq \gamma M$;
(H3) $f(u, |v|) \leq C(|u|)(1 + |v|^p)$ *for all* (x, u, v), *where* $C\colon [0, \infty) \to [0, \infty)$ *is increasing.*

Then, problem (2) *has a positive solution u such that*

$$\delta \leq \|u\|_\infty \leq M \text{ in } \Omega.$$

Remark 6. We would like to observe that the inequality (19) always occurs, if ρ is taken sufficiently small such that

$$\frac{\|\nabla \phi_{\Omega_2}\|_\infty}{\|\phi_{\Omega_2}\|_\infty} \leq \frac{1}{\rho}. \tag{20}$$

In fact, we have the gross estimate

$$\frac{1}{\rho} \leq \gamma_\rho, \quad \text{for any } B_\rho \subset \Omega$$

since $\gamma_\rho = \frac{\|\nabla \phi_\rho\|_\infty}{\|\phi_\rho\|_\infty}$ and

$$\|\phi_\rho\|_\infty = \phi_\rho(0) = -\int_0^\rho \phi'_\rho(s)\, ds = \int_0^\rho |\phi'_\rho(s)|\, ds \leq \rho \|\nabla \phi_\rho\|_\infty.$$

We supposed that the weight function ω has isolated zeroes. As mentioned, this assumption is not necessary.

In Section 5 we give examples of Ω_2 and ρ satisfying (19). There, we consider the cases $\Omega_2 = B_R \supset \Omega$ and, supposing Ω convex, $\Omega_2 = \Omega$. Moreover, we present better estimates than (20) to choose ρ.

The obtention of a sub-solution for problem (2) is based on the following general result:

Lemma 7. *Let Ω and Ω_1 be smooth domains in $\mathbb{R}^N$ $(N > 1)$, with $\Omega_1 \subset \Omega$. Let $u_1 \in C^{1,\alpha}\left(\overline{\Omega_1}\right)$ be a positive solution of*

$$\begin{cases} -\Delta_p u_1 = f_1(x, u_1, \nabla u_1) & \text{in } \Omega_1, \\ u_1 = 0 & \text{on } \partial\Omega_1, \end{cases}$$

where the nonnegative nonlinearity f_1 is continuous.

Suppose also that

$$Z_1 = \{x \in \Omega_1 : \nabla u_1 = 0\}$$

is a finite set of points.

Then the extension

$$\underline{u}(x) = \begin{cases} u_1(x), & \text{if } x \in \overline{\Omega_1}, \\ 0, & \text{if } x \in \overline{\Omega} \setminus \Omega_1 \end{cases}$$

is a sub-solution of

$$\begin{cases} -\Delta_p u = f(x,u,\nabla u) & \text{in } \Omega, \\ u = 0 & \text{on } \partial\Omega \end{cases}$$

for all continuous nonlinearities $f \geq 0$ *such that* $f_1(x,u_1,\nabla u_1) \leq f(x,u_1,\nabla u_1)$ *in* Ω_1.

Proof. This proposition is a consequence of the Divergence Theorem combined with the Hopf's lemma. (See [3] for details.) □

Remark 8. The hypothesis on Z_1 can be obtained if we suppose, for instance, $0 \leq f(x,t,v)$ and that $\{(x,v) : f(x,t,v) = 0\}$ is a finite set of points, for all $t > 0$. (Of course, the more interesting case occurs when $f(x,0,v) = 0$.)

Proof of the theorem. From Remark 1 follows that $k_1(\Omega_2) \leq k_1(B_\rho)$. So, if f satisfies the hypotheses (H1) and (H2), it also satisfies the hypotheses of Theorem 4. By applying Theorem 4, there exists a positive radial function $u_\rho \in C^{1,\alpha}\left(\overline{B_\rho}\right)$ such that

$$\begin{cases} -\Delta_p u_\rho = \omega_\rho(|x-x_0|)f(u_\rho,|\nabla u_\rho|) & \text{in } B_\rho(x_0), \\ u_\rho = 0 & \text{on } \partial B_\rho(x_0). \end{cases}$$

Moreover, the only critical point of u_ρ occurs at $x = x_0$.

It follows from Lemma 7 that

$$\underline{u}(x) = \begin{cases} u_\rho(x), & \text{if } x \in B_\rho, \\ 0, & \text{if } x \in \Omega \setminus B_\rho \end{cases}$$

is a sub-solution of problem (2).

Define

$$\overline{u} = M\frac{\phi_{\Omega_2}}{\|\phi_{\Omega_2}\|_\infty}.$$

Of course, $\overline{u} \leq M$ and $\|\nabla\overline{u}\|_\infty = M\frac{\|\nabla\phi_{\Omega_2}\|_\infty}{\|\phi_{\Omega_2}\|_\infty} \leq \gamma_\rho M$, by hypothesis. So, it follows from (H1) that $f(\overline{u},|\nabla\overline{u}|) \leq k_1(\Omega_2)M^{p-1}$. Thus,

$$-\Delta_p\overline{u} = -\Delta_p\left(M\frac{\phi_{\Omega_2}}{\|\phi_{\Omega_2}\|_\infty}\right) = k_1(\Omega_2)M^{p-1}\|\omega\|_\infty \geq f(\overline{u},|\nabla\overline{u}|)\omega, \tag{21}$$

and, since $\overline{u} > 0$ on $\partial\Omega$, $\overline{u}$ is a super-solution of (2).

Moreover, the pair $(\underline{u},\overline{u})$ is ordered. In fact, if $x \in \Omega\setminus B_\rho$ the result is immediate. Otherwise we know that,

$$\underline{u} = u_\rho \in C = \left\{u \in C^1\left(\overline{B_\rho}\right) : 0 \leq u \leq M, \text{ and } \|\nabla u\|_\infty \leq \gamma_\rho M\right\},$$

and therefore, by (H1), $f(u_\rho,|\nabla u_\rho|) \leq k_1(\Omega_2)M^{p-1}$ and then

$$-\Delta_p\underline{u} = \omega_\rho f(u_\rho,|\nabla u_\rho|) \leq k_1(\Omega_2)M^{p-1}\|\omega\|_\infty = -\Delta_p\left(M\frac{\phi_{\Omega_2}}{\|\phi_{\Omega_2}\|_\infty}\right) = -\Delta_p\overline{u}.$$

We have

$$u_\rho = 0 \leq M\frac{\phi_{\Omega_2}}{\|\phi_{\Omega_2}\|_\infty} = \overline{u}$$

on ∂B_ρ. We are done, since follows from the comparison principle that $\underline{u} \leq \overline{u}$ in $B_\rho \subset \Omega$. □

5. Applications

In this section we choose two concrete domains Ω_2 for application of Theorem 5. In the first example, we consider a ball $B_R(x_1) = \Omega_2$ so that $\Omega \subset B_R$. In the second, supposing Ω convex, we consider $\Omega_2 = \Omega$ and use a result by Payne and Philippin [9].

5.1. Radial supersolution

For all $x \in \Omega$, let $d(x) = \operatorname{dist}(x, \partial\Omega)$. We denote by $r_* = \sup_{x\in\Omega} d(x)$. Let B_{r_*} be a ball with center at $x_0 \in \Omega$ such that $B_{r_*} \subset \Omega$.

Choose R such that $\Omega \subset B_R$, where B_R is a ball with center at $x_1 \in \Omega$ and let $\phi_R \in C^{1,\alpha}(\overline{B_R(x_1)}) \cap W_0^{1,p}(B_R(x_1))$ be the unique positive solution of

$$\begin{cases} -\Delta_p \phi_R = \|\omega\|_\infty & \text{in } B_R(x_1), \\ \phi_R = 0 & \text{on } \partial B_R(x_1), \end{cases} \tag{22}$$

and consider the positive constant $k_1(\Omega_2) = k_1(B_R) = \|\phi_R\|_\infty^{-(p-1)}$.

We define, as in Theorem 5,

$$\overline{u} := M\frac{\phi_R}{\|\phi_R\|_\infty} \in C^{1,\alpha}\left(\overline{B_R(x_1)}\right) \cap W_0^{1,p}(B_R(x_1)).$$

Of course, $0 < \overline{u} \leq M$. We have

$$\begin{aligned} \phi_R(r) &= \|\omega\|_\infty^{\frac{1}{p-1}} \int_r^R \varphi_q\left(\frac{1}{\theta^{N-1}}\int_0^\theta s^{N-1}ds\right)d\theta \\ &= \frac{p-1}{p}\left(\frac{\|\omega\|_\infty}{N}\right)^{\frac{1}{p-1}}\left(R^{\frac{p}{p-1}} - r^{\frac{p}{p-1}}\right). \end{aligned} \tag{23}$$

On the other hand, we have $\nabla\phi_R(x) = \phi_R'(r)\frac{x-x_0}{r}$, from what follows $|\nabla\phi_R(x)| = |\phi_R'(r)|$. Thus,

$$\|\nabla\phi_R\|_\infty = |\phi_R'(R)| = \left(\int_0^R \left(\frac{s}{R}\right)^{N-1}\|\omega\|_\infty ds\right)^{\frac{1}{p-1}} = \left(\frac{\|\omega\|_\infty}{N}\right)^{\frac{1}{p-1}} R^{\frac{1}{p-1}}$$

and

$$\frac{\|\nabla\phi_R\|_\infty}{\|\phi_R\|_\infty} = \frac{p}{p-1}R^{\frac{1}{p-1}-\frac{p}{p-1}} = \frac{q}{R}. \tag{24}$$

So, we need to choose $\rho > 0$ such that $B_\rho \subset \Omega$ and

$$\frac{q}{R} < \gamma_\rho,$$

in order to have

$$0 \leq |\nabla\overline{u}| = M\frac{|\nabla\phi_R|}{\|\phi_R\|_\infty} \leq M\frac{\|\nabla\phi_R\|_\infty}{\|\phi_R\|_\infty} = \frac{q}{R}M \leq \gamma_\rho M.$$

To choose ρ, let us consider the possibilities

(i) $r_* \leq \frac{R}{q}$ $(< R)$.

We choose $\rho = r_*$, because

$$\frac{\|\nabla \phi_R\|_\infty}{\|\phi_R\|_\infty} = \frac{q}{R} \leq \frac{1}{r_*} = \frac{1}{\rho} \leq \gamma_\rho.$$

(ii) $\frac{R}{q} \leq r_*$ $(< R)$.
We choose $\rho = \frac{R}{q}$, since

$$\frac{\|\nabla \phi_R\|_\infty}{\|\phi_R\|_\infty} = \frac{q}{R} = \frac{1}{\rho} \leq \gamma_\rho.$$

In the special case $\omega_\rho \equiv 1$, we can always choose $\rho = r_*$, since

$$\frac{\|\nabla \phi_R\|_\infty}{\|\phi_R\|_\infty} = \frac{q}{R} \leq \frac{q}{r_*} = \frac{q}{\rho} = \gamma_\rho.$$

This value of ρ corresponds to the smallest values of $k_2(B_\rho)$ and γ_ρ. The best value for $k_1(B_R)$ is obtained when R is the smallest radius such that $B_R(x_1) \supset \Omega$ for $x_1 \in \Omega$.

Example 9. We consider the problem

$$\begin{cases} -\Delta_p u &=& \lambda u(x)^{q-1}(1+|\nabla u(x)|^p) & \text{on } \Omega, \\ u &=& 0 & \text{on } \partial\Omega, \end{cases} \tag{25}$$

where Ω is a smooth, bounded domain in $\mathbb{R}^N$, $1 < q < p$, and λ a positive parameter. Problem (25) is sub-linear at the origin.

To solve problem (25) we consider B_ρ as the largest open ball contained in Ω and B_R such that $\Omega \subseteq B_R$. Since $\omega(x) \equiv 1$ in the case of the nonlinearity $\lambda u(x)^{q-1}(1+|\nabla u(x)|^p)$, the constants in hypotheses (H1) and (H2) are given by

$$k_1 := k_1(B_R) = \|\phi_R\|_\infty^{-(p-1)} = \left(\frac{p-1}{p}\right)^{1-p} NR^{-p}, \tag{26}$$

$$k_2 := k_2(B_\rho) = \begin{cases} \left[\dfrac{p-1}{p}\left(\dfrac{p}{N}\right)^{\frac{N}{N-p}}\right]^{1-p} \dfrac{N}{\rho^p}, & \text{if } N \neq p, \\ \left(\dfrac{p-1}{ep}\right)^{1-p} \dfrac{p}{\rho^p}, & \text{if } N = p, \end{cases} \tag{27}$$

and

$$\gamma = \gamma_\rho = \frac{p}{p-1}\frac{1}{\rho}. \tag{28}$$

From now on, k_1 and k_2 denote the constants (26) and (27), respectively. According to Lemma 2, we have $k_1 < k_2$.

Of course, the nonlinearity $\lambda u(x)^{q-1}(1+|\nabla u(x)|^p)$ satisfies (H3) for any value of λ.

By defining the function $H\colon [0,\infty) \to [0,\infty]$ by $H(M) = M^{q-p}(1+\mu^p M^p)$, we see that condition (H1) is satisfied if $H(M) \leq \frac{k_1}{\lambda}$. It is not difficult to verify that H has a unique critical point M_*, given by

$$\mu^p M_*^p = \frac{p}{q} - 1,$$

where H assumes its minimum value

$$H(M_*) = M_*^{q-p}(1+\mu^p M_*^p) = \frac{1}{\mu^{q-p}}\left(\frac{p}{q}-1\right)^{\frac{q-p}{p}}\left(\frac{p}{q}\right) = \frac{p}{q}M_*^{p-q}. \tag{29}$$

So, taking $M := M_*$ and defining

$$\lambda^* = \frac{k_1}{H(M_*)}, \tag{30}$$

hypothesis (H1) is verified for any $0 < \lambda \leq \lambda^*$. The choice $M = M_*$ makes λ^* to be the best possible value of the parameter such that Theorem 5 guarantees the existence of a positive solution for problem (25).

Now, for any fixed $\lambda \in (0, \lambda^*]$, we try to verify (H2). For this, we consider the function $G\colon (0,\infty) \to [0,\infty)$ given by

$$G(x) = x^{q-p}. \tag{31}$$

We clearly have $G(x) \leq H(x)$ for any $x \in (0,\infty)$ and (H2) is verified if

$$\lambda G(\delta_\lambda) \geq k_2, \tag{32}$$

that is,

$$\delta_\lambda \leq \left(\frac{\lambda}{k_2}\right)^{\frac{1}{p-q}}. \tag{33}$$

So, for any $\lambda \in (0, \lambda^*]$, (H2) is satisfied if we take $\delta_\lambda > 0$ verifying the above inequality. Observe that the same value of δ_λ is valid for any $\tilde{\lambda} \in [\lambda, \lambda^*]$.

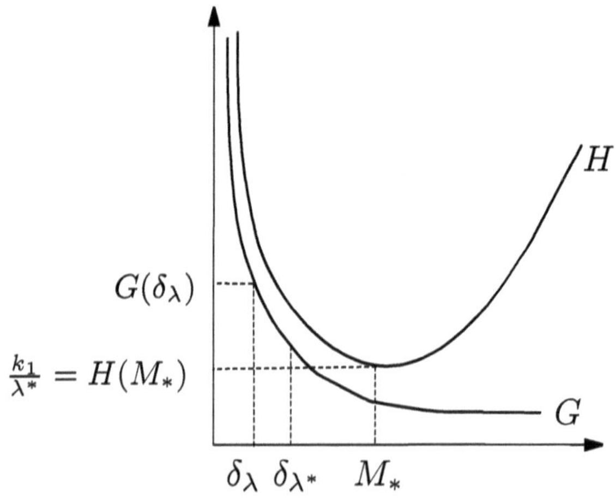

FIGURE 2. The graphs of H and G.

Since $0 < \lambda \leq \lambda^*$, the largest value of δ_λ is attained at λ^*. So, by (29), the condition $\delta_\lambda < M_*$ always holds:

$$\delta_\lambda \leq \left(\frac{\lambda^*}{k_2}\right)^{\frac{1}{p-q}} \leq \left(\frac{\lambda^*}{k_1}\right)^{\frac{1}{p-q}} = \left(\frac{1}{H(M_*)}\right)^{\frac{1}{p-q}} = M_* \left(\frac{q}{p}\right)^{\frac{1}{p-q}} < M_*.$$

5.2. Applying a maximum principle of Payne and Phillipin

If we choose $\Omega_2 = \Omega$, we need to suppose that Ω is convex to control the quotient (19).

To handle this case, we consider the torsional creep problem

$$\begin{cases} -\Delta_p \psi_\Omega = 1 \text{ in } \Omega, \\ \psi_\Omega = 0 \text{ on } \partial\Omega. \end{cases} \tag{34}$$

In order to estimate the quotient (19), we state a maximum principle of Payne and Philippin [9], which was proved for non-degenerate operators.

Theorem 10 (Payne–Philippin). *Let $\Omega \subset \mathbb{R}^N$ be a convex domain such that $\partial\Omega$ is a $C^{2,\alpha}$ surface. If $u = \text{const.}$ on $\partial\Omega$, then*

$$\Phi(x) = 2\frac{p-1}{p}|\nabla\psi_\Omega|^p + 2\psi_\Omega \tag{35}$$

takes its maximum value at a critical point of ψ_Ω.

Regularization methods and the results of Lieberman [8] permit us to apply it to the p-Laplacian:[2]

Lemma 11. *If the bounded, smooth domain Ω is convex, then*

$$\|\nabla\psi_\Omega\|_\infty \leq (q\|\psi_\Omega\|_\infty)^{\frac{1}{p}},$$

what yields

$$\frac{\|\nabla\psi_\Omega\|_\infty}{\|\psi_\Omega\|_\infty} \leq \frac{q^{\frac{1}{p}}}{\|\psi_\Omega\|_\infty^{\frac{1}{q}}}.$$

An immediate consequence of Lemma 11 is an estimate of the quotient (19) in the case Ω convex by taking $\Omega = \Omega_2$: we have

$$\frac{\|\nabla\phi_\Omega\|_\infty}{\|\phi_\Omega\|_\infty} \leq \frac{(q\|\omega\|_\infty)^{\frac{1}{p}}}{\|\phi_\Omega\|_\infty^{\frac{1}{q}}}. \tag{36}$$

We observe that the quotient (19) was controlled for any convex domain $\Omega_2 \supset \Omega$. So, for instance, we can take $\Omega_2 = co(\Omega)$, the convex hull of Ω.

As in the Subsection 5.1, let B_{r_*} be a ball with larger radius such that $B_{r_*} \subset \Omega$. We consider the solution ϕ_* of the problem

$$\begin{cases} -\Delta_p \phi_* = \|\omega\|_\infty \text{ in } B_{r_*}, \\ \phi_* = 0 \qquad \text{on } \partial B_{r_*}. \end{cases}$$

[2] See [3] for details.

Since $B_{r_*} \subset \Omega$, from the comparison principle follows that $\|\phi_*\|_\infty \le \|\phi_\Omega\|_\infty$. But

$$\|\phi_*\|_\infty = \left(\frac{\|\omega\|_\infty}{N}\right)^{\frac{1}{p-1}} \int_0^{r_*} \theta^{\frac{1}{p-1}}\,d\theta = \left(\frac{\|\omega\|_\infty}{N}\right)^{\frac{q}{p}} \frac{r_*^q}{q},$$

thus yielding

$$\frac{\|\nabla\phi_\Omega\|_\infty}{\|\phi_\Omega\|_\infty} \le \frac{(q\|\omega\|_\infty)^{\frac{1}{p}}}{\|\phi_\Omega\|_\infty^{\frac{1}{q}}} \le \frac{q^{\frac{1}{p}+\frac{1}{q}}}{r_*}\|\omega\|_\infty^{\frac{1}{p}} \left(\frac{N}{\|\omega\|_\infty}\right)^{\frac{1}{p}} = \sqrt[p]{N}\frac{q}{r_*}.$$

We now choose ρ given by

$$\rho = \frac{r_*}{q\sqrt[p]{N}} = \frac{p-1}{p\sqrt[p]{N}} r_* \; (< r_*).$$

Then, we have

$$\frac{\|\nabla\phi_\Omega\|_\infty}{\|\phi_\Omega\|_\infty} \le \frac{1}{\rho} \le \gamma_\rho.$$

In the special case $\omega \equiv 1$, we can take ρ such that

$$\frac{q}{\rho} = \frac{q\sqrt[p]{N}}{r_*},$$

since $\gamma_\rho = q/\rho$. It follows

$$\rho = \frac{r_*}{\sqrt[p]{N}} < r_* \quad \text{and} \quad \frac{\|\nabla\phi_\Omega\|_\infty}{\|\phi_\Omega\|_\infty} \le \frac{q}{\rho} = \gamma_\rho.$$

6. Fast growing gradient

In this section we consider the existence of positive solutions for the following problem in two positive parameters in the bounded, smooth domain $\Omega \subset \mathbb{R}^N$, $N \ge 2$:

$$\begin{cases} -\Delta_p u = \lambda h(x,u) + \beta f(x,u,\nabla u) & \text{in } \Omega \\ u = 0 & \text{on } \partial\Omega, \end{cases} \tag{37}$$

where $\Delta_p u := \operatorname{div}(|\nabla u|^{p-2}\nabla u)$ is the p-Laplacian operator, $p > 1$, and h, f are continuous nonlinearities satisfying

(H4) $0 \le \omega_1(x)u^{q-1} \le h(x,u) \le \omega_2(x)u^{q-1}$, $1 < q < p$;

(H5) $0 \le f(x,u,v) \le \omega_3(x)u^a|v|^b$, $a, b > 0$,

and $\omega_i \colon \overline{\Omega} \to [0,\infty)$, $1 \le i \le 3$, are positive continuous *weights*.

The combined effects of the sublinear and superlinear terms make possible the definition of a fixed point operator for each (λ,β) in a region $\mathcal{D}$ of the $\lambda\beta$-plane and use a global estimate for the solution of the Poisson equation $-\Delta_p u = g$ with homogeneous Dirichlet boundary conditions on Ω to obtain an invariant subset by this operator. Hence, by applying Schauder's fixed point theorem we prove the existence of at least one positive solution for the Dirichlet problem above if $(\lambda,\beta) \in \mathcal{D}$. Details of this result can be found in [5].

We now state a consequence of the global regularity results by Lieberman (see [8]).

Lemma 12. *Let Ω be a bounded, smooth domain of $\mathbb{R}^N$ and $g \in L^\infty(\Omega)$. Assume that $u \in W_0^{1,p}(\Omega)$ is a weak solution of*

$$\begin{cases} -\Delta_p u = g \text{ in } \Omega, \\ \quad u = 0 \text{ on } \partial\Omega. \end{cases} \tag{38}$$

Then there exists a positive constant $\mathcal{K}$, depending only on p, N and Ω, such that

$$\|\nabla u\|_\infty \le \mathcal{K}(\|g\|_\infty)^{\frac{1}{p-1}}. \tag{39}$$

To solve problem (37) we define

$$r := a + b + 1, \quad \omega(x) := \max_{i \in \{1,2,3\}} \omega_i(x)$$

and denote by λ_1 and u_1 the first eigenpair of the p-Laplacian with weight ω_1, that is,

$$\begin{cases} -\Delta_p u_1 = \lambda_1 \omega_1 u_1^{p-1} & \text{in } \Omega, \\ \quad u_1 = 0 & \text{on } \partial\Omega, \end{cases}$$

with u_1 positive satisfying $\|u_1\|_\infty = 1$.

Let also $\phi \in W_0^{1,p}(\Omega) \cap C^{1,\alpha}(\overline{\Omega})$ be the solution of the problem

$$\begin{cases} -\Delta_p \phi = \omega \text{ in } \Omega \\ \quad \phi = 0 \text{ on } \partial\Omega \end{cases}$$

and define

$$\gamma := (\mathcal{K}\, \|\omega\|_\infty^{\frac{1}{p-1}}) / \|\phi\|_\infty,$$

where $\mathcal{K}$ satisfies (39). We stress that γ depends only on ω, p, N and Ω.

Lemma 13. *There exists a region $\mathcal{D}$ in the $\lambda\beta$-plane such that, if $(\lambda, \beta) \in \mathcal{D}$ then*

$$\lambda M^{q-1} + \beta\gamma^b M^{a+b} \le (M/\|\phi\|_\infty)^{p-1}, \tag{40}$$

for some positive constant M.

Proof. The inequality (40) can be written as

$$\Phi(M) := \lambda A M^{q-p} + \beta B M^{r-p} \le 1,$$

where the coefficients $A = \|\phi\|_\infty^{p-1}$ and $B := \mathcal{K}^b \|\phi\|_\infty^{p-1-b} \|\omega\|_\infty^{\frac{b}{p-1}}$ clearly depend only on ω, p and Ω.

In order to determine an adequate value for M, we consider the possibilities for the sign of $r - p$.

In the case $r - p > 0$, Φ has an unique critical point and (40) is satisfied if

$$\lambda^{r-p}\beta^{p-q} \le \left(\frac{r-p}{A}\right)^{r-p} \left(\frac{p-q}{B}\right)^{p-q} \frac{1}{(r-q)^{r-q}} =: K. \tag{41}$$

Thus, if the positive parameters λ and β satisfy (41), we conclude that $\overline{u} := (M/\|\phi\|_\infty)\phi$ is a super-solution for (46), where M is the minimum value of Φ.

In the case $r-p=0$, to have $\Phi(M)\le 1$ for some $M>0$ it is necessary that $\beta B<1$, that is

$$\text{if } \lambda>0 \text{ and } \beta<B^{-1} \tag{42}$$

we can take $M>0$ such that $\Phi(M)=1$. Thus, if λ and β satisfy (42) then $\overline{u}=(M/\|\phi\|_\infty)\phi$, where M is the solution of $\Phi(M)=1$.

If $r-p<0$, for any positive parameters λ and β, there always exists $M>0$ such that $\Phi(M)=\lambda AM^{q-p}+\beta BM^{r-p}=1$ and for such an M the function $\overline{u}=(M/\|\phi\|_\infty)\phi$ is a super-solution of (46).

Summarizing, there exists a positive constant M satisfying (40) whenever the pair (λ,β) belongs to the set $\mathcal{D}$ defined by:

$$\mathcal{D}=\begin{cases}\{\lambda,\beta>0:\lambda^{r-p}\beta^{p-q}\le K\} & \text{if } \; r-p>0,\\ \{\lambda,\beta>0:\beta<B^{-1}\} & \text{if } \; r-p=0,\\ \{\lambda,\beta>0\} & \text{if } \; r-p<0,\end{cases} \tag{43}$$

where K is given by (41). □

For each $u\in C^1(\overline{\Omega})$ we define the continuous nonlinearity $F^u:\overline{\Omega}\times\mathbb{R}\to\mathbb{R}$ by

$$F^u(x,\xi):=\lambda\omega_1\xi^{q-1}+\lambda\left(h(x,u(x))-\omega_1 u(x)^{q-1}\right)+\beta f(x,u(x),\nabla u(x)) \tag{44}$$

and observe that $F^u(x,u)=\lambda h(x,u)+\beta f(x,u,\nabla u)$.

Our main result of existence of solution for problem (37) is given by

Theorem 14. *Assume that h and f are continuous and satisfy* (H1) *and* (H2). *There exists a region $\mathcal{D}$ in the $\lambda\beta$-plane such that if $(\lambda,\beta)\in\mathcal{D}$ the Dirichlet problem* (37) *has at least one positive solution u satisfying, for some positive constants ϵ and M:*

$$\epsilon u_1\le u\le(M/\|\phi\|_\infty)\phi \quad \textit{and} \quad \|\nabla u\|_\infty\le\gamma M.$$

Proof. Let $(\lambda,\beta)\in\mathcal{D}$ where the region $\mathcal{D}$ is defined by (43) and take $M>0$ satisfying (40) from Lemma 13. Let us define the subset

$$\mathcal{F}:=\left\{u\in C^1(\overline{\Omega}):\epsilon u_1\le u\le(M/\|\phi\|_\infty)\phi \text{ and } \|\nabla u\|_\infty\le\gamma M\right\}\subset C^1(\overline{\Omega}) \tag{45}$$

where $0<\epsilon\le\min\left\{(\lambda/\lambda_1)^{\frac{1}{p-q}},(M\lambda_1^{-\frac{1}{p-1}})/\|\phi\|_\infty\right\}$.

It is not difficult to see that, for each $u\in\mathcal{F}$, there exists a unique positive solution U of the problem

$$\begin{cases}-\Delta_p U=F^u(x,U) & \text{in } \Omega\\ U=0 & \text{on } \partial\Omega\end{cases} \tag{46}$$

satisfying $\epsilon u_1\le u\le(M/\|\phi\|_\infty)\phi$. Uniqueness follows directly from [6]. Existence follows from the fact that the functions $\underline{u}:=\epsilon u_1$ and $\overline{u}:=(M/\|\phi\|_\infty)\phi$ constitute an ordered pair of sub- and super-solutions of (46). This fact implies, by applying a standard iteration process, that there exists a weak solution U of (46) satisfying $\underline{u}\le U\le\overline{u}$.

By proving that $|\nabla U|\le\gamma M$, we see that $U\in\mathcal{F}$.

The regularity $U \in C^{1,\alpha}(\overline{\Omega})$ for some $0 < \alpha < 1$ uniform with respect to $u \in \mathcal{F}$ follows from the uniform boundedness of both U and $|\nabla U|$. We emphasize that the bounds for U and $|\nabla U|$ are determined by the positive constant M which, in its turn, is fixed according with the pair $(\lambda, \beta) \in \mathcal{D}$.

So, it follows that the operator

$$\begin{array}{rcl} T\colon \mathcal{F} \subset C^1(\overline{\Omega}) & \longrightarrow & C^{1,\alpha}(\overline{\Omega}) \cap W_0^{1,p}(\Omega) \subset C^1(\overline{\Omega}) \\ u & \longrightarrow & U, \end{array}$$

is well defined, U being the unique positive solution of (46). Moreover, the compactness of the immersion $C^{1,\alpha}(\overline{\Omega}) \hookrightarrow C^1(\overline{\Omega})$ implies that T is continuous and compact. Thus, since T leaves invariant the set $\mathcal{F}$ defined by (45) and this set is bounded and convex we can apply Schauder's fixed point theorem to obtain a fixed point u for T. Of course, such a fixed point u satisfies (37) since $-\Delta_p u = F^u(x,u) = \lambda h(x,u) + \beta f(x,u,\nabla u)$ in Ω. □

References

[1] A. Ambrosetti, H. Brezis and G. Cerami, *Combined effects of concave and convex nonlinearities in some elliptic problems*, J. Functional Analysis **122** (1994), 519–543.

[2] L. Boccardo, F. Murat and J.-P. Puel, *Résultats d'existence pour certains problèmes elliptiques quasilinéaires.* Ann. Scuola Norm. Sup. Pisa Cl. Sci. (4) **11** (1984), 213–235.

[3] H. Bueno, G. Ercole, W. Ferreira and A. Zumpano, *Existence and multiplicity of positive solutions for the p-Laplacian with nonlocal coefficient*, J. Math. Anal. Appl. **343** (2008), 151–158.

[4] H. Bueno, G. Ercole and A. Zumpano, *Positive solutions for the p-Laplacian and bounds for its first eigenvalue*, Advanced Nonlinear Studies **9** (2009), 313–338.

[5] H. Bueno and G. Ercole, *A quasilinear problem with fast growing gradient*, Appl. Math. Lett. **26** (2013), 520–523.

[6] J.I. Díaz and J.E. Saa, *Existence et unicité de solutions positives pour certaines équations elliptiques quasilinéaires* C. R. Acad. Sci. Paris **305** Série I (1987) 521–524.

[7] G. Ercole and A. Zumpano, *Existence of positive radial solutions for the n-dimensional p-Laplacian*, Nonlinear Analysis TMA **44** (2001), 355–360.

[8] G.M. Lieberman, *Boundary regularity for solutions of degenerate elliptic equations*, Nonlinear Analysis TMA **12** (1988), 1203–1219.

[9] L.E. Payne and G.A. Philippin, *Some maximum principles for nonlinear elliptic equations in divergence form with applications to capillarity surfaces and to surfaces of constant mean curvature*, Nonlinear Analysis TMA **3**, no. 2 (1979), 193–211.

H. Bueno and G. Ercole
Depto. de Matemática
Universidade Federal de Minas Gerais
Belo Horizonte, 30.123-970, Brazil
e-mail: hamilton@mat.ufmg.br
grey@mat.ufmg.br

Progress in Nonlinear Differential Equations
and Their Applications, Vol. 85, 163–174

Some Weighted Inequalities of Trudinger–Moser Type

Marta Calanchi

Dedicated to Bernhard Ruf and Daniela Lupo with respect, friendship and warmth

Abstract. We discuss some extensions of the Trudinger–Moser inequality in a special case of weighted Sobolev spaces

Mathematics Subject Classification (2010). Primary 46E35; Secondary 35J60, 35J20.

Keywords. Trudinger–Moser type inequalities, critical exponents.

1. Introduction: the state of the art

The classical Trudinger–Moser inequality concerns the limiting case $p = N$ of the well-known Sobolev embeddings

$$W^{1,p}(\Omega) \hookrightarrow L^{p^*}(\Omega) \quad p^* = \frac{pN}{N-p}$$

where Ω is a bounded smooth domain in $\mathbb{R}^N$.

Here we denote by $W^{1,p}(\Omega)$ the usual Sobolev space of L^p-functions with derivatives in L^p endowed by the norm $||u||_{1,p} := \left(\int_\Omega |u|^p dx + \int_\Omega |\nabla u|^p dx\right)^{1/p}$ and by $W_0^{1,p}(\Omega)$ the completion of $C_0^\infty(\Omega)$ in the norm $||u||_{1,p}$.

If $p = N$, p^* becomes infinity, the Sobolev space $W_0^{1,N}(\Omega)$ embeds into any $L^q(\Omega)$, but $W_0^{1,N}(\Omega) \not\subseteq L^\infty(\Omega)$ as the following simple example shows

Example 1. $N = 2$, $\Omega = B_1(0)$. Then $u(x) = \log(1 - \log|x|)$ belongs to $H_0^1(\Omega) = W_0^{1,2}(\Omega)$ but u is not bounded.

The question is to find the maximal growth function $\phi : \mathbb{R} \to \mathbb{R}^+$ such that

$$\text{if } u \in W_0^{1,N}(\Omega), \text{ then } \int_\Omega \phi(u)dx \quad \text{is finite.}$$

Roughly speaking one can express this fact in terms of an embedding by introducing the Orlicz space (which is a generalization of L^p): find a continuous convex function ϕ such that if

$$L_\phi := \text{vector space generated by } \{u : \int_\Omega \phi(u) < +\infty\}$$

then

$$W_0^{1,N}(\Omega) \hookrightarrow L_\phi.$$

S.I. Pohozaev (1965, [12]) and N. Trudinger (1967, [14]) showed independently that this maximal growth is given by

$$\phi(t) = e^{|t|^{\frac{N}{N-1}}} - 1.$$

This growth is optimal in the sense that for any higher growth the integral may become infinite. The proof is based upon an expansion in power series of the exponential function, and on a control of the L^p norm of each term of the series. This result was improved in 1970 by Moser ([11]), who showed that the supremum on the unitary ball in $W_0^{1,N}(\Omega)$

$$\sup_{\int_\Omega |\nabla u|^N dx \leq 1} \int_\Omega \mathrm{e}^{\alpha u^{\frac{N}{N-1}}} dx$$

is bounded if and only if $\alpha \leq \alpha_N = N\omega^{\frac{1}{N-1}}$, where ω_{N-1} is the $N-1$-dimensional surface of the unit sphere. The integral on the left actually is finite for any positive α, but if $\alpha > \alpha_N$ it can be made arbitrarily large by a suitable choice of u.

From now on we will consider the two-dimensional case. Let us recall Moser's result in this case

Theorem 2 (Moser 1970 [11], $N = 2$).

$$(TM) \qquad \sup_{\int_\Omega |\nabla u|^2 dx \leq 1} \int_\Omega e^{\alpha u^2} \leq \begin{cases} C|\Omega| & \textit{if} \quad \alpha \leq 4\pi \\ +\infty & \textit{if} \quad \alpha > 4\pi. \end{cases} \tag{1}$$

The proof relies in an essential way on symmetrization. Indeed, one can substitute u by a radial function u^* on the ball $B_R(0)$ whose sub-levels are balls with the same measure of the corresponding sub-levels of $|u|$. One has the following properties: the "mass" doesn't change, i.e., for all continuous functions G

$$\int_{B_R(0)} G(u^*)dx = \int_\Omega G(u)dx$$

and for the gradient there is the Pólya–Szegö inequality

$$\int_{B_R(0)} |\nabla u^*|^2 dx \leq \int_\Omega |\nabla u|^2 dx$$

which implies

$$\sup_{\int_\Omega |\nabla u|^2 dx \leq 1} \int_\Omega e^{\alpha |u|^2} \leq \sup_{\int_{B_R} |\nabla u^*|^2 dx \leq 1} \int_{B_R} e^{\alpha |u^*|^2},$$

and hence it is sufficient to consider the radial case.

Let u be a radial function. By the change of variable

$$|x| = Re^{-t/2} \qquad u(x) = \sqrt{4\pi}w(t)$$

one has

$$\int_{B_R} |\nabla u^*|^2 dx = \int_0^{+\infty} |w'(t)|^2 dt$$

and

$$\int_{B_R} e^{\alpha|u^*|^2} dx = 2\pi R^2 \int_0^{+\infty} e^{\frac{\alpha}{4\pi}|w(t)|^2 - t} dt.$$

So the problem reduces to

$$\sup_{\int_0^{+\infty} |w'|^2 dt \le 1} \int_0^{+\infty} e^{\frac{\alpha}{4\pi}|w(t)|^2 - t} dt < +\infty.$$

For every C^1-functions w such that $\int_0^{+\infty} |w'|^2 dt \le 1$ and $w(0) = 0$ one has that w is controlled by $t^{1/2}$; indeed

$$|w(t)| = \left| \int_0^t w'(t) dt \right| \le t^{1/2} \left(\int_0^t |w'(t)|^2 \right)^{1/2} \le t^{1/2}.$$

Therefore if $\alpha < 4\pi$ it is easy to prove that the supremum is finite and if $\alpha > 4\pi$ it is sufficient to test the functional on the Moser sequence

$$w_k(t) = \begin{cases} \dfrac{t}{k^{\frac{1}{2}}} & t \le k \\ k^{\frac{1}{2}} & t \ge k \end{cases}$$

to obtain that the supremum goes to infinity. The critical case $\alpha = 4\pi$ is more complicated and we will return to it later in a more general case.

2. Weighted TM inequalities

We point out that, beginning from the celebrated work of Caffarelli–Kohn–Nirenberg [3], the weighted Poincaré–Sobolev inequalities and fundamental questions concerning these inequalities (such as best embedding constants, existence/non-existence, symmetry properties of extremal functions) have attracted a lot of attention in the literature. Only recently limiting embeddings with weights have been considered. We mention for instance [9], [7], [6], [8].These papers threat embeddings of Sobolev spaces in weighted Orlicz spaces: they consider the weight only acting on the functional and they are principally interested in characterizing weights which do not change an exponential Orlicz space up to equivalence of norms.

On the other hand, we consider a weighted version of Moser' s theorem, for which the presence of the weight can change the range of the exponent for which the supremum is still finite. More precisely: let $w = w(x)$ and $v(x)$ two radial weights on the unitary ball (let's say two nonnegative L^1-functions).

Let $\Omega = B$ the unitary ball in $\mathbb{R}^2$ centered in the origin. We denote with $H_0^1(B,w)$ the Sobolev space given by the completion of $C_0^\infty(B)$ with respect to the weighted norm

$$||u||_w := \left(\int_B |\nabla u|^2 \; w(x) \; dx\right)^{1/2} \tag{2}$$

and with $H^1_{0,\text{rad}}(B,w)$ the corresponding subspace of radial functions.

Consider the following problem: find the best exponential maximal growth $F(t)$ for which

$$\sup_{||u||_w \leq 1} \int_B v(x) \; F(u) dx \text{ is finite.}$$

Quite recently M. Calanchi and E. Terraneo [4] investigated this problem in the case $w \equiv 1$ and $v(x) = |x|^\delta$ with $\delta > 0$ (Hénon weight). They concentrated the attention on the following functional $F_\gamma : H_0^1(B) \to \mathbb{R}$

$$F_\gamma(u) = \int_B |x|^\delta \; \left(\mathrm{e}^{p|u|^\gamma} - 1 - p|u|^\gamma\right) \; dx \tag{3}$$

where $\delta > 0$, $p > 0$ and $1 < \gamma \leq 2$. They were first interested in understanding for which values of γ, p and δ the supremum of $F(u)$ on the set $\{u \in H_0^1 : \; \|u\|_{H_0^1} \leq 1\}$ is finite and attained, and secondly their purpose was essentially prove a symmetry breaking result.

They proved that for $0 < \gamma < 2$, or $\gamma = 2$ and $0 < p \leq 4\pi + 2\pi\delta$ the supremum over the subspace of *radial functions* in H_0^1 is finite.

Then they analysed the case $\gamma = 2$ and $p > 4\pi$ (*supercritical*), and they showed that the supremum over the whole space H_0^1 is not finite. It is enough to evaluate the functional F on a suitable family of Moser type functions that concentrate on the boundary. In this way the effect of the weight $|x|^\delta$ becomes negligible. In fact they established the following result:

Theorem 3 (M. Calanchi–E. Terraneo, [4]). *Let*

$$S^R_{\delta,p} = \sup_{u \; radial, ||\nabla u||_2 \leq 1} \int_B |x|^\delta \; \left(\mathrm{e}^{p|u|^2} - 1 - p|u|^2\right) \; dx$$

the supremum taken on the subspace of radial functions and

$$S_{\delta,p} = \sup_{||\nabla u||_2 \leq 1} \int_B |x|^\delta \; \left(\mathrm{e}^{p|u|^2} - 1 - p|u|^2\right) \; dx$$

the supremum taken on the whole space. Then

i) $S^R_{\delta,p} < +\infty \iff p \leq 4\pi + 2\pi\delta$ *and* ii) $S_{\delta,p} < +\infty \iff p \leq 4\pi$

Moreover if $p < 4\pi$ *there exists* $\delta_0 > 0$ *such that*

$$S^R_{\delta,p} < S_{\delta,p} \quad \forall \delta \geq \delta_0.$$

This second result says that if the problem is subcritical, the so-called *ground state solution* (the solution that maximizes the functional) is not radial. This is an extension to the case of exponential growth of a result due to Smets, Su and Willem ([13]) related to the Hénon equation. In order to prove this they give the asymptotic behavior of the radial level $S^R_{\delta,p}$ as $\delta \to +\infty$, and test the level on the whole space along functions which concentrate near the boundary.

To prove part i) the transformation

$$u(|x|^{\frac{2}{\delta+2}}) = \sqrt{\frac{2}{\delta+2}}\, w(|x|) \tag{4}$$

is considered. By an easy computation one has

$$\int_B |x|^\delta \left(\mathrm{e}^{p|u|^2} - 1 - p|u|^2\right)\, dx = \frac{2}{\delta+2}\int_B \left(\mathrm{e}^{\frac{2p}{\delta+2}|w|^\gamma} - 1 - \frac{2p}{\delta+2}|w|^\gamma\right) dx;$$

and

$$\int_B |\nabla u(x)|^2\, dx = \int_B |\nabla w(x)|^2\, dx \leq 1.$$

So one obtains

$$S^R_{\delta,p} = \sup_{\|w\|\leq 1,\ w\ rad} \frac{2}{\delta+2}\int_B \left(\mathrm{e}^{\frac{2p}{\delta+2}|w|^2} - 1 - \frac{2p}{\delta+2}|w|^2\right) dx$$

and one can conclude using the standard Trudinger–Moser inequality.

A similar argument is used by Adimurthy and Sandeep (see [2]) in order to check the critical exponent in the singular case (Hardy weight). Their starting point is to establish some interpolation inequalities between the Hardy inequality and the Sobolev inequality in the limit case. Even if the authors consider the problem in general dimension, for simplicity we describe here only the two-dimensional case. They consider an embedding of the form

$$u \to \int_\Omega \frac{e^{\alpha u^2}}{|x|^\delta (\log(e/|x|))^\gamma}\, dx$$

and the sharpness of constants α, δ, γ for which the supremum (on the unitary ball) of the functional on the right is finite. They first observe that $\gamma = 0$ is the optimal choice and prove the following

Theorem 4 (Adimurthi–K. Sandeep, 2007, [2]). *Let $u \in H^1_0(\Omega)$. Then for every $\alpha > 0$ and $\delta \in [0,2)$*

$$\int_\Omega \frac{e^{\alpha u^2}}{|x|^\delta} dx < +\infty. \tag{5}$$

Moreover

$$\sup_{||\nabla u||_2 \leq 1} \int_\Omega \frac{e^{\alpha u^2}}{|x|^\delta} dx < +\infty \iff \alpha \leq 4\pi - 2\pi\delta. \tag{6}$$

Since in this case the radial weight is decreasing, they can reduce the problem to a radial one, by standard Schwartz symmetrization. Moreover by the same transformation as in (4) they can use Moser's result.

3. Trudinger–Moser inequality with weighted Sobolev spaces

In this section we deal with some results concerning Trudinger–Moser type inequalities with logarithmic weights.

Let $w(x) = w_1(x) = \left(\log\left(\frac{e}{|x|}\right)\right)^\beta$, or $w(x) = w_0(x) = \left(\log\left(\frac{1}{|x|}\right)\right)^\beta$, $v(x) = |x|^\delta$, $\delta > -2$ and the functional $G_\delta : H_0^1(B) \to \mathbb{R}$

$$G_\delta(u) = \int_B |x|^\delta \, e^{\alpha|u|^\gamma} \, dx.$$

Consider the following problem: find γ, δ, α such that

$$\sup_{||u||_w \le 1} \int_B |x|^\delta \, e^{\alpha|u|^\gamma} \, dx < +\infty$$

(we recall that $||u||_w = \left(\int_B |\nabla u|^2 w(x) dx\right)^{1/2}$).

In a previous paper we considered the case $v(x) \equiv 1$ (i.e., $\delta \equiv 0$), and proved

Theorem 5 (M. Calanchi–B. Ruf [5]). *Let* $\beta \in [0, 1)$ *and* $w_0(x) = \left(\log \frac{1}{|x|}\right)^\beta$ *or* $w_1(x) = \left(\log \frac{e}{|x|}\right)^\beta$. *Then*

$$\int_{B_1(0)} e^{|u|^\gamma} dx < +\infty, \quad \textit{for all} \quad u \in H^1_{0,\mathrm{rad}}(B_1, w) \ , \quad \textit{iff} \quad \gamma \le \overline{\gamma} := \frac{2}{1-\beta} \, ,$$

and

$$\sup_{\|u\|_w \le 1, \mathrm{rad}} \int_{B_1(0)} e^{\alpha|u|^{\frac{2}{1-\beta}}} dx < +\infty$$

if and only if

$$\alpha \le \alpha_\beta = 2\,[2\pi(1-\beta)]^{\frac{1}{1-\beta}} \quad \textit{(critical growth)}.$$

Here, considering a Hardy or Hénon weight in the functional, we prove the following

Theorem 6. *Let* $\delta > -2$, *and* w *as in the previous theorem. Then*

$$\text{(i)} \int_{B_1(0)} |x|^\delta e^{|u|^\gamma} dx < +\infty, \quad \textit{for all} \quad u \in H^1_{0,rad}(B_1, w) \ , \quad \textit{iff} \quad \gamma \le \overline{\gamma} := \frac{2}{1-\beta} \, ,$$

and

$$\text{(ii)} \quad \sup_{||u||_w \le 1, rad} \int_{B(0,1)} |x|^\delta \, e^{\alpha|u|^{\frac{2}{1-\beta}}} \, dx < +\infty$$

if and only if

$$\alpha \le \left(1 + \frac{\delta}{2}\right) \alpha_\beta = \left(1 + \frac{\delta}{2}\right) 2\,[2\pi(1-\beta)]^{\frac{1}{1-\beta}} \quad \textit{(critical growth)}.$$

Remark 7. For $\beta = 0$ one has the same exponent as in [4] for the Hénon case and as in [2] for the Hardy case.

We first prove this preliminary result (Radial Lemma).

Lemma 8 ([5]). *Let u a radially symmetric C_0^1 function on $B = B_1(0)$. Then one has for $w = w_0(x) = |\log|x||^\beta$*

$$\text{(i)}\quad |u(x)| \le \frac{|\log|x||^{\frac{1-\beta}{2}}}{\sqrt{2\pi(1-\beta)}}\,||u||_w, \quad \forall \beta < 1,$$

while for $w = w_1(x) = |\log(e/|x|)|^\beta$

$$\text{(ii)}\quad |u(x)| \le \frac{\left|[\log(e/|x|)]^{1-\beta} - 1\right|^{1/2}}{\sqrt{2\pi|1-\beta|}}\,||u||_w, \quad \beta \ne 1.$$

Proof. Since u is radial, let $v(|x|) = u(x)$. Then we have (for $w(x) = |\log(e/|x|)|^\beta$, $\beta \ne 1$)

$$||u||_w = \left(2\pi \int_0^1 |v'(t)|^2\, t\,|\log e/t|^\beta dt\right)^{1/2}.$$

Moreover

$$\begin{aligned} |u(x)| &= |v(|x|) - v(1)| = \left|\int_1^{|x|} v'(t)dt\right| \\ &\le \int_{|x|}^1 |v'(t)|t^{1/2}\,|\log e/t|^{\beta/2}\,t^{-1/2}\,|\log e/t|^{-\beta/2}dt \\ &\le \left(\int_{|x|}^1 |v'(t)|^2\, t\,|\log e/t|^\beta dt\right)^{1/2}\left(\int_{|x|}^1 \frac{1}{t|\log e/t|^\beta}dt\right)^{1/2} \\ &\le \frac{\left|[\log(e/|x|)]^{1-\beta} - 1\right|^{1/2}}{\sqrt{2\pi|1-\beta|}}\,||u||_w, \quad \beta \ne 1. \end{aligned}$$

For $w(x) = |\log|x||$ the procedure is similar. □

Proof of Theorem 6. We first consider the case $w = w_0(x) = |\log|x||^\beta$. We may assume that $u \ge 0$ (one can replace u by $|u|$). Since the problem is radially symmetric we introduce the variable t by

$$|x| = e^{-t/2},$$

and set

$$\psi(t) = 2^{\frac{1-\beta}{2}}\,[2\pi(1-\beta)]^{1/2}\,u(x). \tag{7}$$

Then

$$\int_0^{+\infty} \frac{\dot\psi^2 t^\beta}{1-\beta}\,dt = \left(\int_{B_1(0)} |\nabla u|^2 |\log|x||^\beta\,dx\right). \tag{8}$$

It is sufficient to estimate

$$\int_0^{+\infty} e^{\bar\alpha\psi^\gamma - (1+\frac{\delta}{2})t}\,dt = \frac{1}{m(B)}\int_B |x|^\delta e^{\alpha u^\gamma}\,dx, \tag{9}$$

where

$$\bar{\alpha} = \frac{\alpha}{2\left[2\pi(1-\beta)\right]^{\frac{1}{1-\beta}}}.$$

We first prove that the condition $\gamma \leq \overline{\gamma} = \frac{2}{1-\beta}$ is necessary. Let $\gamma = \overline{\gamma} + \epsilon$, with $\epsilon > 0$. It is sufficient to test the first integral in equation (9) on the following function

$$\psi_\eta(t) = \begin{cases} t^{\frac{1}{\overline{\gamma}}-\eta} = t^{\frac{1-\beta}{2}-\eta} & t \geq 1 \\ t & 0 \leq t \leq 1 \end{cases}$$

where $\eta > 0$ can be chosen such that

$$(\overline{\gamma}+\epsilon)\left(\frac{1}{\overline{\gamma}} - \eta\right) = \left(\frac{2}{1-\beta} + \epsilon\right)\left(\frac{1-\beta}{2} - \eta\right) =: 1 + \bar{\eta} > 1.$$

It is not difficult to prove that

$$\int_0^{+\infty} \frac{\dot{\psi}_\eta^2 t^\beta}{1-\beta}\, dt < C \text{ and } \int_0^{+\infty} e^{\bar{\alpha}\psi_\eta^\gamma - (1+\frac{\delta}{2})t}\, dt = +\infty.$$

Indeed

$$\int_0^{+\infty} \dot{\psi}_\eta^2 t^\beta\, dt = \int_0^1 t^\beta + \int_1^{+\infty} \left(\frac{1-\beta-2\eta}{2}\right)^2 t^{-1-2\eta}\, dt < +\infty$$

and

$$\int_0^{+\infty} e^{\bar{\alpha}\psi_\eta^\gamma - (1+\frac{\delta}{2})t}\, dt \geq \int_1^{+\infty} e^{\bar{\alpha}t^{1+\bar{\eta}} - (1+\frac{\delta}{2})t}\, dt = +\infty.$$

Again from (8) and (9), the sufficient condition can be rewritten as

$$\int_0^{+\infty} e^{\bar{\alpha}|\psi|^{\frac{2}{1-\beta}} - (1+\frac{\delta}{2})t}\, dt < +\infty, \quad \text{for all } \psi \text{ such that } \int_0^{+\infty} \frac{|\psi'|^2 t^\beta}{1-\beta}\, dt < +\infty.$$

We proceed as in [11]. For all $\varepsilon > 0$ there exists $T = T(\varepsilon)$ such that $\int_T^{+\infty} \frac{|\psi'|^2 t^\beta}{1-\beta}\, dt < \varepsilon^2$. Hence, for the Cauchy–Schwarz inequality

$$\begin{aligned}
\psi(t) &= \psi(T) + \int_T^t \psi'(s)\, ds = \psi(T) + \int_T^t |\psi'(s)|^2\, s^{\beta/2} s^{-\beta/2} ds \\
&\leq \psi(T) + \left(\int_T^t |\psi'(s)|^2\, s^\beta\, ds\right)^{1/2} \left(\int_T^t s^{-\beta}\, ds\right)^{1/2} \\
&= \psi(T) + \left(\int_T^t \frac{|\psi'(s)|^2\, s^\beta}{1-\beta}\, ds\right)^{1/2} \left(t^{1-\beta} - T^{1-\beta}\right)^{1/2} \\
&\leq \psi(T) + \varepsilon\left(t^{1-\beta} - T^{1-\beta}\right)^{1/2} \text{ for all } t \geq T\,.
\end{aligned}$$

This implies that there exists $\bar{T}$ such that $\bar{\alpha}\psi^{\frac{2}{1-\beta}}(t) \le (1+\frac{\delta}{2})t$ for all $t \ge \bar{T}$, and this is sufficient to guarantee the existence of the integral

$$\int_0^{+\infty} e^{\bar{\alpha}|\psi|^{\frac{2}{1-\beta}} - (1+\frac{\delta}{2})t}\, dt.$$

We proceed with giving an idea of proof for (ii). By the Radial Lemma we know that

$$\int_0^{+\infty} \frac{\dot{\psi}^2 t^\beta}{1-\beta}\, dt \le 1 \Rightarrow \psi(t) \le t^{\frac{1-\beta}{2}}. \tag{10}$$

Then if $\gamma < \frac{2}{1-\beta}$ or $\gamma = \frac{2}{1-\beta}$ and $\bar{\alpha} < 1 + \delta/2$ it easily follows that

$$\int_0^{+\infty} e^{\bar{\alpha}\psi^\gamma - (1+\frac{\delta}{2})t}\, dt \le \int_0^{+\infty} e^{\bar{\alpha}t - (1+\frac{\delta}{2})t}\, dt < +\infty.$$

The proof of (ii) for the critical exponent is much more delicate and it is an adapted version of Moser's proof. The result of Moser is based on the observation that relation (10) is actually a rather strong inequality. We refer the reader to [5] for a complete proof.

This concludes the case of w_0. In order to prove the assertion for $w_1(x) = \big(\log(e/|x|)\big)$ it is sufficient to observe that, for $\beta \in [0, 1)$

$$H_0^1(B, w_1) \hookrightarrow H_0^1(B, w_0).$$

(*Sharpness*) Now we prove that the theorem is sharp in the sense that if $\alpha > (1+\delta/2)2\,[2\pi(1-\beta)]^{\frac{1}{1-\beta}}$, then the supremum is infinite.

It is sufficient to consider the case $w(x) = |\log(e/|x|)|^\beta$.

Now it is sufficient to test

$$\int_0^{+\infty} e^{(\bar{\alpha}\psi^\gamma - (1+\delta/2)t)}\, dt, \quad \gamma = \frac{2}{1-\beta}$$

on the family of functions (here the change of variables is $t = 2 - 2\log|x|$)

$$\psi_k(t) = \begin{cases} \dfrac{t^{1-\beta} - 2^{1-\beta}}{((k+2)^{1-\beta} - 2^{1-\beta})^{1/2}} & 2 \le t \le k+2 \\ ((k+2)^{1-\beta} - 2^{1-\beta})^{1/2} & t \ge k+2 \end{cases}$$

$$\int_2^{+\infty} e^{(\bar{\alpha}\psi_k^\gamma - (1+\delta/2)t)}\, dt \ge e^{\bar{\alpha}[(k+2)^{1-\beta} - 2^{1-\beta}]^{\frac{1}{1-\beta}}} \int_{k+2}^{+\infty} e^{-(1+\delta/2)t}\, dt$$
$$= \frac{1}{1+\delta/2} e^{\{\bar{\alpha}[(k+2)^{1-\beta} - 2^{1-\beta}]^{\frac{1}{1-\beta}} - (1+\delta/2)(k+2)\}} \to +\infty \text{ if } \bar{\alpha} > 1 + \frac{\delta}{2}. \qquad \square$$

3.1. Remark: the weight is not effective

In this section we consider all the functions in $H_0^1(B_1)$. We observe that in the supercritical case with respect to the Moser inequality, i.e., $\gamma = 2$, $\alpha > 4\pi$, the weight is not effective and the supremum in the whole space $H_0^1(B_1)$ is infinite.

Proposition 9. *Suppose that $\alpha > 4\pi$. Then*

$$\sup_{\{u \in H_0^1(B_1), \|u\|_w \le 1\}} \int_{B_1} |x|^\delta e^{\alpha |u|^2} dx = +\infty.$$

Proof. Case I) $w(x) = |\log |x||^\beta$. In tho case we evaluate the functional on some particular functions obtained by a suitable translation and dilation of Moser's functions in a region of $B(0,1)$ far from the origin and far from the boundary where the presence of $\left|\log |x|\right|^\beta$ can be "neglected".

Consider the following family of functions

$$w_{k,a}(x) = \frac{1}{\sqrt{2\pi}} \begin{cases} \sqrt{\log k} & |x - x_0| < \dfrac{a}{k} \\ \dfrac{\log(\frac{a}{|x - x_0|})}{\sqrt{\log k}} & \dfrac{a}{k} \le |x - x_0| < a \\ 0 & |x - x_0| \ge a \end{cases} \tag{11}$$

where $x_0 = (\frac{1}{2}, 0)$ and $a < \frac{1}{2}$ will be chosen later.

Since $\alpha > 4\pi$, we can write $\alpha = 4\pi(1+\epsilon)^\beta$, with $\varepsilon > 0$. Let

$$u_{k,a}(x) = \left|\log\left(\frac{1}{2} - a\right)\right|^{-\beta/2} w_{k,a}(x)\ .$$

Then, since $|x| \ge \frac{1}{2} - a$ in $B_a(x_0)$ and $\left|\log |x|\right|$ is decreasing, one has

$$\left(\int_{B_1(0)} |\nabla u_{k,a}|^2 \left|\log |x|\right|^\beta dx\right)^{1/2} = \left(\int_{B_a(x_0)} |\nabla w_{k,a}|^2 \frac{\left|\log |x|\right|^\beta}{\left|\log|\frac{1}{2} - a|\right|^\beta} dx\right)^{1/2} \le 1.$$

We evaluate the functional on this sequence and obtain

$$\int_{B_1(0)} |x|^\delta e^{4\pi(1+\varepsilon)^\beta u_{k,a}^2} dx = \int_{B_1(0)} |x|^\delta e^{4\pi(1+\varepsilon)^\beta \left|\log\left(\frac{1}{2} - a\right)\right|^{-\beta} w_{k,a}^2} dx.$$

Now, choosing $0 < a < \frac{1}{2} - \frac{1}{e^{1+\varepsilon}}$, we have $\frac{1+\varepsilon}{\left|\log\left(\frac{1}{2} - a\right)\right|} > 1$.

Let $\bar{\varepsilon} = (1+\varepsilon)^\beta \left|\log\left(\frac{1}{2}-a\right)\right|^{-\beta} - 1 > 0$, we can conclude, as $k \to +\infty$

$$\int_{B_1(0)} |x|^\delta e^{4\pi(1+\varepsilon)^\beta u_{k,a}^2} dx$$

$$\geq \begin{cases} \left(\frac{1}{2}-a\right)^\delta \int_{B_{\frac{a}{k}}(x_0)} e^{4\pi(1+\bar{\varepsilon})w_{k,a}^2} dx = \left(\frac{1}{2}-a\right)^\delta \pi a^2 k^{2\bar{\varepsilon}} \to +\infty, \ (\delta \geq 0), \\ \left(\frac{1}{2}+a\right)^\delta \int_{B_{\frac{a}{k}}(x_0)} e^{4\pi(1+\bar{\varepsilon})w_{k,a}^2} dx = \left(\frac{1}{2}+a\right)^\delta \pi a^2 k^{2\bar{\varepsilon}} \to +\infty, \ (\delta < 0). \end{cases}$$

Case II) $w(x) = \left(\log \frac{e}{|x|}\right)^\beta$. For this case it is necessary to "concentrate" Moser's sequence near the boundary where the weight is almost 1. Let

$$z_{k,a}(x) = \frac{1}{\sqrt{2\pi}} \begin{cases} \sqrt{\log k} & |x-x_a| < \frac{a}{k} \\ \dfrac{\log(\frac{a}{|x-x_a|})}{\sqrt{\log k}} & \frac{a}{k} \leq |x-x_a| < a \\ 0 & |x-x_a| \geq a \end{cases} \tag{12}$$

where $x_a = (1-a, 0)$, $0 < a < 1/2$, $k > 2$. Here it is sufficient to choose $a < \left(\frac{1}{2} - \frac{1}{e^\varepsilon}\right)$ and test the following sequence

$$u_{k,a}(x) = \left|\log\left(\frac{e}{1-2a}\right)\right|^{-\beta/2} w_{k,a}(x) \ .$$

on the functional. □

Acknowledgement

The author is grateful to the anonymous referee for his helpful comments concerning a prior version of this work.

References

[1] D.R. Adams, *A Sharp Inequality of J. Moser for Higher Order Derivatives* Annals of Mathematics, (**128**), no. 2 (1988), 385–398.

[2] Adimurthi, K. Sandeep, *A singular Moser–Trudinger embedding and its applications*, NoDEA, Nonlinear Differ. Equ. Appl. 13, (2007), 585–603.

[3] L. Caffarelli, R. Kohn, L. Nirenberg, *First order interpolation inequalities with weights*, Compositio Math., **53**, (1984), 259–275.

[4] M. Calanchi, E. Terraneo, *Non-radial Maximizers For Functionals With Exponential Non-linearity in* $\mathbb{R}^2$, Advanced Nonlinear Studies **5** (2005), 337–350.

[5] M. Calanchi, B. Ruf, *On Trudinger–Moser Type Inequalities with Logarithmic Weights*, preprint (2013).

[6] J.F. De Oliveira, J.M. do Ó, *Trudinger–Moser type inequalities for weighted spaces involving fractional dimensions*, to appear in Proceedings of the AMS.

[7] D.E. Edmunds, H.Hudzik, M. Krbec, On weighted critical imbeddings of Sobolev spaces. Math. Z. 268 (2011), no. 1-2, 585–592.

[8] H. Hudzik, M. Krbec, *On non-effective weights in Orlicz spaces* Indag. Mathem., N.S., **18** (2007) (2), 215–231.

[9] M. Krbec, T. Schott, *Embeddings of weighted Sobolev spaces in the borderline case*, Real Anal. Exchange **23** (1997/98), no. 2, 395–420.

[10] D.E. Marshall, *A new proof of a sharp inequality concerning the Dirichlet integral*, Ark. Mat. **27** (1989), no. 1, 131–137.

[11] J. Moser, *A sharp form of an inequality by N. Trudinger*, Indiana Univ. Math. J. **20** (1970/71), 1077–1092.

[12] S.I. Pohozaev, *The Sobolev embedding in the case* $pl = n$, Proc. of the Technical Scientific Conference on Advances of Scientific Research 1964–1965, Mathematics Section, (Moskov. Energet. Inst., Moscow), (1965), 158–170.

[13] Smets, D., Su, J., Willem, M., *Non radial ground states for the Hénon equation*, Commun. Contemp. Math. 4(3), (2002), 467–480.

[14] N.S. Trudinger, *On imbeddings into Orlicz spaces and some applications*, J. Math. Mech., **17** (1967), 473–483.

Marta Calanchi
Dipartimento di Matematica
Università degli Studi di Milano
via C. Saldini 50
I-20133 Milano, Italy
e-mail: `marta.calanchi@unimi.it`

Progress in Nonlinear Differential Equations
and Their Applications, Vol. 85, 175–186

Multiple Solutions for p-Laplacian Type Problems with an Asymptotically p-linear Term

Anna Maria Candela and Giuliana Palmieri

Dedicated to Bernhard Ruf on the occasion of his 60th Birthday

Abstract. The aim of this paper is studying the asymptotically p-linear problem

$$\begin{cases} -\mathrm{div}(A(x,u)|\nabla u|^{p-2}\nabla u) + \frac{1}{p}A_t(x,u)|\nabla u|^p & \\ \qquad = \lambda|u|^{p-2}u + g(x,u) & \text{in } \Omega, \\ u = 0 & \text{on } \partial\Omega, \end{cases}$$

where $\Omega \subset \mathbb{R}^N$ is an open bounded domain and $p > N \geq 2$. Suitable assumptions both at infinity and in the origin on the even function $A(x,\cdot)$ and the odd map $g(x,\cdot)$ allow us to prove the existence of multiple solutions by means of variational tools and the pseudo-index theory related to the genus in $W_0^{1,p}(\Omega)$.

Mathematics Subject Classification (2010). 35J60, 35J35, 47J30, 58E05.

Keywords. p-Laplacian type problem, asymptotically p-linear problem, genus, pseudo-index theory, quasi-eigenvalue.

1. Introduction

Let us consider the Dirichlet problem

$$(P) \qquad \begin{cases} -\mathrm{div}(A(x,u)|\nabla u|^{p-2}\nabla u) + \frac{1}{p}A_t(x,u)|\nabla u|^p = f(x,u) & \text{in } \Omega, \\ u = 0 & \text{on } \partial\Omega, \end{cases}$$

Partially supported by Research Funds PRIN2009 and *Fondi d'Ateneo* 2011.

where $\Omega \subset \mathbb{R}^N$ is an open bounded domain, $N \geq 2$, and A, $f : \Omega \times \mathbb{R} \to \mathbb{R}$ are given functions such that the partial derivative $A_t(x,t) = \frac{\partial A}{\partial t}(x,t)$ exists for a.e. $x \in \Omega$, all $t \in \mathbb{R}$.

If we assume $F(x,t) = \int_0^t f(x,s)ds$, at problem (P) we can associate the functional $\mathcal{J} : \mathcal{D} \subset W_0^{1,p}(\Omega) \to \mathbb{R}$ defined as

$$\mathcal{J}(u) = \frac{1}{p}\int_\Omega A(x,u)|\nabla u|^p dx - \int_\Omega F(x,u)dx.$$

In general, $\mathcal{J}$ is not C^1 in the Sobolev space $W_0^{1,p}(\Omega)$ but, under the following conditions:

(H_0) A, A_t are Carathéodory functions on $\Omega \times \mathbb{R}$ such that

$$\sup_{|t|\leq r} |A(\cdot,t)| \in L^\infty(\Omega), \quad \sup_{|t|\leq r} |A_t(\cdot,t)| \in L^\infty(\Omega) \quad \text{for any } r \geq 0;$$

(h_0) f is a Carathéodory function on $\Omega \times \mathbb{R}$ such that

$$\sup_{|t|\leq r} |f(\cdot,t)| \in L^\infty(\Omega) \quad \text{for any } r \geq 0;$$

it is surely well defined in the Banach space

$$X := W_0^{1,p}(\Omega) \cap L^\infty(\Omega), \qquad \|u\|_X := \|u\| + |u|_\infty,$$

with $\|\cdot\|$, respectively $|\cdot|_\infty$, classical norm of $W_0^{1,p}(\Omega)$, respectively $L^\infty(\Omega)$, i.e.,

$$\|u\|^p = \int_\Omega |\nabla u|^p dx, \quad |u|_\infty = \operatorname{ess\,sup}_{x\in\Omega} |u(x)|.$$

Here, our aim is investigating the existence of multiple weak (bounded) solutions of (P) when it is an *elliptic asymptotically p-linear* problem, i.e., A and f satisfy the following hypotheses:

(H_1) there exists $\alpha_0 > 0$ such that

$$A(x,t) \geq \alpha_0 \quad \text{a.e. in } \Omega, \text{ for all } t \in \mathbb{R};$$

(H_2) there exists $A^\infty \in L^\infty(\Omega)$ such that

$$\lim_{|t|\to+\infty} A(x,t) = A^\infty(x) \quad \text{uniformly a.e. in } \Omega;$$

(h_1) there exist $\lambda \in \mathbb{R}$ and $g : \Omega \times \mathbb{R} \to \mathbb{R}$ such that

$$f(x,t) = \lambda|t|^{p-2}t + g(x,t) \quad \text{for a.e. } x \in \Omega, \text{ for all } t \in \mathbb{R},$$

$$\lim_{|t|\to+\infty} \frac{g(x,t)}{|t|^{p-2}t} = 0 \quad \text{uniformly a.e. in } \Omega. \tag{1.1}$$

If conditions (H_0)–(H_2), (h_0)–(h_1) hold, problem (P) reduces to

$$(P_\lambda) \qquad \begin{cases} -\operatorname{div}(A(x,u)|\nabla u|^{p-2}\nabla u) + \dfrac{1}{p}A_t(x,u)|\nabla u|^p \\ \qquad\qquad = \lambda|u|^{p-2}u + g(x,u) & \text{in } \Omega, \\ u = 0 & \text{on } \partial\Omega, \end{cases}$$

and $\mathcal{J}$ is a C^1 functional on X (see [10, Proposition 3.1]); whence, (P_λ) has a variational structure and its weak bounded solutions are critical points of $\mathcal{J}$ in the Banach space X. Then, variational and topological tools may be applied but the presence of the L^∞-norm makes it difficult. In fact, if $N \geq p$, the classical Palais–Smale condition may not occur as it requires the convergence not only in the $W_0^{1,p}$-norm but also in the L^∞ one. Thus, some weak versions of the Palais–Smale condition are required so to obtain a suitable version of the Deformation Lemma; anyway they are not enough for distinguishing multiple critical points at the same critical level so multiplicity results follow from the existence of multiple distinct critical levels (see [8, 9]).

This problem does not arise if $N < p$ as the embedding $W_0^{1,p}(\Omega) \hookrightarrow L^\infty(\Omega)$ implies $X = W_0^{1,p}(\Omega)$, so we can consider the usual $W_0^{1,p}$-norm and the Palais–Smale condition can be proved (see Section 2).

On the other hand, in the hypotheses (h_0) and (h_1), if $p = 2$ and $A(x,t) \equiv 1$ problem (P_λ) reduces to the asymptotically linear one which has been widely investigated (see [1, 4] and references therein). On the contrary, in spite of the large amount of papers dealing with this kind of nonlinearities in the semilinear case, only a few results have been obtained when $p \neq 2$. Namely, some existence results can be found in [2, 3, 5, 11, 12, 14, 15, 17, 18] if $A(x,t) \equiv 1$, but, to our knowledge, there is no result of this kind with a coefficient $A(x,t)$, which depends on t, up to [10]. If $p > 1$ is any, the main difficulty is that, while the structure of the spectrum of $-\Delta$ in $H_0^1(\Omega)$ is known, the full spectrum of $-\Delta_p$ is still unknown, even if various authors have introduced different characterizations of eigenvalues and definitions of quasi-eigenvalues. Furthermore, in our setting we have also to consider the asymptotic behaviour both at the origin and at infinity of the coefficient $A(x,t)$, namely the operators $\mathcal{A}_p^0$, $\mathcal{A}_p^\infty : W_0^{1,p}(\Omega) \to W_0^{-1,p'}(\Omega)$ defined as $\mathcal{A}_p^0 u = -\mathrm{div}(A^0(x)|\nabla u|^{p-2}\nabla u)$, $\mathcal{A}_p^\infty u = -\mathrm{div}(A^\infty(x)|\nabla u|^{p-2}\nabla u)$, with $A^0(x) = A(x,0)$ and A^∞ as in (H_2) (see Section 2) as it is made in [10] but using a cohomological index.

Here, considering the particular case $p > N \geq 2$ and by means of the pseudo-index theory related to the genus in $W_0^{1,p}(\Omega)$, our aim is investigating the existence of multiple solutions of (P_λ) when the parameter λ in (h_1) interacts with sequences of quasi-eigenvalues related to the operators $\mathcal{A}_p^0$ and $\mathcal{A}_p^\infty$ which are defined according to the approach in [7] and [16]. Thus, essentially following the ideas in [5], we are able to prove our main result (for the complete statement, see Theorem 3.1).

Main Theorem. *Assume that $p > N \geq 2$, (H_0)–(H_2), (h_0)–(h_1) hold and the parameter λ is not an eigenvalue of the operator $\mathcal{A}_p^\infty$. If, furthermore, $A(x,\cdot)$ is even and $g(x,\cdot)$ is odd and they satisfy further suitable assumptions in $\Omega \times \mathbb{R}$, then the number of the weak solutions of (P_λ) depends on the interaction of λ, the asymptotic behaviour of g in the origin and the quasi-eigenvalues related to the operators $\mathcal{A}_p^0$ and $\mathcal{A}_p^\infty$ in $W_0^{1,p}(\Omega)$.*

2. Abstract tools and some technical remarks

In order to state the abstract multiplicity theorem we will apply to our problem, we recall some main definitions on the pseudo-index theory related to the genus on a Banach space B for a C^1 even functional $J : B \to \mathbb{R}$ with symmetry group $\mathbb{Z}_2 = \{\mathrm{id}, -\mathrm{id}\}$. Here, we denote $\mathbb{N} = \{1, 2, \dots\}$.

Define

$$\Sigma = \Sigma(B) = \{A \subseteq B : A \text{ closed and symmetric with respect to the origin, i.e., } -u \in A \text{ if } u \in A\}$$

and $\mathcal{H} = \{h \in C(B, B) : h \text{ odd}\}$.

Taking $A \in \Sigma$, $A \neq \emptyset$, the *genus* of A is

$$\gamma(A) = \inf\{k \in \mathbb{N} : \exists \psi \in C(A, \mathbb{R}^k \setminus \{0\}) \text{ s.t. } \psi(-u) = -\psi(u) \text{ for all } u \in A\},$$

if such an infimum exists, otherwise $\gamma(A) = +\infty$. Assume $\gamma(\emptyset) = 0$.

The index theory $(\Sigma, \mathcal{H}, \gamma)$ related to $\mathbb{Z}_2$ is also called *genus* (for more details, we refer to [19, Section 1] or [21, Section II.5]).

According to [6], the *pseudo-index* related to the genus and to a symmetric subset $S \in \Sigma$ is the triplet $(S, \mathcal{H}^*, \gamma^*)$ such that $\mathcal{H}^*$ is a group of odd homeomorphisms, eventually related to an even functional J, and $\gamma^* : \Sigma \longrightarrow \mathbb{N} \cup \{+\infty\}$ is the map defined by

$$\gamma^*(A) = \min_{h \in \mathcal{H}^*} \gamma(h(A) \cap S) = \min_{h \in \mathcal{H}^*} \gamma(A \cap h(S)).$$

The following *mini-max* theorem can be proved (see [4, Theorem 2.9] in the setting of Hilbert spaces; but the same proof holds on Banach spaces, just taking into account [20, Theorem A.4]).

Theorem 2.1. *Let $J : B \to \mathbb{R}$ be a C^1 even functional on a Banach space B and, taking $a, b, c_0, c_\infty \in \bar{\mathbb{R}}$, $-\infty \le a < c_0 < c_\infty < b \le +\infty$, consider the pseudo-index theory $(S, \mathcal{H}^*, \gamma^*)$ related to the genus $(\Sigma, \mathcal{H}, \gamma)$ on B, the functional J and the subset $S \in \Sigma$, with*

$$\mathcal{H}^* = \{h \in \mathcal{H} : h \text{ bounded homeomorphism s.t. } h(u) = u \text{ if } u \notin J^{-1}(]a, b[)\}.$$

Assume that:

(i) *the functional J satisfies the Palais–Smale condition in $]a, b[$;*
(ii) *$J(u) \ge c_0$ for all $u \in S$;*
(iii) *there exist $\tilde{k} \in \mathbb{N}$ and $\tilde{A} \in \Sigma$ such that*

$$J(u) \le c_\infty \quad \text{for all } u \in \tilde{A} \quad \text{and} \quad \gamma^*(\tilde{A}) \ge \tilde{k}.$$

Then the numbers

$$c_i = \inf_{A \in \Sigma_i^*} \sup_{u \in A} J(u), \qquad i \in \{1, \dots, \tilde{k}\},$$

with $\Sigma_i^* = \{A \in \Sigma : \gamma^*(A) \ge i\}$, are critical values for J and

$$c_0 \le c_1 \le \dots \le c_{\tilde{k}} \le c_\infty.$$

Furthermore, if $c = c_i = \cdots = c_{i+r}$, with $i \geq 1$ and $i + r \leq k$, then $\gamma(K_c) \geq r + 1$, with $K_c = \{u \in B : \ J(u) = c, \ \mathrm{d}J(u) = 0\}$.

In order to apply the theorem above, we need the following result, which allows us to obtain a lower bound for the pseudo-index of a suitable $\tilde{A}$ as in (iii) (for more details, see [4, Theorem A.2] or [5, Theorem 2.7]).

Proposition 2.2. *Let $(\Sigma, \mathcal{H}, \gamma)$ be the genus theory on B and V, W two closed subspaces of B. Assume that*

$$\dim V \ < \ +\infty \quad \text{and} \quad \mathrm{codim} W \ < \ +\infty.$$

Then, for every odd bounded homeomorphism h on B and every open bounded symmetric neighbourhood $\mathcal{N}$ of 0 in B, it results

$$\gamma(V \cap h(\partial\mathcal{N} \cap W)) \ \geq \ \dim V - \mathrm{codim} W. \tag{2.1}$$

Till the end of this section, assume that (H_0), (h_0) and (h_1) hold, thus g in (h_1) is a Carathéodory function on $\Omega \times \mathbb{R}$ such that

$$\sup_{|t|\leq r} |g(\cdot, t)| \in L^\infty(\Omega) \quad \text{for any } r \geq 0. \tag{2.2}$$

Then, if $G(x,t) = \int_0^t g(x,s)ds$, the functional $\mathcal{J}$ reduces to

$$J_\lambda(u) \ = \ \frac{1}{p} \int_\Omega (A(x,u)|\nabla u|^p - \lambda |u|^p) dx \ - \ \int_\Omega G(x,u) dx.$$

As already remarked, in the hypothesis $p > N$ the functional J_λ is C^1 on $W_0^{1,p}(\Omega)$ and for all u, $\varphi \in W_0^{1,p}(\Omega)$ it results

$$\begin{aligned} \langle dJ_\lambda(u), \varphi \rangle = & \int_\Omega A(x,u)|\nabla u|^{p-2}\nabla u \cdot \nabla\varphi \ dx \ + \ \frac{1}{p} \int_\Omega A_t(x,u)\varphi|\nabla u|^p dx \\ & - \ \lambda \int_\Omega |u|^{p-2}u\varphi \ dx \ - \ \int_\Omega g(x,u)\varphi \ dx. \end{aligned}$$

In the further assumptions (H_1) and (H_2), we can consider the function $A^\infty \in L^\infty(\Omega)$ such that $A^\infty(x) \geq \alpha_0$ a.e. in Ω; hence, the functional

$$I^\infty(u) \ = \ \int_\Omega A^\infty(x)|\nabla u|^p dx$$

is a weighted norm equivalent to the usual one in $W_0^{1,p}(\Omega)$ and its differential is the operator $A_p^\infty : W_0^{1,p}(\Omega) \to W_0^{-1,p'}(\Omega)$ so that

$$\langle A_p^\infty u, \varphi \rangle \ = \ \int_\Omega A^\infty(x)|\nabla u|^{p-2}\nabla u \cdot \nabla\varphi \ dx \quad \text{for all } u, \ \varphi \in W_0^{1,p}(\Omega).$$

Let $\sigma(A_p^\infty)$ denote the set of the *eigenvalues* of the elliptic operator A_p^∞, i.e., of the parameters $\mu \in \mathbb{R}$ such that the Dirichlet problem

$$\begin{cases} -\mathrm{div}(A^\infty(x)|\nabla u|^{p-2}\nabla u) \ = \ \mu|u|^{p-2}u & \text{in } \Omega, \\ u = 0 & \text{on } \partial\Omega, \end{cases}$$

admits non-trivial weak solutions in $W_0^{1,p}(\Omega)$.

Now, let us consider the following further conditions on $A(x,t)$:

(H_3) there exists

$$\lim_{|t|\to+\infty} A_t(x,t)t \;=\; 0 \qquad \text{uniformly a.e. in } \Omega;$$

(H_4) there exists $\alpha_1 > 0$ such that

$$A(x,t) + \frac{1}{p}A_t(x,t)t \geq \alpha_1 A(x,t) \quad \text{a.e. in } \Omega, \text{ for all } t \in \mathbb{R}.$$

Then the following result can be proved (for all the details, see [10, Proposition 3.5]).

Proposition 2.3. *Assume that the hypotheses (H_0)–(H_4), (h_0)–(h_1) hold and $p > N$. Then if $\lambda \notin \sigma(A_p^\infty)$ the functional J_λ satisfies the Palais–Smale condition in $W_0^{1,p}(\Omega)$.*

Taking $A^0(x) = A(x,0)$, from (H_0) and (H_1) it follows $A^0 \in L^\infty(\Omega)$ and $A^0(x) \geq \alpha_0$ a.e. in Ω; hence, also the functional

$$I^0(u) \;=\; \int_\Omega A^0(x)|\nabla u|^p dx$$

is a weighted norm equivalent to the usual one in $W_0^{1,p}(\Omega)$ and its differential is the operator $A_p^0 : W_0^{1,p}(\Omega) \to W_0^{-1,p'}(\Omega)$ so that

$$\langle A_p^0 u, \varphi\rangle \;=\; \int_\Omega A^0(x)|\nabla u|^{p-2}\nabla u\cdot\nabla\varphi\, dx \quad \text{for all } u,\, \varphi \in W_0^{1,p}(\Omega).$$

Furthermore, from (H_3) it follows

$$|A_t(x,t)| \;\leq\; b_0 \quad \text{a.e. in } \Omega, \text{ for all } t \in \mathbb{R},$$

for a suitable positive constant b_0; hence,

$$\lim_{t\to 0} A(x,t) \;=\; A^0(x) \quad \text{uniformly a.e. in } \Omega. \tag{2.3}$$

Now, taking $\sharp = 0$ or $\sharp = \infty$, we introduce the definitions of sequences of pseudo-eigenvalues related to $A_p^\sharp$, or equivalently its potential $I^\sharp$, that we need in the statement of our main result.

Firstly, taking $\mathcal{S} = \{u \in W_0^{1,p}(\Omega) : \ |u|_p = 1\}$, with $|u|_p^p = \int_\Omega |u|^p dx$, we define

$$\eta_1^\sharp \;=\; \inf_{u\in\mathcal{S}} \int_\Omega A^\sharp(x)|\nabla u|^p dx.$$

By assumptions, $0 < \alpha_0\lambda_1 \leq \eta_1^\sharp \leq \lambda_1|A^\sharp|_\infty$, with λ_1 first (positive, simple, isolated) eigenvalue of the p-Laplacian $-\Delta_p$ in $W_0^{1,p}(\Omega)$. Standard arguments allow us to prove the existence of $\psi_1^\sharp \in \mathcal{S}$ such that $I^\sharp(\psi_1^\sharp) = \eta_1^\sharp$. Thus, reasoning as in [7, Section 5], but replacing $I(u) = \|u\|^p$ with $I^\sharp$ and starting from $\eta_1^\sharp$ at the place of λ_1, we can prove the existence of a sequence $(\eta_k^\sharp)_k$ of positive real numbers with corresponding functions $(\psi_k^\sharp)_k$ such that:

(a) $0 < \eta_1^\sharp \leq \eta_2^\sharp \leq \cdots \leq \eta_k^\sharp \leq \cdots$ with $\eta_k^\sharp \nearrow +\infty$;

(b) $\psi_i^\sharp \neq \psi_j^\sharp$ if $i \neq j$;

(c) $(\psi_k^\sharp)_k$ generates the whole space $W_0^{1,p}(\Omega)$ and

$$W_0^{1,p}(\Omega) = V_h^\sharp \oplus W_h^\sharp \quad \text{for all } h \in \mathbb{N}, \tag{2.4}$$

where $V_h^\sharp = \text{span}\{\psi_1^\sharp, \ldots, \psi_h^\sharp\}$ and its complement $W_h^\sharp$ can be explicitly described. By definition, $\text{codim} W_h^\sharp = \dim V_h^\sharp = h$.

Moreover, for all $h \in \mathbb{N}$ on the infinite-dimensional subspace $W_h^\sharp$ the following inequality holds:

$$\eta_{h+1}^\sharp \int_\Omega |w|^p dx \ \le \ \int_\Omega A^\sharp(x)|\nabla w|^p dx \quad \text{for all } w \in W_h^\sharp \tag{2.5}$$

(cf. [7, Lemma 5.4]).

On the contrary, in order to have a reversed inequality on finite-dimensional subspaces, we reason according to the arguments introduced in [16] but, also in this case, replacing I with $I^\sharp$ and making $\psi_1^\sharp$ play the role of the "first" eigenfunction of $-\Delta_p$ in $W_0^{1,p}(\Omega)$. More precisely, for all $k \in \mathbb{N}$ we consider

$$\mathbb{W}_k^\sharp \ = \ \{V : \ V \text{ is a subspace of } W_0^{1,p}(\Omega), \ \psi_1^\sharp \in V \text{ and } \dim V \ge k\} \tag{2.6}$$

and

$$\nu_k^\sharp \ = \ \inf_{V \in \mathbb{W}_k^\sharp} \sup_{u \in V \cap S} \int_\Omega A^\sharp(x)|\nabla u|^p \, dx. \tag{2.7}$$

At last let us remark that, starting from the genus $(\Sigma, \mathcal{H}, \gamma)$ defined on the Banach space $B = W_0^{1,p}(\Omega)$, we can define a sequence of eigenvalues $(\lambda_k)_k$ of $-\Delta_p$ as

$$\lambda_k \ = \ \inf_{A \in \Sigma_k} \sup_{u \in A \cap S} \int_\Omega |\nabla u|^p \, dx,$$

with $\Sigma_k = \{A \in \Sigma : \ \gamma(A) \ge k\}$, and $\lambda_k \nearrow +\infty$ (see [13]).

Thus, from the properties of the genus, for all $k \in \mathbb{N}$ it is

$$\{V \cap S : V \in \mathbb{W}_k^\sharp\} \subset \Sigma_k$$

and then $\alpha_0 \lambda_k \le \nu_k^\sharp$; whence, $\nu_k^\sharp \nearrow +\infty$.

3. The main result

Now, we can state our main result.

Theorem 3.1. *Taking $p > N \ge 2$, assume that (H_0)–(H_4), (h_0)–(h_1) hold and $\lambda \notin \sigma(A_p^\infty)$. Moreover, suppose that*

(H_5) *$A(x,-t) = A(x,t)$ for all $t \in \mathbb{R}$, a.e. $x \in \Omega$;*

(h_2) *there exist $\lambda^0 \in \mathbb{R}$ such that*

$$\lim_{t \to 0} \frac{g(x,t)}{|t|^{p-2}t} \ = \ \lambda^0 \quad \textit{uniformly a.e. in } \Omega;$$

(h_3) *$g(x,-t) = -g(x,t)$ for all $t \in \mathbb{R}$, a.e. $x \in \Omega$.*

If $\lambda^0 < 0$ *and there exist* $k,\ h \in \mathbb{N}$, $k \geq h$, *such that*

$$\lambda + \lambda^0 \ < \ \eta_h^0 \ \leq \ \nu_k^\infty \ < \ \lambda, \tag{3.1}$$

then problem (P_λ) *has at least* $k-h+1$ *distinct pairs of weak non-trivial solutions.*

Proof. From the hypotheses and Proposition 2.3 it follows that J_λ is a C^1 even functional which satisfies the Palais–Smale condition in $W_0^{1,p}(\Omega)$.

Now, in order to apply Theorem 2.1, we have to prove the "geometric" assumptions related to the genus theory $(\Sigma, \mathcal{H}, \gamma)$ defined on $W_0^{1,p}(\Omega)$.

To this aim, let us point out that from (1.1) it follows

$$\lim_{|t|\to+\infty} \frac{G(x,t)}{|t|^p} \ = \ 0 \quad \text{uniformly a.e. in } \Omega,$$

while from (h_2) we have

$$\lim_{t\to 0} \frac{G(x,t)}{|t|^p} \ = \ \frac{\lambda^0}{p} \quad \text{uniformly a.e. in } \Omega;$$

whence, fixing any $\sigma > 0$, there exist R_σ, $\delta_\sigma > 0$ (without loss of generality $R_\sigma \geq 1$) such that

$$|G(x,t)| \leq \frac{\sigma}{p}\,|t|^p \qquad \text{if } |t| > R_\sigma, \text{ for a.e. } x \in \Omega, \tag{3.2}$$

$$\left|G(x,t) - \frac{\lambda^0}{p}|t|^p\right| \leq \frac{\sigma}{p}|t|^p \qquad \text{if } |t| < \delta_\sigma, \text{ for a.e. } x \in \Omega. \tag{3.3}$$

Moreover, by (H_0), taking any $s > 0$, there exists $a_{\sigma,1} > 0$, depending on σ and s, such that,

$$|G(x,t)| \ \leq \ a_{\sigma,1}|t|^{s+p} \qquad \text{if } \delta_\sigma \leq |t| \leq R_\sigma, \text{ for a.e. } x \in \Omega. \tag{3.4}$$

On the other hand, (H_2) and (2.3) imply that R_σ^∞, $\delta_\sigma^0 > 0$ exist such that

$$|A(x,t) - A^\infty(x)| < \sigma \qquad \text{if } |t| > R_\sigma^\infty, \text{ for a.e. } x \in \Omega, \tag{3.5}$$

$$|A(x,t) - A^0(x)| < \sigma \qquad \text{if } |t| < \delta_\sigma^0, \text{ for a.e. } x \in \Omega. \tag{3.6}$$

As $\lambda^0 < 0$ then (3.2)–(3.4) imply

$$G(x,t) \ \leq \ \frac{\sigma + \lambda^0}{p}\ |t|^p + a_{\sigma,2}|t|^{p+s}$$

for a suitable positive constant $a_{\sigma,2}$ depending also on $|\lambda^0|$. Hence, by the Sobolev Embedding Theorem we have

$$\int_\Omega G(x,u)\ dx \leq \ \frac{\sigma + \lambda^0}{p} \int_\Omega |u|^p dx + a_\sigma \|u\|^{p+s} \quad \text{for all } u \in W_0^{1,p}(\Omega).$$

Then, taking any $u \in W_0^{1,p}(\Omega)$, we have

$$\begin{aligned} J_\lambda(u) &\geq \frac{1}{p}\int_\Omega A(x,u)|\nabla u|^p dx - \frac{\lambda+\lambda^0+\sigma}{p}|u|_p^p - a_\sigma\|u\|^{s+p} \\ &= \frac{1}{p}\int_\Omega A^0(x)|\nabla u|^p dx + \frac{1}{p}\int_\Omega (A(x,u)-A^0(x))|\nabla u|^p dx \\ &\quad - \frac{\lambda+\lambda^0+\sigma}{p}|u|_p^p - a_\sigma\|u\|^{s+p}, \end{aligned}$$

where from (3.6), (H_1), $p > N$ and direct computations it follows

$$\begin{aligned} &\frac{1}{p}\int_\Omega (A(x,u)-A^0(x))|\nabla u|^p dx \\ &\quad \geq -\frac{\sigma}{p}\int_{\Omega_\sigma^0} |\nabla u|^p dx - \frac{1}{p}\int_{\Omega\setminus\Omega_\sigma^0} A^0(x)|\nabla u|^p dx \\ &\quad \geq -\frac{\sigma}{p}\int_\Omega |\nabla u|^p dx - \frac{|A^0|_\infty}{p}\int_{\Omega\setminus\Omega_\sigma^0}\left(\frac{|u|}{\delta_\sigma^0}\right)^s |\nabla u|^p dx \\ &\quad \geq -\frac{\sigma}{\alpha_0 p}\int_\Omega A^0(x)|\nabla u|^p dx - \frac{|A^0|_\infty}{p}\left(\frac{|u|_\infty}{\delta_\sigma^0}\right)^s\int_\Omega |\nabla u|^p dx \\ &\quad \geq -\frac{\sigma}{\alpha_0 p}\int_\Omega A^0(x)|\nabla u|^p dx - a'_\sigma\|u\|^{p+s}, \end{aligned}$$

with $\Omega_\sigma^0 = \{x \in \Omega : \ |u(x)| \leq \delta_\sigma^0\}$, for a suitable positive constant $a'_\sigma = a'_\sigma(\sigma, s)$. Whence,

$$J_\lambda(u) \geq \left(\frac{1}{p} - \frac{\sigma}{\alpha_0 p}\right)\int_\Omega A^0(x)|\nabla u|^p dx - \frac{\lambda+\lambda^0+\sigma}{p}|u|_p^p - a''_\sigma\|u\|^{p+s}, \qquad (3.7)$$

with $a''_\sigma = a_\sigma + a'_\sigma$.
Now, from (3.1), fixing $\sigma > 0$ such that

$$\lambda + \lambda^0 + \left(1 + \frac{\eta_h^0}{\alpha_0}\right)\sigma \ < \ \eta_h^0, \qquad (3.8)$$

from (2.4) and (2.5) referred to W_{h-1}^0, we have $\operatorname{codim} W_{h-1}^0 = h-1$ and, for all $u \in W_{h-1}^0$, (3.7) and (H_1) imply

$$\begin{aligned} J_\lambda(u) &\geq \frac{1}{p}\left(1 - \frac{\sigma}{\alpha_0} - \frac{\lambda+\lambda^0+\sigma}{\eta_h^0}\right)\int_\Omega A^0(x)|\nabla u|^p dx - a''_\sigma\|u\|^{p+s} \\ &\geq \frac{\alpha_0}{p}\left(1 - \frac{\sigma}{\alpha_0} - \frac{\lambda+\lambda^0+\sigma}{\eta_h^0}\right)\|u\|^p - a''_\sigma\|u\|^{p+s}. \end{aligned}$$

Thus, if $\varrho > 0$ is small enough there exists $c_0 > 0$ such that

$$J_\lambda(u) \ \geq \ c_0 \qquad \text{for all } u \in S_\varrho \cap W_{h-1}^0, \qquad (3.9)$$

with $S_\varrho = \{u \in W_0^{1,p}(\Omega) : \|u\| = \varrho\}$.

On the other hand, (2.2) implies

$$\sup_{|t|\leq r} |G(\cdot,t)| \in L^\infty(\Omega) \quad \text{for any } r \geq 0,$$

then from (3.2) a constant $L_\sigma > 0$ exists such that

$$|G(x,t)| \;\leq\; \frac{\sigma}{p}|t|^p + L_\sigma \quad \text{for a.e. } x \in \Omega, \text{ all } t \in \mathbb{R};$$

whence,

$$\left|\int_\Omega G(x,u)\, dx\right| \;\leq\; \frac{\sigma}{p}|u|_p^p + b_\sigma \quad \text{for all } u \in W_0^{1,p}(\Omega),$$

with $b_\sigma = L_\sigma \text{meas}(\Omega)$ (here, $\text{meas}(\cdot)$ is the Lebesgue measure in $\mathbb{R}^N$). Thus,

$$J_\lambda(u) \;\leq\; \frac{1}{p}\int_\Omega A(x,u)|\nabla u|^p dx - \frac{\lambda-\sigma}{p}|u|_p^p + b_\sigma$$

From (2.6) and (2.7) there exists $V_k^\sigma \in \mathbb{W}_k^\infty$ such that

$$\int_\Omega A^\infty(x)|\nabla u|^p dx \;\leq\; (\nu_k^\infty + \sigma)|u|_p^p \quad \text{for all } u \in V_k^\sigma,$$

where, without loss of generality, we can assume $\dim V_k^\sigma = k$. Hence, as all the norms are equivalent on a finite-dimensional space, from the compactness of $\mathcal{S}\cap V_k^\sigma$ and the assumptions on A it follows

$$\int_\Omega A(x,u)|\nabla u|^p dx \;\leq\; a_0\sigma|u|_p^p + \int_\Omega A^\infty(x)|\nabla u|^p dx$$

for all $u \in V_k^\sigma$ with $|u|_p$ large enough, where a_0 is a suitable positive constant independent of σ and u (for more details, see [10]). Thus, if $\lambda - \sigma > 0$, for all $u \in V_k^\sigma$ with $\|u\|$ large enough, we have

$$J_\lambda(u) \;\leq\; \frac{1}{p}\left(1 - \frac{\lambda-(a_0+1)\sigma}{\nu_k^\infty+\sigma}\right)\int_\Omega A^\infty(x)|\nabla u|^p dx + b_\sigma. \tag{3.10}$$

From (3.1), there exists σ small enough such that

$$1 - \frac{\lambda-(a_0+1)\sigma}{\nu_k^\infty+\sigma} \;<\; 0, \tag{3.11}$$

whence (3.10) implies

$$J_\lambda(u) \;\to\; -\infty \quad \text{if } u \in V_k^\sigma,\ \|u\| \to +\infty.$$

Thus, $c_\infty > c_0$ exists such that

$$J_\lambda(u) \;\leq\; c_\infty \quad \text{for all } u \in V_k^\sigma. \tag{3.12}$$

At last, if $\sigma \in\,]0,\lambda[$ is such that both (3.8) and (3.11) are satisfied, then both (3.9) and (3.12) hold and, considered the pseudo-index theory $(S_\varrho \cap W_{h-1}^0, \mathcal{H}^*, \gamma^*)$ related to the genus theory $(\Sigma, \mathcal{H}, \gamma)$ on $W_0^{1,p}(\Omega)$ with

$$\mathcal{H}^* = \{h \in \mathcal{H} : h \text{ bounded homeomorphism}\},$$

from (2.1) in Proposition 2.2 it follows $\gamma^*(V_k^\sigma) \geq k - h + 1$, thus Theorem 2.1 applies and J_λ admits at least $k - h + 1$ distinct pairs of critical points whose critical levels are strictly positive (while $J_\lambda(0) = 0$). □

Finally, here we give an example which points out as some inequalities in (3.1) can be always true.

Example 3.2. Let us define

$$A(x,t) = \left(1 + \mathrm{arctg}(a(x)t^2)\right)^\alpha \quad \text{for a.e. } x \in \Omega, \text{ all } t \in \mathbb{R},$$

with $\alpha \geq 0$ and $a : \Omega \to \mathbb{R}$ is a measurable function such that

$$k_0 \leq a(x) \leq k_1 \quad \text{for a.e. } x \in \Omega$$

for some k_0, $k_1 > 0$. By direct computations for a.e. $x \in \Omega$ and all $t \in \mathbb{R}$ we have $1 \leq A(x,t) \leq (1 + \frac{\pi}{2})^\alpha$ and

$$A_t(x,t)t = \alpha\left(1 + \mathrm{arctg}(a(x)t^2)\right)^{\alpha-1} \frac{2a(x)t^2}{1 + (a(x)t^2)^2};$$

hence, conditions (H_0)–(H_5) hold with $A(x,0) = 1$ and

$$\lim_{|t| \to +\infty} A(x,t) = \left(1 + \frac{\pi}{2}\right)^\alpha$$

uniformly a.e. in Ω. Thus, if $(\eta_h)_h$ and $(\nu_k)_k$ represent the sequences of quasi-eigenvalues of the p-Laplacian $-\Delta_p$ in $W_0^{1,p}(\Omega)$, i.e., such that (2.5) and (2.7) hold with $A^\sharp \equiv 1$, then in this setting we have

$$\eta_h^0 = \eta_h \text{ for all } h \in \mathbb{N} \quad \text{and} \quad \nu_k^\infty = \left(1 + \frac{\pi}{2}\right)^\alpha \nu_k \text{ for all } k \in \mathbb{N}.$$

Thus, the inequality $\eta_h^0 \leq \nu_k^\infty$ is not an assumption but holds for all $k \geq h$ as $\eta_k \leq \nu_k$ for all $k \in \mathbb{N}$ (see [5, Proposition 2.9]).

References

[1] H. Amann and E. Zehnder, *Nontrivial solutions for a class of nonresonance problems and applications to nonlinear differential equations*, Ann. Scuola Norm. Sup. Pisa Cl. Sci. (4) **7** (1980), 539–603.

[2] A. Anane and J.P. Gossez, *Strongly nonlinear elliptic problems near resonance: a variational approach*, Comm. Partial Differential Equations **15** (1990), 1141–1159.

[3] D. Arcoya and L. Orsina, *Landesman–Lazer conditions and quasilinear elliptic equations*, Nonlinear Anal. **28** (1997), 1623–1632.

[4] P. Bartolo, V. Benci and D. Fortunato, *Abstract critical point theorems and applications to some nonlinear problems with "strong" resonance at infinity*, Nonlinear Anal. **7** (1983), 981–1012.

[5] R. Bartolo, A.M. Candela and A. Salvatore, *p-Laplacian problems with nonlinearities interacting with the spectrum*, NoDEA Nonlinear Differential Equations Appl. (to appear). DOI:10.1007/s00030-013-0226-1.

[6] V. Benci, *On the critical point theory for indefinite functionals in the presence of symmetries*, Trans. Am. Math. Soc. **274** (1982), 533–572.

[7] A.M. Candela and G. Palmieri, *Infinitely many solutions of some nonlinear variational equations*, Calc. Var. Partial Differential Equations **34** (2009), 495–530.

[8] A.M. Candela and G. Palmieri, *Some abstract critical point theorems and applications*, in: "Dynamical Systems, Differential Equations and Applications" (X. Hou, X. Lu, A. Miranville, J. Su & J. Zhu, eds), Discrete Contin. Dynam. Syst., AIMS Press (2009), 133–142.

[9] A.M. Candela and G. Palmieri, *An abstract three critical points theorem and applications*, Proceedings of Dynamic Systems and Applications **6** (2012), 70–77.

[10] A.M. Candela, G. Palmieri and K. Perera, *Multiple solutions for p-Laplacian type problems with asymptotically p-linear terms via a cohomological index theory.* Preprint 2013.

[11] D.G. Costa and C.A. Magalhães, *Existence results for perturbations of the p-Laplacian*, Nonlinear Anal. **24** (1995), 409–418.

[12] P. Drábek and S. Robinson, *Resonance problems for the p-Laplacian*, J. Funct. Anal. **169** (1999), 189–200.

[13] P. Lindqvist, *On a nonlinear eigenvalue problem*, Berichte Univ. Jyväskylä Math. Inst. **68** (1995), 33–54.

[14] S. Liu and S. Li, *Existence of solutions for asymptotically 'linear' p-Laplacian equations*, Bull. London Math. Soc. **36** (2004), 81–87.

[15] G. Li and H.S. Zhou, *Asymptotically linear Dirichlet problem for the p-Laplacian*, Nonlinear Anal. **43** (2001), 1043–1055.

[16] G. Li and H.S. Zhou, *Multiple solutions to p-Laplacian problems with asymptotic nonlinearity as u^{p-1} at infinity*, J. London Math. Soc. **65** (2002), 123–138.

[17] K. Perera, R.P. Agarwal and D. O'Regan, *Morse Theoretic Aspects of p-Laplacian Type Operators*, Math. Surveys Monogr. **161**, Amer. Math. Soc., Providence RI, 2010.

[18] K. Perera and A. Szulkin, *p-Laplacian problems where the nonlinearity crosses an eigenvalue*, Discrete Contin. Dyn. Syst. **13** (2005), 743–753.

[19] P.H. Rabinowitz, *Variational Methods for Nonlinear Eigenvalues Problems* (G. Prodi, ed.), Edizioni Cremonese, Roma, 1974, 141–195.

[20] P.H. Rabinowitz, *Minimax Methods in Critical Point Theory with Applications to Differential Equations*, CBMS Reg. Conf. Ser. Math. **65**, Providence, 1986.

[21] M. Struwe, *Variational Methods. Applications to Nonlinear Partial Differential Equations and Hamiltonian Systems*, 4th edition, Ergeb. Math. Grenzgeb. (3) **34**, Springer-Verlag, Berlin, 2008.

Anna Maria Candela and Giuliana Palmieri
Dipartimento di Matematica
Università degli Studi di Bari "A. Moro"
Via E. Orabona 4
I-70125 Bari, Italy
e-mail: annamaria.candela@uniba.it
giuliana.palmieri@uniba.it

Progress in Nonlinear Differential Equations
and Their Applications, Vol. 85, 187–206

Nonlocal Dynamic Problems with Singular Nonlinearities and Applications to MEMS

D. Cassani, L. Fattorusso and A. Tarsia

Dedicated to Bernhard Ruf on the occasion of his 60th birthday

Abstract. We establish existence and regularity results for a time dependent fourth-order integro-differential equation with a possibly singular nonlinearity which has applications in designing MicroElectroMechanicalSystems. The key ingredient in our approach, besides basic theory of hyperbolic equations in Hilbert spaces, exploits the Near Operators Theory introduced by Campanato.

Mathematics Subject Classification (2010). 35A01, 35G30, 35L75.

Keywords. Singular nonlinear problems, nonlocal differential equations, higher-order hyperbolic PDE, regularity results, Steklov boundary conditions, near operators theory, MEMS.

1. Introduction

In this paper we study a time dependent nonlocal fourth-order equation which is a model for describing electrostatic actuation in MEMS devices. From the mathematical point of view, we can think of a plate problem set on a micro-scale in which usual first-order approximations, acceptable in the standard "visible" scale, loose their validity and where one needs to take into account nonlocal effects which in this context are not negligible. Precisely, we consider the following problem

$$\begin{cases} \Delta^2 u + c(x,t)\, u' + u'' = G(\beta,\gamma,u) + H(\lambda(t),\chi,p(x),u), \text{ in } \Omega\times[0,T] \\ 0 \le u(x,t) < 1, \quad \text{in } \Omega\times(0,T] \\ u(x,0) = u_0, \quad x\in\Omega \\ u'(x,0) = 0, \quad x\in\Omega \\ u(x,t) = 0, \quad \Delta u(x,t) - d\dfrac{\partial u(x,t)}{\partial\nu} = 0, \quad \text{on } \partial\Omega\times[0,T] \end{cases} \tag{1}$$

where $\Omega \subset \mathbb{R}^N$, $1 \leq N \leq 3$, is a bounded domain with sufficiently smooth boundary (ν denotes the outward pointing normal to $\partial\Omega$ whereas the derivative with respect to time is denoted by $'$) and

$$G(\beta, \gamma, u) := \left[\beta \int_\Omega |\nabla u(x,t)|^2 \, dx + \gamma\right] \Delta u$$

$$H(\lambda(t), \chi, p(x), u) := \frac{\lambda(t)p(x)}{[1 - u(x,t)]^\sigma \left[1 + \chi \displaystyle\int_\Omega \frac{1}{[1 - u(x,t)]^{\sigma-1}} \, dx\right]^\sigma}.$$

We assume $\sigma \geq 2$ (in the case of a Coulomb potential in the capacitor one has $\sigma = 2$, see [19, 15, 20, 16]), for constants $\beta, \gamma, \chi \geq 0$ which are respectively connected to self-stretching forces, tension forces and capacitance properties of the MEMS device, and for bounded real functions c, p, λ which are respectively related to anisotropic damping phenomena, permittivity profile of the constitutive material and the drop voltage applied between the ground plate at height one and the plate whose displacement is governed by the function $u(x,t)$: we refer to [19, 6] and reference therein for the physical aspects and deduction of (1). We assume in (1) Steklov boundary conditions, with nonnegative parameter d, accordingly to applications which demand more flexible conditions than Navier's, corresponding to $d = 0$ and Dirichlet conditions $u = u_\nu = 0$, obtained formally by setting $d = \infty$.

Existence of steady states for problem (1) have been established in [5] for drop voltage λ below the so-called pull-in voltage λ^*, a critical value which accordingly to the Euclidean space dimension may produce instability, namely solutions u^* such that $\|u^*\|_\infty = 1$, which corresponds to the physical situation in which the deflecting MEMS' plate touches the ground plate, and this actually occurs in dimension higher than the so-called critical dimension N^*, see [4, 7]. As far as we know, for the dynamic version (1) no results in this direction are available at the moment and this work is a first step towards a deeper understanding of those problems. In [6] the dynamic is considered from the point of view of the inverse problem of identifying unknown coefficients under additional information on the solution (which therefore is assumed to exist). Here we consider a variant of the nonlocal contribution due to $\chi > 0$, which generalizes the first-order approximation model (in Taylor's expansion, see [15]), in case of non-constant capacitance, corresponding to $\sigma = 2$. However, we will see that our approach allows more general nonlocal effects than the one considered here and in previous works [5, 6, 16, 20, 13, 14]. We mention that evolution MEMS equations have been previously handled by different methods in [18, 11, 12], where existence results are obtained avoiding nonlocal contributions. More recently, results for nonlocal parabolic problems are obtained in [13] whereas the second-order hyperbolic nonlocal MEMS equation is studied in the one-dimensional case in [14].

For $\beta, \chi > 0$, nonlocal perturbations destroy the variational structure of problem (1) and we investigate existence of weak solutions by exploiting in Sections 3 and 4 a near operator theorem in the sense of Campanato [3, 22]. This approach enables us to prove existence and uniqueness of the solution locally in time but globally in the physical parameters involved in the problem: a key ingredient in our approach relies on a penalization technique. Differently from the stationary case covered in [5] here the problem is somehow delicate as it manifests itself through an hyperbolic nature. Then we are concerned with proving regularity of solutions by adapting and further developing abstract results of [1] and [10]. In this respect, it is worth to mention that standard interpolation theory does not suite optimal regularity results even with the aid of higher-order (operator) perturbations.

Our main results are the following:

Theorem 1. *Let $\Omega \subset \mathbb{R}^N$, $1 \leq N \leq 3$, be a bounded domain with sufficiently small diameter, $\sigma \geq 2$, nonnegative constants β, γ, χ and $0 \leq d < d_0$, where d_0 is the first boundary eigenvalue of the biharmonic operator subject to Steklov boundary conditions. Let also p, c be bounded functions and $\lambda \in C^1((0,T); L^2(\Omega))$ such that $\|\lambda\|_\infty < \lambda^*$, $u_0 \in H^2 \cap H_0^1(\Omega)$ (satisfying suitable compatibility conditions) and $u_1 \in L^2(\Omega)$. Then, problem (1) possesses a unique solution $u \in C^0([0,T]; H^2(\Omega)) \cap C^1([0,T]; L^2(\Omega))$. The same conclusion holds if $d = \infty$ and Ω is a ball.*

Theorem 2. *Let $u \in C^0([0,T]; H_0^2(\Omega)) \cap C^1([0,T]; L^2(\Omega))$ be the solution to problem (1) given by Theorem 1. Assume $u_0, u_1 \in H^2 \cap H_0^1(\Omega)$ and $c \in W^{1,\infty}((0,T); L^2(\Omega))$. Then, the solution enjoys the following regularity:*

$$u \in C^0([0,T]; H^4(\Omega)) \cap C^1([0,T]; H^2 \cap H_0^1(\Omega)) \cap C^2([0,T]; L^2(\Omega)).$$

2. Preliminaries

Next we recall some basic facts in the abstract setting which will be used in the sequel. Let V, H be Hilbert spaces such that $V \hookrightarrow H \hookrightarrow V'$ with continuous and dense embeddings. Let $\mathbf{A}$ be a linear operator such that $\mathbf{A} : V \longrightarrow V'$ and which enjoys the following properties:

$$\exists \nu > 0 \text{ such that } \langle \mathbf{A}u, u \rangle \geq \nu \|u\|_V^2, \qquad \forall u \in V \tag{2}$$

$$\exists M_1 > 0 \text{ such that } |\langle \mathbf{A}u, v \rangle| \leq M_1 \|u\|_V \|v\|_{V'}, \qquad \forall u \in V, \forall v \in V' \tag{3}$$

$$\langle \mathbf{A}u, v \rangle = \langle \mathbf{A}v, u \rangle, \qquad \forall u, v \in V. \tag{4}$$

Here we denote by $\langle \cdot, \cdot \rangle$ the duality pairing.

Let $T > 0$ and for all $t \in [0,T]$ let $\mathbf{R}(t) : V \longrightarrow H$ be a linear operator such that $\mathbf{R} \in L^\infty(0,T)$ and there exists $M_2 > 0$ such that

$$|(\mathbf{R}(t)u, v)_H| \leq M_2 \|u\|_V \|v\|_H, \quad \forall t \in [0,T], \ \forall u \in V, \, v \in H. \tag{5}$$

Let $\mathbf{C}(t) : H \longrightarrow H$ be a linear operator such that $\mathbf{C} \in L^\infty(0,T)$ and there exists $M_3 > 0$ such that

$$|(\mathbf{C}(t)u, v)_H| \leq M_3 \|u\|_H \|v\|_H, \quad \forall t \in [0,T], \ \forall u, v \in H. \tag{6}$$

Consider the following Cauchy problem

$$\begin{cases} \mathbf{A}u(t) + \mathbf{R}(t)u(t) + \mathbf{C}(t)u'(t) + u''(t) = f(t), \quad t \in [0,T] \\ u(0) = u_0 \\ u'(0) = u_1 \end{cases} \tag{7}$$

where $f \in L^2((0,T);H)$.

Definition 1. As a solution of problem (7) we mean

$$u \in L^2((0,T);V) \cap H^1((0,T);H) \cap H^2((0,T);V')$$

such that the following holds

$$\begin{cases} \displaystyle\int_0^T \langle \mathbf{A}u(t), v(t)\rangle + (\mathbf{R}(t)u(t), v(t))_H + (\mathbf{C}(t)u'(t), v(t))_H \\ \quad +\langle u''(t), v(t)\rangle \, dt = \displaystyle\int_0^T (f(t), v(t))_H \, dt \\ u(0) = u_0 \end{cases} \tag{8}$$

for all $v \in L^2((0,T);V) \cap H^1((0,T);H)$ such that $v(T) = 0$.

Remark 1. Notice that

$$\begin{aligned} \int_0^T \langle u''(t), v(t)\rangle \, dt &= \int_0^T \frac{d}{dt}\langle u'(t), v(t)\rangle - \int_0^T \langle u'(t), v'(t)\rangle \, dt \\ &= \langle u'(T), v(T)\rangle - \langle u'(0), v(0)\rangle - \int_0^T \langle u'(t), v'(t)\rangle \, dt \\ &= -(u'(0), v(0))_H - \int_0^T (u'(t), v'(t))_H \, dt \\ &= -(u_1, v(0))_H - \int_0^T (u'(t), v'(t))_H \, dt. \end{aligned}$$

Hence (8) is equivalent to requiring

$$\begin{cases} \displaystyle\int_0^T [\langle \mathbf{A}u(t), v(t)\rangle + (\mathbf{R}(t)u(t), v(t))_H + (\mathbf{C}(t)u'(t), v(t))_H \\ \quad -(u'(t), v'(t))_H] \, dt = \displaystyle\int_0^T (f(t), v(t))_H \, dt + (u_1, v(0))_H \\ u(0) = u_0 \end{cases}$$

for all $v \in L^2((0,T);V) \cap H^1((0,T);H)$ such that $v(T) = 0$.

As a consequence of [17, Chap. 3.8] and [1], we have the following

Theorem 3 ([17, 1]). *Let $f \in L^1((0,T);H)$, $u_0 \in V$ and $u_1 \in H$, then there exists a unique solution $u \in L^2((0,T);V) \cap H^1((0,T);H) \cap H^2((0,T);V')$ to problem (7).*

Moreover, the solution $u \in C^0([0,T];V) \cap C^1((0,T);H)$ *and the following energy identity holds*

$$\begin{aligned} \langle \mathbf{A}u(t), u(t)\rangle &+ \int_0^t [(\mathbf{R}(s)u(s), u'(s))_H + (\mathbf{C}(s)u'(s), u'(s))_H]\, ds + \|u'(t)\|_H^2 \\ &= \langle \mathbf{A}(0)u_0, u_0\rangle + \|u_1\|_H^2 + \int_0^t \langle \mathbf{A}'(s)u(s), u(s)\rangle\, ds \\ &+ \int_0^t (f(s), u'(s))_H\, ds, \quad t \in [0,T]. \end{aligned} \tag{9}$$

Remark 2. Notice that in particular if we set

$$\Theta u = (\mathbf{A}u + \mathbf{R}(t)u + \mathbf{C}(t)u' + u'', u(0), u'(0))$$

and

$$\tilde{Y} = \{u \,|\, u \in C^0([0,T];V) \cap C^1([0,T];H),\ \Theta u \in L^1((0,T);V') \times V \times H\}$$

then $\Theta_{|\tilde{Y}}$ is an isomorphism of $\tilde{Y}$ onto $L^1(0,T,L^2(\Omega)) \times V \times H$; see also Remark 4.4 in [1].

Define the function space $\mathcal{H}_d$ as the Sobolev space $H^2(\Omega) \cap H_0^1(\Omega)$ endowed with the scalar product

$$(v,w)_d := \int_\Omega \Delta v \Delta w\, dx - d\int_{\partial\Omega} v_\nu w_\nu\, dS$$

which induces on $H^2 \cap H_0^1(\Omega)$ a norm which is equivalent to the standard Sobolev norm, provided $d < d_0$, the first simple boundary eigenvalue of the biharmonic operator subject to Steklov boundary conditions, see [9]:

$$d_0 := \inf_{H^2 \cap H_0^1(\Omega) \setminus H_0^2(\Omega)} \frac{\int_\Omega |\Delta u|^2\, dx}{\int_{\partial\Omega} |u_\nu|^2\, dS}.$$

In particular, the operator Δ^2 yields an isomorphism of $\mathcal{H}_d$ onto $L^2(\Omega)$. In the case $d = \infty$ in which the Dirichlet boundary condition $u = u_\nu = 0$ is considered, we set $\mathcal{H}_\infty := H_0^2(\Omega)$ endowed with the scalar product

$$(v,w)_\infty := \int_\Omega \Delta v \Delta w\, dx.$$

Problem (1) enters this abstract framework by choosing $V = \mathcal{H}_d$, $H = L^2$,

$$\begin{aligned} \langle \mathbf{A}u, v\rangle &:= \int_\Omega \Delta u(x,t)\Delta v(x,t)\, dx - d\int_{\partial\Omega} u_\nu(x,t)v_\nu(x,t)\, dS \\ \mathbf{C}(t) &:= c(x,t)I \\ f(x,t) &:= G(x,t) + H(x,t) \end{aligned}$$

where $I : H \longrightarrow H$ is the identity map, so that

$$u \in L^2((0,T);\mathcal{H}_d) \cap H^1((0,T);L^2(\Omega)) \cap H^2((0,T);\mathcal{H}_d')$$

is a solution of (1) provided

$$\begin{cases} \int_0^T \int_\Omega [\Delta u(x,t)\Delta v(x,t) + c(x,t)u'(x,t)v(x,t) - u'(x,t)v'(x,t)]\, dx \\ \quad -d \int_{\partial\Omega} u_\nu(x,t) v_\nu(x,t)\, dS\, dt \\ = \int_0^T \int_\Omega [G(\beta,\gamma,u(x,t)) + H(\lambda(t),\chi,p(x),u(x,t))] v(x,t)\, dx\, dt \\ \quad + \int_\Omega u_1(x)\, v(x,0)\, dx \\ 0 \le u(x,t) < 1, \qquad \text{in } \Omega \times [0,T] \\ u(x,0) = u_0 \in H^4 \cap \mathcal{H}_d(\Omega), \quad 0 < u_0 < 1 \end{cases}$$

for all $v \in L^2((0,T);\mathcal{H}_d(\Omega)) \cap H^1((0,T);L^2(\Omega))$ such that $v(x,T) = 0$ (notice that in the case $d = \infty$ it is enough to assume $u_0 \in \mathcal{H}_d$). Moreover, from the energy identity (9) we have

$$\begin{aligned} &\int_\Omega |\Delta u(x,t)|^2\, dx \; - \; d \int_{\partial\Omega} |u_\nu|^2\, dS \\ &\qquad + \int_0^t \int_\Omega c(x,s)|u'(x,s)|^2\, dx\, ds \; + \; \int_\Omega |u'(x,t)|^2\, dx \\ &\quad = \int_\Omega |\Delta u_0(x)|^2\, dx \; - \; d \int_{\partial\Omega} |(u_0)_\nu|^2\, dS \; + \; \int_\Omega |u_1(x)|^2\, dx \\ &\qquad + \int_0^t \int_\Omega f(x,s)\, u'(x,s)\, dx\, ds, \end{aligned}$$

from which we get

$$\begin{aligned} &\int_\Omega |\Delta u(x,t)|^2\, dx \; - \; d \int_{\partial\Omega} |u_\nu|^2\, dS \; + \; \int_\Omega |u'(x,t)|^2\, dx \\ &\quad \le \int_\Omega |\Delta u_0(x)|^2\, dx \; - \; d \int_{\partial\Omega} |(u_0)_\nu|^2\, dS \; + \; \int_\Omega |u_1(x)|^2\, dx \\ &\qquad + \int_0^t \left(\int_\Omega |f(x,s)|^2\, dx\right)^{\frac{1}{2}} \left(\int_\Omega |u'(x,s)|^2\, dx\right)^{\frac{1}{2}} ds \\ &\qquad + \|c\|_\infty \int_0^t \int_\Omega |u'(x,s)|^2\, dx\, ds. \end{aligned} \tag{10}$$

Remark 3. Note that condition $0 \le u < 1$ holds pointwise, if $N < 4$, by Sobolev's embedding of H^2 into the space of continuous functions.

Remark 4. The second-order boundary condition involved in (1) needs to be legitimated since in the Sobolev space $H^2 \cap H^1_0(\Omega)$ second-order derivatives do not have, in general, trace on $\partial\Omega$. However, by elliptic regularity theory, the weak solution to the problem

$$\begin{cases} \Delta^2 u = f, & \text{in } \Omega \\ u = 0 \text{ and } u_\nu = g, & \text{on } \partial\Omega \end{cases}$$

belongs to $H^4(\Omega)$ provided $f, g \in L^2$. Let us rewrite (1) in the following form

$$\Delta^2 u = -c(x,t)\, u' - u'' + G(\beta, \gamma, u) + H(\lambda(t), \chi, p(x), u) \tag{11}$$

so that, for all $t \in [0,T]$, the trace of the Laplacian $\Delta u(x,t)$ turns out to be well defined on $\partial\Omega$ once that the right-hand side in (11) belongs to L^2. This will be a consequence of the a priori estimate proved in Theorem 2.

The following version of Gronwall's lemma (proved in [21] actually for $\delta = 1$ but the argument trivially extends to any $\delta \geq 0$) will be used in the next section.

Lemma 1. *Let $w \in C^0([0,T])$, $w(t) \geq 0$ for all $t \in [0,T]$ and $w(0) = w_0 \geq 0$ such that the following holds*

$$w(t) \leq \delta\, w_0 + \int_0^t [h(s) + g(s)\, w^\beta(s)]\, ds, \qquad \forall t \in [0,T]$$

with $\delta \geq 0$, $\beta \in [0,1)$ and $g, h \in L^1(0,T)$ such that $h, g \geq 0$, for almost all $t \in [0,T]$. Then the following inequality holds

$$w(t) \leq \frac{1}{1-\beta}\left(\delta\, w_0 + \int_0^t h(s)\, ds\right) + \left(\int_0^t g(s)\, ds\right)^{\frac{1}{1-\beta}}, \ \forall t \in [0,T].$$

Lemma 1 applied to (10) with

$$\begin{aligned} w(t) &= \int_\Omega |\Delta u(x,t)|^2\, dx - d\int_{\partial\Omega} |u_\nu|^2\, dS + \int_\Omega |u'(x,t)|^2\, dx \\ w_0 &= \int_\Omega |\Delta u_0(x)|^2\, dx - d\int_{\partial\Omega} |(u_0)_\nu|^2\, dS + \int_\Omega |u_1(x)|^2\, dx \\ h(t) &= \|c\|_\infty \int_\Omega |u'(x,t)|^2\, dx, \;\; g(t) = \left(\int_\Omega |f(x,t)|^2\, dx\right)^{\frac{1}{2}}, \;\; \beta = \frac{1}{2}, \; \delta = 1 \end{aligned}$$

$$\begin{aligned} &\int_\Omega |\Delta u(x,t)|^2\, dx - d\int_{\partial\Omega} |u_\nu|^2\, dS + \int_\Omega |u'(x,t)|^2\, dx \\ &\quad \leq 2\int_\Omega |\Delta u_0(x)|^2 - d\int_{\partial\Omega} |(u_0)_\nu|^2\, dS + |u_1(x)|^2\, dx \\ &\qquad + \int_0^t \int_\Omega 2\|c\|_\infty |u'(x,t)|^2\, dx\, dt + \left\{\int_0^t \left[\int_\Omega |f(x,t)|^2\, dx\right]^{\frac{1}{2}} dt\right\}^2. \end{aligned}$$

Finally by the classical Gronwall inequality we get

$$\begin{aligned} &\int_\Omega |\Delta u(x,t)|^2\,dx - d\int_{\partial\Omega} |u_\nu|^2\,dS + \int_\Omega |u'(x,t)|^2\,dx \\ &\leq e^{2T\|c\|_\infty}\Big(2\int_\Omega |\Delta u_0(x)|^2 - d\int_{\partial\Omega} |(u_0)_\nu|^2\,dS + |u_1(x)|^2\,dx \\ &+\Big\{\int_0^t \Big[\int_\Omega |f(x,t)|^2\,dx\Big]^{\frac{1}{2}}\,dt\Big\}^2\Big). \end{aligned} \tag{12}$$

3. Existence via Campanato's method: a penalization approach

The proof of existence in Theorem 1 is achieved by extending to the dynamic setting the idea introduced in [5]. We buy the line of the stationary case whose key-ingredient is an abstract result of [22]. The argument belongs to the so-called Near Operators Theory introduced by S. Campanato in [3] within the framework of non-variational nonlinear elliptic systems and which we will use in the following form due to the third named author:

Theorem 4 (Theorem 2.1 in [22]). *Let X be a topological space, Y a set, Z a Banach space and the following mappings $\mathbf{F} : X \times Y \longrightarrow Z$, $\mathbf{B} : Y \longrightarrow Z$. Assume that:*

(i) *there exists $(\mathbf{x}_0, y_0) \in X \times Y$ such that $\mathbf{F}(\mathbf{x}_0, y_0) = 0$;*
(ii) *the map $\mathbf{x} \mapsto \mathbf{F}(\mathbf{x}, y_0)$ is continuous at $\mathbf{x}_0$;*
(iii) *there exist $k_1 > 0, k_2 \in (0,1)$ and a neighborhood of $\mathbf{x}_0$, $U(\mathbf{x}_0) \subset X$, such that for all $y_1, y_2 \in Y$ and for all $\mathbf{x} \in U(\mathbf{x}_0)$ we have*
$$\|\mathbf{B}(y_1) - \mathbf{B}(y_2) - k_1[\mathbf{F}(\mathbf{x}, y_1) - \mathbf{F}(\mathbf{x}, y_2)]\|_Z \leq k_2\|\mathbf{B}(y_1) - \mathbf{B}(y_2)\|_Z.$$
(iv) *$\mathbf{B}$ is injective;*
(v) *$\mathbf{B}(Y)$ is a neighborhood of $z_0 = \mathbf{B}(y_0)$.*

Then, there exists a ball $S(z_0, r) \subset \mathbf{B}(Y)$ and a neighborhood of $\mathbf{x}_0$, $V(\mathbf{x}_0) \subset U(\mathbf{x}_0)$, such that the following problem:
$$\begin{cases} \mathbf{F}(\mathbf{x}, y(\mathbf{x})) = 0, & \forall \mathbf{x} \in V(\mathbf{x}_0) \\ y(\mathbf{x}_0) = y_0 \end{cases}$$
possesses a unique solution $y : V(\mathbf{x}_0) \longrightarrow \mathbf{B}^{-1}(S(z_0, r))$. Moreover, if condition (iii) holds for all $\mathbf{x} \in X$, then the solution $y = y(\mathbf{x})$ turns out to be defined in the whole X.

In our context we choose $X = \mathbb{R}^+ \times \mathbb{R}^+ \times \mathbb{R}^+ \times \Phi \times \Lambda$ where
$$\Phi = \{f \in L^\infty(\Omega) : |x : f(x) > 0| \neq 0\}$$
and
$$\Lambda = \{\lambda \in L^\infty[0,T] : 0 < \lambda < \lambda^*, \quad \lambda^* \in \mathbb{R}^+\}$$

Here we take the opportunity to better explain one of the argument used in [5] and for which we realized the need of giving more details. Indeed, in order to verify condition (v) in Theorem 4 we need to perform a penalization procedure. Namely let $\varepsilon > 0$ and consider the following penalized problem

$$\begin{cases} \Delta^2 u_\varepsilon + c(x,t)u'_\varepsilon + u''_\varepsilon = G(\beta,\gamma,u_\varepsilon) + H_\varepsilon(\lambda,\chi,p(x),u_\varepsilon), \ \text{in } \Omega\times[0,T] \\ 0 < u_\varepsilon(x,t) < 1, \quad \text{in } \Omega\times[0,T] \\ u_\varepsilon(x,t) = \varepsilon, \quad \Delta u_\varepsilon(x,t) - d\dfrac{\partial u_\varepsilon(x,t)}{\partial\nu} = 0, \quad \text{on } \partial\Omega\times[0,T] \\ u_\varepsilon(x,0) = u_0 + \varepsilon, \ u'_\varepsilon(x,0) = 0, \ \text{on } \Omega. \end{cases} \tag{13}$$

where

$$H_\varepsilon(\lambda(t)\,\chi, p(x), u) := \frac{\lambda(t)\,p(x,t)}{[1+\varepsilon-u(x,t)]^\sigma\left[1+\chi\displaystyle\int_\Omega \frac{1}{[1-u(x,t)]^{\sigma-1}}\,dx\right]^\sigma}.$$

For problem (13) the following holds

Lemma 2. *Let* $\varepsilon_0 = \sup_\Omega v_\varepsilon$ *where* v_ε *is the solution to the stationary problem. Under the assumption of Theorem 1, for all* $\varepsilon \in (0, \frac{1-\varepsilon_0}{2})$ *problem* (13) *admits a unique solution* $u_\varepsilon \in C^0([0,T];H^2(\Omega)) \cap C^1([0,T];L^2(\Omega))$.

Define Y_ε as the set of functions $y \in C^0([0,T];\mathcal{H}_{d,\varepsilon}(\Omega)) \cap C^1([0,T];L^2(\Omega))$, where $\mathcal{H}_{d,\varepsilon}(\Omega)$ is the set of function y such that $y-\varepsilon$ belongs to $\mathcal{H}_d(\Omega)$, such that:

$$\Delta^2 y(x,t) + c(x,t)y'(x,t) + y''(x,t) \in L^1((0,T);L^2(\Omega)) \tag{14}$$

$$0 < y(x,t) < 1, \ \text{in } \overline{\Omega}\times[0,T], \quad y'(x,0) = 0 \text{ in } \Omega \tag{15}$$

$$\int_\Omega \frac{1}{[1-y(x,t)]^{4(\sigma-1)}}\,dx < M_1, \quad \forall t\in[0,T] \tag{16}$$

$$\int_\Omega |\Delta y(x,t)|^2\,dx < M_2, \quad \forall t\in[0,T] \tag{17}$$

for positive constants M_1, M_2. Set also $Z = L^1((0,T);L^2(\Omega)) \times \mathcal{H}_{d,\varepsilon}$ and finally set $\mathbf{x} = (\beta,\gamma,\chi,p,\lambda)$ to denote an element of the space X. Define

$$\begin{aligned} \mathbf{F}_\varepsilon(\mathbf{x},y) &:= (F_\varepsilon(x,y), y(x,0), y'(x,0)) \\ &= \Big(\Delta^2 y(x,t) + c(x,t)y'(x,t) + y''(x,t) \\ &\quad - G(\beta,\gamma,y(x,t)) - H_\varepsilon(\lambda(t),\chi,p(x),y(x,t)), \\ &\quad (y(x,0)-\varepsilon-u_0)|\lambda(t)-\lambda_0|, y'(x,0)\Big), \end{aligned}$$

$$\begin{aligned} \mathbf{B}(y) &:= (B(y), y(x,0)) \\ &= \left(\Delta^2 y(x,t) + c(x,t)y'(x,t) + y''(x,t),\ y(x,0)\right). \end{aligned}$$

Remark 5. For parameters $\beta,\chi,\gamma = 0$, the existence of the solution v_ε to the stationary problem related to (13), follows from the existence of the solution v_0 corresponding to $\varepsilon = 0$, obtained in [4, 2], by taking $0 < \varepsilon < (1-\varepsilon_0)/2$ and setting $v_\varepsilon = v_0 + \varepsilon$.

4. Nearness estimates: proof of Lemma 2 and Theorem 1

In this section we show that assumptions (i)–(v) of Theorem 4 are fulfilled for the penalized problem (13). As a consequence of [5] we have proved that for any $(\beta_0, \gamma_0, \chi_0, p_0) \in \mathbb{R}^+ \times \mathbb{R}^+ \times \mathbb{R}^+ \times \Phi$ and for any $t_0 \in [0, T]$ and $\lambda_0 \in (0, \lambda^*)$, there exists a unique stationary solution $v_\varepsilon(x, t_0) \in \mathcal{H}_d(\Omega) \cap H^4(\Omega)$ to the problem obtained from (13) by freezing the time variable at $t = t_0$, which satisfies $0 < v_\varepsilon(x, t_0) < 1$ together with

$$\int_\Omega \frac{1}{[1 - v_\varepsilon(x, t_0)(x)]^{4(\sigma-1)}}\, dx < M_1, \qquad \int_\Omega |\Delta v_\varepsilon(x, t_0)(x)|^2 dx < M_2$$

provided the diameter of Ω is sufficiently small and either $0 \leq d < d_0$, the first simple boundary eigenvalue of the biharmonic operator subject to Steklov boundary conditions, or $d = \infty$ and Ω is a ball. Both conditions on the parameter d are required in order to have a positive preserving property to hold for the biharmonic operator (see [9]) and which is used to prove the existence of λ^*, the so-called pull-in voltage; estimates on λ^* can be found in [4, 2].

Remark 6. Actually the nonlocal contribution due to $\chi > 0$ considered here is slight different from the one in [5]: however it is readily seen that calculations adapt to this case with minor changes and moreover that, beyound physical motivations, our argument allows quite general nonlocal "capacitance" effects.

Thus we set

$$\mathbf{x_0} = (\beta_0, \gamma_0, \chi_0, p_0, \lambda_0) \quad \text{and} \quad y_{0,\varepsilon} = v_\varepsilon(x, t_0)$$

to have

$$\mathbf{F}_\varepsilon(\mathbf{x_0}, y_{0,\varepsilon}) = 0.$$

Let us verify (ii):

$$\begin{aligned}
&\|\mathbf{F}_\varepsilon(\mathbf{x}, y_{0,\varepsilon}) - \mathbf{F}_\varepsilon(\mathbf{x}_0, y_{0,\varepsilon})\|_Z \\
&= \int_0^T \int_\Omega |F_\varepsilon(\beta, \gamma, \chi, p(x), \lambda(t), v_\varepsilon(x, t_0)) \\
&\quad - F_\varepsilon(\beta_0, \gamma_0, \chi_0, p_0(x), \lambda_0, v_\varepsilon(x, t_0))|^2 \, dx \; dt \\
&\leq 2 \int_0^T \int_\Omega \Bigg| \left[\beta \int_\Omega |\nabla v_\varepsilon(x, t_0)|^2 \, dx + \gamma\right] \Delta v_\varepsilon(x, t_0) \\
&\quad - \left[\beta_0 \int_\Omega |\nabla v_\varepsilon(x, t_0)|^2 \, dx + \gamma_0\right] \Delta v_\varepsilon(x, t_0)\Bigg|^2 dx\, dt \\
&\quad + 2 \int_0^T \int_\Omega \Bigg| \frac{\lambda(t) p(x)}{[1 + \varepsilon - v_\varepsilon(x, t_0)]^\sigma \, [1 + \chi\, h(v_\varepsilon)]^\sigma} \\
&\quad - \frac{\lambda_0\, p_0(x)}{[1 + \varepsilon - v_\varepsilon(x, t_0)]^\sigma \, [1 + \chi_0\, h(v_\varepsilon)]^\sigma}\Bigg|^2 dx\, dt \\
&= I_1 + I_2
\end{aligned}$$

where we set for simplicity

$$h(u) = \int_\Omega \frac{1}{[1-u(x,t)]^{\sigma-1}}\,dx.$$

We have

$$I_1 \le 4T\left[\,|\beta-\beta_0|^2|\Omega|\left(\int_\Omega |\nabla v_\varepsilon(x,t_0)|^2\,dx\right)^2 + |\gamma-\gamma_0|^2\right] T\int_\Omega |\Delta v_\varepsilon(x,t_0)|^2\,dx$$

whereas the second integral can be estimated as follows

$$\begin{aligned}
I_2 \le 4\int_0^T\int_\Omega & \frac{1}{[1+\varepsilon - v_\varepsilon(x,t_0)]^{2\sigma}}\left[\left|\frac{\lambda(t)[p(x)-p_0(x)]}{[1+\chi\,h(v_\varepsilon)]^\sigma}\right|^2\right.\\
&+\left|\frac{[\lambda(t)-\lambda_0]p_0(x)}{[1+\chi\,h(v_\varepsilon)]^\sigma}\right|^2\\
&+\left.\left|\frac{\lambda_0 p_0(x)}{[1+\chi\,h(v_\varepsilon)]^\sigma}-\frac{\lambda_0 p_0(x)}{[1+\chi_0\,h(v_\varepsilon)]^\sigma}\right|^2\right]dx\,dt \qquad (18)\\
\le\, & C(T,\Omega,M_1)\,(\lambda^*)^2\,\|p-p_0\|^2_{L^\infty(\Omega)}\\
&+C(T,\Omega,M_1),\|p_0\|^2_{L^\infty(\Omega)}\,\|\lambda-\lambda_0\|^2_{L^\infty(0,T)}\\
&+C(T,\Omega,M_1,\sigma)\|p_0\|^2_{L^\infty(\Omega)}\,\|\lambda_0\|^2_{L^\infty(\Omega)}|\chi-\chi_0|^2
\end{aligned}$$

where we have used in the last term of (18) the elementary inequality: $|a^s-b^s|\le s(a+b)^{s-1}|a-b|$, which is valid for $a,b\ge 0$ and $s\ge 1$.

Next we prove the nearness estimate (iii) of Theorem 4 in the global form, that is we have to prove that there exists $k_1>0$ and $k_2\in(0,1)$ such that for all $\mathbf{x}\in X$, $y_1,y_2\in Y_\varepsilon$, the following inequality holds

$$\begin{aligned}
\int_0^T\int_\Omega & |(1-k_1)[B(y_1)-B(y_2)]\\
&+k_1[G(\beta,\,\gamma,\,y_1(x,t)) - H_\varepsilon(\lambda(t),\,\chi,\,p(x),\,y_1(x,t))]\\
&-[G(\beta,\,\gamma,\,y_2(x,t)) - H_\varepsilon(\lambda(t),\,\chi,\,p(x),\,y_2(x,t))]|^2\,dx\,dt\\
&+(1-k_1)^2\|\lambda(t)-\lambda\|^2_{\infty,[0,T]}\,\|y_1(x,0)-y_2(x,0)\|^2_{\mathcal{H}_d} \qquad (19)\\
\le k_2 \int_0^T\int_\Omega & |\Delta^2 y_1(x,t) + c(x,t)\,y_1'(x,t) + y_1''(x,t)\\
&-[\Delta^2 y_2(x,t) + c(x,t)\,y_2'(x,t) + y_2''(x,t)]|^2\,dx\,dt\\
&+k_2\|y_1(x,0)-y_2(x,0)\|^2_{\mathcal{H}_d}.
\end{aligned}$$

Observe that initial data comply (19) provided we chose $(1-k_1)\le k_2$.

Next we evaluate the integral part and begin to estimate

$$\begin{aligned}
&\int_0^T \int_\Omega |G(\beta,\,\gamma,\,y_1(x,t)) \;-\; G(\beta,\,\gamma,\,y_2(x,t))|^2\,dx\,dt \\
&= \int_0^T \int_\Omega \left| \left[\beta \int_\Omega |\nabla y_1(x,t)|^2\,dx \,+\, \gamma\right] \Delta y_1(x,t) \right. \\
&\qquad \left. -\left[\beta \int_\Omega |\nabla y_2(x,t)|^2\,dx \,+\, \gamma\right] \Delta y_2(x,t) \right|^2 dx\,dt \\
&\leq 2 \int_0^T \int_\Omega \left| \left[\beta \int_\Omega |\nabla y_1(x,t)|^2\,dx \,+\, \gamma\right] [\Delta y_1(x,t) \;-\; \Delta y_2(x,t)] \right|^2 dx\,dt \\
&\quad + 2\,\beta^2 \int_0^T \int_\Omega \left|\Delta y_2(x,t)\right| \left| \int_\Omega |\nabla y_1(x,t)|^2\,dx \;-\; \int_\Omega \left|\nabla y_2(x,t)|^2\right|\,dx \right|^2 dx\,dt \\
&\leq 2 \int_0^T \int_\Omega \left| \left[\beta \int_\Omega |\nabla y_1(x,t)|^2\,dx \,+\, \gamma\right] [\Delta y_1(x,t) \;-\; \Delta y_2(x,t)] \right|^2 dx\,dt \\
&\quad + 2\,\beta^2 \int_0^T \left[\int_\Omega |\Delta y_2(x,t)|^2\,dx\right] \left[\int_\Omega |\nabla y_1(x,t) \;-\; \nabla y_2(x,t)|^2\,dx\right] \\
&\qquad \cdot \left[\int_\Omega |\nabla y_1(x,t) \,+\, \nabla y_2(x,t)|^2\,dx\right] dt \\
&\leq 2 \int_0^T \left\{ \beta^2 \left[\int_\Omega |\nabla y_1(x,t)|^2\,dx\right]^2 \int_\Omega |\Delta y_1(x,t) - \Delta y_2(x,t)|^2\,dx \right. \\
&\quad + 2\,\beta\,\gamma \left[\int_\Omega |\nabla y_1(x,t)|^2\,dx\right] \int_\Omega |\Delta y_1(x,t) - \Delta y_2(x,t)|^2\,dx \\
&\quad \left. + \gamma^2 \int_\Omega |\Delta y_1(x,t) - \Delta y_2(x,t)|^2\,dx \right\} dt \\
&\quad + 2\,\beta^2 \left\{ \int_0^T \left[\int_\Omega |\Delta y_2(x,t)|^2 dx\right]^2 dt \right\}^{\frac{1}{2}} \\
&\qquad \cdot \left\{ \int_0^T \left[\int_\Omega |\nabla y_1(x,t) \;-\; \nabla y_2(x,t)|^2 dx\right]^4 dt \right\}^{\frac{1}{4}} \\
&\qquad \cdot \left\{ \int_0^T \left[\int_\Omega |\nabla y_1(x,t) \,+\, \nabla y_2(x,t)|^2 dx\right]^4 dt \right\}^{\frac{1}{4}} \\
&= \mathcal{I}G_1 + \mathcal{I}G_2 + \mathcal{I}G_3 + \mathcal{I}G_4.
\end{aligned}$$

Remark 7. By the Poincaré inequality and elliptic regularity theory one has (see, e.g., Chap. 2 of [9] and also the Appendix in [6]) the following gradient estimate

$$\sup_{t\in[0,T]} \int_\Omega |\nabla y_1(x,t)|^2\,dx \,\leq\, C\,d_\Omega^2 \sup_{t\in[0,T]} \int_\Omega |\Delta y_1(x,t)|^2\,dx.$$

where d_Ω denotes the diameter of Ω.

Next by Remark 7 and the Gronwall type estimate (12) we have

$$\begin{aligned}
\mathcal{I}G_1 &\le 2\,C^2\,\beta^2\,M_2^2\,d_\Omega^4 \int_0^T \int_\Omega |\Delta y_1(x,t) \,-\, \Delta y_2(x,t)|^2\,dx \\
&\le 2\,C^2\,\beta^2\,M_2^2\,d_\Omega^4\,e^{2T\|c\|_\infty}\Bigg\{\Bigg[\int_0^T \Bigg(\int_\Omega |\Delta^2 y_1(x,t) \,-\, \Delta^2 y_2(x,t) \\
&\quad + c(x,t)[y_1'(x,t) - y_2'(x,t)] + [y_1''(x,t) - y_2''(x,t)]|^2\,dx\Bigg)^{\frac{1}{2}} dt\Bigg]^2 \\
&\quad + \|y_1(x,0) \,-\, y_2(x,0)\|_{\mathcal{H}_d}^2\Bigg\} \\
&= 2\,C^2\,\beta^2\,M_2^2\,d_\Omega^4\,e^{2T\|c\|_\infty}\Bigg\{\Bigg[\int_0^T \Bigg(\int_\Omega |B(y_1) \,-\, B(y_2)|^2\,dx\Bigg)^{\frac{1}{2}} dt\Bigg]^2 \\
&\quad + \|y_1(x,0) \,-\, y_2(x,0)\|_{\mathcal{H}_d}^2\Bigg\}.
\end{aligned} \tag{20}$$

Similarly one has

$$\begin{aligned}
\mathcal{I}G_2 &\le 2\,C\,\beta\,\gamma\,M_2\,d_\Omega^2\,e^{2T\|c\|_\infty}\Bigg\{\Bigg[\int_0^T \Bigg(\int_\Omega |B(y_1) - B(y_2)|^2 dx\Bigg)^{\frac{1}{2}} dt\Bigg]^2 \\
&\quad + \|y_1(x,0) \,-\, y_2(x,0)\|_{\mathcal{H}_d}^2\Bigg\}
\end{aligned} \tag{21}$$

$$\begin{aligned}
\mathcal{I}G_3 \le\; &\gamma^2\,e^{2T\|c\|_\infty}\Bigg\{\Bigg[\int_0^T \Bigg(\int_\Omega |B(y_1) - B(y_2)|^2\,dx\Bigg)^{\frac{1}{2}} dt\Bigg]^2 \\
&+ \|y_1(x,0) \,-\, y_2(x,0)\|_{\mathcal{H}_d}^2\Bigg\}.
\end{aligned} \tag{22}$$

Finally the last integral can be estimated as follows

$$\begin{aligned}
\mathcal{I}G_4 &\le 2\,C^2\,\beta^2\,M_2\,d_\Omega^4\Bigg\{\int_0^T \Bigg[\int_\Omega |\Delta y_1(x,t) \,-\, \Delta y_2(x,t)|^2\,dx\Bigg]^4 dt\Bigg\}^{\frac{1}{4}} \\
&\quad \cdot\Bigg\{\int_0^T \Bigg[\int_\Omega |\Delta y_1(x,t) \,+\, \Delta y_2(x,t)|^2\,dx\Bigg]^4 dt\Bigg\}^{\frac{1}{4}} \\
&\le 4\,C^2\,\beta^2\,M_2^2\,d_\Omega^4\,e^{2T\|c\|_\infty}\Bigg\{\Bigg[\int_0^T \Bigg(\int_\Omega |B(y_1) - B(y_2)|^2\,dx\Bigg)^{\frac{1}{2}} dt\Bigg]^2 \\
&\quad + \|y_1(x,0) \,-\, y_2(x,0)\|_{\mathcal{H}_d}^2\Bigg\}.
\end{aligned} \tag{23}$$

Joining estimates (20), (21), (22), (23) we get

$$\int_0^T \int_\Omega |G(\beta,\gamma, y_1(x,t)) - G(\beta,\gamma, y_2(x,t))|^2\, dx\, dt \\ \leq C(\beta,\gamma,M_2,d_\Omega)\, e^{2T\|c\|_\infty}\left\{\left[\int_0^T \left(\int_\Omega |B(y_1)-B(y_2)|^2\, dx\right)^{\frac{1}{2}} dt\right]^2 \\ + \|y_1(x,0) - y_2(x,0)\|^2_{\mathcal{H}_d}\right\}. \tag{24}$$

Let us resume nearness estimate (19) and evaluate

$$\int_0^T \int_\Omega |\, H_\varepsilon(\lambda(t), \chi, f(x), y_1(x,t))] - H_\varepsilon(\lambda(t), \chi, f(x), y_2(x,t))|^2\, dx\; dt \\ = \int_0^T \int_\Omega \left| \frac{\lambda(t) f(x)}{[1+\varepsilon - y_2(x,t)]^\sigma\,[1 + \chi\, h(y_1)]^\sigma} \right. \\ \left. - \frac{\lambda(t)\, f(x)}{[1+\varepsilon - y_2(x,t)]^\sigma\,[1 + \chi\, h(y_2)]^\sigma}\right|^2 dx\, dt \\ \leq 2(\lambda^*)^2 \|f\|^2_{L^\infty(\Omega)} \int_0^T \int_\Omega \left| \frac{(1+\varepsilon - y_2)^\sigma - (1+\varepsilon - y_1)^\sigma}{(1+\varepsilon - y_2)^\sigma (1+\varepsilon - y_1)^\sigma}\right|^2 \\ + \left|\frac{[1+\chi h(y_2)]^\sigma - [1+\chi h(y_1)]^\sigma}{(1+\varepsilon-y_2)^\sigma}\right|^2 dx\, dt \\ = 2(\lambda^*)^2\|f\|^2_{L^\infty(\Omega)}\, (\mathcal{I}H_1 + \mathcal{I}H_2)\,.$$

Let us estimate separately the integrals $\mathcal{I}H_1$ and $\mathcal{I}H_2$ as follows

$$\mathcal{I}H_1 \leq \sigma^2 \int_0^T \int_\Omega \frac{|2(1+\varepsilon) - y_1(x,t) - y_2(x,t)|^{2(\sigma-1)} |y_1(x,t) - y_2(x,t)|^2}{[1+\varepsilon - y_1(x,t)]^{2\sigma}\,[1+\varepsilon - y_2(x,t)]^{2\sigma}}\, dx\, dt \\ \leq 36^{(\sigma-1)}\sigma^2 \int_0^T \int_\Omega \frac{[y_1(x,t) - y_2(x,t)]^2}{[1+\varepsilon - y_1(x,t)]^{2\sigma}\,[1+\varepsilon - y_2(x,t)]^{2\sigma}}\, dx\, dt \\ \leq 36^{(\sigma-1)}\sigma^2\, M_1^2\, T\, \|y_1 - y_2\|^2_{L^\infty((0,T)\times\Omega)} \\ \leq 36^{(\sigma-1)}\sigma^2\, M_1^2\, T\, C d_\Omega^{4-N}\, \|\Delta(y_1 - y_2)\|^2_{L^\infty(0,T,L^2(\Omega))} \\ \leq 36^{(\sigma-1)}\sigma^2 M_1^2 T C d_\Omega^{4-N} e^{2T\|c\|_\infty}\left\{\left[\int_0^T \left(\int_\Omega |B(y_1)-B(y_2)|^2\, dx\right)^{\frac{1}{2}} dt\right]^2 \\ + \|y_1(x,0) - y_2(x,0)\|^2_{\mathcal{H}_d}\right\} \tag{25}$$

where we have used the Sobolev embedding inequality for $y \in H^2 \cap H^1_0(\Omega)$

$$\|y\|_\infty \leq C d_\Omega^{2-\frac{N}{2}}\, \|\Delta y\|_2$$

which holds as long as $1 \le N < 4$.

$$\mathcal{I}H_2 \le \sigma^2\chi^2 \int_0^T \int_\Omega |2+\chi(h(y_2)+h(y_1))|^{2(\sigma-1)} \frac{|h(y_2)-h(y_1)|^2}{(1+\varepsilon-y_2)^{2\sigma}}\, dx\, dt$$

$$\le \sigma^2\chi^2 4^{\sigma-1}(1+\chi M_1^{\frac14}|\Omega|^{\frac34})^{2(\sigma-1)}$$

$$\int_0^T \int_\Omega \frac{dx}{[1+\varepsilon-y_2(x,t)]^{2\sigma}} \left|\int_\Omega \frac{1}{[1-y_2(x,t)]^{\sigma-1}} - \frac{1}{[1-y_1(x,t)]^{\sigma-1}}\, dx\right|^2 dt$$

$$\le \sigma^2\chi^2 4^{\sigma-1}(1+\chi M_1^{\frac14}|\Omega|^{\frac34})^{2(\sigma-1)}$$

$$\int_0^T \int_\Omega \frac{dx}{[1+\varepsilon-y_2(x,t)]^{2\sigma}} \left|\int_\Omega \frac{[1-y_1(x,t)]^{\sigma-1}-[1-y_2(x,t)]^{\sigma-1}}{[1-y_1(x,t)]^{\sigma-1}\,[1-y_2(x,t)]^{\sigma-1}}\, dx\right|^2 dt$$

$$\le C(\chi,\sigma,M_1,|\Omega|)T\|y_1-y_2\|^2_{L^\infty((0,T)\times\Omega)}$$

by means of the following bound which is a consequence of condition (16) and Hölder's inequality

$$\int_0^T \left[\int_\Omega \frac{dx}{[1+\varepsilon-y_2(x,t)]^{2\sigma}}\right]^2 \left\{\int_\Omega \frac{dx}{[1-y_1(x,t)]^{\sigma-1}\,[1-y_2(x,t)]^{\sigma-1}}\right\}^2 dt$$

$$\le TC(M_1,|\Omega|).$$

Finally, we may argue as in previous cases to get

$$\mathcal{I}H_2 \le C(\chi,\sigma,M_1,|\Omega|)Td_\Omega^2\,\|\Delta[y_1-y_2]\|^2_{L^\infty((0,T);L^2(\Omega))}$$

$$\le C(\chi,\sigma,M_1,|\Omega|)\,T\,d_\Omega^2\,e^{2T\|c\|_\infty}\left\{\left[\int_0^T\left(\int_\Omega |B(y_1)-B(y_2)|^2\,dx\right)^{\frac12} dt\right]^2 + \|y_1(x,0)-y_2(x,0)\|^2_{\mathcal{H}_d}\right\}$$

which together with (25) proves the nearness estimate (19) provided d_Ω is sufficiently small.

It remains to show that condition (v) of Theorem 4 is satisfied, namely we have to exhibit $\eta > 0$ such that for all $g \in L^1((0,T);L^2(\Omega))$ which satisfies

$$\int_0^T \int_\Omega |g - \mathbf{B}y_{0,\varepsilon}|^2\, dx\, dt \ < \eta^2$$

we can find $u_\varepsilon \in Y_\varepsilon$ such that $Bu_\varepsilon = g$. Since $y_{0,\varepsilon} \in Y_\varepsilon \cap H^4(\Omega)$ and $y'_{0,\varepsilon} = y''_{0,\varepsilon} = 0$ we have $By_{0,\varepsilon} = \Delta^2 y_{0,\varepsilon} \in L^2(\Omega)$. From what recalled in Section (3), we know that the Cauchy–Steklov problem

$$\begin{cases} Bu_\varepsilon = \Delta^2 u_\varepsilon(x,t) + c(x,t)\,u'_\varepsilon(x,t) + u''_\varepsilon(x,t) = g(x,t), & \text{q.o. in } \Omega\times[0,T],\\ u_\varepsilon(x,0) = y_{0,\varepsilon}, & \text{in } \Omega\\ u'_\varepsilon(x,0) = u_1, & \text{q.o. in } \Omega\\ u_\varepsilon(x,t) = \varepsilon,\ \Delta u_\varepsilon - d\frac{\partial u_\varepsilon}{\partial \nu} = 0, & \text{in } \partial\Omega\times[0,T] \end{cases}$$

possesses a unique solution $u_\varepsilon \in C^0((0,T);\mathcal{H}_{d,\varepsilon}) \cap C^1((0,T);L^2(\Omega))$.

We next verify the remaining conditions for functions with membership in the set Y_ε. From the energy estimate (12) applied to $u_\varepsilon - y_{0,\varepsilon}$ we get for $t \in [0,T]$

$$\begin{aligned}\left(1-\frac{d}{d_0}\right)&\int_\Omega |\Delta u_\varepsilon(x,t)-\Delta y_{0,\varepsilon}(x)|^2\,dx\\ &\le \int_\Omega |\Delta u_\varepsilon(x,t)-\Delta y_{0,\varepsilon}(x)|^2\,dx - d\int_{\partial\Omega}\left|\frac{\partial u_\varepsilon}{\partial\nu}-\frac{\partial y_{0,\varepsilon}}{\partial\nu}\right|^2 dS\\ &\le e^{2t\|c\|_\infty}\left\{\left[\int_0^T\left(\int_\Omega |B(u_\varepsilon)-B(y_{0,\varepsilon})|^2\,dx\right)^{\frac12} dt\right]^2\right\}.\end{aligned}$$

Since

$$\int_\Omega |\Delta y_{0,\varepsilon}(x)|^2\,dx < M_2, \quad \forall t\in[0,T]$$

we have also

$$\int_\Omega |\Delta u_\varepsilon(x,t)|^2\,dx < M_2, \quad \forall t\in[0,T]$$

provided η is small enough; this yields condition (17).
Recalling that $1\le N<4$, by the Sobolev embedding theorem we have

$$\begin{aligned}&\sup_\Omega |u_\varepsilon(x,t)-y_{0,\varepsilon}(x)|\\ &\quad\le C(d_\Omega^{2-\frac{N}{2}})\|\Delta[u_\varepsilon(x,t)-y_{0,\varepsilon}(x)]\|_{L^2(\Omega)} < C(d_\Omega^{2-\frac{N}{2}})\eta.\end{aligned} \tag{26}$$

Since $0<y_{0,\varepsilon}(x)<1$, for all $x\in\overline{\Omega}$, there exists $\eta_1>0$ such that for $0<\eta<\eta_1$ we have also $0<u_\varepsilon(x,t)<1$ and condition (15) is satisfied.

Finally from the following inequality

$$\int_\Omega \frac{1}{[1-u_\varepsilon(x,t)]^\sigma}\,dx \le \int_\Omega\left|\frac{[1-u_\varepsilon]^\sigma-[1-y_{0,\varepsilon}]^\sigma}{[1-y_{0,\varepsilon}]^\sigma[1-u_\varepsilon]^\sigma}\right|dx + \int_\Omega\frac{1}{[1-y_{0,\varepsilon}(x)]^\sigma}\,dx$$

and arguments similar to those used in verifying condition (ii), we have

$$\int_\Omega\left|\frac{[1-u_\varepsilon]^\sigma-[1-y_{0,\varepsilon}]^\sigma}{[1-y_{0,\varepsilon}]^\sigma[1-u_\varepsilon]^\sigma}\right|dx \le C\sup_\Omega|u-y_{0,\varepsilon}|$$

and by (26) there exists $\eta_2>0$ such that for $0<\eta<\eta_2$

$$\int_\Omega \frac{1}{[1-u_\varepsilon(x,t)]^{4(\sigma-1)}}\,dx < M_1, \quad \forall t\in[0,T]$$

and hence condition (16). This concludes the proof of Lemma 2.

We next prove Theorem 1 by showing that the solution of the penalized problem converges, as $\varepsilon\to 0$ to the solution of the original problem (1). In order to do this we prove that the family of penalized solutions yields a Cauchy sequence in $C^0([0,T];\mathcal{H}_d)\cap C^1([0,T];L^2(\Omega))$. Indeed, consider two penalized solutions u_ε

and u_{ε_1} corresponding to parameters ε and ε_1 respectively and evaluate

$$\begin{cases} \Delta^2 (u_\varepsilon - u_{\varepsilon_1}) + c(x,t)(u'_\varepsilon - u'_{\varepsilon_1}) + u''_\varepsilon - u''_{\varepsilon_1} = G(\beta,\gamma,u_\varepsilon) - G(\beta,\gamma,u_{\varepsilon_1}) \\ \quad + H_\varepsilon(\lambda,\chi,p(x),u_\varepsilon) - H_{\varepsilon_1}(\lambda,\chi,p(x),u_{\varepsilon_1}), \text{ in } \Omega\times[0,T] \\ u_\varepsilon(x,t) - u_{\varepsilon_1}(x,t) = \varepsilon - \varepsilon_1, \\ \Delta[u_\varepsilon(x,t) - u_{\varepsilon_1}(x,t)] - d\dfrac{\partial[u_\varepsilon(x,t) - u_{\varepsilon_1}(x,t)]}{\partial\nu} = 0, \qquad \text{on } \partial\Omega\times[0,T], \\ u_\varepsilon(x,0) = \varepsilon - \varepsilon_1, \ (u'_\varepsilon - u'_{\varepsilon_1})(x,0) = 0, \text{ on } \Omega. \end{cases} \tag{27}$$

Arguing as in previous cases we have

$$\begin{aligned} &\int_0^t \int_\Omega |G(\beta,\gamma,u_\varepsilon) - G(\beta,\gamma,u_{\varepsilon_1})|^2 dx\, dt \\ &\qquad \le C \int_0^t \int_\Omega |\Delta[u_\varepsilon(x,t) - u_{\varepsilon_1}(x,t)]|^2 dx\, dt, \quad t\in[0,T] \end{aligned}$$

$$\begin{aligned} &\int_0^t \int_\Omega |H_\varepsilon(\lambda,\chi,p(x),u_\varepsilon) - H_{\varepsilon_1}(\lambda,\chi,p(x),u_{\varepsilon_1})|^2 dx\, dt \\ &\qquad \le C_1(\varepsilon-\varepsilon_1)^2 T\,|\Omega| + C_2 \int_0^t \int_\Omega |\Delta[u_\varepsilon(x,t) - u_{\varepsilon_1}(x,t)]|^2 dx\, dt. \end{aligned}$$

Then applying inequality (12) to problem (27) we get for all $t\in(0,T]$

$$\begin{aligned} &\int_\Omega |\Delta[u_\varepsilon(x,t) - u_{\varepsilon_1}(x,t)]|^2 dx \\ &\qquad \le C_1(\varepsilon-\varepsilon_1)^2 T\,|\Omega| + C_2 \int_0^t \int_\Omega |\Delta[u_\varepsilon(x,t) - u_{\varepsilon_1}(x,t)]|^2 dx\, dt \end{aligned}$$

from which Gronwall's lemma yields

$$\int_\Omega |\Delta[u_\varepsilon(x,t) - u_{\varepsilon_1}(x,t)]|^2 dx \le C(\varepsilon-\varepsilon_1)^2 T\,|\Omega|$$

and the proof is now complete. □

Remark 8. We point out that the limiting procedure as $\varepsilon \to 0$ carried out so far, has to be used also in the stationary case [5], by following step by step the above calculations.

5. An abstract result towards regularity: proof of Theorem 2

We have proved so far that problem (1) possesses a unique solution

$$u \in C^0([0,T]; H^2(\Omega)) \cap C^1([0,T]; L^2(\Omega))$$

and in particular, since $u = 0$ on $\partial\Omega$ we have, regardless of higher-order boundary conditions,

$$u \in C^0([0,T]; H^2 \cap H_0^1(\Omega)) \cap C^1([0,T]; L^2(\Omega)).$$

In order to prove the regularity Theorem 2 for the MEMS problem, we need to prove that the nonlocal term $\int_\Omega |\nabla u(x,t)|^2 dx$ belongs to $C^1([0,T])$. This issue is somehow delicate, as one may think of applying interpolation theory via penalization techniques but this would yield just $u \in C^{\frac{1}{2}}([0,T]; H_0^1(\Omega))$. Let us prove the following

Lemma 3. *Let $u \in C^0([0,T]; H^2 \cap H_0^1(\Omega)) \cap C^1([0,T]; L^2(\Omega))$, then the map*

$$t \mapsto \int_\Omega |\nabla u(x,t)|^2 dx$$

belongs to $H^{1,\infty}(0,T)$.

Proof. Consider the following problem

$$\begin{cases} -\Delta u(x,t) + u'(x,t) = f(x,t), \\ u(x,0) = u(x,0) \\ u(x,t) = 0, \text{ on } \partial\Omega \times [0,T] \end{cases}$$

where by assumption $f \in C^0([0,T]; L^2(\Omega))$ and for which the energy identity reads as follows

$$\int_\Omega |\nabla u(x,t)|^2 dx - \int_\Omega |\nabla u(x,s)|^2 dx + 2\int_s^t \int_\Omega |u'(x,\xi)|^2 dx\, d\xi$$
$$= 2\int_s^t \int_\Omega f(x,\xi) u'(\xi,t)\, d\xi, \quad s < t.$$

Hence we get

$$\left| \int_\Omega |\nabla u(x,t)|^2 dx - \int_\Omega |\nabla u(x,s)|^2 dx \right|$$
$$\leq 2\int_s^t \int_\Omega |u'(x,\xi)|^2 dx\, d\xi + 2\int_s^t \int_\Omega |f(x,\xi)|\, |u'(\xi,t)|\, d\xi$$
$$\leq 2|t-s| \left(\sup_{[0,T]} \|u'(x,t)\|^2_{L^2(\Omega)} + \sup_{[0,T]} \|f(x,t)\|_{L^2(\Omega)} \sup_{[0,T]} \|u'(x,t)\|_{L^2(\Omega)} \right)$$

and thus the claim is proved. □

Let us resume equation of problem (1) written in the following form

$$\Delta^2 u + \left[-\beta \int_\Omega |\nabla u(x,t)|^2 dx + \gamma \right] \Delta u + c(x,t)u' + u'' = H(\lambda(t), \chi, p(x), u) \quad (28)$$

As a consequence of Theorem 1 the right-hand side in (28) belongs to the space $C^1([0,T]; L^2(\Omega))$, provided $\lambda \in C^1([0,T]; L^2(\Omega))$, whence by Lemma 3

$$t \mapsto -\beta \int_\Omega |\nabla u(x,t)|^2 dx + \gamma$$

belongs to $H^{1,\infty}(0,T)$.

Now we are in a situation where the following abstract result due to G. Gilardi applies:

Theorem 5. *Let* $f \in W^{1,1}((0,T);H)$, $u_0, u_1 \in V$ *and* $Au_0 \in H$. *Let* $\mathbf{R}(\mathbf{t})$ *and* $\mathbf{C}(t)$ *belong to* $W^{1,1}(0,T)$. *Moreover for any* $v \in V$ *we have* $\|\mathbf{R}'(t)v\|_H \leq c\|v\|_V$ *and for any* $v \in H$ *we have* $\|\mathbf{C}'(t)v\|_H \leq c\|v\|_H$. *Then the solution* u *of problem 7 is in* $C^0((0,T);D(A)) \cap C^1((0,T);V) \cap C^2((0,T);H)$.

which is straightforward from [10, Teorema 4.5], to obtain

$$u \in C^0([0,T];H^4(\Omega)) \cap C^1([0,T];H^2 \cap H_0^1(\Omega)) \cap C^2([0,T];L^2(\Omega))$$

provided $c \in H^{1,\infty}((0,T);L^2(\Omega))$.

References

[1] C. Baiocchi, *Soluzioni ordinarie e generalizzate del problema di Cauchy per equazioni differenzilai astratte del secondo ordine in spazi di Hilbert*, Ricerche Mat. **16** (1967), 27–95.

[2] E. Berchio, D. Cassani and F. Gazzola, *Hardy–Rellich inequalities with boundary remainder terms and applications*, Manuscripta Math. **131** (2010), 427–458.

[3] S. Campanato, *On the condition of nearness between operators*, Ann. Mat. Pura Appl. **167** (1994), 243–256.

[4] D. Cassani, J.M. do Ó and N. Ghoussoub,*On a fourth order elliptic problem with a singular nonlinearity*, Adv. Nonlinear Stud. **9** (2009), 189–209.

[5] D. Cassani, L. Fattorusso and A. Tarsia, *Global existence for nonlocal MEMS*, Nonlinear Analysis **74** (2011), 5722–5726.

[6] D. Cassani, B. Kaltenbacher and A. Lorenzi, *Direct and inverse problems related to MEMS*, Inverse Problems **25** (2009) 105002 (22 pp).

[7] C. Cowan, P. Esposito, N. Ghoussoub and A. Mordifam, *The Critical Dimension for a Fourth Order Elliptic Problem with Singular Nonlinearity*, Arch. Ration. Mech. Anal. **198** (2010), 763–787.

[8] M.G. Crandall, P.H. Rabinowitz and L. Tartar, *On a Dirichlet problem with a singular nonlinearity*, Comm. Partial Differential Equations **2** (1977), 193–222.

[9] F. Gazzola, H.C. Grunau and G. Sweers, *Polyharmonic boundary value problems*, Springer Lecture Notes, 2010.

[10] G. Gilardi, *Teoremi di regolarità per la soluzione di un'equazione differenziale astratta lineare del secondo ordine*, Rend. Ist. Lombardo, Classe Sc. A **106** (1972), 641–675.

[11] Z. Guo and J. Wei, *On the Cauchy problem for a reaction-diffusion equation with a singular nonlinearity*, J. Differential Equations **240** (2007), 279–323.

[12] Y. Guo, *Global solutions of singular parabolic equations arising from electrostatic MEMS*, J. Differential Equations **245** (2008), 809–844.

[13] K.M. Hui, *The existence and dynamic properties of a parabolic nonlocal MEMS equation*, Nonlinear Analysis **74** (2011), 298–316.

[14] N.I. Kavallaris, A.A. Lacey, C.V. Nikolopoulos, and D.E. Tzanetis, *A hyperbolic non-local problem modelling MEMS technology*, Rocky Mountain J. Math. **41** (2011), 505–534.

[15] V. Kempe, *Inertial MEMS, principles and practice*, Cambridge University Press, 2011.

[16] F.H. Lin and Y.S. Yang, *Nonlinear non-local elliptic equation modelling electrostatic actuation*, Proc. R. Soc. London, Ser. A **463** (2007), 1323–1337.

[17] J.L. Lions and E. Magenes, *Problemès aux limites non homogènes et applications*, vol. 1, Dunod, Paris, 1968.

[18] M. Loayza, *The heat equation with singular nonlinearity and singular initial data*, J. Differential Equations **229** (2006), 509–528.

[19] J.A. Pelesko and D.H. Bernstein, *Modeling MEMS and NEMS*, Chapman & Hall/CRC, 2003.

[20] J.A. Pelesko and A.A. Triolo, *Nonlocal problems in MEMS device control*, J. Engrg. Math. **41** (2001), 345–366.

[21] M. Sassetti and A. Tarsia, *Su un'equazione non lineare della corda vibrante*, Annali di Matematica, **CLXI** (1992), 1–42.

[22] A. Tarsia, *Differential equations and implicit functions: a generalization of the near operator theorem*, Topol. Methods Nonlinear Anal. **11** (1998), 115–133.

D. Cassani
Università degli Studi dell'Insubria
Via Valleggio 11
I-22100 Como, Italy
e-mail: daniele.cassani@uninsubria.it

L. Fattorusso
DIMET, Università Mediterranea degli Studi di Reggio Calabria
Via Graziella (loc. Feo di Vito)
I-89100 Reggio Calabria, Italy
e-mail: luisa.fattorusso@unirc.it

A. Tarsia
Dipartimento di Matematica
Università di Pisa
Via F. Buonarroti 2
I-56127 Pisa, Italy
e-mail: tarsia@dm.unipi.it

Progress in Nonlinear Differential Equations
and Their Applications, Vol. 85, 207–230

Existence and Multiplicity Results for Some Scalar Fields Equations

Giovanna Cerami

Abstract. In this paper the results of some researches concerning Scalar Field Equations are summarized. The interest is focused on the question of existence and multiplicity of stationary solutions, so the model equation

$$-\Delta u + a(x)u = |u|^{p-1}u \quad \text{in } \mathbb{R}^N,$$

is considered. The difficulties and the ideas introduced to face them as well as some well known results are discussed. Some recent advances concerning existence and multiplicity of multi-bump solutions are described in more detail.

Mathematics Subject Classification (2010). primary 35J20, secondary 35J60.

Keywords. Elliptic equations in $\mathbb{R}^N$, variational methods, multi-bump solutions, infinitely many positive and nodal solutions.

1. Introduction and survey on well-known and more recent results

The aim of this paper is to describe some recent advances concerning the equation

$$(E) \qquad -\Delta u + a(x)u = |u|^{p-1}u \text{ in } \mathbb{R}^N,$$

where $N \geq 2$, $p > 1$, $p < 2^* - 1 = \frac{N+2}{N-2}$, if $N \geq 3$, and the potential $a(x)$ is a positive function that is not required to possess any symmetry property.

It is well known that a strong motivation for studying such equation is due to its connection with Mathematical Physics. For instance, the search of certain kind of solitary waves (stationary states) in nonlinear equations of the Klein–Gordon or Schrödinger type leads to look for solutions of (E) (see, e.g., [5] and [8] for a detailed discussion of this fact). Moreover, Euclidean scalar fields equations, like (E), appear in several other context of Physics (nonlinear optics, laser propagation, constructive field theory, etc.).

Work supported by the Italian national research project "Metodi Variazionali e topologici nello studio di fenomeni non lineari" and by GNAMPA of INDAM..

On the other hand, it is worth stressing that, besides the importance in the applications, another considerable reason which motivates the researchers interest is the presence, in equations like (E), of some specific mathematical difficulties that make their study challenging.

Therefore, it is easy to understand why this type of equations has been object of extensive studies during last thirty years.

Equation (E) is variational in nature: its finite energy solutions can be searched as critical points of the 'action' functional defined in $H^1(\mathbb{R}^N)$ by

$$I(u) = \frac{1}{2} \int_{\mathbb{R}^N} (|\nabla u|^2 + a(x)u^2)dx - \frac{1}{p+1} \int_{\mathbb{R}^N} |u|^{p+1} dx.$$

However, the usual variational methods cannot be applied in a standard way because of a lack of compactness. The origin of the trouble is, essentially, the invariance of $\mathbb{R}^N$ under the action of the non compact group of the translations and, technically, appears as the non-compactness, whatever p is, of the embedding $j : H^1(\mathbb{R}^N) \to L^p(\mathbb{R}^N)$, and as a failure of the basic Palais–Smale condition at some energy levels.

This difficulty can be overcome when $a(x)$ enjoys some symmetry.

Indeed, first known results (see [6], [14], [15], [19], [21], [5]) have been obtained assuming either $a(x) = a \in \mathbb{R}$ or $a(x) = a(|x|)$. Actually, radial symmetry was formerly used to reduce (E) to one dimension, so that ordinary differential equations methods could be applied. However, the crucial role of symmetry was clear after the observation (due essentially to Strauss [21]) that $H^1_r(\mathbb{R}^N)$, the subspace of $H^1(\mathbb{R}^N)$ consisting of radially symmetric functions, embeds compactly in $L^p(\mathbb{R}^N)$. This device, together with the Palais Symmetric Criticality Principle, allows to prove, by usual variational arguments, the existence of a positive (ground state if $a(x) = a \in \mathbb{R}$) solution and of infinitely many, radially symmetric, changing sign solutions to (E) [5]. Moreover, it must be mentioned that, still under assumption $a(x) = a(|x|)$, one can also show the existence of infinitely many nonradial, changing sign, solutions of (E), under suitable restrictions on the dimension N (see [3], [8] and references therein), and, furthermore, that the existence of infinitely many non radial positive solutions can be proved if, in addition, the assumption that $a(|x|)$ decays at infinity with a prescribed polynomial rate is imposed (see [22]).

When $a(x)$ has no symmetry properties, even the existence question becomes a quite difficult matter, the loss of compactness is severe. Most researches have been concerned with the case in which

$$\lim_{|x|\to\infty} a(x) = a_\infty > 0$$

exists (for some existence and nonexistence results when this condition is not required see, e.g., [20],[9], [10]).

Considering the non symmetric case, the first observation is that, looking for critical points of the functional I, the topological situation appears quite different according that,

$$a(x) \to a_\infty, \qquad \text{as } |x| \to \infty, \qquad \text{from below} \tag{1.1}$$

or not.

Actually, when (1.1) is true, using concentration-compactness arguments it is possible to show (see [17], [20]) that a positive, ground state, solution of (E) exists either when $a_\infty \in \mathbb{R}$, either when $a_\infty = +\infty$ and can be found, for instance, by minimizing the functional I on the Nehari natural constraint $\mathcal{M} := \{u \in H^1(\mathbb{R}^N) : I'(u)[u] = 0\}$.

On the contrary, if (1.1) does not hold, (E) may not have a least energy solution. This is the case when

$$a(x) \geq a_\infty, \quad \forall x \in \mathbb{R}^N, \qquad \text{and} \qquad a(x) \neq a_\infty \quad \text{on a positive measure set.} \tag{1.2}$$

Indeed, when (1.2) holds, denoting the functional related to the limit equation

$$(E_\infty) \qquad -\Delta u + a_\infty u = |u|^{p-1}u \qquad \text{in } \mathbb{R}^N,$$

by

$$I_\infty(u) := \frac{1}{2}\int_{\mathbb{R}^N} (|\nabla u|^2 + a_\infty u^2)dx - \frac{1}{p+1}\int_{\mathbb{R}^N} |u|^{p+1}dx,$$

and by m_∞ the value I_∞ takes at the ground state solution w of (E_∞), it is not difficult to show the equalities

$$\inf\{I(u) : u \in \mathcal{M}\} = \min\{I_\infty(u) : u \in H^1(\mathbb{R}^N), I'_\infty(u)[u] = 0\} = m_\infty \tag{1.3}$$

and that the infimum cannot be achieved (see, e.g., [8], Prop. 3.1).

Nevertheless, this fact does not mean that when (1.1) is not verified there is no hope of finding positive solutions to (E). When (1.2) is true, the existence of a positive, not ground state, solution has been proved in [2] (Sect. VII, see also [1]) when the additional decay condition

$$\int_{\mathbb{R}^N} (a(x) - a_\infty) \exp(\sigma(a_\infty)^{1/2}|x|)|x|^{\frac{N-1}{2}} \in L^1(\mathbb{R}^N) \qquad \text{for some } \sigma > 0 \tag{1.4}$$

is satisfied. The idea of the proof is to look for critical points of I at higher energy levels, using subtle topological tools and minimax methods, and taking advantage of a deep study of the nature of the obstacles to the compactness ([4], [2]).

The subsequent natural question to investigate is under which conditions (E) admits multiple solutions, eventually infinitely many, as in the radially symmetric case. However, facing this question, it is not difficult to guess that the strategy of trying to construct critical levels of I, avoiding the 'bad levels' for the compactness, probably is not the most appropriate. Indeed, the representation theorem of the non-compact Palais–Smale sequences of I (see [4]) supplies the information that 'bad' levels of I can be located by critical values of the limit functional. Therefore, since by the Berestycki–Lions result ([5]) (E_∞) has infinitely many solutions, infinitely many 'dangerous' levels exist.

The existence of infinitely many *changing sign solutions* to (E), when (1.1) holds, has been proved in [9], using a quite natural approach, under the decay condition

$$\lim_{|x|\to+\infty} \frac{\partial a}{\partial \mathbf{x}}(x)e^{\sigma|x|} = +\infty, \qquad \forall\sigma > 0, \quad \text{where } \mathbf{x} = \frac{x}{|x|},$$

assuming, in addition, some stability of the value $\frac{\partial a}{\partial \mathbf{x}}(x)$ with respect to small perturbations of the direction. The proof is based on the idea of approximating the equation in the whole space $\mathbb{R}^N$ by a sequence of problems in balls, $B_{r_n}(0)$, centered at the origin and whose radius r_n is increasing to $+\infty$, as $n \to +\infty$. Clearly, by using standard arguments, it is possible to construct for any such approximating problem in $B_{r_n}(0)$ infinitely many solutions $(u_k^n)_k$. Then, using such families of solutions, infinitely many sequences consisting of approximate solutions having the same topological nature can be built. Finally, the fact that, passing to the limit, these sequences give the desired infinitely many solutions of (E), and do not give rise to non compact sequences, is obtained by very delicate estimates involving Pohozaev type inequalities and Morse index arguments.

The possibility of proving the existence of *infinitely many 'multi-bump' positive solutions* for (E) has been firstly successfully explored in the pioneering paper [16], under periodicity assumptions on the coefficients. This property, of course, plays a crucial role in realizing the project of getting multi-bump solutions by 'gluing' positive bumps of the same nature.

Subsequently, as already mentioned, the existence of infinitely many positive multi-bump solutions has been proved in [22], in a radially symmetric framework, imposing on the potential the decay condition $a(|x|) = a_\infty + \frac{c_1}{|x|^m} + O(\frac{1}{|x|^{m+\sigma}})$, as $|x| \to +\infty$, with a suitable choice of m and σ.

On the other hand, considering the semi-classical equation

$$-\varepsilon^2\Delta u + a(x)u = |u|^{p-1}u, \qquad \text{in } \mathbb{R}^N, \tag{1.5}$$

where ε is a small parameter, it is worth recalling that a lot of work has been done on the question of the positive solutions multiplicity and that (1.5) is strongly related to (E), because, by a change of variables, it becomes

$$-\Delta u + a(\varepsilon x)u = |u|^{p-1}u, \qquad \text{in } \mathbb{R}^N.$$

The number of solutions of (1.5) has been related to the number and/or the type of critical points of $a(x)$ and, also, to the topology of the sublevel sets of $a(x)$. The method mostly used in the proofs has been the so-called method of the 'projections' and a Lyapunov–Schmidt reduction of the problem to a finite-dimensional one. Since it is very difficult to cite all the interesting contributions in this direction without forgetting something, we just refer the interested reader to the latest ones [7] [18] and to references therein. With respect to (1.5), it must be pointed out that even the best results one can obtain in this setting sound, more or less, as follows: under suitable assumptions on the nature (of critical points) of a, for any fixed integer k, there exists $\varepsilon_k > 0$ such that, when $\varepsilon \in (0, \varepsilon_k)$, there are at least k positive (possibly multi-bump) solutions. So, it is not difficult to realize that ε_k goes to 0 as the number k of the desired solutions tends to infinity. Thus, there is no hope of obtaining, by this approach, the existence of infinitely many solutions to (E).

Recently some contributions to the settlement of the question of the existence of infinitely many 'multi-bump' solutions to (E), when no symmetry assumptions on $a(x)$ are available, have been given in [12], [13], [11]. All these results concern the case

$$a(x) \not\equiv a_\infty, \qquad a(x) \to a_\infty, \quad \text{as } |x| \to \infty, \qquad \text{from above} \tag{1.6}$$

and are proved by using purely variational methods.

The rest of the present work is devoted to describe these results, providing some insight on the main ideas.

The first result (see [12]) is stated as follows:

Theorem 1.1. *Let assumptions*

(h_1) $a(x) \longrightarrow a_\infty > 0 \quad$ *as* $\quad |x| \to \infty$,
(h_2) $a(x) \geq a_0 > 0 \quad \forall x \in \mathbb{R}^N$,
(h_3) $a \in L^{N/2}_{\rm loc}(\mathbb{R}^N)$,
(h_4) $\exists\ \bar\eta \in (0, \sqrt{a_\infty}) : \ \lim_{|x|\to+\infty}(a(x) - a_\infty)e^{\bar\eta|x|} = +\infty$

be satisfied.

Then, there exists a positive constant, $\mathcal{A} = \mathcal{A}(N, \bar\eta, a_0, a_\infty) \in \mathbb{R}$, *such that, when*

$$(*) \qquad |a(x) - a_\infty|_{L^{N/2}_{\rm loc}} := \sup_{y\in\mathbb{R}^N} |a(x) - a_\infty|_{L^{N/2}(B_1(y))} < \mathcal{A},$$

equation (E) *has infinitely many positive solutions belonging to* $H^1(\mathbb{R}^N)$.

It is worth making at once some remarks on the above theorem assumptions. These comments are also helpful to understand the reasons of the development of the researches we discuss in this paper.

First of all, let notice that regularity assumption (h_3) on a is very mild and, moreover, that neither $\inf_{x\in\mathbb{R}^N} a(x) = a_\infty$, nor $a(x) \geq a_\infty$ for all $x \in \mathbb{R}^N$ are required.

Assumption (h_4) is a 'slow decay' condition, it can be satisfied when $a(x)$ decays very slowly, although, unlike [22], a suitable exponential decay is allowed. It is interesting to observe that (h_4) is almost complementary of the 'fast decay' condition (1.4) imposed to $a(x)$ in [2] and in [1]). The role played by (h_4) is basic: it is the deep motivation for which the variational argument works. Indeed, as we shall see, the solutions are found by a max-min argument on the action functional I and the procedure is successful because the attractive effect of $a(x)$ on the 'bumps' is dominating on the repulsive disposition, which is of a specified exponential type, of positive masses with respect to each other.

On the contrary, the 'small oscillation' condition $(*)$ on a appears less natural, hence, reasonably, one wonders whether it is necessary or not.

The answer to this question is contained in a theorem, proved in [11], in which the same claim of Theorem 1.1 is obtained without imposing any 'small oscillation' condition on the potential, but asking that the potential has a, suitably large and suitably far away, 'pit around the origin':

Theorem 1.2. *Let $a(x)$ satisfy (h_1), (h_2), (h_3), and (h_4). Let $A \subset \mathbb{R}^N$ be an open, bounded set such that $0 \in A$. Let $\mathcal{N}$ be a bounded neighborhood of ∂A, and let $b : \mathcal{N} \to \mathbb{R}$ be a continuous function, having compact support, such that*

$$\inf_{x \in \partial A} b(x) = b_0 > 0, \qquad \sup_{\mathcal{N}} b(x) < a_0.$$

Let us consider the equation

$$(E_\varepsilon) \qquad -\Delta u + a_\varepsilon(x) u = |u|^{p-1} u \text{ in } \mathbb{R}^N,$$

where $a_\varepsilon(x) = a(x) - b_\varepsilon(x)$ and $b_\varepsilon : \mathbb{R}^N \to \mathbb{R}$ is defined by $b_\varepsilon(x) = b(\varepsilon x)$ if $x \in \mathcal{N}/\varepsilon$, $b_\varepsilon(x) = 0$ if $x \notin \mathcal{N}/\varepsilon$ ($\mathcal{N}/\varepsilon = \{x \in \mathbb{R}^N : \varepsilon x \in \mathcal{N}\}$).

Then, there exists $\bar{\varepsilon} > 0$ such that, for all $\varepsilon \in (0, \bar{\varepsilon})$, equation (E_ε) has infinitely many positive solutions belonging to $H^1(\mathbb{R}^N)$.

It is worth stressing that equations (E_ε) are exactly of the type (E) for any choice of ε and that the claim, unlike the above quoted results for semi-classical equations, gives *for all* ε suitably small *infinitely many positive solutions* to (E_ε).

It is an open question if (*) can be merely dropped.

A further question coming in a quite natural way, thinking out the attractive effect of $a(x)$, is to investigate whether the multi-bump solutions, found in Theorem 1.1, can converge to a solution of (E), when the number of the bumps tends to infinity.

The study of this question has been object of [13] and the positive answer is contained in the following

Theorem 1.3. *Let assumptions of Theorem* 1.1 *be satisfied.*

Then, there exists a solution of (E), $\bar{u} \in H^1_{\text{loc}}(\mathbb{R}^N)$, which has infinitely many positive bumps.

More precisely, $\bar{u}$ is emerging (in the sense specified in first step of Sect. 2*) around an unbounded sequence of points $(\bar{x}_n)_n$, $\bar{x}_n \in \mathbb{R}^N$ ($\bar{x}_n \neq \bar{x}_m$ for $m \neq n$).*

Furthermore $\bar{u}$ and $(\bar{x}_n)_n$ have the following properties:

$$\lim_{n \to \infty} \min\{|\bar{x}_n - \bar{x}_m| : m \in \mathbb{N},\ m \neq n\} = +\infty\,, \tag{1.7}$$

$$\lim_{n \to \infty} \bar{u}(x + \bar{x}_n) = w(x) \qquad \textit{uniformly on all compact subsets of } \mathbb{R}^N.$$

We observe that relation (1.7) indicates the bumps of $\bar{u}$ rarefy, as the distance from the origin increases, giving rise to a quite new phenomenon. Indeed, for instance, multi-bump positive solutions, obtained in [22], when $a(x)$ is radially symmetric, cannot converge as the number of the bumps increases, and, on the contrary the bumps, as their number increases, spread out, going far away each other and far away from the origin in a uniform way.

We remark that the same conclusion of the above theorem can be shown true also for equation (E_ε), for all ε for which the claim of Theorem 1.2 holds (see [11]).

The last question, that naturally appears, is to investigate the possibility, of obtaining *infinitely many multi-bump changing sign solutions*, when (1.6) holds, complementing, in some sense, the result of [9]. A positive answer has been given in [11]:

Theorem 1.4. *Let $a(x)$ satisfy (h_1), (h_2), (h_3), and (h_4). Let $A \Subset B \subset \mathbb{R}^N$, be open, bounded sets such that $0 \in A$. Let $\mathcal{N}_A$ and $\mathcal{N}_B$ be bounded neighborhoods of ∂A and ∂B respectively, such that $\mathcal{N}_A \cap \mathcal{N}_B = \emptyset$. Let be $b : \mathcal{N} \to \mathbb{R}$, $\mathcal{N} = \mathcal{N}_A \cup \mathcal{N}_B$, a continuous function, having compact support, such that*

$$\inf_{x \in \partial A \cup \partial B} b(x) = b_0 > 0, \qquad \sup_{x \in \mathcal{N}} b(x) < a_0.$$

Let $\mathcal{N}/\varepsilon$, b_ε, a_ε and E_ε, be defined as in Theorem 1.2.

Then, there exist $\bar{\varepsilon} > 0$ such that for all $\varepsilon \in (0, \bar{\varepsilon})$ equation (E_ε) has infinitely many nodal solutions belonging to $H^1(\mathbb{R}^N)$.

The remainder part of the paper is organized as follows. Section 2 contains a description of the main steps necessary to prove Theorem 1.1: the new method for finding critical points, that can be called 'of multiple baricenters and multiple local Nehari constraints', is displayed in details, making an attempt of emphasizing the ideas and avoiding technicalities. Section 3 is devoted to an outline of the construction of a solution to (E) having infinitely many bumps. In Section 4 a sketch of the way of proving Theorems 1.2 and 1.4 is exposed, skipping the points in which the arguments are similar to those of Theorem 1.1 and stressing the differences.

2. Infinitely many multi-bump positive solutions

In this section we describe main ideas and arguments used to prove Theorem 1.1; for sake of simplicity in what follows, instead of $(h_1), (h_2)$, we assume (1.2).

Solutions of (E) are searched, by using purely variational methods, as critical points of the functional I in special classes of 'k-bump' functions. This program is carried out in several steps.

First Step: Classes of admissible multi-bump functions

We start by considering a positive number $\delta > 0$ and by defining, for any function $u \in H^1(\mathbb{R}^N)$, its *emerging part* above δ

$$u^\delta(x) := (u - \delta)^+(x)$$

and its *submerged part* under δ.

$$u_\delta(x) := u(x) - u^\delta(x).$$

Then, fixing a suitably small δ and a large ρ ($\rho = \rho(\delta)$ chosen so large that $w(x) < \delta$ outside $B_{\rho/2}(0)$), we say that $u \in H^1(\mathbb{R}^N)$ is *emerging (above δ) around the points $x_1, x_2, \ldots, x_k$ (in k balls of radius ρ)* if

$$u^\delta(x) = \sum_{i=1}^{k} u_i^\delta(x)$$

where, for all $i \in \{1, 2, \ldots, k\}$, $u_i^\delta \geq 0$, $u_i^\delta \not\equiv 0$, $u_i^\delta \in H_0^1(B_\rho(x_i))$, $B_\rho(x_i) \cap B_\rho(x_j) = \emptyset$ if $i \neq j$.

Setting

$$\mathcal{K}_1 = \mathbb{R}^N;$$

$$\mathcal{K}_k = \Big\{(x_1, x_2, \ldots, x_k) \in \big(\mathbb{R}^N\big)^k : |x_i - x_j| \geq 3\rho,\ i, j = 1, 2, \ldots, k,\ i \neq j\Big\} \quad \forall k > 1,$$

we now define, for all $(x_1, x_2, \ldots, x_k) \in \mathcal{K}_k$, the classes:

$$S_{x_1,x_2,\ldots,x_k} = \{u \in H^1(\mathbb{R}^N) : u \geq 0,\ u \text{ emerging around } (x_1, x_2, \ldots, x_k) \in \mathcal{K}_k, \text{ and } I'(u)[u_i^\delta] = 0,\ \beta_i(u) = 0\ \forall\, i = 1, 2, \ldots, k\}$$

where

$$\beta_i(u) \;=\; \frac{1}{|u_i^\delta|_2^2} \int_{\mathbb{R}^N} (x - x_i)(u_i^\delta(x))^2 dx \tag{2.1}$$

are barycenter type maps.

The sets $S_{x_1,x_2,\ldots,x_k}$ defined above are not empty. Indeed, assumptions (h_1), (h_2), and (h_3) guarantee that, for all $u \in H^1(\mathbb{R}^N)$ such that $u^\delta \neq 0$, the function $\mathcal{I}_u : [0, +\infty) \longrightarrow \mathbb{R}$ defined as $\mathcal{I}_u(t) = I(u_\delta + tu^\delta)$ has a unique maximum point $t_u \in (0, +\infty)$, so it makes sense to call the function $u_\delta + t_u u^\delta$ *the projection* of u on the natural nonsmooth constraint $\{u \in H^1(\mathbb{R}^N) : I'(u)[u^\delta] = 0\}$. Then, given $(x_1, x_2, \ldots, x_k) \in \mathcal{K}_k$, an example of function belonging to $S_{x_1,x_2,\ldots,x_k}$ is provided considering:

$$u(x) = \begin{cases} 0 & \forall\, x \in \mathbb{R}^N \setminus (\bigcup_{i=1}^k \text{supp } v_i) \\ \tilde{v}_i(x) & \forall\, x \in \text{supp } v_i,\ i = 1, 2, \ldots, k \end{cases}$$

where $\tilde{v}_i(x)$ is, for all $i \in \{1, 2, \ldots, k\}$, the projection on

$$\{u \in H^1(\mathbb{R}^N) : I'(u)[u^\delta] = 0\}$$

of the function $v_i(x) = v(x - x_i)$, $v \in \mathcal{C}_0^\infty(B_\rho(0))$ being a positive, radially symmetric (around the origin), function such that $v(x) > \delta$ on a positive measure subset of $B_\rho(0)$.

Second step: Minimization between functions emerging around a given set of points

A choice of δ suitably small and assumptions (h_1), (h_2), and (h_3) allow to show that, for all $k \in \mathbb{N} \setminus \{0\}$ and for all $(x_1, x_2, \ldots, x_k) \in \mathcal{K}_k$, the relation

$$u \in S_{x_1,x_2,\ldots,x_k} \Longrightarrow I(u) > I(u_\delta) > 0$$

holds. Thus, obviously,

$$\mu(x_1, x_2, \dots, x_k) := \inf_{S_{x_1,x_2,\dots,x_k}} I(u) \geq 0,$$

and, the following proposition states that more is true:

Proposition 2.1. *Let assumptions* (h_1), (h_2), (h_3) *be satisfied. Then, for all* $k \in \mathbb{N} \setminus \{0\}$, *for all* $(x_1, x_2, \dots, x_k) \in \mathcal{K}_k$, *there exists* $\bar{u} \in S_{x_1,x_2,\dots,x_k}$ *such that*

$$I(\bar{u}) = \inf_{S_{x_1,x_2,\dots,x_k}} I(u) = \mu(x_1, x_2, \dots, x_k).$$

The method used for proving Proposition 2.1 is in the spirit of known arguments exploited to show that some functionals, satisfying suitable assumptions, possess a minimum on their natural Nehari constraint. Nevertheless, the situation, in this setting, is more delicate, because, when we work in $S_{x_1,x_2,\dots,x_k}$, we deal with functions satisfying 'natural constraints' that are only *local.* Therefore, the proof is not standard.

An important effect of the above considered minimization procedure is some remarkable feature, described by two propositions, of any minimizing function. First statement provides a decay estimate on the *submerged part* of a minimizing function and ensures that such minimizing function solves (E) on $\mathbb{R}^N$ except the points of the support of its *emerging part*:

Proposition 2.2. *Let assumptions* (h_1), (h_2), (h_3), *and* (h_4) *be satisfied. Let* k, $\bar{u}$, *and* $(x_1, x_2, \dots, x_k)$ *be as in Proposition* 2.1. *Then* $\bar{u}_\delta$ *solves*

$$(P_\delta) \qquad \begin{cases} -\Delta u + a(x)u = u^p & in \quad \mathbb{R}^N \setminus \text{supp } \bar{u}^\delta, \\ u = \delta & on \quad \text{supp } \bar{u}^\delta \\ u > 0 & in \quad \mathbb{R}^N. \end{cases}$$

Moreover, setting $d(x) = \text{dist}(x, \text{supp } \bar{u}^\delta)$, *the relation*

$$0 < \bar{u}_\delta(x) < C\delta e^{-\bar{\eta} d(x)} \tag{2.2}$$

holds, with $C > 0$ *depending only on* $\bar{\eta}$, a_∞, N.

Latter proposition makes clear the equations that a minimizer (actually its *emerging part*), being a constrained critical point of I, must satisfy:

Proposition 2.3. *Let assumptions* (h_1), (h_2), *and* (h_3) *be satisfied. Then, for all* $k \in \mathbb{N} \setminus \{0\}$, *for all* $(x_1, x_2, \dots, x_k) \in \mathcal{K}_k$, *for all* $\bar{u} \in S_{x_1,x_2,\dots,x_k}$ *such that* $I(\bar{u}) = \mu(x_1, x_2, \dots, x_k)$, *for all* $i \in \{1, 2, \dots, k\}$, *a* $\lambda_i \in \mathbb{R}^N$ *exists so that the relation*

$$I'(\bar{u})[\psi] = \int_{B_\rho(x_i)} \bar{u}^\delta(x)\psi(x)(\lambda_i \cdot (x - x_i))dx \qquad \forall \psi \in H_0^1(B_\rho(x_i)) \tag{2.3}$$

holds true.

Third step: Maximization among minimal functions

The purpose is, now, to show that, when we make the k-tuples $(x_1, x_2, \ldots, x_k)$ vary in the set $\mathcal{K}_k$, the supremum, on the set of minima $\mu(x_1, x_2, \ldots, x_k)$ of I on the admissible classes, is a maximum. We remark that, unlike the previous step, where assumption (h_4) was only used to prove the decay estimate (2.2), now this assumption plays a crucial role.

Setting

$$\mu_k = \sup_{\mathcal{K}_k} \mu(x_1, x_2, \ldots, x_k) = \sup_{\mathcal{K}_k} \min_{S_{x_1,x_2,\ldots,x_k}} I(u),$$

the desired result is stated in the following:

Proposition 2.4. *Let assumptions* (h_1), (h_2), (h_3), (h_4) *be satisfied. Then, for all* $k \in \mathbb{N} \setminus \{0\}$*:*

$$\begin{array}{ll} \text{i)} & \exists\ (\bar{x}_1, \bar{x}_2, \ldots, \bar{x}_k) \in \mathcal{K}_k\ : \mu_k = \mu(\bar{x}_1, \bar{x}_2, \ldots, \bar{x}_k); \\ \text{ii)} & \mu_k + m_\infty < \mu_{k+1}. \end{array} \tag{2.4}$$

The idea underlying the above proposition can be explained, in a rough way, as follows: in order to maximize the functional I among functions having a fixed number of bumps, one needs that the functions (and especially the bumps) feel as much as possible the attractive effect of $a(x)$. On the other hand, the interaction between the 'bumps' let the value of I on a multi-bump function decrease as much as closer the bumps are, giving rise to a tendency of the bumps to escape to infinity. Therefore, the possibility of finding a maximizer is strongly connected to a delicate balance of these two opposite effects and, more precisely, to the possibility that the attractive force of $a(x)$, because of the slow decay, imposed by (h_4), prevails over the repulsive interaction between the bumps (that, by Proposition 2.2, is of exponential type).

The proof of Proposition 2.4 follows an inductive argument, we give here an outline of it when $k = 1$ and $k = 2$.

Case $k = 1$. Writing

$$I(u) = I_\infty(u) + \frac{1}{2}\int_{\mathbb{R}^N} (a(x) - a_\infty) u^2(x)dx$$

it is easy to realize that, being $a(x) - a_\infty > 0$ on a positive measure set, for all $x \in \mathbb{R}^N$

$$\mu_1 \geq \mu(x) > m_\infty,$$

holds true.

To show that μ_1 is achieved, main point is proving that any sequence $(y_n)_n$, $y_n \in \mathbb{R}^N$ such that $\lim_{n\to+\infty} \mu(y_n) = \mu_1$, must be bounded. Intuitively, this fact is true because as $|y_n| \to \infty$ the interaction of minimizing functions u_n 'centered' at y_n with the potential a decreases going to zero, hence $I(u_n)$ approaches more and more $I_\infty(u_n)$, and, then, m_∞. Technically, denoting by $(u_n)_n$ a sequence of functions such that $u_n \in S_{y_n}$ and $I(u_n) = \mu(y_n)$ and by $(\tilde{w}_{y_n})_n$ the sequence of the projections on the local Nehari constraint of $w(\cdot - y_n)$, $w(x)$ being the function realizing (1.3), the idea is formalized by the relations

$$\begin{aligned} \mu(y_n) = I(u_n) \leq I(\tilde{w}_n) &= I^\infty(\tilde{w}_n) + \frac{1}{2}\int_{\mathbb{R}^N}(a(x) - a_\infty)(\tilde{w}_n(x))^2 dx \\ &\leq m_\infty + \frac{1}{2}\int_{\mathbb{R}^N}(a(x) - a_\infty)(\tilde{w}_n(x))^2 dx, \end{aligned} \tag{2.5}$$

and

$$\lim_{n\to+\infty}\frac{1}{2}\int_{\mathbb{R}^N}(a(x) - a_\infty)(\tilde{w}_n(x))^2 = 0.$$

Case $k = 2$. To show the inequality

$$\mu_2 > \mu_1 + m_\infty, \tag{2.6}$$

we consider a sequence of pairs $(\bar{x}, \sigma_n\tau)_n$, with $\sigma_n \in \mathbb{R}$, $\sigma_n \underset{n\to+\infty}{\longrightarrow} +\infty$, $\tau \in \mathbb{R}^N$, $|\tau| = 1$, $\bar{x} \in \mathbb{R}^N$ such that $\mu_1 = \mu(\bar{x})$, and a sequence $(u_n)_n$ of functions so that $u_n \in S_{\bar{x},\sigma_n\tau}$, $I(u_n) = \mu(\bar{x}, \sigma_n\tau)$.

For large n, the emerging parts of u_n (emerging around $\bar{x}$ and $\sigma_n\tau$ respectively) are on opposite hyperspaces with respect to the strip

$$\Sigma_n = \left\{x \in \mathbb{R}^N : \frac{\sigma_n}{2} - 1 < (x\cdot\tau) < \frac{\sigma_n}{2} + 1\right\}.$$

By using a suitable cut-off function, one can set, for all n, u_n equal to 0 on Σ_n. Evaluating I on the sequence, $(v_n)_n$, obtained cutting the functions u_n, one gets:

$$\begin{aligned} I(v_n) &\leq I(u_n) + \frac{1}{2}\bar{c}_1\int_{\Sigma_n}((u_n)_\delta)^2 dx + \bar{c}_2\int_{\Sigma_n}((u_n)_\delta)^{p+1} dx \\ &\leq \mu_2 + O\left(e^{-\bar{\eta}\sigma_n}\right). \end{aligned} \tag{2.7}$$

On the other hand, for large n, v_n can be written as the sum of two functions $v_n^1 \in S_{\bar{x}}$, and $v_n^2 \in S_{\sigma_n\tau}$, for which the relations:

$$I(v_n^1) \geq \mu(\bar{x}) = \mu_1, \qquad \text{and} \qquad I(v_n^2) \geq m_\infty + \frac{1}{2}\int_{\mathbb{R}^N}(a(x) - a_\infty)(v_n^2)^2$$

hold true. Therefore,

$$I(v_n) \geq \mu_1 + m_\infty + \frac{1}{2}\int_{\mathbb{R}^N}(a(x) - a_\infty)(v_n^2)^2. \tag{2.8}$$

Hence, thanks to (h_4), (2.7) together with (2.8) gives (2.6).

Now, proving that μ_2 is achieved is not a difficult matter. Again the main point is showing that a maximizing sequence $((y_n^1, y_n^2))_n$, is bounded.

The argument can be carried out by contradiction. Denoting by $(u_n)_n$, a sequence of functions so that $u_n \in S_{y_n^1,y_n^2}$ and $I(u_n) = \mu(y_n^1, y_n^2)$ assume, for instance, $|y_n^2| \to \infty$. Then, considering $(s_n)_n$, and $(z_n)_n$, two sequences of functions so that $s_n \in S_{y_n^1}$, and $I(s_n) = \mu(y_n^1)$, $z_n \in S_{y_n^2}$, and $I(z_n) = \mu(y_n^2)$, one deduces as in (2.5) that $I(z_n) \leq m_\infty + o(1)$. Moreover, $s_n \vee z_n \in S_{y_n^1,y_n^2}$ and, being $s_n \wedge z_n <$

δ using the convexity and coercivity of I on functions having small L^∞ norm, $I(s_n \wedge z_n) > 0$ can be obtained. Thus,

$$\begin{aligned}\mu_2 = I(u_n) + o(1) \le I(s_n \vee z_n) &= I(s_n) + I(z_n) - I(s_n \wedge z_n) + o(1)\\ &\le I(s_n) + I(z_n) + o(1) \le \mu_1 + m_\infty + o(1)\end{aligned}$$

follows, contradicting (2.6).

Last step: Constrained critical points are free critical points

The functions u_k, found by the above-described max-min procedure are 'good candidates' to be critical points. Since we already know that the submerged parts of such functions are solutions of (E) in $\mathbb{R}^N$ except the supports of their emerging parts, to complete the argument, what is left to show is that the emerging parts too satisfy (E).

First of all, we need to be sure that, at least when $|a - a_\infty|_{L^{N/2}_{\text{loc}}}$ is small, the k-tuples, around which the maximizers are emerging, are contained in the interior part of $\mathcal{K}_k$ and that the supports of the emerging parts do not touch the boundary of the balls in which they are contained. To this end, a deep inspection of the maximizing functions asymptotic properties, as $|a - a_\infty|_{L^{N/2}_{\text{loc}}} \to 0$, is in order.

A careful energy balance, again strongly depending on assumption (h_4), shows that, if the distance of the centers of the emerging parts would not go over any fixed quantity, when $|a - a_\infty|_{\frac{N}{2},\text{loc}}$ decreases going to 0, then, the energy lowering, due to the interaction of the bumps, could exceed the attractive effect of a, yielding a contradiction with relation (2.6) (that holds for all $k \in \mathbb{N}$ and for all a satisfying (h_1)–(h_4)).

Denoting by $\mathcal{F}$ the family of functions a satisfying (h_1)–(h_4), for each $a \in \mathcal{F}$ by I^a, S^a, μ_k^a the corresponding functional, classes, max-min values, and considering functions u_k^a emerging around points $(x_1^a, x_2^a, \dots, x_k^a)$ such that $u_k^a \in S^a_{x_1^a,\dots,x_k^a}$, and $I^a(u_k^a) = \mu_k^a$ the above reasoning is summarized by the relation:

$$\min\left\{|x_i^a - x_j^a| : i \neq j,\ \ i,j = 1,2,\dots,k\right\} \to +\infty \qquad \text{as } |a - a_\infty|_{L^{N/2}_{\text{loc}}} \to 0. \tag{2.9}$$

Furthermore, taking advantage of the uniqueness (up to translations) of the ground state solution w of (E_∞), it is possible to describe the asymptotic shape of the solutions emerging parts:

$$\sup\left\{|u_k^a(x + x_i^a) - w(x)| :\ i = 1,2,\dots,k,\ |x| \le r\right\} \to 0 \quad \text{as } |a - a_\infty|_{L^{N/2}_{\text{loc}}} \to 0, \tag{2.10}$$

for all $r > 0$.

So, as consequence of the exponential decay of $w(x)$ and of the choice of ρ the desired relation:

$$\text{supp } (u_k^a)_i^\delta \Subset B_\rho(x_i^a) \qquad \forall i \in \{1,2,\dots,k\}, \quad \text{for small } |a - a_\infty|_{L^{N/2}_{\text{loc}}}$$

can be deduced, thanks to the choice of ρ.

Now, the proof is completed as the following claim is shown true: when $|a - a_\infty|_{L^{N/2}_{\rm loc}}$ is small enough, for all $k \in \mathbb{N} \setminus \{0\}$, the Lagrange multipliers appearing in relations (2.3) are equal to zero.

Here the arguments become quite delicate and technical, nevertheless the underlying idea is simple and can be roughly summarized as follows: if some Lagrange multiplier related to a maximizing k-tuple would be nonzero, then, moving the points of that k-tuple, a little bit, along the directions of the nonzero Lagrange multipliers, we could get a contradiction proving that the energy of minimizers related to the 'new' k-tuple is greater than the energy of the minimizer related to the maximizing k-tuple.

Technically one argues as follows: if the claim is false, a sequence $(a_n)_n$ of potentials satisfying (h_1)–(h_4), a sequence of natural numbers $(k_n)_n$ and a sequence of functions $(u_{k_n})_n$ exist so that (denoting by I_n, $\mathcal{K}^n_{k_n}$, S^n, $\mu^n_{k_n}$ the corresponding functional, sets, classes, max-min values…)

$$|a_n - a_\infty|_{L^{N/2}_{\rm loc}} \underset{n\to+\infty}{\longrightarrow} 0\,; \quad u_{k_n} \in S^n_{x_1^n, x_2^n, \ldots, x^n_{k_n}}\,; \quad I_n(u_{k_n}) = \mu^n_{k_n} \ \text{ and } \ I_n'(u_{k_n}) \neq 0.$$

Therefore, according to Proposition 2.3, some $\lambda_i^n \neq 0$ exists for which

$$I_n'(u_{k_n})[\psi] = \int_{B_\rho(x_i^n)} u^\delta_{k_n}(x)\psi(x)\ (\lambda_i^n \cdot (x - x_i^n))dx \qquad \forall \psi \in H^1_0(B_\rho(x_i^n))$$

holds true. We can assume, e.g., that $|\lambda_1^n| \neq 0$, for all n and, up to a subsequence, $\lim_{n\to+\infty} \lambda_1^n/|\lambda_1^n| = \lambda$. Then, we build, for all $n \in \mathbb{N} \setminus \{0\}$, another k-tuple of points in $\mathbb{R}^N$, $(y_1^n, y_2^n, \ldots, y^n_{k_n}) \in \mathcal{K}^n_{k_n}$, setting

$$y_i^n = \begin{cases} x_i^n & i \neq 1 \\ x_1^n + \varepsilon_n \lambda & i = 1, \end{cases}$$

($(\varepsilon_n)_n$ being a sequence of real, positive, suitably small numbers).

To $(y_1^n, y_2^n, \ldots, y^n_{k_n})$ there corresponds a sequence of functions $(v_n)_n$, made up by minimizers of I_n in $S^n_{y_1^n, y_2^n, \ldots, y^n_{k_n}}$, which, by construction, satisfy the inequalities

$$I_n(v_n) \le \mu^n_{k_n} = I_n(u_{k_n}) \qquad \forall n \in \mathbb{N}. \tag{2.11}$$

On the other hand, writing the variation of I, passing from u_{k_n} to v_n, by means of a Taylor expansion we obtain

$$I_n(v_n) - I_n(u_{k_n}) = I_n'(u_{k_n})[v_n - u_{k_n}] + \frac{1}{2}\int_{\mathbb{R}^N} |\nabla(v_n - u_{k_n})|^2 dx$$
$$+ \frac{1}{2}\int_{\mathbb{R}^N} (a_n)(v_n - u_{k_n})^2 dx - \frac{p}{2}\int_{\mathbb{R}^N} [u_{k_n} + \tilde{\omega}_n\ (v_n - u_{k_n})]^{p-1}(v_n - u_{k_n})^2 dx$$

where $\tilde{\omega}_n(x) \in [0,1]$.

Very careful estimates of the expansion terms, involving relations (2.9) and (2.10) and other consequences of assumption $(*)$, show that, for large n,

$$I_n(v_n) - I_n(u_{k_n}) > 0$$

contradicting (2.11) and completing the proof.

3. A solution with infinitely many bumps

This section is devoted to describe how the existence of a positive solution to (E) having infinitely many bumps can be obtained.

This kind of solution is searched as limit, as the number k of the bumps goes to $+\infty$, of a sequence $(\bar{u}_k)_k$ of multi-bumps solutions given by Theorem 1.1. Therefore, unlike the functions $\bar{u}_k$, this new solution has no a variational characterization.

To carry out this program, in view of what seen in Section 2, it is useful to complete the statement of Theorem 1.1 as follows:

Theorem 3.1. *Let assumptions of Theorem* 1.1 *be satisfied. Then, for all* $k \in \mathbb{N} \setminus \{0\}$, *there exists (at least) one solution* $\bar{u}_k$ *of* (E) *which is emerging around* k *points* $(\bar{x}_1^k, \dots, \bar{x}_k^k) \in \mathcal{K}_k$.

Moreover, $\bar{u}_k$ *is found as critical point of* I *and is characterized as*

$$I(\bar{u}_k) = \min_{S_{\bar{x}_1^k,\dots,\bar{x}_k^k}} I(u) = \max_{\mathcal{K}_k} \min_{S_{x_1,\dots,x_k}} I(u).$$

The proof of Theorem 1.3 is divided in two steps.

Step 1. We begin constructing a sequence $(\bar{u}_k)_k$ whose elements $\bar{u}_k$ are, for all k, solutions of (E) having exactly k bumps and, then, proving a relation that, in some sense, provides a bound from below to the number of emerging parts that can be contained in balls centered at the origin:

Proposition 3.2. *Let* $(\bar{u}_k)_k$ *be a sequence of solutions to* (E) *obtained as described in Theorem* 3.1. *For all real number* $r > 0$, *let us denote by* $\nu(\bar{u}_k, r)$ *the number of points around which* $\bar{u}_k$ *is emerging and that are contained in* $B_r(0)$.

Then, for all $h \in \mathbb{N}$ *there exist a real number* $r_h > 0$ *and a number* $k_h \in \mathbb{N}$ *such that*

$$\nu(\bar{u}_k, r_h) \geq h, \qquad \forall k > k_h. \tag{3.1}$$

Relation (3.1) is a basic ingredient for getting the desired result; its truth is founded on the slow decay assumption (h_4). The proof follows this scheme: assuming false (3.1) means admitting the possibility of constructing a sequence of solutions to (E), increasing with respect to the number of bumps, and a sequence of 'bumps', belonging to these solutions, centered at points that go to infinity. Then, one shows this construction is, for large n, in contrast with the tendency of the bumps of solutions not to go too far away from origin, in order to maximize the energy, feeling the effect of the potential a.

Technically, one assumes there exist $h \in \mathbb{N}$ and sequences of numbers $(r_n)_n$, $r_n \in \mathbb{R}^+ \setminus \{0\}$, $(k_n)_n$, $k_n \in \mathbb{N}$, such that

$$r_n \to +\infty, \ k_n \to +\infty, \ \text{as } n \to +\infty, \text{ and } \ \nu(\bar{u}_{k_n}, r_n) < h, \ \forall n \in \mathbb{N}.$$

Then, denoting by $(\bar{x}_1^n, \dots, \bar{x}_{k_n}^n)$ the points around which $\bar{u}_{k_n}$ is emerging and passing, if necessary, to a subsequence, one can assume that for some $j < h$,

$$\bar{x}_i^n \underset{n\to+\infty}{\longrightarrow} \bar{x}_i \qquad \forall i \leq j, \quad \text{and} \quad \bar{x}_i^n \underset{n\to+\infty}{\longrightarrow} +\infty \qquad \forall i > j.$$

Now, in view of the fact that $|\bar{x}^n_{j+1}| \underset{n\to+\infty}{\longrightarrow} +\infty$, keeping in mind the arguments to obtain (2.5), one can prove the inequality

$$\limsup_{n\to+\infty} \left[\mu(\bar{x}^n_1,\dots,\bar{x}^n_{k_n}) - \mu(\bar{x}^n_{j+2},\dots,\bar{x}^n_{k_n})\right] \le \mu(\bar{x}_1,\dots,\bar{x}_j) + m_\infty, \tag{3.2}$$

and, on the other hand, developing an argument similar to that used to prove (2.6), the strict inequality

$$\max_{m\in\mathbb{R}^+} \mu(\bar{x}_1,\dots,\bar{x}_j, m\frac{\bar{x}_{j+1}}{|\bar{x}_{j+1}|}) > \mu(\bar{x}_1,\dots,\bar{x}_j) + m_\infty,$$

can be shown true. Therefore, the existence follows of some $\bar{y} = \bar{m}\frac{\bar{x}_{j+1}}{|\bar{x}_{j+1}|}$, so that the estimate

$$\begin{aligned}\lim_{n\to+\infty} &\left[\mu(\bar{x}^n_1,\dots,\bar{x}^n_j,\bar{y},\bar{x}^n_{j+2},\dots,\bar{x}^n_{k_n}) - \mu(\bar{x}^n_{j+2},\dots,\bar{x}^n_{k_n})\right] \\ &= \mu(\bar{x}_1,\dots,\bar{x}_j,\bar{y}) > \mu(\bar{x}_1,\dots,\bar{x}_j) + m_\infty,\end{aligned} \tag{3.3}$$

hold true, for large n.

Then the desired contradiction is reached combining (3.2) and (3.3), because one gets, for large n, the relation

$$\mu(\bar{x}^n_1,\dots,\bar{x}^n_{k_n}) < \mu(\bar{x}^n_1,\dots,\bar{x}^n_j,\bar{y},\bar{x}^n_{j+2},\dots,\bar{x}^n_{k_n}),$$

which contradicts the maximality of $\mu(\bar{x}^n_1,\dots,\bar{x}^n_{k_n})$.

Once obtained (3.1) one observes that, for all h, an upper bound to $\nu(\bar{u}_k, r_h)$ is also available, because points around which any $\bar{u}_k$ is emerging have interdistances greater or equal than 3ρ.

Thus, for all $h \in \mathbb{N}\setminus\{0\}$, r_h, k_h, and H_h exist so that

$$h \le \nu(\bar{u}_k, r_h) \le H_h, \quad \forall k > k_h.$$

The above inequalities, clearly, imply,

$$\lim_{h\to+\infty} r_h = +\infty,$$

furthermore, up to a subsequence, $r_h \le r_{h+1}$, for all $h \in \mathbb{N}$, can be assumed.

Step 2. A family of subsequences of $(\bar{u}_k)_k$ is defined as follows:

$(\bar{u}^1_{k_n})_n$ is a subsequence of $(\bar{u}_k)_k$ such that, $\forall n$, $\nu(\bar{u}^1_{k_n}, r_1) = h_1$, with $1 \le h_1 \le H_1$

$(\bar{u}^2_{k_n})_n$ is a subsequence of $(\bar{u}^1_{k_n})_n$ such that, $\forall n$, $\nu(\bar{u}^2_{k_n}, r_2) = h_2$, with $2 \le h_2 \le H_2$

......

$(\bar{u}^m_{k_n})_n$ is a subsequence of $(\bar{u}^{m-1}_{k_n})_n$ so that $\forall n$, $\nu(\bar{u}^m_{k_n}, r_m) = h_m$ with $m \le h_m \le H_m$

......

and, after that, the 'diagonal' sequence

$$\bar{u}^1_{k_1}, \bar{u}^2_{k_2}, \ldots, \bar{u}^n_{k_n}, \ldots \tag{3.4}$$

(definitively) subsequence of $(\bar{u}^m_{k_n})_n$, for all m, is considered.

The desired solution of (E) will be obtained as limit, in a suitable sense, of the sequence (3.4).

Indeed, first of all we observe that, for all $m \in \mathbb{N}$, the sequences of h_m-tuples $((\bar{x}^n_1)^m, \ldots, (\bar{x}^n_{h_m})^m)_n$, consisting of points, contained in $B_{r_m}(0)$, around which the functions of the sequence $(\bar{u}^m_{k_n})_n$ are emerging, are bounded and, then, up to subsequences, converging as n goes to $+\infty$. Thus, the sequences consisting of points, around which the functions of $\bar{u}^n_{k_n}$ are emerging, are converging as n goes to infinity. Furthermore, the interdistances between these limit points are greater or equal than 3ρ, because the interdistances between the centers of bumps of functions $\bar{u}^n_{k_n}$ are greater or equal than 3ρ. As consequence, these limit points form an unbounded numerable subset $L := \{\bar{x}_n : n \in \mathbb{N}\}$ of $\mathbb{R}^N$.

On the other hand, for all h, $B_{r_h}(0)$, contains only a finite number of 'emerging parts' of the functions $\bar{u}^n_{k_n}$, hence a control from above can be obtained on $I((\bar{u}^n_{k_n})_{|B_{r_h}(0)})$ and, this yields the boundedness of $\|\bar{u}^n_{k_n}\|_{H^1(B_{r_h}(0))}$, by using regularity arguments.

Therefore, we are in position to conclude. Indeed, being, for all n, $\bar{u}^n_{k_n}$ a solution of (E), since $r_h \to +\infty$ as $h \to +\infty$, we can infer that, up to a subsequence, $(\bar{u}^n_{k_n})_n$ uniformly converges on every compact set of $\mathbb{R}^N$ to a function $\bar{u} \in H^1_{\rm loc}(\mathbb{R}^N)$, which is a solution of (E) and has, for all h, at least h emerging parts around points belonging to $B_{r_h}(0)$.

We end this section giving an idea of the way in which relation (1.7) can be proved. Once again the cause of this rarefaction phenomenon is the attractive effect of the potential, due to (h_4), and the nature of energy maximizers that the solutions $\bar{u}_k$ have.

Actually, if relation (1.7) would be false, two subsequences of $(\bar{x}_n)_n$, $(b_n)_n$ and $(\bar{b}_n)_n$, would exist so that, for all n, $\bar{b}_n \neq b_n$ and $(|\bar{b}_n - b_n|)_n$ would be bounded. This fact would imply the existence of a sequence $(\bar{u}_{k_n})_n$, of solutions of (E), emerging around k_n points $(\bar{x}^n_1, \ldots, \bar{x}^n_{k_n})$, so that

$$I(\bar{u}_{k_n}) = \mu_{k_n}\,, \qquad \sup_{B_R(b_n)\cup B_R(\bar{b}_n)} |\bar{u}_{k_n} - \bar{u}| < \frac{1}{n} \quad \text{for some fixed R} \tag{3.5}$$

and, moreover, possessing two sequences of centers of bumps, for instance $(\bar{x}^n_1)_n$ and $(\bar{x}^n_2)_n$, having bounded interdistances.

Now, in computing $I(\bar{u}_{k_n})$, the effect of the interaction between the masses corresponding to $(\bar{x}^n_1)_n$ and $(\bar{x}^n_2)_n$, could exceed, for large n, the attractive effect of the potential producing an energy drop described by

$$\limsup_{n\to+\infty}[\mu(\bar{x}^n_1, \ldots, \bar{x}^n_{k_n}) - \mu(\bar{x}^n_2, \ldots, \bar{x}^n_{k_n})] < m_\infty. \tag{3.6}$$

But, this relation contrasts the maximizers character of functions $\bar{u}_{k_n}$. Indeed, similarly to what already seen proving (3.3), for any n a y_n can be found such that the relation

$$\mu(y_n, \bar{x}_2^n, \ldots, \bar{x}_{k_n}^n) \geq \mu(\bar{x}_2^n, \ldots, \bar{x}_{k_n}^n) + m_\infty - \frac{1}{n} \tag{3.7}$$

holds. Hence, combining (3.5), (3.6), and (3.7), for large n, the impossible relation would follow

$$\mu_{k_n} = I(\bar{u}_{k_n}) = \mu(\bar{x}_1^n, \ldots, \bar{x}_{k_n}^n) < \mu(y_n, \bar{x}_2^n, \ldots, \bar{x}_{k_n}^n) \leq \mu_{k_n},$$

yielding the desired conclusion.

4. Multiple solutions when the potential has a 'pit'

In this section a sketch of the proof of Theorems 1.2 and 1.4 is proposed. Theorem 1.2 is considered in part A of section, while part B is concerned with Theorem 1.4.

A) *Infinitely many positive solutions*

We start observing that the continuity of b allows to find a real number $\sigma > 0$ such that, setting $\tilde{\mathcal{N}} := \{x \in \mathbb{R}^N : \operatorname{dist}(x, \partial A) \leq \sigma\}$, $\tilde{\mathcal{N}} \subset \mathcal{N}$ follows and $\inf_{\tilde{\mathcal{N}}} b(x) \geq b_0/2$.

The variational framework to prove Theorem 1.2 is similar to that considered for Theorem 1.1. Solutions of (E_ε) are searched as critical points of

$$I_\varepsilon(u) = \frac{1}{2}\int_{\mathbb{R}^N} (|\nabla u|^2 + a_\varepsilon(x)u^2)dx - \frac{1}{p+1}\int_{\mathbb{R}^N} |u|^{p+1}dx$$

on suitable classes of multi-bump functions.

Setting $A/\varepsilon = \{x \in \mathbb{R}^N : \varepsilon x \in A\}$, $\mathcal{H}_1^\varepsilon = \mathbb{R}^N \setminus (A/\varepsilon)$, and, $\forall k > 1$

$$\mathcal{H}_k^\varepsilon = \{(x_1, x_2, \ldots, x_k) \in (\mathbb{R}^N)^k : |x_i - x_j| \geq 3\rho, \\ x_i, x_j \notin A/\varepsilon,\ i,j = 1,2,\ldots,k,\ i \neq j\},$$

we define for all $(x_1, x_2, \ldots, x_k) \in \mathcal{H}_k^\varepsilon$, the classes:

$$S^\varepsilon_{x_1,x_2,\ldots,x_k} = \{u \in H^1(\mathbb{R}^N) : u \geq 0,\ u \text{ emerging around } (x_1, x_2, \ldots, x_k) \in \mathcal{H}_k^\varepsilon, \\ \text{and } I'_\varepsilon(u)[u_i^\delta] = 0,\ \beta_i(u) = 0\ \forall\, i = 1, 2, \ldots, k\},$$

where $\beta_i(u)$ are barycenter type maps defined as in (2.1).

Working, for all ε, with the functional I_ε on the classes $S^\varepsilon_{x_1,x_2,\ldots,x_k}$ and using arguments similar to those displayed in the second step of Section 2, it is possible to show that

$$\mu_\varepsilon(x_1, x_2, \ldots, x_k) := \inf_{S^\varepsilon_{x_1,x_2,\ldots,x_k}} I_\varepsilon(u) > 0,$$

and that it is achieved. Furthermore, defining

$$\mu_{\varepsilon,k} = \sup_{\mathcal{H}_k^\varepsilon} \mu_\varepsilon(x_1, x_2, \ldots, x_k) = \sup_{\mathcal{H}_k^\varepsilon} \min_{S^\varepsilon_{x_1,x_2,\ldots,x_k}} I_\varepsilon(u).$$

the existence of $(\bar{x}_{1,\varepsilon}, \bar{x}_{2,\varepsilon}, \dots, \bar{x}_{k,\varepsilon}) \in \mathcal{H}_k^\varepsilon$ so that

$$\mu_{\varepsilon,k} = \mu_\varepsilon(\bar{x}_{1,\varepsilon}, \bar{x}_{2,\varepsilon}, \dots, \bar{x}_{k,\varepsilon}),$$

follows as in the third step of Section 2.

Once the max-min procedure has been accomplished, some work is still needed to show that, at least for small ε, the k-tuples around which the maximizers are emerging are contained in the interior part of $\mathcal{H}_k^\varepsilon$. This means to show that, for all $i \in \{1, \dots, k\}$, the relation $\bar{x}_{i,\varepsilon} \notin \partial(A/\varepsilon)$ holds true.

Keeping in mind the max-min variational method used to produce candidate critical points and the definition of a_ε, one understands that the possibility that some point of the maximizing k-tuple could belong to the boundary of A/ε is excluded, when ε is small, by the presence of the potential 'pit' around the boundary of A/ε. Indeed, when a component of a k-tuple comes in $\tilde{\mathcal{N}}/\varepsilon$, the interaction of the potential with the bump, corresponding to this component, makes the functional I_ε loose energy.

We give, now, a little more detailed, even if simplified, outline of the proof of this fact. Arguing by contradiction, the existence is assumed of a sequence $(\varepsilon_n)_n$, $\varepsilon_n > 0$, $\varepsilon_n \to 0$, a sequence $(k_n)_n$, $k_n \in \mathbb{N}$, a sequence of functions $(u_{\varepsilon_n,k_n})_n$, emerging around $(x_{1,\varepsilon_n}, \dots, x_{k_n,\varepsilon_n})$, realizing μ_{ε_n,k_n} (i.e., $I_{\varepsilon_n}((u_{\varepsilon_n,k_n})) = \mu_{\varepsilon_n,k_n}$) and such that, for all n, a $j_n \in \{1, \dots, k_n\}$ exists for which $x_{j_n,\varepsilon_n} \in \partial(A/\varepsilon_n)$.

For sake of simplicity, we consider only the case $k_n = 1$, for all $n \in \mathbb{N}$. So, we denote by $(u_n)_n$ a sequence consisting of functions emerging around one point, $x_n \in \partial(A/\varepsilon_n)$, realizing the values $\mu_{\varepsilon_{n,1}}$. Denoting by $\tilde{w}_n$ the sequence made up projections of $w_n(x) := w(x - x_n)$ on the sets $\{u \in H^1(\mathbb{R}^N) : I'_{\varepsilon_n}(u)[u^\delta] = 0\}$, arguing as in (2.5), we infer:

$$\begin{aligned}\mu_{\varepsilon_{n,1}} = I_{\varepsilon_n}(u_n) \le I_{\varepsilon_n}(\tilde{w}_n) &= I_\infty(\tilde{w}_n) + \frac{1}{2}\int_{\mathbb{R}^N}(a_{\varepsilon_n} - a_\infty)(\tilde{w}_n)^2 dx \\ &\le m_\infty + C\int_{\mathbb{R}^N}(a_{\varepsilon_n} - a_\infty)(w_n)^2 dx,\end{aligned}$$

$C > 0$ constant. Now, writing

$$\begin{aligned}&\int_{\mathbb{R}^N}(a_{\varepsilon_n} - a_\infty)(w_n)^2 dx \\ &\quad = \int_{\mathbb{R}^N\setminus B_R(x_n)}(a_{\varepsilon_n} - a_\infty)(w_n)^2 dx + \int_{B_R(x_n)}(a_{\varepsilon_n} - a_\infty)(w_n)^2 dx,\end{aligned} \tag{4.1}$$

the first integral in the right-hand side of (4.1) becomes as small as one wants as R increases, namely, for all $\eta > 0$ a real positive number R, suitably big and not depending on ε_n, can be found so that, for large n,

$$\int_{\mathbb{R}^N\setminus B_R(x_n)}(a_{\varepsilon_n} - a_\infty)(w_n)^2 dx \le \eta.$$

On the other hand, the second addend in right-hand side of (4.1) can be controlled from above by a negative quantity depending neither on η nor on R. Indeed, for large n, the inclusion $B_R(x_n) \subset (\tilde{\mathcal{N}}/\varepsilon_n)$, allows to obtain:

$$\int_{B_R(x_n)} (a_{\varepsilon_n} - a_\infty)(w_n)^2 dx \leq -\frac{b_0}{4} \int_{B_1(0)} (w(x))^2 dx.$$

As a consequence, one gets

$$\mu_{\varepsilon_n,1} < m_\infty$$

contradicting the relation

$$\mu_{\varepsilon_n,1} > m_\infty,$$

that is nothing but a particular case of (2.4).

The final step is, as in section 2, to show that the functions $u_{\varepsilon,k}$, realizing the values $\mu_{\varepsilon,k}$ and 'candidates' to be critical points of I_ε, are actually critical points when ε is suitably small. Here, instead of working with potentials having 'smaller and smaller oscillation', we consider potentials $a_\varepsilon(x)$ having a pit around the origin that becomes larger and larger and more and more far away as ε becomes small. As effect of this behaviour and of the definition of $\mathcal{H}_k^\varepsilon$, when ε goes to zero, the emerging parts of the candidate solutions are 'pushed' in regions of $\mathbb{R}^N$ in which $a_\varepsilon(x) - a_\infty$, is smaller and smaller.

The asymptotic properties of the maximizing functions $u_{\varepsilon,k}$ are investigated and the results are summarized by two relations, the first concerning the points around which they are emerging, the latter concerning the asymptotic shape of the emerging parts and showing that it approaches the ground state solution of the limit problem (E_∞) :

$$\min\{|x_{\varepsilon,i} - x_{\varepsilon,j}| : i \neq j, \quad i,j = 1,2,\dots,k\} \to +\infty \qquad \text{as } \varepsilon \to 0, \tag{4.2}$$

and

$$\sup\{|u_{\varepsilon,k}(x + x_{\varepsilon,i}) - w(x)| : \ i = 1,2,\dots,k, \ |x| \leq r\} \to 0 \quad \text{as } \varepsilon \to 0, \tag{4.3}$$

for all $r > 0$.

Hence

$$\operatorname{supp}(u_{\varepsilon,k})_i^\delta \Subset B_\rho(x_{\varepsilon,j}), \qquad \forall i = 1,2,\dots,k \qquad \text{for small } \varepsilon$$

can be deduced. Thus, considering the equations which must be satisfied by a function $u_{\varepsilon,k}$ (obtained by the max-min procedure)

$$I'_\varepsilon(u_{\varepsilon,k})\psi = \int_{B_\rho(x_{\varepsilon,i})} u_{\varepsilon,k}^\delta \psi(x)(\lambda_i(x + x_{\varepsilon,i}))dx \qquad \forall \psi \in H_0^1(B_\rho(x_i)),$$

to complete the proof one has to show that the Lagrange multipliers $\lambda_i \in \mathbb{R}^N$, are equal to zero.

This last information is obtained by a quite analogous method to that indicated in the last step of Section 2. Of course, in this case the analysis must be based on relations (4.2) and (4.3) and on the smallness of $a_\varepsilon(x) - a_\infty$ outside A/ε, when ε is small.

B) *Infinitely many multi-bump nodal solutions*
As for Theorem 1.2, we look for critical points of functionals I_ε (with a_ε as described in Theorem 1.4). Nevertheless, since the desired solutions are changing sign, we need to define new classes of functions having different features with respect to those considered in the proofs of Theorems 1.1 and 1.3.

For fixed $\delta > 0$, suitably small, we write any function $u \in H^1(\mathbb{R}^N)$ as

$$u = u_\delta + (u^+)^\delta - (u^-)^\delta$$

where

$$(u^+)^\delta = (u - \delta)^+$$

is the *positive emerging part of* u,

$$(u^-)^\delta = (-u - \delta)^+$$

is the *negative emerging part of* u, and

$$u_\delta = (u \wedge \delta) \vee (-\delta)$$

is the *middle part of* u.

Then, fixed a suitably small $\delta > 0$ and a large $\rho > 0$ (depending on δ), we say that a function $u \in H^1(\mathbb{R}^N)$ is *positively emerging around* $x_1, \dots, x_h$ and *negatively emerging around* $y_1, \dots, y_k$ in balls of radius ρ if

$$(u^+)^\delta = \sum_{i=1}^{h} (u_i^+)^\delta, \ \ (u_i^+)^\delta \in H_0^1(B_\rho(x_i)), \ \ (u_i^+)^\delta \geq 0, \ \ (u_i^+)^\delta \not\equiv 0, \ \ \forall i \in \{1, \dots, h\}$$

$$(u^-)^\delta = \sum_{i=1}^{k} (u_i^-)^\delta, \ \ (u_i^-)^\delta \in H_0^1(B_\rho(y_i)), \ \ (u_i^-)^\delta \geq 0, \ \ (u_i^-)^\delta \not\equiv 0, \ \ \forall i \in \{1, \dots, k\}$$

where $B_\rho(x_i) \cap B_\rho(x_m) = \emptyset$, $B_\rho(y_j) \cap B_\rho(y_l) = \emptyset$, $B_\rho(x_i) \cap B_\rho(y_j) = \emptyset$, $i \neq m$, $j \neq l$, $\forall\, i, m \in \{1, \dots, h\}$, $\forall\, j, l \in \{1, \dots, k\}$.

The solutions whose existence is claimed in Theorem 1.4 are searched fixing $h \in \mathbb{N}$ and looking for solutions of (E_ε), in special classes of functions having at most h positive emerging parts and an arbitrarily large number of negative emerging parts. In this framework, Theorem 1.4 can be more precisely stated as follows:

Theorem 4.1. *Let assumptions of Theorem* 1.4 *be satisfied. Then, for all* $h \in \mathbb{N}$ *there exists* $\varepsilon_h > 0$ *such that for all* $\varepsilon \in (0, \varepsilon_h)$, *for all* $k \in \mathbb{N} \setminus \{0\}$, *and for all* $j \in \mathbb{N}$, $0 \leq j \leq \min(h, k)$, *there exists a solution of* (E_ε) *having* j *positive emerging parts and* $k - j$ *negative emerging parts.*

Remark 4.2. We remark that Theorem 4.1 setting $h = 0$ yields a result analogous to that of Theorem 1.2: the existence of infinitely many multi-bump negative solutions for (E_ε).

We, also, notice that the statement of Theorem 4.1 holds true exchanging the roles of positive and negative emerging parts.

In order to prove the above theorem, we set

$$\mathcal{H}_1^\varepsilon = [(\bar{B} \setminus A)/\varepsilon)], \qquad \mathcal{K}_1^\varepsilon = \mathbb{R}^N \setminus (B/\varepsilon);$$

and, for all $m > 1$

$$\mathcal{H}_m^\varepsilon = \big\{(x_1, x_2, \dots, x_m) \in \left(\mathbb{R}^N\right)^m : |x_i - x_l| \geq 3\rho, \\ x_i,\, x_l \in (\bar{B}/\varepsilon) \setminus (A/\varepsilon),\ i, l = 1, 2, \dots, m,\ i \neq l\big\},$$

and

$$\mathcal{K}_m^\varepsilon = \big\{(y_1, y_2, \dots, y_m) \in \left(\mathbb{R}^N\right)^m : |y_i - y_l| \geq 3\rho, \\ \mathrm{dist}(y_i, B/\varepsilon) \geq 3\rho\ ,\ i, l = 1, 2, \dots, m,\ \ l \neq i\big\}.$$

Obviously, for all $m \in \mathbb{N} \setminus \{0\}$, there exists $\varepsilon_1 = \varepsilon_1(m)$ such that $\varepsilon \in (0, \varepsilon_1)$ implies $\mathcal{H}_m^\varepsilon \neq \emptyset$, moreover, from $\mathcal{H}_m^\varepsilon \neq \emptyset,\ \ \mathcal{H}_i^\varepsilon \neq \emptyset,\ \forall i = 1, 2, \dots, m$, follows.

In view of Remark 4.2, in what follows we suppose $h \in \mathbb{N}, h \geq 1$. We define for all $h,\ k \in \mathbb{N} \setminus \{0\}$ and for $0 \leq j \leq \min(h, k)$, the classes:

$$S^\varepsilon_{x_1,\dots,x_j,y_1,\dots,y_{k-j}} = \big\{u \in H^1(\mathbb{R}^N) :\ u \text{ positively emerging around} \\ (x_1, \dots, x_j) \in \mathcal{H}_j^\varepsilon,\ \text{negatively emerging around} \\ (y_1, \dots, y_{k-j}) \in \mathcal{K}_{k-j}^\varepsilon,\ \beta_i(u) = 0,\ \beta_{j+l}(u) = 0, \\ I'_\varepsilon(u)[u_i^+]^\delta = 0,\ I'_\varepsilon(u)[u_l^-]^\delta = 0,\ i = 1, \dots, j, \\ l = 1, \dots, k - j\big\}$$

and

$$S^\varepsilon_{x_1,\dots,x_j} = \big\{u \in H^1(\mathbb{R}^N) :\ u \text{ positively emerging around} \\ (x_1, \dots, x_j) \in \mathcal{H}_j^\varepsilon, \beta_i(u) = 0,\ I'_\varepsilon(u)[u_i^+]^\delta = 0,\ i = 1, \dots, j.\big\}$$

Working as indicated in Step 1 of Section 2 it is not difficult to see that the above-defined classes are not empty. Furthermore, arguments similar to those described in the second step of Section 2, allow us to conclude that

$$\inf_{S^\varepsilon_{x_1,\dots,x_j}} I_\varepsilon(u) > 0\ ; \qquad \inf_{S^\varepsilon_{x_1,\dots,x_j,y_1,\dots,y_{k-j}}} I_\varepsilon(u) > 0$$

and the infima are achieved.

The middle part of any (changing sign) minimizer $\hat{u}_\varepsilon$ of I_ε on $S^\varepsilon_{x_1,\dots,x_j,y_1,\dots,y_{k-j}}$ has good properties. Next proposition states that it solves (E_ε) in $\mathbb{R}^N$ except the points belonging to the supports of the (positive and negative) emerging parts of $\hat{u}_\varepsilon$, and that its decay can be estimated:

Proposition 4.3. *Let assumptions of Theorem* 1.4 *be satisfied. Let* h *and* $k \in \mathbb{N} \setminus \{0\}$, $j \in \{1, \dots, \min(h, k)\}$, $\varepsilon \in (0, \varepsilon_1)$, *and let* $\hat{u}_\varepsilon$ *be a minimizer for* I_ε *on*

$S^{\varepsilon}_{x_1,\dots,x_j,y_1,\dots,y_{k-j}}$. *Then* $(\hat{u}_\varepsilon)_\delta$ *solves*

$$(P_\delta) \qquad \begin{cases} -\Delta u + a_\varepsilon(x)u = |u|^{p-1}u & in \quad \mathbb{R}^N \setminus (\text{supp } (\hat{u}^+_\varepsilon)^\delta \cup \text{supp } (\hat{u}^-_\varepsilon)^\delta) \\ u = \delta & on \quad \text{supp } (\hat{u}^+_\varepsilon)^\delta \\ u = -\delta & on \quad \text{supp } (\hat{u}^-_\varepsilon)^\delta. \end{cases}$$

Moreover, setting $d_\varepsilon(x) = \text{dist}(x, [\text{supp } (\hat{u}^+_\varepsilon)^\delta \cup \text{supp}(\hat{u}^-_\varepsilon)^\delta]$, *the relation*

$$|(\hat{u}_\varepsilon)_\delta(x)| < C\delta e^{-\bar{\eta} d_\varepsilon(x)} \tag{4.4}$$

holds, with $C > 0$ *depending only on* $\bar{\eta}$, a_∞, *and* N.

Now, we first define

$$\mu^\varepsilon_j := \sup_{\mathcal{H}^\varepsilon_j} \min_{S^\varepsilon_{x_1,\dots,x_j}} I_\varepsilon(u),$$

and we observe that, thanks to the compactness of the set $\mathcal{H}^\varepsilon_j$ and to the continuity of the map $(x_1,\dots,x_j) \to \min_{S^\varepsilon_{x_1,\dots,x_j}} I_\varepsilon(u)$ on $\mathcal{H}^\varepsilon_m$, the existence of $(\bar{x}_{1,\varepsilon},\dots,\bar{x}_{j,\varepsilon})$ such that $\mu^\varepsilon_j = \min\left\{ I_\varepsilon(u) : u \in S^\varepsilon_{\bar{x}_{1,\varepsilon},\dots,\bar{x}_{j,\varepsilon}} \right\}$, easily follows. We then set

$$\mu^\varepsilon_{j,k-j} = \sup_{\mathcal{H}^\varepsilon_j \times \mathcal{K}^\varepsilon_{k-j}} \min_{S^\varepsilon_{x_1,\dots,x_j,y_1,\dots,y_{k-j}}} I_\varepsilon(u).$$

The fact that $\mu^\varepsilon_{j,k-j}$ are realized is stated in the following:

Proposition 4.4. *Let assumptions of Theorem* 1.4 *be satisfied. Let* k, h, j, ε_1 *be as in Proposition* 4.3. *Then for all* $\varepsilon \in (0, \varepsilon_1)$ *the following relation holds*

$$\mu^\varepsilon_{j,k-j} > \mu^\varepsilon_{j,k-j-1} + m_\infty \tag{4.5}$$

and $(\tilde{x}_{1,\varepsilon},\dots,\tilde{x}_{j,\varepsilon}) \in \mathcal{H}^\varepsilon_j$, $(\tilde{y}_{1,\varepsilon},\dots,\tilde{y}_{k-j,\varepsilon}) \in \mathcal{K}^\varepsilon_{k-j}$ *exist so that*

$$\mu^\varepsilon_{j,k-j} = \max_{\mathcal{H}^\varepsilon_j \times \mathcal{K}^\varepsilon_{k-j}} \min_{S^\varepsilon_{x_1,\dots,x_j,y_1,\dots,y_{k-j}}} I_\varepsilon(u) = \min_{S^\varepsilon_{\tilde{x}_{1,\varepsilon},\dots,\tilde{x}_{j,\varepsilon},\tilde{y}_{1,\varepsilon},\dots,\tilde{y}_{k-j,\varepsilon}}} I_\varepsilon(u).$$

The proof of the crucial energy estimate (4.5), as well as the maximizer existence are obtained by an inductive method similar to that used in third step of Section 2. We remark that in this proof the slow decay of the potential and the decay estimate (4.4) have a decisive role. Nevertheless, once proved Proposition 4.4, some work is still needed to show that the found functions are 'good' candidates to be critical points of I_ε and to be able to control the interaction between the positive and the negative bumps.

Arguments similar to that displayed in Part A of this section show that an $\varepsilon_2 \leq \varepsilon_1$ exists so that, for $\varepsilon \in (0, \varepsilon_2)$, the k-tuples around which the maximizers are emerging are such that $\tilde{x}_{i,\varepsilon} \notin \partial[(\bar{B}\setminus A)/\varepsilon]$ and $\tilde{y}_{l,\varepsilon} \notin \bar{B}/\varepsilon$, for all $i \in \{1,\dots,j\}$, $l \in \{1,\dots,k-j\}$. Then the proof can be completed as in Subsection A, after proving for the 'candidate' critical points $\tilde{u}_{\varepsilon,k}$ the asymptotic relations:

$$\min\{|\tilde{x}_{\varepsilon,i}-\tilde{x}_{\varepsilon,l}| : i\neq l,\quad i,l=1,2,\ldots,j\}\to+\infty \quad \text{as } \varepsilon\to 0,$$
$$\min\{|\tilde{y}_{\varepsilon,i}-\tilde{y}_{\varepsilon,l}| : i\neq l,\quad i,l=1,2,\ldots,k-j\}\to+\infty \quad \text{as } \varepsilon\to 0,$$
$$\min\{|\tilde{x}_{\varepsilon,i}-\tilde{y}_{\varepsilon,l}| : i=1,2,\ldots,j,\ l=1,2,\ldots,k-j\}\to+\infty \quad \text{as } \varepsilon\to 0,$$

and, for all $r>0$,

$$\sup\{|\tilde{u}_{\varepsilon,k}(x+\tilde{x}_{\varepsilon,i})-w(x)| :\ i=1,2,\ldots,j,\ |x|\leq r\}\to 0 \quad \text{as } \varepsilon\to 0,$$
$$\sup\{|\tilde{u}_{\varepsilon,k}(x+\tilde{y}_{\varepsilon,i})-w(x)| :\ i=1,2,\ldots,k-j,\ |x|\leq r\}\to 0 \quad \text{as } \varepsilon\to 0.$$

References

[1] A. Bahri, Y.Y. Li, *On a min-max procedure for the existence of a positive solution for certain scalar field equations in R^N*. Rev. Mat. Iberoamericana **6** (1990), 1–15.

[2] A. Bahri, P.L. Lions, *On the existence of a positive solution of semilinear elliptic equations in unbounded domains.* Ann. Inst. H. Poincaré Anal. Non Linéaire **14** (1997), 365–413.

[3] T. Bartsch, M. Willelm *Infinitely many nonradial solutions of an Euclidean scalar field equation.* J. Funct. Anal., **117**, (1993), 447–460.

[4] V. Benci, G. Cerami, *Positive solutions of some nonlinear elliptic problem in exterior domains.* Arch. Rat. Mech. Anal. **99** (1987), 283–300.

[5] H. Berestycki, P.L. Lions, *Nonlinear scalar fields equations – I. Existence of a ground-state; II Existence of infinitely many solutions.* Arch. Rational Mech. Anal., **82** (1983), 313–346; 347–376.

[6] M.S. Berger *On the existence and structure of stationary states for a nonlinear Klein–Gordon equation.* J. Funct. Anal., **9**, (1972), 249–261.

[7] J.Y. Byeon, K. Tanaka *Semiclassical standing waves with clustering peaks for nonlinear Schrödinger equations.* Memoirs Amer. Math. Soc., to appear.

[8] G. Cerami, *Some nonlinear elliptic problems in unbounded domains.* Milan J. Math., **74**, (2006), 47–77.

[9] G. Cerami, G. Devillanova, S. Solimini, *Infinitely many bound states for some nonlinear scalar field equations.* Calc. Var. Partial Differential Equations, **23**, (2005), 139–168.

[10] G. Cerami, R. Molle *On some Schrödinger equations with nonregular potential at infinity.* Disc. Cont. Dyn. Systems, **28**, (2010), 827–844

[11] G. Cerami, R. Molle, D. Passaseo *Multiple positive and nodal solutions for some nonlinear scalar field equations.* Preprint.

[12] G. Cerami, D. Passaseo, S. Solimini, *Infinitely many positive solutions to some scalar field equations with non-symmetric coefficients.* Comm. Pure. Appl. Math. **66**, (2013), 372–413.

[13] G. Cerami, D. Passaseo, S. Solimini, *Nonlinear Scalar Field Equations: Existence of a solution with infinitely many bumps.* Ann. Inst. H. Poincaré Anal. Non Linéaire, to appear.

[14] C.V. Coffman *Uniqueness of the ground state solution for* $\Delta u - u + u^3 = 0$ *and a variational characterization of other solutions.* Arch. Rat. Mech. Anal., **46**, (1972), 81–95.

[15] S. Coleman, V. Glazer, A. Martin *Action minima among solutions to a class of Euclidean scalar field equations.* Comm. Math. Phys., **58**, (1978), 211–221.

[16] V. Coti Zelati, P. Rabinowitz, *Homoclinic type solutions for a semilinear elliptic PDE on* $\mathbb{R}^N$. Comm. Pure. Appl. Math., **10**, (1992), 1217–1269.

[17] P.L. Lions, *The concentration compactness principle in the calculus of variations, Parts I and II.* Ann. Inst. H. Poincaré – Analyse Non Linéaire. **1**, (1984), 109–145, 223–283.

[18] L. Lin, Z. Liu *Multi-bump solutions and multi-tower solutions for equations in* $\mathbb{R}^N$. J. Funct. Anal., **257** (2009), 485–505.

[19] Z. Nehari *On a nonlinear differential equation arising in nuclear physics.* Proc. Roy. Irish. Acad., **62**, (1963), 117–135.

[20] P.H. Rabinowitz *On a class of nonlinear Schrödinger equations.* Z. Angew. Math. Phys., **43**, (1992), 270–291.

[21] W. Strauss, *Existence of solitary waves in higher dimensions.* Comm. Math. Phys., **55**, (1977), 149–162.

[22] J. Wei, S. Yan, *Infinitely many positive solutions for the nonlinear Schrödinger equations in* $\mathbb{R}^N$. Calc. Var. Partial Differential Equations, **37**, (2010), 423–439

Giovanna Cerami
Dipartimento di Matematica
Politecnico di Bari, Campus Universitario
Via Orabona 4
I-70125 Bari, Italia
e-mail: cerami@poliba.it

Progress in Nonlinear Differential Equations
and Their Applications, Vol. 85, 231–240

A Supercritical Elliptic Problem in a Cylindrical Shell

Mónica Clapp and Andrzej Szulkin

To Bernhard Ruf on his birthday, with our friendship and great esteem

Abstract. We consider the problem

$$-\Delta u = |u|^{p-2} u \ \text{ in } \Omega, \quad u = 0 \ \text{ on } \partial\Omega,$$

where $\Omega := \{(y, z) \in \mathbb{R}^{m+1} \times \mathbb{R}^{N-m-1} : 0 < a < |y| < b < \infty\}$, $0 \leq m \leq N-1$ and $N \geq 2$. Let $2^*_{N,m} := 2(N-m)/(N-m-2)$ if $m < N-2$ and $2^*_{N,m} := \infty$ if $m = N-2$ or $N-1$. We show that $2^*_{N,m}$ is the true critical exponent for this problem, and that there exist nontrivial solutions if $2 < p < 2^*_{N,m}$ but there are no such solutions if $p \geq 2^*_{N,m}$.

Mathematics Subject Classification (2010). 35J61, 35J20, 35B33, 35B07.

Keywords. Nonlinear elliptic equation, supercritical nonlinearity, unbounded domain.

1. Introduction

Consider the Lane–Emden–Fowler problem

$$-\Delta u = |u|^{p-2} u \ \text{ in } \mathcal{D}, \qquad u = 0 \ \text{ on } \partial\mathcal{D}, \tag{1.1}$$

where $\mathcal{D}$ is a smooth domain in $\mathbb{R}^N$ and $p > 2$.

If $\mathcal{D}$ is bounded it is well known that this problem has at least one positive solution and infinitely many sign changing solutions when p is smaller than the critical Sobolev exponent 2^*, defined as $2^* := \frac{2N}{N-2}$ if $N \geq 3$ and as $2^* := \infty$ if $N = 1$ or 2. In contrast, the existence of solutions for $p \geq 2^*$ is a delicate issue. Pohozhaev's identity [12] implies that problem (1.1) has no nontrivial solution if the domain $\mathcal{D}$ is strictly starshaped. On the other hand, Bahri and Coron [2]

This research is partially supported by CONACYT grant 129847, UNAM-DGAPA-PAPIIT grant IN106612 (México), and the Swedish Research Council.

proved that a positive solution to (1.1) exists if $p = 2^*$ and $\mathcal{D}$ is bounded and has nontrivial reduced homology with $\mathbb{Z}/2$ coefficients.

One may ask whether this last statement is also true for $p > 2^*$. Passaseo showed in [10, 11] that this is not so: for each $1 \leq m < N-2$ he exhibited a bounded smooth domain $\mathcal{D}$ which is homotopy equivalent to the m-dimensional sphere, in which problem (1.1) has infinitely many solutions if $p < 2^*_{N,m} := \frac{2(N-m)}{N-m-2}$ and does not have a nontrivial solution if $p \geq 2^*_{N,m}$. Examples of domains with richer homology were recently given by Clapp, Faya and Pistoia in [3]. Wei and Yan established in [17] the existence of infinitely many positive solutions for $p = 2^*_{N,m}$ in some bounded domains. For p slightly below $2^*_{N,m}$ solutions concentrating along an m-dimensional manifold were recently obtained in [1, 4]. Note that $2^*_{N,m}$ is the critical Sobolev exponent in dimension $N-m$. It is called the $(m+1)$-st critical exponent for problem (1.1).

The purpose of this note is to exhibit unbounded domains in which this problem has the behavior described by Passaseo.

We consider the problem

$$\begin{cases} -\Delta u = |u|^{p-2}u & \text{in } \Omega, \\ u = 0 & \text{on } \partial\Omega, \\ |\nabla u|^2, |u|^p \in L^1(\Omega), \end{cases} \tag{1.2}$$

in a cylindrical shell

$$\Omega := \{x = (y,z) \in \mathbb{R}^{m+1} \times \mathbb{R}^{N-m-1} : a < |y| < b\}, \qquad 0 < a < b < \infty,$$

for $p > 2$.

If $m = N-1$ or $N-2$, we set $2^*_{N,m} := \infty$. First note that if $m = N-1$ then $\Omega = \{x \in \mathbb{R}^N : a < |x| < b\}$, and a well-known result by Kazdan and Warner [9] asserts that (1.2) has infinitely many radial solutions for any $p > 2$. In the other extreme case, where $m = 0$, the domain Ω is the union of two disjoint strips $(a,b) \times \mathbb{R}^{N-1}$ and $(-b,-a) \times \mathbb{R}^{N-1}$. Each of them is starshaped, so there are no solutions for $p \geq 2^*_{N,0} = 2^*$. Esteban showed in [5] that there are infinitely many solutions in $(a,b) \times \mathbb{R}^{N-1}$ if $N \geq 3$ and $p < 2^*$, and one positive solution if $N = 2$ (in fact, she considered a more general problem). These solutions are axially symmetric, i.e., $u(y,z) = u(y,|z|)$ for all $(y,z) \in \Omega$.

Here we study the remaining cases, i.e., $1 \leq m \leq N-2$. Our first result states the nonexistence of solutions other than $u = 0$, if $p \geq 2^*_{N,m}$.

Theorem 1.1. *If $1 \leq m < N-2$ and $p \geq 2^*_{N,m}$, then problem (1.2) does not have any nontrivial solution $u \in \mathcal{C}^2(\Omega) \cap \mathcal{C}^1(\overline{\Omega})$.*

Our next result shows that solutions $u \neq 0$ do exist if $2 < p < 2^*_{N,m}$.

As usual, we write $O(k)$ for the group of linear isometries of $\mathbb{R}^k$ (represented by orthogonal $k \times k$-matrices). Recall that if G is a closed subgroup of $O(N)$ then a subset X of $\mathbb{R}^N$ is G-invariant if $gX = X$ for every $g \in G$, and a function $u : X \to \mathbb{R}$ is called G-invariant provided $u(gx) = u(x)$ for all $g \in G$, $x \in X$.

Note that Ω is $[O(m+1) \times O(N-m-1)]$-invariant for the obvious action given by $(g,h)(y,z) := (gy, hz)$ for all $g \in O(m+1)$, $h \in O(N-m-1)$, $y \in \mathbb{R}^{m+1}$, $z \in \mathbb{R}^{N-m-1}$.

Theorem 1.2.

(i) *If $1 \leq m < N-2$ and $2 < p < 2^*_{N,m}$, then problem* (1.2) *has infinitely many $[O(m+1) \times O(N-m-1)]$-invariant solutions and one of these solutions is positive.*

(ii) *If $1 \leq m = N-2$ and $2 < p < \infty$, then problem* (1.2) *has a positive $[O(N-1) \times O(1)]$-invariant solution.*

In Section 2 we prove Theorem 1.1. Theorem 1.2 is proved in Section 3. We conclude the paper with a multiplicity result and an open question in Section 4.

2. A Pohožaev identity and the proof of Theorem 1.1

We prove Theorem 1.1 by adapting Passaseo's argument in [10, 11], see also [3]. The proof relies on the following special case of a Pohožaev type identity due to Pucci and Serrin [13].

For $(u,v) \in \mathbb{R} \times \mathbb{R}^N$ we set

$$\phi(u,v) := \frac{1}{2}|v|^2 - \frac{1}{p}|u|^p.$$

Lemma 2.1. *If $u \in C^2(\Omega)$ satisfies $-\Delta u = |u|^{p-2}u$ in Ω then, for every $\chi \in C^1(\overline{\Omega}, \mathbb{R}^N)$, the equality*

$$(\operatorname{div}\chi)\,\phi(u,\nabla u) - D\chi\,[\nabla u]\cdot\nabla u = \operatorname{div}\left[\phi(u,\nabla u)\chi - (\chi\cdot\nabla u)\nabla u\right] \tag{2.1}$$

holds true.

Proof. Put $\chi = (\chi_1, \ldots, \chi_N)$, denote the partial derivative with respect to x_k by ∂_k and let LHS and RHS denote the left- and the right-hand side of (2.1). Then

$$\text{LHS} = (\operatorname{div}\chi)\,\phi(u,\nabla u) - \sum_{j,k}\partial_k\chi_j\,\partial_j u\,\partial_k u$$

and

$$\begin{aligned}\text{RHS} &= (\operatorname{div}\chi)\,\phi(u,\nabla u) + \sum_{j,k}\chi_k\,\partial_j u\,\partial^2_{jk}u - |u|^{p-2}u\,\nabla u\cdot\chi \\ &\quad - (\nabla u\cdot\chi)\Delta u - \sum_{j,k}\partial_k\chi_j\,\partial_j u\,\partial_k u - \sum_{j,k}\chi_j\,\partial_k u\,\partial^2_{jk}u \\ &= (\operatorname{div}\chi)\,\phi(u,\nabla u) - (\nabla u\cdot\chi)(\Delta u + |u|^{p-2}u) - \sum_{j,k}\partial_k\chi_j\,\partial_j u\,\partial_k u.\end{aligned}$$

Since $-\Delta u = |u|^{p-2}u$, the conclusion follows. □

Using a well-known truncation argument, we can now prove the following result.

Proposition 2.2. *Assume that* $\chi \in \mathcal{C}^1(\overline{\Omega}, \mathbb{R}^N)$ *has the following properties:*
(a) $\chi \cdot \nu$ *is bounded on* $\partial\Omega$, *where* $\nu(s)$ *is the outer unit normal at* $s \in \partial\Omega$,
(b) $|\chi(x)| \leq |x|$ *for every* $x \in \Omega$,
(c) $\operatorname{div}\chi$ *is bounded in* Ω,
(d) $|D\chi(x)\xi \cdot \xi| \leq |\xi|^2$ *for all* $x \in \Omega$, $\xi \in \mathbb{R}^N$.
Then every solution $u \in \mathcal{C}^2(\Omega) \cap \mathcal{C}^1(\overline{\Omega})$ *of* (1.2) *satisfies*

$$\frac{1}{2}\int_{\partial\Omega} |\nabla u|^2 \chi \cdot \nu = -\int_\Omega (\operatorname{div}\chi)\,\phi(u,\nabla u) + \int_\Omega D\chi\,[\nabla u]\cdot\nabla u. \tag{2.2}$$

Proof. Choose $\psi \in \mathcal{C}^\infty(\mathbb{R})$ such that $0 \leq \psi(t) \leq 1$, $\psi(t) = 1$ if $|t| \leq 1$ and $\psi(t) = 0$ if $|t| \geq 2$. For each $k \in \mathbb{N}$ define

$$\psi_k(x) := \psi\left(\frac{|x|^2}{k^2}\right) \qquad \text{and} \qquad \chi^k(x) := \psi_k(x)\chi(x).$$

Note that there is a constant $c_0 > 0$ such that

$$|x|\,|\nabla\psi_k(x)| \leq c_0 \qquad \text{for all } x \in \mathbb{R}^N,\ k \in \mathbb{N}. \tag{2.3}$$

Next, choose a sequence of bounded smooth domains $\Omega_k \subset \Omega$ such that

$$\Omega_k \supset \Omega \cap \overline{B_{2k}(0)}. \tag{2.4}$$

Integrating (2.1) with $\chi := \chi^k$ in Ω_k and using the divergence theorem and Lemma 2.1 we obtain

$$\begin{aligned}&\int_{\Omega_k} (\operatorname{div}\chi^k)\,\phi(u,\nabla u) - \int_{\Omega_k} D\chi^k\,[\nabla u]\cdot\nabla u \\ &\quad = \int_{\partial\Omega_k} \left[\phi(u,\nabla u)\,(\chi^k\cdot\nu^k) - (\chi^k\cdot\nabla u)\,(\nabla u\cdot\nu^k)\right],\end{aligned}$$

where ν^k is the outer unit normal to Ω_k. Property (2.4) implies that $\chi^k = 0$ in $\overline{\Omega \smallsetminus \Omega_k}$, so we may replace Ω_k by Ω, $\partial\Omega_k$ by $\partial\Omega$ and ν^k by ν in the previous identity. Moreover, since $u = 0$ on $\partial\Omega$, we have that

$$\nabla u = (\nabla u \cdot \nu)\,\nu \ \text{ on } \partial\Omega.$$

Therefore,

$$\begin{aligned}&\int_{\Omega} (\operatorname{div}\chi^k)\,\phi(u,\nabla u) - \int_{\Omega} D\chi^k\,[\nabla u]\cdot\nabla u \\ &\quad = \int_{\partial\Omega} \left[\phi(u,\nabla u)\,(\chi^k\cdot\nu) - (\chi^k\cdot\nabla u)\,(\nabla u\cdot\nu)\right] \\ &\quad = \int_{\partial\Omega} \left[\phi(u,\nabla u) - |\nabla u|^2\right](\chi^k\cdot\nu) \\ &\quad = -\frac{1}{2}\int_{\partial\Omega} |\nabla u|^2\,\psi_k(x)\,(\chi\cdot\nu)\,.\end{aligned} \tag{2.5}$$

Since $\operatorname{div}\chi^k = \psi_k \operatorname{div}\chi + \nabla\psi_k\cdot\chi$, using (2.3) and properties (b) and (c) we obtain

$$\left|\operatorname{div}\chi^k\right| \leq |\operatorname{div}\chi| + |\nabla\psi_k|\,|\chi| \leq |\operatorname{div}\chi| + c_0 \leq c_1 \ \text{ in } \Omega. \tag{2.6}$$

Similarly, since

$$D\chi^k(x)\xi \cdot \xi = \psi_k(x) D\chi(x)\xi \cdot \xi + (\nabla\psi_k \cdot \xi)(\chi \cdot \xi),$$

property (d) yields

$$\left|D\chi^k(x)\xi \cdot \xi\right| \leq (1 + c_0)\,|\xi|^2 \qquad \text{for all } x \in \Omega,\ \xi \in \mathbb{R}^N. \tag{2.7}$$

Inequalities (2.6), (2.7) and property (a) allow us to apply Lebesgue's dominated convergence theorem to the left- and the right-hand side of (2.5) to obtain

$$\int_\Omega (\operatorname{div}\chi)\,\phi(u, \nabla u) - \int_\Omega D\chi\,[\nabla u] \cdot \nabla u = -\frac{1}{2}\int_{\partial\Omega} |\nabla u|^2\,(\chi \cdot \nu),$$

as claimed. □

Proof of Theorem 1.1. Let $\varphi(t) = \frac{1}{m+1}\left[1 - (\frac{a}{t})^{m+1}\right]$ be the solution to the boundary value problem

$$\begin{cases} \varphi'(t)t + (m+1)\varphi(t) = 1, & t \in (0, \infty), \\ \varphi(a) = 0. \end{cases}$$

Define

$$\chi(y, z) := (\varphi(|y|)y, z). \tag{2.8}$$

Then, if ν denotes the outer unit normal on $\partial\Omega$,

$$(\chi \cdot \nu)(y, z) = \begin{cases} 0 & \text{if } |y| = a, \\ \frac{1}{m+1}\left[1 - (\frac{a}{b})^{m+1}\right] b & \text{if } |y| = b. \end{cases} \tag{2.9}$$

So property (a) of Proposition 2.2 holds. Clearly, (b) holds. Now,

$$\operatorname{div}\chi(y, z) = [\varphi'(|y|)\,|y| + (m+1)\varphi(|y|)] + N - m - 1 = N - m. \tag{2.10}$$

In particular, (c) holds. To prove (d) notice that χ is $O(m+1)$-equivariant, i.e.,

$$\chi(gy, z) = g\chi(y, z) \qquad \text{for every } g \in O(m+1).$$

Therefore, $g \circ D\chi(y, z) = D\chi(gy, z) \circ g$ and, hence,

$$\langle D\chi\,(y, z)\,[\xi]\,, \xi\rangle = \langle g\,(D\chi\,(y, z)\,[\xi])\,, g\xi\rangle = \langle D\chi\,(gy, z)\,[g\xi], g\xi\rangle$$

for all $\xi \in \mathbb{R}^N$. Thus, it suffices to show that the inequality (d) holds for $y = (t, 0, \ldots, 0)$ with $t \in (a, b)$. A straightforward computation shows that, for such y, $D\chi(y)$ is a diagonal matrix whose diagonal entries are $a_{11} = 1 - m\varphi(t)$, $a_{jj} = \varphi(t)$ for $j = 2, \ldots, m+1$, and $a_{jj} = 1$ for $j = m+2, \ldots, N$. Since $a_{jj} \in (0, 1]$,

$$0 < \langle D\chi\,(y, z)\,[\xi]\,, \xi\rangle \leq |\xi|^2 \quad \text{for all } \xi \in \mathbb{R}^N \smallsetminus \{0\} \tag{2.11}$$

and (d) follows. From (2.9), (2.2), (2.11) and (2.10) we obtain

$$\begin{aligned}0 &< \frac{1}{2}\int_{\partial\Omega} |\nabla u|^2 \chi\cdot\nu \\ &= -\int_\Omega (\operatorname{div}\chi)\,\phi(u,\nabla u) + \int_\Omega D\chi\,[\nabla u]\cdot\nabla u \\ &\le (N-m)\int_\Omega \left[\frac{1}{p}|u|^p - \frac{1}{2}|\nabla u|^2\right] + \int_\Omega |\nabla u|^2 \\ &= (N-m)\left(\frac{1}{p}-\frac{1}{2}+\frac{1}{N-m}\right)\int_\Omega |\nabla u|^2.\end{aligned}$$

The first (strict) inequality follows from the unique continuation property [8, 7]. This immediately implies that $p < 2^*_{N,m}$. □

3. The proof of Theorem 1.2

An $O(m+1)$-invariant function $u(y,z) = v(|y|,z)$ solves problem (1.2) if and only if $v = v(r,z)$ solves

$$\begin{cases} -\Delta v - \frac{m}{r}\frac{\partial v}{\partial r} = |v|^{p-2}v & \text{in } (a,b)\times\mathbb{R}^{N-m-1} =: \mathcal{S}, \\ v = 0 & \text{on } \{a,b\}\times\mathbb{R}^{N-m-1} = \partial\mathcal{S}, \end{cases} \tag{3.1}$$

and $|\nabla v|^2, |v|^p \in L^1(\mathcal{S})$. Problem (3.1) can be rewritten as

$$-\operatorname{div}(r^m\nabla v) = r^m|v|^{p-2}v \quad \text{in } \mathcal{S}, \qquad v = 0 \quad \text{on } \partial\mathcal{S}. \tag{3.2}$$

By Poincaré's inequality (see Lemma 3 in [5]) and since $a < r < b$, the norms

$$\|v\|_m := \left(\int_{\mathcal{S}} r^m |\nabla v|^2\right)^{1/2} \quad \text{and} \quad |v|_{m,p} := \left(\int_{\mathcal{S}} r^m |v|^p\right)^{1/p} \tag{3.3}$$

are equivalent to those of $H_0^1(\mathcal{S})$ and $L^p(\mathcal{S})$ respectively.

Consider the functional $I(v) := \|v\|_m^2$ restricted to

$$M := \{v \in H_0^1(\mathcal{S}) : |v|_{m,p} = 1\}.$$

Then M is a C^2-manifold, and v is a critical point of $I|_M$ if and only if $v \in H_0^1(\mathcal{S})$ and $\|v\|_m^{2/(p-2)}v$ is a nontrivial solution to (3.2). Note that $I|_M$ is bounded below by a positive constant.

Proof of Theorem 1.2 (i). Assume that $1 \le m < N-2$ and $2 < p < 2^*_{N,m}$. Set $G := O(N-m-1)$ and denote by $H_0^1(\mathcal{S})^G$ and $L^p(\mathcal{S})^G$ the subspaces of $H_0^1(\mathcal{S})$ and $L^p(\mathcal{S})$ respectively, consisting of functions v such that $v(r,gz) = v(r,z)$ for all $g \in G$. Esteban and Lions showed in [6] that, for these values of m and p, $H_0^1(\mathcal{S})^G$ is compactly embedded in $L^p(\mathcal{S})^G$ (see also Theorem 1.24 in [18]). So $H_0^1(\mathcal{S})^G$ is compactly embedded in $L^p(\mathcal{S})^G$ for the norms (3.3) as well.

Let

$$M^G := \{v \in H_0^1(\mathcal{S})^G : |v|_{m,p} = 1\}.$$

It follows from the principle of symmetric criticality [18, Theorem 1.28] that the critical points of $I|_{M^G}$ are also critical points of $I|_M$. The manifold M^G is radially diffeomorphic to the unit sphere in $H_0^1(\mathcal{S})^G$, so its Krasnoselskii genus is infinite. A standard argument, using the compactness of the embedding $H_0^1(\mathcal{S})^G \hookrightarrow L^p(\mathcal{S})^G$ for the norms (3.3), shows that $I|_{M^G}$ satisfies the Palais–Smale condition. Hence $I|_{M^G}$ has infinitely many critical points (see, e.g., Theorem II.5.7 in [15]). It can also be shown by a well-known argument that the critical values of $I|_{M^G}$ tend to infinity (see, e.g., Proposition 9.33 in [14]).

It remains to show that (3.2) has a positive solution. The argument is again standard: since $I|_{M^G}$ satisfies the Palais–Smale condition,

$$c_0^G := \inf\{I(v) : v \in M^G\}$$

is attained at some v_0. Since $I(v) = I(|v|)$ and $|v| \in M^G$ if $v \in M^G$, we have that $I(|v_0|) = c_0^G$ and we may assume $v_0 \geq 0$. The maximum principle applied to the corresponding solution u_0 of (1.2) implies $u_0 > 0$. □

If $m = N - 2$, then $G = O(1)$ and it is easy to see that the space $H_0^1(\mathcal{S})^G$ is not compactly embedded in $L^p(\mathcal{S})^G$. So part (ii) of Theorem 1.2 requires a different argument.

Proof of Theorem 1.2 (ii). Assume that $1 \leq m = N - 2$ and $2 < p < \infty$. We shall show that

$$c_0 := \inf\{I(v) : v \in M\}$$

is attained. Clearly, a minimizing sequence (v_n) is bounded, so we may assume that $v_n \rightharpoonup v$ weakly in $H_0^1(\mathcal{S})$. According to P.-L. Lions' lemma [18, Lemma 1.21] either $v_n \to 0$ strongly in $L^p(\mathcal{S})$, which is impossible because $v_n \in M$, or there exist $\delta > 0$ and $(r_n, z_n) \in [a, b] \times \mathbb{R}$ such that, after passing to a subsequence if necessary,

$$\int_{B_1(r_n, z_n)} v_n^2 \geq \delta. \tag{3.4}$$

Here $B_1(r_n, z_n)$ denotes the ball of radius 1 and center at (r_n, z_n). Since the problem is invariant with respect to translations along the z-axis, replacing $v_n(r, z)$ by $v_n(r, z + z_n)$, we may assume the center of the ball above is $(r_n, 0)$. It follows that for this – translated – sequence the weak limit v cannot be zero due to (3.4) and the compactness of the embedding of $H_0^1(\mathcal{S})$ in $L^2_{loc}(\mathcal{S})$. Passing to a subsequence once more, we have that $v_n(x) \to v(x)$ a.e. It follows from the Brezis–Lieb lemma [18, Lemma 1.32] that

$$1 = |v_n|^p_{m,p} = \lim_{n\to\infty} |v_n - v|^p_{m,p} + |v|^p_{m,p}.$$

Using this identity and the definition of c_0 we obtain

$$\begin{aligned} c_0 &= \lim_{n\to\infty} \|v_n\|_m^2 = \lim_{n\to\infty} \|v_n - v\|_m^2 + \|v\|_m^2 \\ &\geq c_0 \left(\lim_{n\to\infty} |v_n - v|_{m,p}^2 + |v|_{m,p}^2 \right) \\ &= c_0 \left((1 - |v|_{m,p}^p)^{2/p} + (|v|_{m,p}^p)^{2/p} \right) \\ &\geq c_0 (1 - |v|_{m,p}^p + |v|_{m,p}^p)^{2/p} = c_0. \end{aligned}$$

Since $v \neq 0$, it follows that $|v_n - v|_{m,p} \to 0$ and $|v|_{m,p} = 1$. So $v \in M$ and, as $c_0 = \lim_{n\to\infty} I(v_n) \geq I(v)$, we must have $I(v) = c_0$.

So the infimum is attained at v and using the moving plane method [18, Appendix C], we may assume, after translation, that $v(r, -z) = v(r, z)$, i.e., $v \in H_0^1(\mathcal{S})^{O(1)}$. As in the preceding proof, replacing v by $|v|$, we obtain a positive solution. □

4. Further solutions and an open question

If $1 \leq m = N - 2$ and $p \in (2, 2^*_{N,m})$, the method we have used to prove Theorem 1.2 only guarantees the existence of two solutions to problem (1.2), one positive and one negative, up to translations along the z-axis. However, if $p \in (2, 2^*)$, then it is possible to show that there are infinitely many solutions, which are not radial in y, but have other prescribed symmetry properties.

Write $y = (y^1, y^2) \in \mathbb{R}^2 \times \mathbb{R}^{m-1} \equiv \mathbb{R}^{m+1}$ and identify $\mathbb{R}^2$ with the complex plane $\mathbb{C}$. Following [16], we denote by G_k, $k \geq 3$, the subgroup of $O(2)$ generated by two elements α, β which act on $\mathbb{C}$ by

$$\alpha y^1 := \mathrm{e}^{2\pi i/k} y^1, \qquad \beta y^1 := \mathrm{e}^{2\pi i/k} \overline{y^1},$$

i.e., α is the rotation in $\mathbb{C}$ by the angle $2\pi/k$ and β is the reflection in the line $y_2^1 = \tan(\pi/k) y_1^1$, where $y^1 = y_1^1 + \mathrm{i} y_2^1 \in \mathbb{C}$. Observe that α, β satisfy the relations $\alpha^k = \beta^2 = e$, $\alpha\beta\alpha = \beta$. Let G_k act on $\mathbb{R}^N$ by $gx = (gy^1, y^2, z)$.

Theorem 4.1. *If $1 \leq m \leq N - 2$ and $2 < p < 2^*$ then, for each $k \geq 3$, problem (1.2) has a solution u_k which satisfies*

$$u_k(x) = \det(g) u_k(g^{-1}x) \qquad \text{for all } g \in G_k, \tag{4.1}$$

and $u_k \neq u_j$ if $k \neq j$.

Proof. Since the approach is taken from [16], we give only a brief sketch of the proof here and refer to Section 2 of [16] for more details.

The group G_k acts on $H_0^1(\Omega)$ by

$$(gu)(x) := \det(g) u(g^{-1}x),$$

where $\det(g)$ is the determinant of g. Let

$$H_0^1(\Omega)^{G_k} := \{u \in H_0^1(\Omega) : u(gx) = \det(g) u(x) \text{ for all } g \in G_k\}$$

be the fixed point space of this action, and define $I(u) := \int_\Omega |\nabla u|^2$ and

$$M^{G_k} := \{u \in H_0^1(\Omega)^{G_k} : |u|_p = 1\}.$$

By the principle of symmetric criticality the critical points of $I|_{M^{G_k}}$ are nontrivial solutions to problem (1.2) which satisfy (4.1). Now we can see as in the proof of part (ii) of Theorem 1.2 that there exists a minimizer u_k for I on the manifold M^{G_k}. Moreover, we may assume that u_k has exactly $2k$ nodal domains, see Corollary 2.7 in [16]. So in particular, $u_k \neq u_j$ if $k \neq j$. □

The question whether problem (1.2) has infinitely many solutions when $1 \leq m = N-2$ and $p \in [2^*, 2^*_{N,m})$ remains open. We believe that the answer is yes, but the proof would require different methods.

References

[1] N. Ackermann, M. Clapp, A. Pistoia, Boundary clustered layers near the higher critical exponents, *J. Differential Equations* **254** (2013), 4168–4183.

[2] A. Bahri, J.M. Coron, On a nonlinear elliptic equation involving the critical Sobolev exponent: The effect of the topology of the domain, *Comm. Pure Appl. Math.* **41** (1988), 253–294.

[3] M. Clapp, J. Faya, A. Pistoia, Nonexistence and multiplicity of solutions to elliptic problems with supercritical exponents, *Calc. Var. Partial Differential Equations* **48** (2013), 611–623.

[4] M. del Pino, M. Musso, and F. Pacard, Bubbling along boundary geodesics near the second critical exponent, J. Eur. Math. Soc. (JEMS) **12** (2010), 1553–1605.

[5] M.J. Esteban, Nonlinear elliptic problems in strip-like domains: symmetry of positive vortex rings, *Nonlinear Anal.* **7** (1983), 365–379.

[6] M.J. Esteban, P.-L. Lions, A compactness lemma, *Nonlinear Anal.* **7** (1983), 381–385.

[7] N. Garofalo, F.-H. Lin, Unique continuation for elliptic operators: a geometric-variational approach, *Comm. Pure Appl. Math.* **40** (1987), 347–366.

[8] D. Jerison, C.E. Kenig, Unique continuation and absence of positive eigenvalues for Schrödinger operators, *Annals of Math.* **121** (1985), 463–494.

[9] J.L. Kazdan, F.W. Warner, Remarks on some quasilinear elliptic equations, *Comm. Pure Appl. Math.* **28** (1975), 567–597.

[10] D. Passaseo, Nonexistence results for elliptic problems with supercritical nonlinearity in nontrivial domains, *J. Funct. Anal.* **114** (1993), 97–105.

[11] D. Passaseo, New nonexistence results for elliptic equations with supercritical nonlinearity, *Differential Integral Equations* **8** (1995), 577–586.

[12] S.I. Pohožaev, Eigenfunctions of the equation $\Delta u + \lambda f(u) = 0$, *Soviet Math. Dokl.* **6** (1965), 1408–1411.

[13] P. Pucci, J. Serrin, A general variational identity, *Indiana Univ. Math. J.* **35** (1986), 681–703.

[14] P.H. Rabinowitz, "Minimax Methods in Critical Point Theory with Applications to Differential Equations", *CBMS* **65**, Amer. Math. Soc., Providence, R.I. 1986.

[15] M. Struwe, "Variational Methods", Springer-Verlag, Berlin-Heidelberg-New York 1990.

[16] A. Szulkin, S. Waliullah, Sign-changing and symmetry-breaking solutions to singular problems, *Complex Variables Elliptic Eq.* **57** (2012), 1191–1208.

[17] J. Wei, S. Yan, Infinitely many positive solutions for an elliptic problem with critical or supercritical growth, *J. Math. Pures Appl.* **96** (2011), 307–333.

[18] M. Willem, "Minimax Theorems", *PNLDE* **24**, Birkhäuser, Boston-Basel-Berlin 1996.

Mónica Clapp
Instituto de Matemáticas
Universidad Nacional Autónoma de México
Circuito Exterior, C.U.
04510 México D.F., Mexico
e-mail: `monica.clapp@im.unam.mx`

Andrzej Szulkin
Department of Mathematics
Stockholm University
106 91 Stockholm, Sweden
e-mail: `andrzejs@math.su.se`

Progress in Nonlinear Differential Equations
and Their Applications, Vol. 85, 241–266

The Geometric Microlocal Analysis of Generalized Kimura and Heston Diffusions

C.L. Epstein and Rafe Mazzeo

Abstract. In this paper we show how to use geometric microlocal analysis techniques to construct the heat kernel for a class of degenerate diffusions, called Kimura diffusions, which arise as continuum limits of the Wright–Fisher model in Population Genetics. We restrict our attention to the case of a Kimura diffusion on a manifold with boundary, and show that, by changing variables and scaling, we can employ the 0-calculus to construct a parametrix for the heat kernel.

Mathematics Subject Classification (2010). 35K65, 58J65, 35K08, 92D25, 35A27.

Keywords. Degenerate diffusion operators, heat kernel, Kimura diffusion, Wright–Fisher model, population genetics, V_0-calculus.

1. Introduction

Our recent work [6], [7] contains a thorough analysis of the existence and regularity theory for solutions of a certain class of degenerate diffusion operators $\partial_t - L$ on a compact manifold with corners M, for data in anisotropic Hölder spaces. The operators we consider arise naturally in applications to population genetics and mathematical finance, and to acknowledge some of the early and influential work on special operators operators of this form, we call these generalized Kimura diffusions; we also use the moniker Wright–Fisher since the original discrete Markov chain model in population genetics is known as the Wright–Fisher model.

Let p be a point on the boundary of M and let $(r_1, \ldots, r_k, y_1, \ldots, y_\ell)$ be a local coordinate system near a point p. Here each $r_k \geq 0$ and $(y_1, \ldots, y_\ell)$ lies in an open ball in $\mathbb{R}^\ell$, with p corresponding to the origin. (If such coordinates exist, we

The research of the first author was partially supported by NSF grant DMS12-05851 and ARO grant W911NF-12-1-0552.

The research of the second author was partially supported by NSF grant DMS11-05050.

say that p lies on a corner of codimension k). We say that L is of general Kimura type if it takes the form

$$L = \sum_{i=1}^{k} r_i \partial_{r_i}^2 + \sum_{p,q=1}^{k} a_{pq} r_p r_q \partial_{r_p r_q}^2 + \sum_{i=1}^{k} b_i \partial_{r_i} \\ + \sum_{p=1}^{k} \sum_{j=1}^{\ell} e_{pj} r_p \partial_{r_p y_j}^2 + \sum_{j,s=1}^{\ell} c_{js} \partial_{y_j y_s}^2 + \sum_{s=1}^{\ell} f_s \partial_{y_s},$$

where all coefficients a_{pq}, b_i, e_{pj}, c_{js} and f_s are smooth functions of (r, y). It is elliptic in this setting provided these coefficients (with the prefactors of r removed) satisfy a certain definiteness condition which we explain below. The dependence of the coefficients b_i on (r, y) is worthy of special note since this dependence causes many of the technical difficulties in our analysis.

Although the aforementioned references do not explicitly treat other types of degenerate diffusions, we observed in the course of that research that our methods and results can be adapted easily to another interesting class of operators which generalize the so-called Heston operator, a useful model in mathematical finance. The elliptic operators L associated to these generalized Heston diffusions are nearly of the same form as the Kimura operators above except that the coefficients c_{rs} associated to the tangential second-order terms also vanish to first order at the boundaries and corners of M. As in the Kimura setting, there is an adapted notion of ellipticity here.

The monograph [7] discusses a number of approaches that have been used successfully in the past to treat these two types of degenerate parabolic problems; we draw attention in particular to [4], [5] which analyze special cases of these types of operators from a geometric point of view.

Our goal in this paper is to indicate how an analysis of generalized Kimura and Heston operators and their associated heat operators can be addressed using the methods of geometric microlocal analysis. This approach comprises a very robust set of tools, pioneered by Melrose and extensively developed by him and many others, building from the elliptic and parabolic parametrix constructions of classical microlocal analysis with a particular focus on the polyhomogeneous structure of the Schwartz kernels of these parametrices. The advantage of this over the more customary focus on symbol classes, etc., is that parametrices of degenerate operators must incorporate information not only about interior singularities (e.g., along the diagonal) but along the boundaries and corners as well, since that is where the effects of the degeneracy appear. These new boundary singularities are best described using the geometric language of blowups of manifolds with corners and the systematic use of spaces of conormal and polyhomogeneous distributions. There are now quite a few detailed treatments of special cases of this methodology, though unfortunately no general expository source. We refer to [12], [15] and [11] for very detailed explanations of the constructions for particular classes of degenerate elliptic operators; the first and last paper treat operators similar to those considered here. For treatments of the parabolic problems associated to such operators see [1], [13] and the forthcoming expository survey [2].

To keep this exposition brief, we consider only the restricted setting where M is a manifold with boundary. This reduces the complexity of the analysis substantially. It will be possible to handle the general case, allowing corners of arbitrary codimension, using extensions of these methods, but this will take substantial further work.

There is at least one compelling reason to pursue this alternate development to the results of [7], since there is a lacuna in the technical results obtained by those earlier methods, as we now describe. In the original applications to population biology which drew us to these problems, it is important to understand Kimura diffusions as semigroups on the function space $\mathcal{C}^0(M)$. Indeed, this is natural because the dual problem ("Kolmogorov's forward equation") acts on measures. Using the earlier approach, we gave a complete analysis of the semigroup on $\mathcal{C}^0$, and specifically the precise and sharp estimates for the smoothing effect of this diffusion, only for two special cases: the one-dimensional case handled in [6], and the invariant model case (the key building block in the parametrix construction) in [7]. This is possible because there are explicit formulæ for the heat kernels in these model cases, from which the regularity of solutions can be deduced quite easily. In the one-dimensional case one can change variables to put the operator *exactly* into model form near each boundary point. In higher dimensions this analysis also suffices when the second-order part of the operator agrees exactly with the model.

$$\sum\nolimits_j x_j \partial_{x_j}^2 + \sum\nolimits_k \partial_{y_k}^2. \tag{1.1}$$

In the general higher-dimensional case, however, a more elaborate perturbation analysis is required, and these arguments are structured with respect to a given Hölder space, hence precluding direct consideration of $\mathcal{C}^0$. That approach does allow one to deduce precise regularity estimates at times $t > 0$ if the initial data lies in one of these Hölder spaces, but to pass from there to initial data in $\mathcal{C}^0$ requires a limiting argument using the maximum principle, which loses a lot of information. By contrast, the geometric microlocal approach developed here allows us to complete the $\mathcal{C}^0$ semigroup theory for generalized Kimura diffusions, at least in the case of manifolds with boundary but no higher codimensional corners.

Let us now turn to the specific results proved here. Consider the parabolic operator $\partial_t - L$ on a manifold with boundary M where L falls into one of the two following classes of degenerate elliptic operators, the generalized Kimura-type diffusion:

$$\begin{aligned} L_{\mathrm{Kim}} = r\partial_r^2 &+ \sum_j c_j(r,y) r\partial_r\partial_{y_j} \\ &+ \sum_{i,j} a_{ij}(r,y)\partial_{y_i y_j}^2 + b_0(r,y)\partial_r + \sum_j b_j(r,y)\partial_{y_j} + e(r,y), \end{aligned} \tag{1.2}$$

where (a_{ij}) is a positive definite $(n-1)$-by-$(n-1)$ matrix, and where we require that L_{Kim} is elliptic, in the ordinary sense, in the interior of M, and the generalized

Heston-type diffusion:

$$\begin{aligned} L_{\mathrm{Hes}} = \; & x\partial_x^2 + \sum_j c_j(x,y)x\partial_x\partial_{y_j} \\ & + x\sum_{i,j} a_{ij}(x,y)\partial^2_{y_iy_j} + b_0(x,y)\partial_x + \sum_j b_j(x,y)\partial_{y_j} + e(x,y). \end{aligned} \tag{1.3}$$

We have normalized L_{Kim} by requiring the coefficient of ∂_r^2 to equal r; similarly, L_{Hes} is normalized by requiring that the coefficient of ∂_x^2 equals x. With these normalizations, the value of the coefficient b_0 is invariantly defined at $r = 0$ (or $x = 0$). We then define

$$\beta(L_{\mathrm{Kim}}) \text{ or } \beta(L_{\mathrm{Hes}}) = \inf\{b_0(0,y) : \; y \in \partial M\}; \tag{1.4}$$

this number is often just denoted by β. In what follows we restrict attention to the case that $\beta > 0$. The character of the Schwartz kernel of the inverse is quite different when $\beta = 0$, with an interior term and a term localized on the boundary, see [7, 17].

The reason for labeling the normal variable r in (1.2) and x in (1.3) will become apparent momentarily. Ellipticity now requires that the symmetric matrix

$$\begin{pmatrix} 1 & c_1 & \cdots & c_{n-1} \\ c_1 & a_{11} & \cdots & a_{1,n-1} \\ \vdots & & & \vdots \\ c_{n-1} & a_{n-1,1} & \cdots & a_{n-1,n-1} \end{pmatrix} \tag{1.5}$$

is positive definite, and in addition that L_{Hes} is elliptic in the interior of M. Notice that the requirements of positive definiteness of the boundary operator along with the ordinary (but nonuniform) ellipticity in the interior together fix the notions of degenerate ellipticity in each setting.

The starting point for our work is the observation that operators in both classes can be transformed into the well-understood class of elliptic uniformly degenerate operators (also called 0-operators; we use these two monikers interchangeably). By definition, an operator $\mathcal{L}$ is said to be uniformly degenerate if it takes the form

$$\begin{aligned} \mathcal{L} = x^2\partial_x^2 + \sum_j c_j(x,y)x^2\partial^2_{xy_j} + \sum_{i,j} a_{ij}(x,y)x^2\partial^2_{y_iy_j} \\ + b_0(x,y)x\partial_x + \sum_j b_j(x,y)x\partial_{y_j} + e(x,y). \end{aligned} \tag{1.6}$$

Just as for Heston operators, 0-ellipticity requires that the same matrix (1.5) be positive definite and that $\mathcal{L}$ is elliptic in the standard sense in the interior.

The relationships between these types of operators is not difficult to explain. First, after a change of variables, an elliptic Kimura operator L_{Kim} is equivalent to a multiple of an elliptic uniformly degenerate operator. Indeed, replace the defining function r by a new variable $x = \sqrt{r}$; observing that

$$\sqrt{r}\,\partial_r = \frac{1}{2}\partial_x \Longrightarrow r\partial_r^2 = \frac{1}{4}\left(\partial_x^2 - \frac{1}{x}\partial_x\right),$$

we obtain the equivalent expression

$$\begin{aligned} L_{\text{Kim}} = \frac{1}{4}\partial_x^2 + \frac{1}{2}\sum c_j x \partial^2_{xy_j} \\ + \sum a_{ij}\partial^2_{y_iy_j} + \frac{1}{2}\left(b_0 - \frac{1}{2}\right)x^{-1}\partial_x + \sum b_j \partial_{y_j} + e. \end{aligned} \tag{1.7}$$

This is smooth and nondegenerate if and only if $b_0 \equiv \frac{1}{2}$; in general, however, $\mathcal{L} := x^2 L_{\text{Kim}}$ is an elliptic uniformly degenerate operator. Similarly, if L_{Hes} is an elliptic Heston operator, then $\mathcal{L} = xL_{\text{Hes}}$ is again elliptic and uniformly degenerate, without need to change variables.

What these transformations show is that elliptic operators of Kimura or Heston type differ only mildly from elliptic uniformly degenerate operators; furthermore, an uniformly degenerate operator which arises in this way has lowest-order term, $e(x, y)$, which vanishes at $x = 0$. We assume that this last condition holds for all the uniformly degenerate operators we consider below since this simplifies several points in the exposition. In other words, we restrict attention to uniformly degenerate operators of the form

$$\begin{aligned} \mathcal{L} = x^2\partial_x^2 + \sum_j c_j(x,y)x^2\partial^2_{xy_j} + \sum_{i,j} a_{ij}(x,y)x^2\partial^2_{y_iy_j} \\ + b_0(x,y)x\partial_x + \sum_j b_j(x,y)x\partial_{y_j}. \end{aligned} \tag{1.8}$$

The advantage of these transformations is that one has available the calculus of 0-pseudodifferential operators, as defined and developed in [11], which leads to detailed understanding of uniformly degenerate operators. These same techniques may therefore be brought to bear, to deduce mapping properties, fine regularity statements, etc., for elliptic Kimura and Heston operators.

It is less obvious how to use this relationship when studying the associated heat operators, however. For example, multiplying $\partial_t - L_{\text{Kim}}$ by r and then setting $r = x^2$ produces the rather difficult looking operator $x^2\partial_t - \mathcal{L}$, where $\mathcal{L}$ is an elliptic 0-operator, which is (rather seriously) non-parabolic at ∂M. A completely analogous issue occurs for Heston operators. Thus the new contribution of this paper is to explain how to adapt the geometric microlocal parametrix methods to these new types of degenerate parabolic operators. As a prelude to this, the next section contains a somewhat abridged review of the elliptic parametrix construction. This is important in its own right, but the methods provide a very good warm-up to the slightly more complicated constructions needed for the parabolic problem. That parabolic parametrix construction is explained in §4, and following this we explain in §5 how to use this parametrix to deduce the sharp regularity properties of solutions.

2. A review of the elliptic theory

As explained above, we begin with a review of the elliptic parametrix construction for elliptic uniformly degenerate operators. Amongst other things, this provides a clear indication of the role of boundary conditions. In this section we identify the space of pseudodifferential operators acting on functions defined on M which contains the partial inverses of a generalized Kimura and Heston operators, L_{Kim}^{-1} and L_{Hes}^{-1}, as the calculus of uniformly degenerate operators on a manifold with boundary.

It is shown in [7] that under the assumption that $\beta(L_{\text{Kim}}) > 0$, the equation

$$L_{\text{Kim}} u = f \tag{2.1}$$

is solvable for all f in a subspace, of codimension 1, defined by a linear functional of the form

$$\ell(f) = \int_M f \varphi dV. \tag{2.2}$$

Here φdV is the solution to the adjoint equation, $L'_{\text{Kim}}\varphi = 0$, and φ is an integrable, non-negative function which is smooth in the interior of M. There is a similar statement about the solvability of $L_{\text{Hes}} u = f$ which can be proved based on the theory developed in this section.

Our main results in the elliptic case are summarized in the two following theorems.

Theorem 2.1. *Let L_{Kim} be a generalized Kimura operator on a manifold with boundary M, for which the inward pointing part of the first-order term is nowhere vanishing. The partial inverse operator L_{Kim}^{-1} belongs to the space of* 0*-pseudo-differential operators $\Psi_0^{-2,2,0,2\beta-1}(M)$, where $\beta = \beta(L_{\text{Kim}})$ is defined in* (1.4).

Remark 2.2. *We explain the notation $\Psi_0^{-2,2,0,2\beta-1}(M)$ below, but note here that the superscripts denote orders of vanishing of the Schwartz kernels at various boundary faces of the* 0*-double space M_0^2 introduced below. Since we have changed variables, setting $r = x^2$, we should note that these orders of vanishing are with respect to the (x, y) rather than the (r, y) coordinate system, and the non-degenerate measure $dx'dy'$.*

Theorem 2.3. *Let L_{Hes} be a generalized Heston operator on a manifold with boundary M, for which the inward pointing part of the first-order term is nowhere vanishing. The partial inverse operator L_{Hes}^{-1} belongs to the space of* 0*-pseudo-differential operators $\Psi_0^{-2,1,0,\beta-1}(M)$, with $\beta = \beta(L_{\text{Hes}})$ defined in* (1.4).

In the remainder of this section we denote by $\mathcal{L}$ any second-order elliptic uniformly degenerate operator (1.6); for simplicity, assume that $\mathcal{L}$ is scalar, though the generalization of all the material below to systems is straightforward (see [11]). We refer to this paper for further details on all of the notation and ideas discussed in this section. At the end of this section we indicate why the results discussed here imply Theorems 2.1 and 2.3.

2.1. Model operators

One of the first invariants associated to $\mathcal{L}$ is its indicial operator. This is an ordinary differential operator, parameterized by points $y \in \partial M$, and given in local coordinates by

$$I(\mathcal{L})_y = s^2\partial_s^2 + b_0(0,y)s\partial_s.$$

This is obtained by formally replacing the local variable x by a global variable $s \in \mathbb{R}^+$ – there is an invariant way of regarding s as lying in the inward-pointing b-normal bundle to ∂M at y – and then setting all other occurrences of x to 0. In particular, we evaluate the coefficient b_0 at $(0, y) \in \partial M$ and because we are assuming that the lowest-order term e vanishes at $x = 0$, this term disappears in this model operator. The indicial operator is a regular singular operator and can be analyzed by classical methods. Its indicial roots are the values γ such that $I(\mathcal{L})_y s^\gamma = 0$; since $\mathcal{L}$ is second-order scalar, there are only two such values, $\gamma_0 = 0$ and $\gamma_1 = 1 - b_0(y)$, and so

$$I(\mathcal{L})_y(\alpha(y)s^0 + \beta(y)s^{1-b_0(y)}) = 0$$

for all y, at least so long as $b_0(y)$ avoids the non-negative integers. We shall assume for the rest of this paper that $b_0(y) > 0$ for all y; it is possible to treat the case $b_0(y) \equiv 0$ by essentially the same methods, though the final result is rather different. The general mixed case, where b_0 may vanish on some closed subset of ∂M presents many technical difficulties. The main applications from biology and finance prohibit $b_0 < 0$, and in fact often require that $0 \le b_0 \le 1$; we do not insist on this latter restriction, however. Thus we assume only that $b_0 > 0$. The compactness of ∂M implies that $\beta(L_{\text{Kim}})$ defined in (1.4) as the infimum of $b_0(0, y)$ over $y \in \partial M$ is positive.

This family of indicial operators and indicial roots indicates what we should expect of more general solutions of $\mathcal{L}u = 0$. For example, in the special case where b_0 remains constant on ∂M, then one of the main results of [11] states that an arbitrary (local) solution u to $\mathcal{L}u = 0$ has an asymptotic expansion as $x \to 0$ of the form

$$u(x, y) \sim \sum_i x^i u_{0i}(y) + \sum_j x^{1-b_0} u_{1j}(y). \tag{2.3}$$

This expansion must be interpreted properly since in many cases it only holds in a weak, or distributional sense. This means simply that if $\chi(y) \in \mathcal{C}^\infty(\partial M)$, then it is always true that

$$\langle u(x,\cdot), \chi(\cdot)\rangle = \int_{\partial M} u(x,y)\chi(y)\,dy \sim \sum_i x^i \langle u_{0i}, \chi\rangle + \sum_j x^{1-b_0}\langle u_{1j}, \chi\rangle$$

is an asymptotic expansion in the usual (strong) sense, but (2.3) need not hold in this same pointwise strong sense. Actually, it is known that if the leading coefficients u_{00} and u_{10} are both smooth, then all coefficients in the expansion are smooth and (2.3) is a classical asymptotic expansion. It is still possible to work with weak expansions, with some important caveats which are noted below.

Two issues complicate the problems of interest to us. The first is that it is more natural in applications to allow the indicial root $1-b_0(y)$ to vary smoothly as a function on ∂M. This leads to a more complicated regularity theory for general solutions of $\mathcal{L}u = 0$ (or $\mathcal{L}u = f$), see [10] for some recent work on this. The choice of boundary condition we shall impose circumvents this to some extent. Namely, we select only the (unique) solution for which $u_{10} \equiv 0$, and hence which contains no term $u_{1j}(y)x^{1-b_0(y)}$ in its expansion. This is tantamount to choosing a solution $u \in \mathcal{C}^\infty(\overline{M})$ to $\mathcal{L}u = f$, where $f \in \mathcal{C}^\infty(\overline{M})$. Of course part of the problem is to show that such a solution exists and is unique, which is proved in [7]. This choice of boundary condition is of Neumann type, in the sense that, at least when $b_0(y) \in (0,1)$, we are requiring a *non-leading* term in the expansion for u at ∂M to vanish. This introduces another complication, familiar even in the classical nondegenerate case with Neumann boundary conditions, that the construction of a solution operator requires some extra global considerations, compared to the solution for the Dirichlet problem.

One further model operator is needed in the analysis of $\mathcal{L}$: the normal operator

$$N(\mathcal{L})_y = s^2\partial_s^2 + \sum_i c_i(0,y)(s\partial_s)(s\partial_{w_i}) + \sum_{ij} a_{ij}(0,y)s^2\partial_{w_iw_j} + b_0(0,y)s\partial_s + \sum_j b_j(0,y)s\partial_{w_j}.$$

This acts on a half-space $\mathbb{R}_s^+ \times \mathbb{R}_w^{n-1}$, which can be naturally identified as the inward pointing half of T_yM at each $y \in \partial M$. Since $N(\mathcal{L})$ is translation invariant in w and jointly dilation invariant in (s,w), it can be analyzed by first passing to the Fourier transform in w and then rescaling, setting $\sigma = s|\eta|$, where η is the Fourier transform variable dual to w. This leads to an ordinary differential operator, depending parametrically on y and $\hat{\eta} = \eta/|\eta|$. Because $\mathcal{L}$ is second order and scalar, there is a standard classical procedure for writing down the corresponding Green function of this ODE; this Green function may then be rescaled and Fourier transformed back to a Green function for $N(\mathcal{L})_y$ itself. This should be regarded as the 'infinitesimal inverse' for $\mathcal{L}$ at $y \in \partial M$, and is the key new building block in the construction for the actual (generalized) inverse for $\mathcal{L}$.

To understand this in a somewhat broader sense, observe that the standard parametrix construction for approximate inverses of nondegenerate elliptic differential operators in microlocal analysis is simply an elegant way to 'glue together' the family of inverses to each of the constant coefficient models obtained by freezing the coefficients of the differential operator at each point. (More broadly still, many proofs of the classical Schauder estimates proceed by some sort of perturbative argument starting from the exact inverses of these model constant coefficient operators.) In the 0-calculus we can proceed very similarly once we allow that there is a different way to make sense of freezing the coefficients at a boundary point y, thus leading to the normal operator $N(\mathcal{L})_y$, and that there is a good way to describe the inverses of the model operators obtained in this way. Both in the standard setting and for the 0-calculus, these considerations eventually lead to one of the first main qualitative results: if $\mathcal{L}$ is fully elliptic in the sense that its model operator at each point $p \in M$ (in the extended sense described above if $\mathcal{L}$ is uniformly degenerate) is invertible, then there exists right and left parametrices, G_r and G_ℓ, for $\mathcal{L}$ such that the remainder terms $\mathrm{Id} - G_\ell \mathcal{L}$, $\mathrm{Id} - \mathcal{L} G_r$, are compact smoothing operators. The existence of such parametrices can then be used to deduce global Fredholm mapping properties and fine regularity theory.

2.2. The 0-double space

Before describing the parametrix construction itself, we first describe an auxiliary geometric object, which we call the 0-double space, written as M_0^2, which is the actual setting for the parametrix construction. This space is an 'enhancement' of the simple product M^2 in the sense that there is a smooth surjection $M_0^2 \to M^2$, which is the identity over the interior (and indeed over most of the boundary points of M_0^2). The geometric microlocal approach is distinguished by its insistence on the Schwartz kernel of the parametrix as the primary object.

The class of 0-pseudodifferential operators is defined by the requirement that the Schwartz kernel of any such operator enjoys specific and rather simple regularity properties only when lifted to M_0^2. In other words, this double space allows one to efficiently encode the asymptotic properties of these Schwartz kernels in various regimes near the boundary. Said differently, the 0-double-space, M_0^2, gives singular coordinates near the boundary of M^2 in which the asymptotic behavior of the Schwartz kernel is transparent.

The space M_0^2 is obtained by blowing up M^2 along the diagonal in $\partial M \times \partial M$, which we denote $\mathrm{diag}((\partial M)^2)$. In the notation used in [15] this space is denoted $M_0^2 = [M^2; \mathrm{diag}((\partial M)^2)]$. This blowup corresponds to replacing each

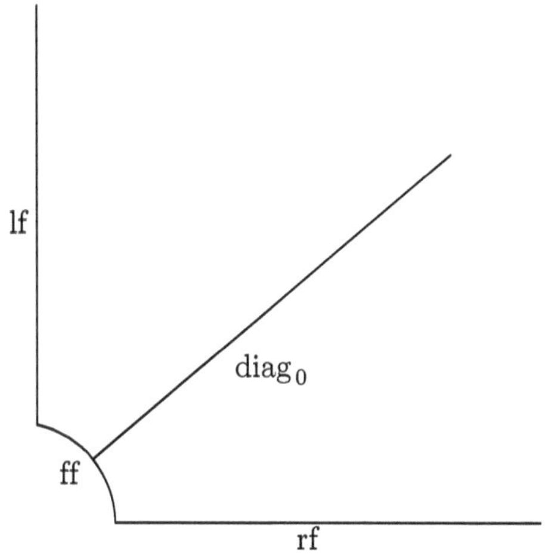

FIGURE 1. A schematic diagram of the 0-double space M_0^2.

point of $\mathrm{diag}(\partial M)^2$ with its inward-pointing spherical normal bundle; equivalently, introduce polar Fermi coordinates around this submanifold and then regard the '$r = 0$' face as a new boundary hypersurface of the blown up space. If $(x_1, y_1; x_2, y_2)$ are coordinates near a boundary point of M^2, then

$$r^2 = x_1^2 + x_2^2 + |y_1 - y_2|^2, \quad \text{and} \quad r, \frac{x_1}{r}, \frac{x_2}{r}, \frac{y_1 - y_2}{r}, \quad \text{along with} \quad y_1 \tag{2.4}$$

provide local coordinates near the new boundary component of M_0^2.

Thus M_0^2 is a manifold with corners, just like M^2, but has one extra boundary hypersurface, called its front face, ff, which is the $r = 0$ face mentioned above, and which 'blows down' to $\mathrm{diag}(\partial M)^2$. It has two other boundary hypersurfaces, the left and right faces, lf and rf, corresponding to $\partial M \times M$ and $M \times \partial M$, respectively; the other distinguished submanifold is the closure of the lift of the diagonal in $\mathrm{int}\, M \times \mathrm{int}\, M$, which is denoted by diag_0. An important advantage of M_0^2 over M^2 is that diag_0 meets only the interior of ff and the intersection is transversal. It does not intersect any other boundary faces, whereas the ordinary diagonal diag intersects the corner of M^2. The Schwartz kernels of the pseudodifferential operators we consider are singular both along diag, and along other components of the boundary of M^2. The blow-up operation physically separates these singularities making their description in M_0^2 much simpler than in M^2.

The class of 0-pseudodifferential operators $\Psi_0^{m,\rho,a,b}(M)$ consists of operators A which are pseudodifferential operators over the interior of M in the classical sense, but which have certain behavior at the boundaries. We require that the Schwartz kernel K_A of any such A, which is a distribution on M^2, lifts to a distribution κ_A on M_0^2 that has the following properties: first, κ_A has a standard classical pseudodifferential singularity along diag_0, and at ff, this conormal singularity is required to be smoothly extendible across ff in the following sense. Namely, (after removing some fixed and explicit singular density factor), we require that κ_A is the restriction from the space obtained by doubling M_0^2 across ff of a distribution which is smooth away from the doubled diagonal and which has a classical pseudodifferential singularity of fixed order uniformly across the 'interface' ff of this double.

We also require that κ_A is conormal at the other two boundary faces, lf and rf; in many cases which arise in applications, it may be polyhomogeneous at one or both of these faces. The indices m, ρ, a, b indicate the orders of vanishing at these various submanifolds and boundary faces: thus m denotes the pseudodifferential order, or (roughly) the rate of blowup of κ_A on approach to diag_0; ρ denotes the rate of vanishing at ff (since we are requiring κ_A to be smooth up to this face, ρ must be a nonnegative integer); finally, a and b are rates of vanishing for the conormal orders along the rf and lf respectively. If κ_A is polyhomogeneous at these faces, then these are lower bounds for the vanishing order of all terms in the expansions. We represent these kernels using the non-degenerate density $dx'dy'$ on the incoming face.

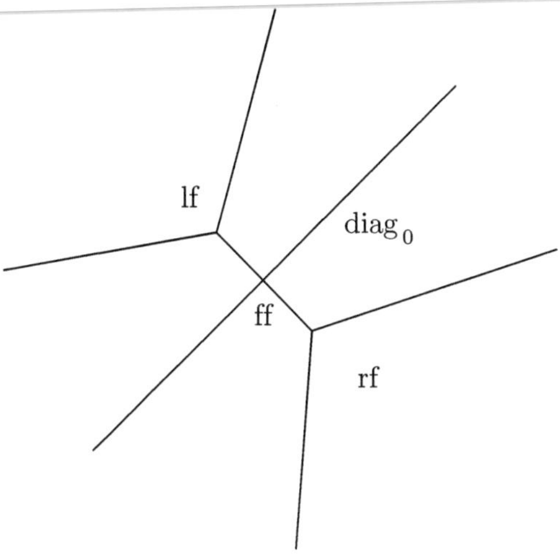

FIGURE 2. A schematic diagram of the double across the ff of M_0^2.

2.3. The elliptic parametrix construction

Having defined the space M_0^2, we now outline the parametrix construction for an elliptic uniformly degenerate operator $\mathcal{L}$. This construction is in two steps: it starts with a 'rough' approximate inverse which is precisely chosen at each of the main boundary faces of M_0^2 and at diag_0, but then extended smoothly away these regions in a fairly arbitrary manner; the second step involves showing that it can be corrected so that the error term is as small as possible. A large part of the work involves showing that this correction preserves certain desirable features of the parametrix, e.g., its conormal or polyhomogeneous structure.

To be definite, suppose that the indicial root structure of $\mathcal{L}$ is as in the discussion above, so one root is identically 0 and the other may vary smoothly but remains strictly less than 1. The first step is to choose an element $G_0 \in \Psi_0^{-2,0,0,\beta'}(M)$, where

$$\beta' = 2\beta(L_{\mathrm{Kim}}) - 3,$$

defined in (1.4). This shift in the order of vanishing at lf (where $x' \to 0$) can be explained as follows. In the x coordinates, $L_{\mathrm{Kim}} = x^{-2}\mathcal{L}$ where $\mathcal{L}$ is uniformly degenerate. Suppose that $\mathcal{G}$ is an inverse (or partial inverse, or at least a sufficiently good parametrix) for $\mathcal{L}$. The order of vanishing of the Schwartz kernel of $\mathcal{G}$ is an indicial root for the *adjoint problem* with respect to the measure $dxdy$. More specifically, working just with indicial operators because these determine the indicial roots,

$$I(\mathcal{L})_y = (x\partial_x)^2 + (2b-2)x\partial_x,$$

hence its adjoint (with respect to this measure) equals

$$I(\mathcal{L})_y^* = (-\partial_x x)^2 + (2b-2)(-\partial_x x) = (x\partial_x)^2 + (4-2b)x\partial_x + 3 - 2b.$$

The indicial roots of this equation satisfy

$$s^2 + (4 - 2b)s + (3 - 2b) = (s + 1)(s - (2b - 3)) = 0,$$

hence these two indicial roots are $s = -1$ and $2b - 3$. The latter corresponds to the order of vanishing of $\mathcal{G}$ along lf. Finally, rewriting $L_{\mathrm{Kim}}G = I$ as

$$x^{-2}\mathcal{L}G = \delta(x - x')\delta(y - y'),$$

we see that $G = \mathcal{G}(x')^2$. This means that G vanishes or blows up like $2b_0 - 1$ along lf (left face). Of course, since b_0 may vary with y, we must take the infimum of this over all $y \in \partial M$.

Similar considerations apply for $L_{\mathrm{Hes}} = x^{-1}\mathcal{L}$, where $I(\mathcal{L})_y = (x\partial_x)^2 + (b_0 - 1)x\partial_x$. The order of vanishing of the inverse for $\mathcal{L}$ at $x' = 0$ is $b_0 - 2$ and hence including the extra factor x' on the right of $\mathcal{G}$ shows that the parametrix for L_{Hes} vanishes like $(x')^{b_0 - 1}$ at lf.

Denoting the lift of the Schwartz kernel of G_0 to M_0^2 by κ_G, then this distribution has a complete classical expansion along diag_0 which is determined using the standard symbol calculus in such a way that $\mathcal{L}G_0 = \mathrm{Id} - Q_0$, where κ_{Q_0} is $\mathcal{C}^\infty$ across diag_0 and up to ff. As part of this, we are also able to demand that the restriction of G_0 to ff equals the inverse to the normal operator $N(\mathcal{L})_y$ over each point $y \in \partial M$. It is precisely at this last step, in choosing the specific inverse of the normal operator family, where we incorporate the choice of boundary conditions. The dual requirements for G_0 along diag_0 and ff are compatible because of the rubric: "the symbol of the normal operator of $\mathcal{L}$ equals the restriction to $\mathrm{diag}_0 \cap \mathrm{ff}$ of the symbol of $\mathcal{L}$".

We make a few observations which expand on this. First, ff is the total space of a fibration: the base is ∂M (via its identification with $\mathrm{diag}((\partial M)^2)$), while the fiber is a closed n-dimensional quarter-sphere (i.e., where two of the coordinates are restricted to be nonnegative). This quarter-sphere is the compactification of the stereographic projection of the half-space which we already encountered as the (s, w)-half plane. The key technical fact that must be proved is that the integral kernel $N(G)_y$ for $(N(\mathcal{L})_y)^{-1}$ has a nice structure on this compactification. More specifically, as we already indicated, we can obtain $N(G)_y$ in rather concrete terms as the inverse Fourier transform of a rescaling ($s = \sigma/|\eta|$) of the Green function of a simple second-order ODE. A priori this is a function of (s, s', w, w'), and still depends parametrically on y. The restriction of G_0 to ff should be chosen on that fiber to equal this function evaluated at $(s', w') = (1, 0)$, which corresponds to the point of intersection of diag_0 with that fiber. This function is $\mathcal{C}^\infty$ on the interior of this fiber except at $(1, 0)$ and has a classical expansion at that point. What must be proved, however, is that it extends as a conormal distribution to the *closed* quarter-sphere and is smooth up to $\mathrm{ff} \cap \mathrm{lf}$ and conormal with vanishing rate β' at $\mathrm{ff} \cap \mathrm{rf}$ (and at the corners).

From the way that G_0 is chosen, we deduce that the initial error term $Q_0 = \mathrm{Id} - \mathcal{L}G_0$ lies in $\Psi_0^{-\infty,1,0,\beta'}$, i.e., κ_{Q_0} is smooth across diag_0 and vanishes to order 1 at ff. It turns out to be easy to add a correction term G_1 to this parametrix so that the resulting error term $Q_1 = \mathrm{Id} - \mathcal{L}(G_0 + G_1)$ lies in $\Psi_0^{-\infty,1,\infty,\beta'}$. In other words, we can also remove the entire expansion of the error term at rf. To see how, note that if we let $\mathcal{L}$ act on the formal expansion of G_1 at rf, then the leading part of this operator is the indicial operator $I(\mathcal{L})_y$. Thus we can successively solve away the terms in the expansion of Q_1 at this face using the inverse $I(\mathcal{L})_y^{-1}$. This step is infinitesimal at each point of rf.

To complete the construction, we would ideally like to remove the error term altogether by multiplying each side of the equation $\mathcal{L}(G_0 + G_1) = \mathrm{Id} - Q_1$ by the inverse $(\mathrm{Id} - Q_1)^{-1}$. Of course, this operator need not be invertible, so instead we multiply 'formally' by the Neumann series $\sum_{j=0}^{\infty} Q_1^j$, or in other words, by the operator $\mathrm{Id} + R$ where R is an asymptotic Borel summation of this Neumann series. The key fact needed to make sense of this Borel sum is the composition law for 0-pseudodifferential operators, which implies in particular that $Q_1^j \in \Psi_0^{-\infty,j,\infty,\beta'}$, or in other words, $\kappa_{Q_1^j}$ vanishes to increasingly higher order at ff. Having formed such an R, we now multiply as intended by $\mathrm{Id} + R$ to obtain an operator $G = (G_0 + G_1) \circ (\mathrm{Id} + R)$ that satisfies

$$\mathcal{L}G = \mathrm{Id} - S, \quad \text{where} \qquad S \in \Psi_0^{-\infty,\infty,\infty,\beta'}(M).$$

The parametrix G itself continues to lie in $\Psi_0^{-2,0,0,\beta'}(M)$.

The Schwartz kernel of this final error term is 'very smoothing', since it is smooth in the interior and vanishes rapidly both along ff and lf. It is straightforward to conclude that this S is compact on all reasonable function spaces. This parametrix G and the structure of S can then be used to deduce not only the global mapping properties of $\mathcal{L}$ on weighted Sobolev and Hölder spaces (and many other natural spaces as well), but also the precise local regularity theory for solutions of $\mathcal{L}u = f$, again in a variety of function spaces. All of this is recorded in detail in [11], to which we refer for complete details of the parametrix construction and its analytic consequences.

To prove Theorem 2.1, we recall what was already described in an earlier part of this proof: namely, if $\mathcal{L}\mathcal{G} = I$, then the (partial) inverse G for $L_{\mathrm{Kim}} = x^{-2}\mathcal{L}$ is equal to $\mathcal{G}(x')^2$. Similarly, the partial inverse G for L_{Hes} equals $\mathcal{G}(x')$. This completes the proofs of Theorems 2.1 and 2.3.

We have admittedly marched through the steps of this elliptic parametrix construction swiftly, but have done so since this construction is recorded carefully elsewhere. These steps are all mirrored in the heat kernel parametrix construction, which is our main goal below.

3. The Kimura and Heston heat kernels

We turn now to the parametrix construction for the heat operators associated to the classes of elliptic Kimura and Heston type operators.

The geometric microlocal construction of parametrices for heat operators is similar to the corresponding construction in the elliptic setting. We refer to [15],

[16], [14], [2], for heat kernel constructions for several other degenerate parabolic equations; the heat kernels for conic and edge problems discussed in all but the first of these sources are analogous to the construction here. In any case, the procedure we follow is substantially the same as in those papers: namely, we focus on the conormal structure of $H(t, z, z')$ on a double heat space which is obtained as a resolution by blowup of $\mathbb{R}^+ \times M^2$. This resolution is constructed to capture the singular structure of H, which occurs along the interior diagonal of M^2 at $t = 0$, the submanifold of the corner at $\{0\} \times \text{diag}(\partial M)^2$ and (to a lesser extent) along $(\mathbb{R}^+ \times M \times \partial M) \cup (\mathbb{R}^+ \times \partial M \times M)$. Thus we first define Kimura and Heston heat spaces, $M^2_{h-\text{Kim}}$ and $M^2_{h-\text{Hes}}$, respectively. As a test, we then show that the explicit solution kernels for the model heat operators are polyhomogeneous on these spaces; this is both useful in the full 'curved' construction, but also indicates that the heat spaces are sufficiently intricate to capture the various types of asymptotic singularities. We then proceed with the iterative parametrix construction. Key components of this analysis, beyond the use of the blown up heat spaces, include a composition formula for the operators represented by Schwartz kernels on these spaces.

Before we begin, recall that our primary objective is to find the solution operators for the Kimura and Heston heat equations. Henceforth we systematically identify these operators with their Schwartz kernels $H(t, z, z')$, the so-called heat kernels. These satisfy

$$(\partial_t - L)H = 0, \text{ where } t > 0, \text{ and } H|_{t=0} = \delta(z - z'),$$

and are unique provided we require that solutions of this problem also satisfy the Neumann-type boundary condition introduced above: namely, $w = H\phi$ must be smooth up to $x = 0$. The delta-function on the right side of the second equation requires some explanation: as an integral kernel, it must be integrated against a density on M, and so it may need an extra factor to compensate for factors on this density. For example, it suffices merely to multiply by a nonvanishing smooth function of z' if we use a density which itself is a nonvanishing smooth function of $dz = dxdy$; on the other hand, if using $drdy = 2xdxdy$, it would be necessary to replace the right side by $\delta(x - x')\delta(y - y')(x')^{-1}$, and so on. We shall therefore agree to fix the volume form $dxdy$ (where $x = \sqrt{r}$ in the Kimura case).

Note that these two defining equations for H remain valid when we multiply the heat operator (not H!) by a prefactor or change variables, provided we compensate with the appropriate Jacobian factor. As already noted in the introduction, it is advantageous to multiply the heat operators by a vanishing prefactor to make the elliptic parts uniformly degenerate. However, at variance with the suggestion there, we multiply here by t rather than x^2 or x.

3.1. Heat spaces

We first review the definition of the blowup of a manifold with corners X along a p-submanifold Y, which was already used (albeit informally) in the last section, and the modification of this definition needed to define blowups with respect to a different homogeneity structure along Y.

Let X be a manifold with corners and $Y \subset X$ a p-submanifold, which means simply that it is a submanifold with the property that for any $q \in Y$, there is a neighborhood $\mathcal{U}$ of q in X which is a product, $\mathcal{U}' \times \mathcal{U}''$, where $\mathcal{U}'$ is a relatively open neighborhood in Y and $\mathcal{U}''$ is a neighborhood of 0 in a manifold with corners of complementary dimension (invariantly, a neighborhood of the 0-section of the normal bundle of Y in X). The blowup $[X; Y]$ is the union of $X \setminus Y$ and the interior spherical normal bundle of Y in X. This union is given the unique minimal differentiable structure so that the lifts of smooth functions on X and polar Fermi coordinates around Y in X are both smooth. It is called the normal blowup of X around Y because it respects the homogeneous dilation structure in directions normal to Y in X. If $Y_1 \subset Y_2 \subset X$ is an inclusion of p-submanifolds, we define the iterated blowup $[X; Y_2; Y_1]$ in a straightforward way.

Next, considering only a special case of a more general inhomogeneous blowup construction, suppose that $Y \subset \{0\} \times X \subset \mathbb{R}^+_t \times X$. The parabolic blowup of Y in $\mathbb{R}^+ \times X$, denoted $[\mathbb{R}^+ \times X; Y; dt]$, consists of equivalence classes of curves where two curves are equivalent if they are tangent to higher order than expected with respect to the parabolic dilations $(t, z, y) \mapsto (\lambda^2 t, \lambda z, y)$, where z lies in the normal bundle to $Y \subset X$ and $y \in Y$. We refer to [8] for the precise definition, see also [15], [2]. This has $\mathcal{C}^\infty$ structure which can be defined using 'parabolic polar coordinates' (see below) around $0 \times Y$. That the homogeneities in the time and spatial directions should differ is essentially a consequence of the fact that the Euclidean heat kernel is a function of $|x - x'|^2/t$; if z is identified with $x - x'$, then this quantity is invariant under the parabolic dilations defined above.

We now define the Kimura and Heston heat spaces using a sequence of two blowups:

$$M^2_{h-\mathrm{Kim}} = [\mathbb{R}^+ \times M^2; \{0\} \times \operatorname{diag}(\partial M)^2, dt; \{0\} \times \operatorname{diag}(M^2), dt], \tag{3.1}$$

$$M^2_{h-\mathrm{Hes}} = [\mathbb{R}^+ \times M^2; \{0\} \times \operatorname{diag}(\partial M)^2; \{0\} \times \operatorname{diag}(M^2), dt]. \tag{3.2}$$

Thus the *only* difference between these spaces is that $\{0\} \times \partial\,\mathrm{diag}$ is blown up parabolically in the first case, but only normally in the second; this reflects the different relative homogeneities of t and x in the two settings. However, on a qualitative level, the two spaces are almost identical.

Each of these spaces has five boundary hypersurfaces: there is the original 'bottom face' tb at $t = 0$ (away from the diagonal and corner), the lifts of the left and right faces lf and rf, corresponding to $\mathbb{R}^+ \times M \times \partial M$ and $\mathbb{R}^+ \times \partial M \times M$, respectively, the 'temporal face' tf which is the lift of the diagonal in M^2 at $t = 0$, and finally the front face ff, which is the lift of the diagonal of the corner at $t = 0$.

To get a better feeling for the geometry of these spaces, it is helpful to introduce various coordinate systems. To be specific, consider first $M^2_{h-\mathrm{Kim}}$. Near tf, but away from the corner, we are near the diagonal in M^2, hence can use two copies w and w' of the same interior coordinate system; we then introduce

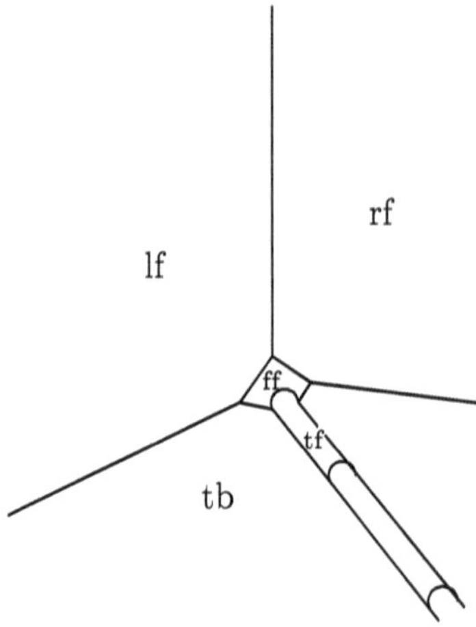

FIGURE 3. A schematic diagram of the blown up heat space $M^2_{h-\mathrm{Kim}}$.

parabolic polar coordinates

$$R = \sqrt{t + |w - w'|^2}, \quad \theta = \left(\frac{t}{R^2}, \frac{w - w'}{R} \right),$$

to fill out a complete nonsingular coordinate system (R, θ, w') near this face (but away from ff). It is not simple to make computations in these coordinates however, and so we define a convenient set of projective coordinates:

$$\tau = \sqrt{t}, \quad W = \frac{w - w'}{\sqrt{t}}. \tag{3.3}$$

Thus $\tau \geq 0$ and $W \in \mathbb{R}^n$. It is quite important that the submanifold described by $\tau = 0$ at a fixed w' is identified in this way with $\mathbb{R}^n$, with Euclidean coordinate W, and in fact this choice of Euclidean coordinate is projectively natural (and $W = 0$ corresponds to the intersection of tf with the diagonal $\{w = w'\}$ at $t = 0$). In particular, tf is the total space of a fibration over $\mathrm{diag}(M^2)$, where each fiber is identified with a 'parabolic' hemisphere, and its interior is projectively identified with $\mathbb{R}^n$. The projective naturality means that a different choice of local coordinate w leads to a new projective coordinate $\widetilde{W}$ which is projectively equivalent to W. These projective coordinates are singular near tb; this turns out not to be important since the kernels vanish to infinite order at this bottom face anyway. Finally, let us point out that the structure of these heat spaces near tf is 'universal' in the sense that it is insensitive to the degeneracies of L at the boundary and captures the standard interior, local, short-time structure of any nondegenerate heat kernel.

There are similar types of polar and projective coordinates near ff as well. We begin with coordinates $(t; x, y; x', y')$ for $M^2 \times \mathbb{R}_+$; in which the boundaries are $x = 0, x' = 0$ and $t = 0$. We are blowing up the submanifold where $x = x' = 0, y = y'$ and $t = 0$. We do not use the polar coordinates at all, so let us consider immediately the projective coordinate system (T, s, u, x', y'), where

$$T = \frac{t}{(x')^2}, \quad s = \frac{x}{x'}, \quad u = \frac{y - y'}{x'}. \tag{3.4}$$

This coordinate system is singular near lf, where $x' = 0$, but there is an alternate good projective coordinate system near that face which is obtained by dividing by x rather than x'. A third useful projective coordinate system is obtained by dividing by $\sqrt{t}$: thus, introduce $(\xi, \xi', \sigma, \upsilon, y')$, where

$$\xi = \frac{x}{\sqrt{t}}, \quad \xi' = \frac{x'}{\sqrt{t}}, \quad \sigma = \sqrt{t}, \quad \upsilon = \frac{y - y'}{\sqrt{t}}. \tag{3.5}$$

These are valid away from $\mathrm{tf} \cup \mathrm{tb}$.

Just as for tf, the front face ff is also the total space of a fibration; this time the base space is $\mathrm{diag}(\partial M)^2$ and the fiber is a parabolic quarter-sphere, parabolically blown up at a boundary point. Using the first projective coordinate system, there is a projectively natural identification of the interior of each of these quarter-sphere fibers with a quarter-space (T, s, u) where $T, s > 0$ and $u \in \mathbb{R}^{n-1}$.

The Heston heat space $M^2_{h-\mathrm{Hes}}$ differs from $M^2_{h-\mathrm{Kim}}$ only in that the blowup of $\mathrm{diag}(\partial M)^2$ at $t = 0$ is normal rather than parabolic. This changes the local coordinate systems near ff in an obvious way: specifically, we replace the coordinates above by

$$\begin{aligned} T &= \frac{t}{x'}, \quad s = \frac{x}{x'}, \quad u = \frac{y - y'}{x'}, \qquad \text{and} \\ \xi &= \frac{x}{t}, \quad \xi' = \frac{x'}{t}, \quad \sigma = \sqrt{t}, \quad \upsilon = \frac{y - y'}{t}. \end{aligned} \tag{3.6}$$

We leave it to the reader to track the corresponding minor changes.

3.2. Model operators

The calculation which justifies the introduction of these heat spaces is the fact that the restrictions of the lifts of the Kimura or Heston heat operators to the boundary faces tf and ff of the corresponding heat spaces are comprehensible model operators.

The model operators on the fibers of tf are, as noted earlier, universal in that away from $x = 0$ there is nothing to distinguish L_{Kim} from L_{Hes} or any other nondegenerate elliptic operator in the interior. The lift of ∂_t, for example, near this face is singular, so to compensate for this, we premultiply the entire operator $\partial_t - L$ by the factor t. This yields an operator which lifts to be nonsingular on M^2_h (we omit the Kim or Hes subscript for the moment) near ff. In fact, as local computations show, this lift is an operator which acts tangentially on this face.

To be clear in these computations, we consider $t\partial_t - tL$ acting on $\mathbb{R}^+_t$ and the first (left) factor of M in M^2, and then lift to the blown up heat space. Working in the projective coordinate system (τ, W, w'), we see that

$$\partial_{w_j} = \sum_k \left(\tau^{-1}(\delta_{jk} + \alpha_{jk})\partial_{W_k} + \beta_{jk}\partial_{w'_k}\right) = \frac{1}{\tau}(\partial_{W_j} + \tau V_j)$$

where V_j is smooth up to tf. Since we can always choose the local coordinate w so that the second-order part of L is the standard Euclidean Laplacian at a given point p, which we take as $w' = 0$, we then compute that the lift of tL equals

$$\sum_{j=1}^n \partial^2_{W_j} + \tau P_0 + \sum w'_j P_j,$$

where the P_i are second-order differential operators on the blown up space which are smooth up to tf. A similar computation, which can either be done in the parabolic polar coordinate system above, or a different suitable projective coordinate system, shows that these are smooth at tf$\cap$tb too. On the other hand, we also have

$$t\partial_t = \frac{1}{2}\tau\partial_\tau - \frac{1}{2}\sum_{j=1}^n W_j \partial_{W_j}.$$

Putting these together, we see that the 'error terms' vanish when restricting the lift of $t\partial_t - tL$ to the fiber of tf over p, and so this restricted operator equals

$$\frac{1}{2}\tau\partial_\tau - \Delta_W + \frac{1}{2}W\cdot\partial_W. \tag{3.7}$$

Note, in particular, that this acts tangent to the hemisphere fibers of tf. Slightly more generally, if we apply $t\partial_t - tL$ to any term of the form $\tau^k H_k$ and then restrict to $\tau = 0$, we get

$$\tau^k\left(\Delta_W - \frac{1}{2}W\partial_W - \frac{k}{2}\right)H_k. \tag{3.8}$$

The operators we have identified here are the model problems for $t(\partial_t - L)$ at order k along ff. These are shifts of operators equivalent to the harmonic oscillator, and hence are invertible (for certain values of k only off a finite rank subspace) on the Schwartz space $\mathcal{S}(\mathbb{R}^n_W)$. The solvability of these model problems has been dealt with even in the earliest papers on heat trace expansions, starting with the work of Minakshisundaram and Pleijel in the 1950s. We refer to [3] for a modern treatment of this.

Now consider the lift of $t\partial_t - tL$ near ff. To be definite, let us focus on Kimura operators. Using the projective coordinates (T, s, u, x', y'),

$$t\partial_t = T\partial_T, \quad t\partial_x^2 = T\partial_s^2, \quad Tx^{-1}\partial_x = Ts^{-1}\partial_s, \quad \text{and} \quad t\partial^2_{y_iy_j} = T\partial^2_{u_iu_j}.$$

Suppose that L_{Kim} as expressed in (1.7), and choose local coordinates y on ∂M so that the matrix $a_{ij}(0,y) = \delta_{ij}$ at $y = y_0$. Then

$$t(\partial_t - L_{\mathrm{Kim}}) = T\partial_T - T\left(\frac{1}{4}\partial_s^2 + \Delta_u + \frac{1}{2}(b_0(0,y_0) - \frac{1}{2})s^{-1}\partial_s + \mathcal{O}(x')\right). \tag{3.9}$$

After removing the overall factor T, the restriction of this to ff (i.e., setting $x' = 0$) is simply the heat operator corresponding to $s^{-2}N(x^2 L_{\mathrm{Kim}})_{y_0}$. Abusing language slightly, we call this latter object the normal operator for L_{Kim}.

The minor modifications of these statements for L_{Hes} are straightforward and left to the reader

To summarize, what we have achieved is that these heat spaces have the property that the lifts of the Kimura and Heston heat operators (multiplied by t), restricted to either tf or ff, act tangentially to the fibers of these spaces, and these restrictions are naturally associated with the model heat problems: the model at each fiber of ff is the heat operator associated to the normal operator, while the model along each fiber of tf is a universal elliptic operator (of Ornstein–Uhlenbeck type) on a projectively equivalent Euclidean space.

3.3. Model heat kernels

A preliminary test for whether these heat spaces are suitable for describing the precise asymptotic structure of the Kimura and Heston heat kernels is to see whether the heat kernels of the model operators of each of these types is polyhomogeneous on the associated heat space. We shall do this in the Kimura setting using an explicit formula for the Kimura heat kernel taken from [6]. We do not do this here for Heston operators simply because we do not know a similar explicit expression (although it is highly likely that such an expression exists).

In any case, consider the model operator, written in terms of $x = \sqrt{r}$ as

$$\partial_t - \left(\frac{1}{4}\partial_x^2 + \frac{1}{2}\left(b_0 - \frac{1}{2}\right)x^{-1}\partial_x + \Delta_y\right), \tag{3.10}$$

where $b_0 > 0$ is a constant. Then according to (6.13) in [6], the heat kernel for this operator which corresponds to the choice of boundary condition which omits the term r^{1-b_0} in the expansion of solutions is given by

$$\begin{aligned} H_{\mathrm{Kim}}(t, r, y, r', y') &= \frac{1}{(4\pi t)^{(n-1)/2}}\frac{1}{t}\left(\frac{r}{r'}\right)^{\frac{1-b_0}{2}} e^{-\frac{r+r'}{t}} I_{b_0-1}\left(2\sqrt{\frac{rr'}{t^2}}\right) e^{-\frac{|y-y'|^2}{2t}} \\ &= (4\pi)^{-(n-1)/2} t^{-(n+1)/2}\left(\frac{x}{x'}\right)^{1-b_0} e^{-\frac{x^2+(x')^2}{t}} I_{b_0-1}\left(2\frac{xx'}{t}\right) e^{-\frac{|y-y'|^2}{2t}}. \end{aligned} \tag{3.11}$$

Here, $I_{b_0-1}(z)$ is the modified Bessel (or Macdonald) function, which is asymptotic to cz^{b_0-1} for $z \searrow 0$ and which grows exponentially like $e^z/\sqrt{z}$ as $z \nearrow \infty$. In addition, $y, y' \in \mathbb{R}^{n-1}$ and the contribution from these variables is the standard Gaussian because the heat kernel is multiplicative with respect to Riemannian products.

To check that this kernel lifts to a polyhomogeneous function on $M^2_{h-\mathrm{Kim}}$, we first observe that

$$H_{\mathrm{Kim}}(\lambda^2 t, \lambda x, \lambda y, \lambda x', \lambda y') = \lambda^{-n-1} H_{\mathrm{Kim}}(t, x, y, x', y').$$

The fact that H_{Kim} is homogeneous under this dilation could have been predicted from the fact that (3.10) is homogeneous of degree -1 and the boundary condition is also invariant under this dilation. However, the boundary condition, that H_{Kim} equals $\delta(x-x')\delta(y-y')(x')^{-1}$ at $t=0$, is homogeneous of degree $-n-1$, which implies that H_{Kim} inherits this same degree of homogeneity. The upshot is that H_{Kim} can be written as $(x')^{-n-1}\mathcal{H}$ where $\mathcal{H}$ is homogeneous of degree 0, and hence smooth, up to ff.

We can see directly, using the asymptotics of the Bessel function as $z \to 0$, that H_{Kim} is smooth up to rf, i.e., as $x \searrow 0$, and is smooth up to lf as well provided we remove the factor $(x')^{2b_0-1}$. To understand its behavior near tf, use the coordinates $(\xi, \xi', \sigma, v, y')$ to get

$$H_{\mathrm{Kim}} = (4\pi)^{(1-n)/2}\sigma^{-n-1}\left(\frac{\xi}{\xi'}\right)^{1-b_0} e^{-(\xi^2+(\xi')^2)} I_{b_0-1}\left(2\frac{\xi\xi'}{\sigma}\right) e^{-|v|^2/2}.$$

Using the Bessel asymptotics as $z \searrow \infty$ shows that H_{Kim} vanishes to infinite order along tb and blows up like σ^{-n} on tf. Looking closer, it is also apparent that the restriction of the leading coefficient at $\sigma=0$ is (up to a constant) $e^{-(\xi^2+|\sigma|^2)/2} = e^{-|W|^2/2}$, which is in the nullspace of (3.8).

Again, it should be possible to understand the Heston heat kernel in similarly explicit terms, but we do not pursue this here. Note that a posteriori from the construction below, since the heat kernel for any Heston-type operator must live as a reasonable distribution on $M^2_{h-\mathrm{Hes}}$, the same is obviously true for the model Heston operator, and using homogeneity considerations, we can conclude that this model Heston heat kernel is polyhomogeneous on this Heston heat space.

3.4. The heat parametrix construction

Following the definitions and calculations above, we now proceed with the construction of heat kernel parametrices. As usual, we explain this carefully only for the Kimura case, since the Heston case is completely analogous.

The first approximation to the parametrix is a Schwartz kernel H_0 which will be chosen carefully so that its asymptotic structure is correct to all orders along tf and to first order at ff. More specifically, we wish to choose H_0 so that $t(\partial_t - L_{\mathrm{Kim}})H_0$ vanishes to infinite order at tf and blows up only to order $-n+1$ at ff (which is better than expected since $t(\partial_t - L)$ is homogeneous of degree 0 and we shall choose H_0 to blow up to order $-n$ at ff.

To arrange matters along tf, it suffices to observe that if we expand $H_0 \sim \sum \tau^{-n+k} H_{0k}$ as $\tau \searrow 0$, then $t\partial_t - tL$ acts as the model operator (3.8) of order k on H_{0k}. As already noted there (and proved carefully in [3], see also [2]), this operator is invertible for $k>0$ and has a one-dimensional nullspace when $k=0$ consisting of the function $e^{-|W|^2/2}$. Thus if we choose H_{00} to equal this Gaussian, then the expansion of $t(\partial_t - L)\tau^{-n}H_{00}$ (and later $t(\partial_t - L)\tau^{-n+j}H_{0j}$) produces error terms which blow up or decay like τ^{-n+k} for $k>0$. We then regard these as inhomogeneous terms and choose H_{0k} to solve away all such inhomogeneous terms produced by earlier steps of the construction. We can then take a Borel sum of all of these Taylor coefficients. This determines H_0 near tf. While this is described in rather different language, what we are doing here is no more nor less than the classical interior parametrix construction for heat kernels.

One aspect of this behavior at tf worth mentioning explicitly is the fact that the leading term is, up to a dimensional constant, $\tau^{-n}e^{-|W|^2/2} = t^{-n/2}e^{-|y-y'|^2/2t}$. This guarantees that H has the correct initial condition,

$$\lim_{t\to 0+} \int_M H(t,x,y,x',y')\phi(x',y')\,dx'dy' = \phi(x,y).$$

Indeed, this is true for the Euclidean heat kernel, which is equal in its entirety to this leading term, and the higher order (in τ) terms in the expansion of more general heat kernels do not contribute to this small t limit.

On the other hand, near ff we solve away only the first term. Specifically, writing $H_0 = (x')^{-n}H_0'$ and applying $t(\partial_t - L)$ to this, the leading term of the resulting function on any fiber of ff is simply the model heat operator for L at the corresponding boundary point $y \in \partial M$ applied to the restriction of H_0' to this face. We should clearly choose this restriction to equal the model heat kernel, and so we do.

The only thing to check is that the singularity in this model heat kernel at $\mathrm{ff}\cap\mathrm{tf}$ is the same as the rate of blowup of H_0 along tf, but a moment's thought shows that this is indeed true. This simply reflects the fact that the singularity at $T=0$ of the model heat operator is the limit of the singularities at $T=0$ of the family of nearby interior problems (transverse to the diagonal) which limit to it.

To recapitulate, then, we choose H_0 so that $t(\partial_t - L_{\mathrm{Kim}})H_0$ vanishes to infinite order along $\mathrm{tf}\cup\mathrm{tb}$, and blows up to order $-n+1$ at ff.

We have not yet discussed the behavior of H_0 along the side faces rf and lf. These are governed by the behavior of the model heat kernel along the intersections $\mathrm{ff}\cap\mathrm{rf}$ and $\mathrm{ff}\cap\mathrm{lf}$. The first of these corresponds to letting $s\to 0$ for $T>0$, and the boundary condition we are imposing dictates precisely that this model heat kernel is smooth at this face. On the other hand, over each fiber of ff along the corner $\mathrm{ff}\cap\mathrm{lf}$, the model heat kernel is polyhomogeneous with leading order $(s')^{b_0} = (x'/x)^{b_0}$. Unfortunately this exponent may vary with $y\in\partial M$, and so the cumulative regularity of this family of model heat kernels is only conormal with vanishing order β, but not polyhomogeneous (unless b_0 is constant in y).

Denote by $\Psi^{\ell,0,\beta}_{h-\mathrm{Kim}}$ the space of all Schwartz kernels on $M^2_{h-\mathrm{Kim}}$ which are smooth in the interior and up to rf, and conormal of order β at lf, vanish to all orders at $\mathrm{tf}\cup\mathrm{tb}$, and which blow up (or decay) like $-n+\ell$ at ff. At this stage we make a slight shift and consider these Schwartz kernels acting on functions $f(t,w)$, rather than functions depending only on w, via the usual formula

$$H\star f(t,w) = \int_0^t\int_M H(t-s,w,w')f(s,w')\,dsdw',$$

with the same choice of volume forms as we used before. The point of doing this is to be able to write the other main technical ingredient of this parametrix construction, which is the composition formula

$$H_j \in \Psi^{\ell_j,0,\beta'}_{h-\mathrm{Kim}}, j = 1,2, \Longrightarrow H_1 \star H_2 \in \Psi^{\ell_1+\ell_2,0,\beta'}_{h-\mathrm{Kim}}. \tag{3.12}$$

for any $\beta' \in \mathbb{R}$, and $\ell_1, \ell_2 \in \mathbb{N}$. When thinking of these as operators in (t, w) in this way, we write H as $H\star$.

Notice that the true heat kernel satisfies $(\partial_t - L_{\mathrm{Kim}})H_{\mathrm{Kim}}\star = \mathrm{Id}$, while the parametrix we have now chosen satisfies only

$$t(\partial_t - L_{\mathrm{Kim}})H_0\star = \mathrm{Id} - K\star \tag{3.13}$$

for some $K \in \Psi^{1,0,2\beta-1}_{h-\mathrm{Kim}}$. The explanation for the rate of vanishing or blow-up at the left face $x' = 0$ is explained using the initial condition $H|_{t=0} = \delta(x - x')\delta(y - y')$. We argue just as in the elliptic case that this vanishing order is determined by a (slight shift of a) indicial root of a related uniformly degenerate operator. Using the composition formula we see that $(K\star)^j \in \Psi^{j,0,\beta}_{h-\mathrm{Kim}}$, or in other words, its Schwartz kernel vanishes like $-n+j$ at ff. Hence we can build the Borel sum of the Neumann series

$$(\mathrm{Id} - K\star)^{-1} \sim \sum_{j=0}^{\infty} (K\star)^j = \mathrm{Id} + \mathcal{K}\star.$$

The proof of the composition formula is somewhat laborious, and can be done by fairly direct computation. There is a quicker and more elegant way to do this using the pushforward theorem of Melrose. The formula above can be derived in precisely the same manner as the corresponding composition formula in [14], for example.

In any case, we now compose (3.13) on the right with $\mathrm{Id} + \mathcal{K}\star$, to get

$$t(\partial_t - L_{\mathrm{Kim}})H_0(\mathrm{Id} + \mathcal{K}\star) = \mathrm{Id} + \mathcal{S},$$

where $\mathcal{S} \in \Psi^{\infty,0,2\beta-1}_{h-\mathrm{Kim}}$.

It is now standard that because of its very rapid vanishing at all faces as $t \searrow 0$, $(\mathrm{Id} + \mathcal{S}\star)^{-1}$ has a Neumann series which converges in the appropriate conormal or polyhomogeneous topology.

We now appeal to the uniqueness of the solution operator for $\partial_t - L_{\mathrm{Kim}}$ which satisfies the boundary condition of smoothness at the outgoing $(x = 0)$ face and the initial condition

$$\lim_{t \searrow 0} \int_M H(t, z, z')\phi(z')\, dz' = \phi(z).$$

This means that we can identify

$$H_{\mathrm{Kim}} = H_0(\mathrm{Id} + \mathcal{K}\star)(\mathrm{Id} + \mathcal{S}\star),$$

and from this we conclude finally that $H_{\mathrm{Kim}} \in \Psi^{0,0,2\beta-1}_{h-\mathrm{Kim}}$, which is the main structural theorem we are after.

Theorem 3.1. *Let L_{Kim} be a generalized Kimura operator on a manifold with boundary M, for which the inward pointing part of the first-order term is nowhere vanishing. The Schwartz kernel for the solution operator of the heat equation*

$$\partial_t w = L_{\text{Kim}} w \tag{3.14}$$

defines an element of the space of pseudodifferential operators $\Psi^{0,0,2\beta-1}_{h-\text{Kim}}(\mathbb{R}_+ \times M)$. Here β is defined in (1.4).

We have not stated the corresponding theorem for the heat operator $\partial_t - L_{\text{Hes}}$ because we have not yet determined a specific formula for the corresponding model heat kernel. If this were available, the structure of H_{Hes} could then be determined using the methods of this section.

4. Mapping properties and regularity theory

In this final section we exploit the structure of Kimura and Heston heat kernels to extend some of the regularity properties of solutions of the homogeneous and inhomogeneous heat equations

$$(\partial_t - L)w = 0, \qquad w\big|_{t=0} = \phi \tag{4.1}$$

$$(\partial_t - L)u = f, \qquad u\big|_{t=0} = 0, \tag{4.2}$$

where $L = L_{\text{Kim}}$ or L_{Hes}.

As explained in the introduction, the nature of the perturbative arguments used in [7] made it necessary to work there with functions ϕ, f, w and u lying in two scales of anisotropic Hölder spaces $\mathcal{C}^{k,\alpha}_{\text{WF}}$ and $\mathcal{C}^{k,2+\alpha}_{\text{WF}}$ (defined either over M or $\mathbb{R}^+ \times M$). These spaces reflect the homogeneity structure of L_{Kim}. The space $\mathcal{C}^{0,\gamma}_{\text{WF}}$ is the standard Hölder space with respect to the variables $(\sqrt{r}, y)$. (The other spaces have a hybrid nature so this simple coordinate transformation does not provide a complete characterization.) There is a simple technical reason for the need to work on spaces for which one has good elliptic or parabolic estimates: the difference E between a general Kimura operator L_{Kim} and the 'constant coefficient' model Kimura operator at a point $p \in \partial M$ is an operator where the coefficients of the second-order terms are small in a small neighborhood of p. If we use the model heat kernel $H^0(t, z, z')$ as an initial parametrix in that neighborhood, then we require estimates of the norm of the error term EH^0. But one can only show that the operator norm of this term is bounded, let alone small, if one has some form of parabolic Schauder estimates for the problem, and these fail for data lying in $\mathcal{C}^0$. However, for data in the appropriate Hölder spaces, these local parametrices can be patched together to obtain a global approximation to the heat inverse for which the error term does have small norm. From this the true heat kernel can be obtained by a (convergent!) Volterra series.

These arguments are supplemented by the robust collection of maximum principles proved in [7]. If $\phi \in \mathcal{C}^0(M)$, for example, and if $\phi_j \in \mathcal{C}^{0,\gamma}_{\text{WF}}(M)$ is a sequence of functions converging to ϕ uniformly, then the homogeneous solutions

$w_j = H\phi_j$ converge uniformly to a continuous function w. Unfortunately, without a better understanding of the a priori parabolic regularity theory, it is impossible to conclude that this limit w is smooth near ∂M when $t > 0$.

These arguments have only been written out for Kimura heat operators, but we claim that this same chain of reasoning applies mutatis mutandis for Heston operators. The necessary maximum principles are contained in [9]. The one gap is that, as remarked earlier, we do not know explicit expressions for the model Heston heat kernels. However, given these, it would be straightforward to develop results paralleling all those in [7] for general Heston heat equations on manifolds with corners.

Our aim in this final section is to point out that the geometric microlocal parametrices constructed here accomplish the perturbation theory in a more refined way and make evident the full parabolic smoothing effect for initial data in $\mathcal{C}^0$. It is possible to use the structure theory of these parametrices to recapture the full regularity theory for the Kimura heat kernel on the 'WF'-Hölder spaces, as developed in [7]. This would require sufficient extra space that we defer this to elsewhere.

One of the key strengths of the geometric microlocal method is that it expedites the passage from the model solution operator to the solution operator for the more general variable coefficient problem. The method from [7] outlined above accomplishes this partly at the level of solution operators but partly on the level of solutions. However, the insistence of considering only solution operators and their parametrices has significant advantages: namely, one can do the perturbation analysis on the solution kernels directly, and since these objects are (at least when viewed correctly) fundamentally smooth (or at least polyhomogeneous) objects, the issues above about the difference between $\mathcal{C}^{0,\gamma}_{\mathrm{WF}}$ and $\mathcal{C}^0$ dissolve. More specifically, the Neumann series argument used to pass from the parametrix to the exact heat kernel relies on the composition calculus for the 0 heat calculus recorded in (3.12), and the proof of this formula, in turn, is simplified by the infinite regularity of the factors.

To illustrate this, consider the heat kernel H_{Kim} constructed in the last section, and fix $\phi \in \mathcal{C}^0(M)$. Let w be the unique solution to (4.1) satisfying the natural (smooth) boundary condition at ∂M; thus

$$w(t,x,y) = \int_M H_{\mathrm{Kim}}(t,x,y,x',y')\phi(x',y')\, x'dx'dy'.$$

We have shown that $H_{\mathrm{Kim}}(t,x,y,x',y')$ is *smooth* as $x \searrow 0$ when $t > 0$, uniformly in all other variables (and in any compact interval $0 < t_0 \leq t \leq t_1$). From this it is clear that this integral produces a function which is also smooth in $x \geq 0$ for $t > 0$. The final thing to check is that $w \to \phi$ *uniformly* as $t \searrow 0$. This can be done by standard arguments involving this integral formula. Since w solves the equation $\partial_t w = L_{\mathrm{Kim}} w$, it is also an immediate consequence of the maximum principle proved in [7] for the Kimura-type operators. This completes the proof of the following theorem:

Theorem 4.1. *Let M be a compact manifold with boundary and L_{Kim} a generalized Kimura operator for which the inward pointing part of the first-order term is nowhere vanishing. If $f \in \mathcal{C}^0(M)$, then the unique regular solution to the homogeneous initial value problem*

$$\begin{aligned} \partial_t w &= L_{\text{Kim}} w \text{ in } (0,\infty) \times M \\ \lim_{t \to 0^+} w(t,x,y) &= \phi(x,y) \quad \text{for } (x,y) \in M \end{aligned} \tag{4.3}$$

belongs to $\mathcal{C}^0([0,\infty) \times M) \cap \mathcal{C}^\infty((0,\infty) \times M)$.

We have discussed this only for Kimura operators. However, the Heston heat kernel is also smooth as $x \to 0$ for $t > 0$, so we deduce the smoothing effect for Heston operators with initial data in $\mathcal{C}^0$ in exactly the same way. There are many further regularity theorems for these equations, including the precise mapping properties of H_{Kim} and H_{Hes} on various types of adapted Hölder spaces. We hope to address these elsewhere.

References

[1] P. Albin, "A renormalized index theorem for some complete asymptotically regular metrics: The Gauss-Bonnet theorem," Adv. Math. **213** (2007), no. 1, 1–52.

[2] P. Albin and R. Mazzeo, "Geometric microlocal constructions of heat kernels: a user's guide," in preparation.

[3] N. Berline, E. Getzler and M. Vergne, "Heat kernels and Dirac operators," Grundlehren der Mathematischen Wissenschaften **298**, Springer-Verlag, Berlin (1992).

[4] P. Daskalopoulos and R. Hamilton, "Regularity of the free boundary for the porous medium equation," J. Amer. Math. Soc. 11 (1998), no. 4, 899–965.

[5] P. Daskalopoulos and R. Hamilton, "The free boundary in the Gauss curvature flow with flat sides," J. Reine Angew. Math. 510 (1999), 187–227.

[6] C.L. Epstein and R. Mazzeo, "Wright–Fisher diffusion in one dimension," *SIAM Journal on Mathematical Analysis* **42** no. 2, (2010), pp. 568–608.

[7] C.L. Epstein and R. Mazzeo, "Degenerate diffusion operators arising in population biology," Annals of Mathematical Studies, no. 185, Princeton University Press (2013).

[8] C.L. Epstein, R.B. Melrose and G. Mendoza, "Resolvent of the Laplacian on strictly pseudoconvex domains," Acta Math. **167** (1991), no. 1-2, 1–106.

[9] P. Feehan, "Partial differential operators with non-negative characteristic form, maximum principles, and uniqueness for boundary value and obstacle problems," arXiv:1204.6613. To appear, Comm. P.D.E.

[10] T. Krainer and G. Mendoza, "The kernel bundle of a holomorphic Fredholm family," Comm. P.D.E., **38** (2013), 2107–2125.

[11] R. Mazzeo, "Elliptic theory of differential edge operators I," Comm. P.D.E. **16** no. 10, (1991), pp. 1616–1664.

[12] R. Mazzeo and R.B. Melrose, "Meromorphic extension of the resolvent for spaces with asymptotically constant negative curvature," Jour. Func. Anal, **75** (1987), pp. 260–310.

[13] R. Mazzeo, Y. Rubinstein and N. Sesum, "Ricci flow on conic surfaces," arXiv: 1306.6688.

[14] R. Mazzeo and B. Vertman, "Analytic torsion on manifolds with edges," *Advances in Mathematics* **231** no. 2 (2012), pp. 1000–1040.

[15] R.B. Melrose, "The Atiyah-Patodi-Singer index theorem," A.K. Peters, 1993.

[16] E. Mooers, "Heat kernel asymptotics on manifolds with conic singularities," J. Anal. Math. **78** (1999), 1–36.

[17] N. Shimakura, "Formulas for diffusion approximations of some gene frequency models," J. Math. Kyoto Univ., 21 (1981), pp. 19–45.

C.L. Epstein
Depts. of Mathematics and Radiology
University of Pennsylvania
209 South 33rd Street
Philadelphia, PA 19104, USA
e-mail: cle@math.upenn.edu

Rafe Mazzeo
Dept. of Mathematics, Building 380
Stanford University
Stanford, CA 94305, USA
e-mail: mazzeo@math.stanford.edu

Progress in Nonlinear Differential Equations
and Their Applications, Vol. 85, 267–276

On a Resonant Lane–Emden Problem

Grey Ercole

Abstract. We study the asymptotic behavior, as $q \to p$, of the positive solutions of the Lane–Emden problem $-\Delta_p u = \lambda_p |u|^{q-2} u$ in Ω, $u = 0$ on $\partial\Omega$, where $\Omega \subset \mathbb{R}^N$ is a bounded and smooth domain, $N \geq 2$ and λ_p is the first eigenvalue of the p-Laplacian operator Δ_p, $p > 1$. We prove that any family of positive solutions of this problem converges in $C^1(\overline{\Omega})$ to the function $\theta_p e_p$ when $q \to p$, where e_p is the positive and L^∞-normalized first eigenfunction of the p-Laplacian and $\theta_p := \exp\left(\|e_p\|_{L^p(\Omega)}^{-p} \int_\Omega e_p^p |\ln e_p| \, dx\right)$.

Mathematics Subject Classification (2010). 35B35, 35B40, 35J92.

Keywords. Asymptotic behavior, blow-up technique, first eigenpair, ground states, Lane–Emden, Picone's inequality, p-Laplacian.

1. Introduction

Consider the Lane–Emden problem

$$\begin{cases} -\Delta_p u = \lambda |u|^{q-2} u & \text{in } \Omega, \\ u = 0 & \text{on } \partial\Omega, \end{cases} \tag{1}$$

where $\lambda > 0$, $\Omega \subset \mathbb{R}^N$ is a bounded and smooth domain, $N \geq 2$, $\Delta_p u := \operatorname{div}\left(|\nabla u|^{p-2} \nabla u\right)$ is the p-Laplacian operator, $p > 1$, and $1 < q < p^\star$, with $p^\star$ denoting the Sobolev critical exponent defined by $p^\star = Np/(N-p)$, if $1 < p < N$, and $p^\star = \infty$, if $p \geqslant N$.

The existence of positive weak solutions for this problem is a well-known fact. Moreover, such solutions are bounded (in the L^∞ norm) and hence (as consequence of classical regularity results) belong to $C^{1,\alpha}(\overline{\Omega})$ for some $0 < \alpha < 1$.

When $q = p$ we have the eigenvalue problem for the p-Laplacian, whose first eigenvalue λ_p is positive, isolated and simple. Moreover, associated eigenfunctions do not change sign in Ω.

The author thanks the support of FAPEMIG and CNPq, Brazil.

In the sub-linear case $1 < q < p$, the positive weak solutions are unique (see [6]). However, in the super-linear case $p < q < p^\star$ this fact does not happen, in general. Non-uniqueness of positive weak solutions of (1) occurs for ring-shaped domains when q is close to $p^\star$ (see [7, 10]) or when $q > p$ and Ω is a sufficiently thin annulus (see [14]). On the other hand, when Ω is a ball, (1) has a unique positive weak solution (see [1]). For the Laplacian ($p = 2$) and a general bounded domain, uniqueness happens if q is sufficiently close to 2 (see [5, Lemma 1]).

With different goals, asymptotics of solutions of the Lane–Emden problem (1) have been studied by many authors since the 1990s. For example, in [10] for $p < N$, $\lambda = 1$ and $q \to p^\star$; or in [15] for $p = N$, $\lambda = 1$ and $q \to \infty$. Recently, in [11], the asymptotic behavior in $W_0^{1,p}(\Omega)$ of the ground state solutions (i.e., positive weak solutions that minimize the energy functional among all possible weak solutions) as $q \to p^+$, was described for all positive values of λ. More recently, the asymptotic behavior with $q \to p^-$ in $W_0^{1,p}(\Omega)$ was described in [3]. Some these asymptotics had already appeared in [12], for $\lambda \neq \lambda_p$.

However, up to our knowledge, only in [11] and [3] the resonant problem, that is, when $\lambda = \lambda_p$, was dealt with, but the asymptotic behavior of its positive solutions was not fully determined. Indeed, although the families of solutions were known to have a subsequence converging in $W_0^{1,p}(\Omega)$ to a first eigenfunction, the correct first eigenfunction was unknown; in principle, distinct first eigenfunctions (each one multiple of the other, of course) could be obtained as limits of different subsequences of these families. Moreover, in the super-linear case, the known results are valid only for ground state families. Therefore, nothing was known about the asymptotic behavior (as $q \to p^+$) of other (eventually existing) families of positive solutions.

In the present work we first consider the resonant Lane–Emden problem

$$\begin{cases} -\Delta_p u = \lambda_p |u|^{q-2} u \text{ in } \Omega, \\ \quad u = 0 \qquad\qquad \text{on } \partial\Omega, \end{cases} \tag{2}$$

and an arbitrary family $\{u_q\}_{q\in[1,p)\cup(p,p^\star)}$ of positive solutions of this problem (not necessarily ground states, in the super-linear case). Our main result is the convergence $u_q \to \theta_p e_p$ in $C^1(\overline{\Omega})$, as $q \to p$, where

$$\theta_p := \exp\left(\frac{\int_\Omega e_p^p |\ln e_p| \, dx}{\int_\Omega e_p^p dx}\right)$$

and e_p is the first positive eigenfunction such that $\|e_p\|_\infty = 1$. (From now on $\|v\|_r$ stands for the usual L^r norm of v.)

By a scaling argument this result also determines the exact asymptotic behavior, as $q \to p$, of positive solutions of the Lane–Emden problem (1), for any $\lambda > 0$. Moreover, it implies the differentiability at $q = p$ of the function $q \in [1, p^\star) \mapsto \lambda_q \in \mathbb{R}$, where λ_q denotes the minimum on $W_0^{1,p}(\Omega)\backslash\{0\}$ of the Rayleigh quotient $\mathcal{R}_q$ defined by $\mathcal{R}_q(u) := \|\nabla u\|_p^p / \|u\|_q^p$.

A third consequence of our main result is that, for each $\lambda > 0$, $1 \le s \le \infty$ and for any sequence $q_n \to p$ one has:

$$\lim_{q_n \to p} \left(\lambda \left\| u_{\lambda,q_n} \right\|_s^{q_n - p} \right) = \lambda_p \qquad \text{and} \qquad \frac{u_{\lambda,q_n}}{\left\| u_{\lambda,q_n} \right\|_s} \to \frac{e_p}{\left\| e_p \right\|_s}$$

the last convergence being in the $C^1(\overline{\Omega})$ space. This might be useful for numerical computation of the first eigenvalue of the p-Laplacian (see [4]) taking into account that λ does not need to be close to λ_p and that the sequence q_n tending to p can be arbitrarily chosen.

2. Asymptotic behavior of the resonant problem

In this section we consider the resonant Lane–Emden problem

$$\begin{cases} -\Delta_p u = \lambda_p \left| u \right|^{q-2} u \text{ in } \Omega \\ \quad u = 0 \qquad\qquad \text{on } \partial\Omega. \end{cases} \tag{3}$$

Our goal is to completely determine the asymptotic behavior of the weak positive solutions of this problem, as $q \to p$. (Some proofs in this section were omitted or just sketched, but all of them are available in [8].)

The weak solutions of (3) are the critical points of the energy functional $I_q : W_0^{1,p}(\Omega) \longrightarrow \mathbb{R}$ defined by

$$I_q(u) := \frac{1}{p} \int_\Omega \left| \nabla u \right|^p dx - \frac{\lambda_p}{q} \int_\Omega \left| u \right|^q dx.$$

Furthermore, a family $\{v_q\}_{q \in [1,p) \cup (p,p^\star)}$ of positive weak solutions of (3) is obtained from minimizers of the Rayleigh quotient

$$\mathcal{R}_q(u) := \frac{\int_\Omega \left| \nabla u \right|^p dx}{\left(\int_\Omega \left| u \right|^q dx \right)^{\frac{p}{q}}}$$

in $W_0^{1,p}(\Omega) \setminus \{0\}$.

In fact, as it is well known, the compactness of the immersion $W_0^{1,p}(\Omega) \hookrightarrow L^q(\Omega)$ for $1 \le q < p^\star$ implies that $\mathcal{R}_q : W_0^{1,p}(\Omega) \setminus \{0\} \longrightarrow \mathbb{R}$ attains a positive minimum at a positive and L^q-normalized function $w_q \in W_0^{1,p}(\Omega) \cap C^{1,\alpha}(\overline{\Omega})$:

$$\left\| w_q \right\|_q = 1 \text{ and } \lambda_q := \min \left\{ \mathcal{R}_q(u) : u \in W_0^{1,p}(\Omega) \setminus \{0\} \right\} = \mathcal{R}_q(w_q). \tag{4}$$

(We remark that this notation is coherent with the case $q = p$, since the first eigenvalue λ_p is also characterized as the minimum of $\mathcal{R}_p$ on $W_0^{1,p}(\Omega) \setminus \{0\}$.)

It is straightforward to verify that w_q is a weak solution of

$$\begin{cases} -\Delta_p u = \lambda_q \left| u \right|^{q-2} u \text{ in } \Omega \\ \quad u = 0 \qquad\qquad \text{on } \partial\Omega \end{cases}$$

and hence that

$$v_q = \left(\frac{\lambda_q}{\lambda_p}\right)^{\frac{1}{q-p}} w_q \tag{5}$$

is a positive weak solution of (3) for each $q \in [1,p) \cup (p,p^\star)$.

Since $\|w_q\|_q = 1$ one has

$$\|v_q\|_q = \left(\frac{\lambda_q}{\lambda_p}\right)^{\frac{1}{q-p}}. \tag{6}$$

In the sub-linear case $1 \le q < p$ the function v_q is the only critical point of I_q. Moreover, this function minimizes the energy functional I_q on $W_0^{1,p}(\Omega) \setminus \{0\}$, that is

$$I_q(v_q) = \min\left\{I_q(v) : v \in W_0^{1,p}(\Omega) \setminus \{0\}\right\}. \tag{7}$$

This property can also be directly proved using (4) and (6).

In the super-linear case $1 < p < q < p^\star$ the energy functional is not bounded from below. However, the weak positive solution v_q minimizes both, the energy functional I_q and the L^q norm, in the Nehari manifold

$$\mathcal{N}_q := \left\{v \in W_0^{1,p}(\Omega) \setminus \{0\} : \int_\Omega |\nabla v|^p \, dx = \lambda_p \int_\Omega |v|^q \, dx\right\}.$$

Therefore, since any nontrivial solution of (3) belongs to $\mathcal{N}_q$, it follows that $v_q \in \mathcal{N}_q$ and also that v_q is a ground state.

Since no general uniqueness result is known for the super-linear case, the existence of multiple ground states for (3) is possible, at least in principle, for each fixed $q \in (p,p^\star)$. However, all of them must have the same energy and also the same L^q norm.

In the remaining of this section we denote by v_q the function defined by (5) and by u_q any positive solution of the resonant Lane–Emden (2). Obviously, in the sub-linear case we must have $u_q = v_q$.

Lemma 1. *Let $\{u_q\}_{q\in[1,p)\cup(p,p^\star)}$ be a family of positive solutions of the Lane–Emden problem* (3). *One has*

$$0 < C_1 \le \|u_q\|_\infty^{q-p} \le C_2$$

for all $q \in [1,p) \cup (p,p+\epsilon)$, where $\epsilon > 0$ and the constants C_1 and C_2 do not depend on $q \in [1,p) \cup (p,p+\epsilon)$.

In the sub-linear case C_1 is obtained after testing (7) with the function e_p, while C_2 is obtained from a simple comparison principle involving the p-torsion function $\phi_p \in W_0^{1,p}(\Omega)$, that is, $-\Delta_p\phi_p = 1$ in Ω.

In the super-linear case, C_1 can be taken as 1 and the constant C_2 follows after combining a blow-up argument with Picone's inequality (as in Lemma 2.1 of [13]).

Lemma 2. *Let* $\{u_q\}_{q\in[1,p)\cup(p,p^\star)}$ *be a family of positive solutions of the Lane–Emden problem* (3) *and define, for each* $q \in [1,p)\cup(p,p^\star)$, *the function* $U_q := \dfrac{u_q}{\|u_q\|_\infty}$. *Then* U_q *converges to* e_p *in* $C^1(\overline{\Omega})$ *as* $q \to p$. *Moreover,*

$$\int_\Omega \frac{U_q^p - U_q^q}{q-p}dx \to \int_\Omega e_p^p \left|\ln e_p\right| dx \quad \text{as } q \to p. \tag{8}$$

Proof. It is easy to verify that

$$\begin{cases} -\Delta_p U_q = \lambda_p \left\|u_q\right\|_\infty^{q-p} U_q^{q-1} & \text{in } \Omega, \\ U_q = 0 & \text{on } \partial\Omega. \end{cases} \tag{9}$$

It follows from Lemma 1 that the right-hand side of the equation in (9) is uniformly bounded with respect to $q \in [1,p)\cup(p,p+\epsilon)$. Therefore, global Hölder regularity implies that U_q converges in $C^1\left(\overline{\Omega}\right)$ to a function $U \geq 0$ (as $q \to p$) with $\|U\|_\infty = 1$. It also holds $\lambda_p \left\|u_q\right\|_\infty^{q-p} \to c \in (\lambda_p C_1, \lambda_p C_2)$.

Taking the limit $q \to p$ in the weak formulation of (9) with $\lambda = \lambda_p \left\|u_q\right\|_\infty^{q-p}$, one obtains

$$\int_\Omega |\nabla U|^{p-2}\, \nabla U \cdot \nabla\varphi dx = c\int_\Omega |U|^{p-2}\, U\varphi dx$$

for any test function $\varphi \in W_0^{1,p}(\Omega)$, which proves that U is a nonnegative eigenfunction associated with the eigenvalue c and such that $\|U\|_\infty = 1$. But this fact necessarily implies that $c = \lambda_p$ and $U = e_p$. Thus, the uniqueness of the limits $\lambda_p \left\|u_q\right\|_\infty^{q-p} \to \lambda_p$ and $U_q \to e_p$ show that these convergences do not depend on subsequences. Therefore, $\left\|u_q\right\|_\infty^{q-p} \to 1$ and $U_q \to e_p$ in $C^1\left(\overline{\Omega}\right)$.

In order to prove (8) we first observe that $\dfrac{U_q^p - U_q^q}{q-p}$ is uniformly bounded with respect to q close to p with

$$\limsup_{q\to p}\left|\frac{U_q^p - U_q^q}{q-p}\right| \leq \lim_{q\to p}\frac{1}{p}\left(\frac{p}{q}\right)^{\frac{q}{q-p}} = \frac{1}{p\exp(1)}.$$

Now, taking into account the convergence $U_q \to e_p$ in $C^1(\overline{\Omega})$, (8) follows from Lebesgue's dominated convergence theorem if we prove that

$$\frac{1-U_q^{q-p}}{q-p} \to \left|\ln e_p\right| \quad \text{as } q \to p^+ \quad \text{a.e. in } \Omega$$

and

$$\frac{U_q^{p-q}-1}{q-p} \to \left|\ln e_p\right| \quad \text{as } q \to p^- \quad \text{a.e. in } \Omega.$$

So, let $\mathcal{K} \subset \Omega$ compact and $0 < \delta < \min_{\mathcal{K}} e_p$. Then

$$0 < \min_{\mathcal{K}} e_p - \delta < e_p - \delta \leq U_q \leq e_p + \delta \quad \text{in } \mathcal{K}$$

for all q sufficiently close to p. Hence, in $\mathcal{K}$ one has

$$-\ln(e_p+\delta) \le \liminf_{q\to p^+} \frac{1-U_q^{q-p}}{q-p} \le \limsup_{q\to p^+} \frac{1-U_q^{q-p}}{q-p} \le -\ln(e_p-\delta), \tag{10}$$

since

$$\lim_{q\to p^+} \frac{1-(e_p+\delta)^{q-p}}{q-p} = -\ln(e_p+\delta) \quad \text{and} \quad \lim_{q\to p^+} \frac{1-(e_p-\delta)^{q-p}}{q-p} = -\ln(e_p-\delta).$$

Therefore, making $\delta \to 0^+$ in (10) we conclude that

$$\lim_{q\to p^+} \frac{1-U_q^{q-p}}{q-p} = -\ln e_p = |\ln e_p| \quad \text{in } \mathcal{K}.$$

Analogously we prove that

$$\lim_{q\to p^-} \frac{U_q^{p-q}-1}{q-p} = |\ln e_p| \ \text{ in } \mathcal{K}.$$

□

Lemma 3. *Let* $\{u_q\}_{q\in[1,p)\cup(p,p^\star)}$ *be a family of positive weak solutions of the Lane–Emden problem* (3). *Then,*

$$\limsup_{q\to p^-} \|u_q\|_\infty \le \exp\left(\frac{\int_\Omega e_p^p |\ln e_p|\, dx}{\int_\Omega e_p^p dx}\right) \le \liminf_{q\to p^+} \|u_q\|_\infty\,. \tag{11}$$

Proof. Applying Picone's inequality (see [2]) to $U_q = \dfrac{u_q}{\|u_q\|_\infty}$ and e_p one has

$$\int_\Omega |\nabla U_q|^p\, dx \ge \int_\Omega |\nabla e_p|^{p-2}\, \nabla e_p \cdot \nabla\left(\frac{U_q^p}{e_p^{p-1}}\right) dx. \tag{12}$$

(Hopf's boundary lemma implies that $U_q^p/e_p^{p-1} \in W_0^{1,p}(\Omega)$.) Therefore, it follows from (9) that

$$\lambda_p \|u_q\|_\infty^{q-p} \int_\Omega U_q^q dx \ge \lambda_p \int_\Omega e_p^{p-1} \frac{U_q^p}{e_p^{p-1}} dx = \lambda_p \int_\Omega U_q^p dx$$

and from this we obtain

$$\frac{\|u_q\|_\infty^{q-p}-1}{q-p} \int_\Omega U_q^q dx \ge \int_\Omega \frac{U_q^p - U_q^q}{q-p} dx \ \text{ if } \ p<q<p^\star \tag{13}$$

and

$$\frac{\|u_q\|_\infty^{q-p}-1}{q-p} \int_\Omega U_q^q dx \le \int_\Omega \frac{U_q^p - U_q^q}{q-p} dx \ \text{ if } \ 1<q<p.$$

Let us suppose, in the case $q\to p^+$, that there exist $L<\theta_p$ and a sequence $q_n \to p^+$ such that $\|u_{q_n}\|_\infty \le L$. Then (13) and Lemma 2 yield

$$\begin{aligned}\int_\Omega e_p^p |\ln e_p|\, dx &= \lim \int_\Omega \frac{U_{q_n}^p - U_{q_n}^{q_n}}{q_n-p} dx \\ &\le \lim \frac{L^{q_n-p}-1}{q_n-p} \int_\Omega U_{q_n}^{q_n} dx = \ln L \int_\Omega e_p^p dx,\end{aligned}$$

that is, $\theta_p \le L$, thus reaching a contradiction. We have proved the second inequality in (11).

The case $q \to p^-$ is analogous. □

Lemma 4. *Let* $\{u_q\}_{q\in[1,p)\cup(p,p^\star)}$ *be a family of positive weak solutions of the Lane–Emden problem* (3)*. Then,*

$$\limsup_{q\to p^+} \|u_q\|_\infty \le \exp\left(\frac{\int_\Omega e_p^p |\ln e_p|\, dx}{\int_\Omega e_p^p dx}\right) \le \liminf_{q\to p^-} \|u_q\|_\infty .$$

Proof. By applying Picone's inequality again, but interchanging U_q with e_q in (12), the lemma follows similarly. □

Theorem 5. *Let* $\{u_q\}_{q\in[1,p)\cup(p,p^\star)}$ *be a family of positive weak solutions of the Lane–Emden problem* (3)*. Then* u_q *converges in* $C^1(\overline{\Omega})$ *to* $\theta_p e_p$ *as* $q \to p$*, where*

$$\theta_p := \exp\left(\frac{\int_\Omega e_p^p |\ln e_p|\, dx}{\int_\Omega e_p^p dx}\right).$$

Proof. Lemmas 3 and 4 imply that

$$\lim_{q\to p} \|u_p\|_\infty \to \theta_p. \tag{14}$$

Thus, the right-hand side of (3) is bounded for all q sufficiently close to p. This fact, combined with the global Hölder regularity ensures that u_q converges in $C^1(\overline{\Omega})$ to a positive first eigenfunction $u \in C^1(\overline{\Omega}) \cap W_0^{1,p}(\Omega)$ when $q \to p$. Thus, $u = ke_p$ for some $k > 0$. But, according to (14) $k = \theta_p$, implying that the limit function is always $\theta_p e_p$ (that is, it does not depend on subsequences). Therefore, $u_q \to \theta_p e_p$ in $C^1(\overline{\Omega})$ as $q \to p$. □

3. Applications

A consequence of Theorem 5 is the differentiability of the function $q \in [1, p^\star) \mapsto \lambda_q$ at $q = p$, where λ_q is defined by (4). We remark that this function is, in fact, differentiable almost everywhere since the function $q \in [1, p^\star) \longmapsto |\Omega|^{\frac{p}{q}} \lambda_q$ is strictly decreasing (see [9]).

Corollary 6. *It holds*

$$\lim_{q\to p} \frac{\lambda_q - \lambda_p}{q-p} = \lambda_p \ln(\theta_p \|e_p\|_p). \tag{15}$$

Proof. We recall that for each $q \in [1,p) \cup (p,p^\star)$ the function $v_q = \left(\frac{\lambda_q}{\lambda_p}\right)^{\frac{1}{q-p}} w_q$ is a positive weak solution of the resonant Lane–Emden problem (3)), where $w_q \in W_0^{1,p}(\Omega) \cap C^{1,\alpha}(\overline{\Omega})$ satisfies $\|w_q\|_q = 1$ and $\mathcal{R}_q(w_q) = \lambda_q$.

Thus, it follows from Theorem 5 that

$$\theta_p \|e_p\|_p = \lim_{q\to p} \|v_q\|_q = \lim_{q\to p} \left(\frac{\lambda_q}{\lambda_p}\right)^{\frac{1}{q-p}} = \exp\left(\lim_{q\to p} \frac{\ln \lambda_q - \ln \lambda_p}{q-p}\right).$$

But this is equivalent to differentiability of λ_q at $q = p$ with $\frac{d}{dq}[\lambda_q]_{q=p}$ given by (15). □

Another consequence of Theorem 5 is the complete description, in the $C^1(\overline{\Omega})$ space, of the asymptotic behavior for the positive solutions of the non-resonant problem ($0 < \lambda \neq \lambda_p$):

$$\begin{cases} -\Delta_p u = \lambda |u|^{q-2} u \text{ in } \Omega \\ u = 0 \qquad\qquad \text{on } \partial\Omega. \end{cases} \tag{16}$$

Corollary 7. *Let* $\{u_{\lambda,q}\}_{q\in[1,p)\cup(p,p^\star)}$ *be a family of positive solutions of* (16). *Then*

$$\lim_{q\to p^-} \|u_{\lambda,q}\|_{C^1} = \begin{cases} 0 & \text{if } \lambda < \lambda_p \\ \infty & \text{if } \lambda > \lambda_p \end{cases}$$

and

$$\lim_{q\to p^+} \|u_{\lambda,q}\|_{C^1} = \begin{cases} \infty & \text{if } \lambda < \lambda_p \\ 0 & \text{if } \lambda > \lambda_p. \end{cases}$$

Proof. The proof follows directly from Theorem 5 after noticing that

$$u_{\lambda,q} := \left(\frac{\lambda}{\lambda_p}\right)^{\frac{1}{p-q}} u_q, \tag{17}$$

where u_q is a positive solution of the resonant Lane–Emden problem (3). □

These results generalize those in [3] and in [11] to C^1 norm. Note that in the super-linear case, our results are really more general than those in [11] since they do apply to arbitrary families of positive solutions and not only for ground states, as in [11].

A third consequence of Theorem 5 is that it provides a theoretical method for obtaining approximations for a first eigenpair of the p-Laplacian by solving a non-resonant problem (16) with $\lambda > 0$ arbitrary and q close to p. In fact, we have the following corollary.

Corollary 8. *For* $1 \le s \le \infty$ *and* $\lambda > 0$ *fixed let* $U_{\lambda,q} := \dfrac{u_{\lambda,q}}{\|u_{\lambda,q}\|_s}$ *and* $\mu_{\lambda,q} := \lambda \|u_{\lambda,q}\|_s^{q-p}$. *Then, as* $q \to p$:

$$\mu_{\lambda,q} \to \lambda_p \quad \text{and} \quad U_q \to \frac{e_p}{\|e_p\|_s} \quad \text{in } C^1(\overline{\Omega}).$$

Proof. The proof follows directly from Theorem 5 after noticing from (17) that

$$U_{\lambda,q} = \frac{u_q}{\|u_q\|_s} \quad \text{and} \quad \mu_{\lambda,q} := \lambda \|u_{\lambda,q}\|_s^{q-p} = \lambda \frac{\lambda_p}{\lambda} \|u_q\|_s^{q-p} = \lambda_p \|u_q\|_s^{q-p}.$$ □

Corollary 8 provides a method for obtaining numerical approximations of the first eigenpair $(\lambda_p, \frac{e_p}{\|e_p\|_s})$. In fact, in a first step one can compute a numerical solution of problem (16) with q close to p and hence, after L^s-normalization, one obtains approximations for λ_p and $\frac{e_p}{\|e_p\|_s}$ simultaneously.

Of course, a numerical solution of the nonlinear problem (16), for some $\lambda > 0$ fixed, is easier to obtain than directly compute the first eigenpair of the p-Laplacian (by solving the corresponding eigenvalue problem). As previously mentioned, the advantage here is that λ can be chosen arbitrarily in computational implementations of (16) and does not need to be close to λ_p. A similar approach was recently used in [4], where the iterative sub- and super-solution method was applied to compute the positive solutions of the sub-linear problem.

We emphasize that this approach is well supported by the results in this work also for the super-linear case, since it does apply to any family of positive solutions. It is worth noticing that since the previously known results are valid only for ground state families, the application of this method (up to now) would be unviable if one takes into account the necessity of proving that a numerical solution of the super-linear problem is in fact a ground state.

References

[1] Adimurthi R. and Yadava S.L., An elementary proof of the uniqueness of positive radial solutions of a quasilinear Dirichlet problem, *Arch. Rational Mech. Anal.* **127** (1994), 219–229.

[2] Allegretto W. and Huang Y.X., A Picone's identity for the p-Laplacian and applications, *Nonlinear Anal.* **32** (1998), 819–830.

[3] Anello G., On the Dirichlet problem involving the equation $-\Delta_p u = \lambda u^{s-1}$, *Nonlinear Anal.* **70** (2009), 2060–2066.

[4] Biezuner R., Brown J., Ercole G. and Martins E., Computing the first eigenpair of the p-Laplacian via inverse iteration of sub-linear supersolutions, *J. Sci. Comput.* **52** (2012), 180–201.

[5] Dancer E.N., Real analyticity and non-degeneracy, *Math. Ann.* **325** (2003), 369–392.

[6] Díaz J.I. and Saa J.E., Existence et unicité de solutions positives pour certaines équations elliptiques quasilinéaires, *C. R. Acad. Sci. Paris (Série I)* **305** (1987), 521–524.

[7] Drábek P., A note on the nonuniqueness for some quasilinear eigenvalue problem, *Appl. Math. Lett.* **13** (2000), 39–41.

[8] Ercole, G., On the resonant Lane–Emden problem for the p-Laplacian, *Commun. Contemp. Math.*, to appear.

[9] Ercole, G., Absolute continuity of the best Sobolev constant, *J. Math. Anal. Appl.* **404** (2013), 420–428.

[10] García Azorero J.P. and Peral Alonso I., On limits of solutions of elliptic problems with nearly critical exponent, *Comm. Part. Diff. Eq.* **17** (1992), 2113–2126.

[11] Grumiau C. and Parini E., On the asymptotics of solutions of the Lane–Emden problem for the p-Laplacian, *Arch. Math.* **91** (2008), 354–365.

[12] Huang Y.X., A note on the asymptotic behavior of positive solutions for some elliptic equation, *Nonlinear Anal.* **29** (1997), 533–537.

[13] Iturriaga L., Lorca S., Sanchez J., Existence and multiplicity results for the p-Laplacian with a p-gradient term, *NoDEA Nonlinear Differential Equations Appl.* **15** (6) (2008), 729–743.

[14] Kawohl B., Symmetry results for functions yielding best constants in Sobolev-type inequalities. *Discrete Contin. Dynam. Systems*, **6** (2000), 683–690.

[15] Ren X. and Wei J., Counting peaks of solutions to some quasilinear elliptic equations with large exponents, *J. Differential Equations* **117** (1995), 28–55.

Grey Ercole
Departamento de Matemática – ICEx
Universidade Federal de Minas Gerais
Av. Antônio Carlos 6627
Caixa Postal 702
30161-970, Belo Horizonte, MG, Brazil
e-mail: grey@mat.ufmg.br

Progress in Nonlinear Differential Equations
and Their Applications, Vol. 85, 277–286

A Note on the Existence of a Positive Solution for a Non-autonomous Schrödinger–Poisson System

Marcelo F. Furtado, Liliane A. Maia and Everaldo S. Medeiros

Abstract. We consider the system

$$\begin{cases} -\Delta u + V(x)u + K(x)\phi(x)u = a(x)|u|^{p-1}u, & x \in \mathbb{R}^3, \\ -\Delta \phi = K(x)u^2, & x \in \mathbb{R}^3, \end{cases} \tag{S}$$

where $3 < p < 5$ and the potentials $K(x), a(x)$ and $V(x)$ has finite limits as $|x| \to +\infty$. By imposing some conditions on the decay rate of the potentials we obtain the existence of a ground state solution. In the proof we apply variational methods.

Mathematics Subject Classification (2010). 35J20, 35J60, 35B38.

Keywords. Schrödinger–Poisson system; ground state solution; variational methods.

1. Introduction

In this note we are concerned with the existence of a positive solution for the nonlinear system

$$\begin{cases} -\Delta u + V(x)u + K(x)\phi(x)u = a(x)|u|^{p-1}u, & x \in \mathbb{R}^3, \\ -\Delta \phi = K(x)u^2, & x \in \mathbb{R}^3, \end{cases} \tag{S}$$

where $3 < p < 5$ and the potentials $K(x), a(x)$ and $V(x)$ satisfy some basic assumptions.

As quoted in the paper [4], this system arises in many interesting physical context. According to a classical model, the interaction of a charge particle with an electromagnetic field can be described by coupling the nonlinear Schrödinger and the Maxwell equations. In particular, if one is looking for electrostatic-type

The three authors were partially supported by CNPq/Brazil. The first two authors were partially supported by PROEX/CAPES, UnB..

solutions, it is natural to solve (S). In many papers the potential V has been supposed constant or radial (see for instance [1, 2, 8] and references therein). Here, motivated by the recent results by G. Cerami and G. Vaira [6] we will assume the following hypotheses:

(H_1) there exist c_K, $\alpha > 0$ such that

$$0 \leq K(x) \leq c_K e^{-\alpha|x|}, \text{ for a.e. } x \in \mathbb{R}^3;$$

(H_2) a, $V \in C(\mathbb{R}^3, \mathbb{R})$ are positive continuous functions such that

$$\lim_{|x|\to+\infty} V(x) = V_\infty > 0, \qquad \lim_{|x|\to+\infty} a(x) = a_\infty > 0. \tag{1.1}$$

Furthermore, it is necessary to have some control on the asymptotic behavior of the potentials V and a. So, we also assume that

(H_3) there exist c_V, c_a, γ, $\theta > 0$ such that, for each $x \in \mathbb{R}^3$, there hold

$$V(x) \leq V_\infty + c_V e^{-\gamma|x|}, \qquad a(x) \geq a_\infty + c_a e^{-\theta|x|}, \tag{1.2}$$

with $\theta < \min\{\gamma, \alpha\} \leq \max\{\gamma, \alpha\} < 2\sqrt{V_\infty}$.

Our main result can be stated as follows:

Theorem 1.1. *If (H_1)–(H_3) hold, then the system (S) has a positive ground state solution.*

For the proof, we use an approach similar to that of [6]. It consists in applying the Mountain Pass Theorem together with some sort of Splitting Lemma. This former result enables us to overcome the lack of compactness of the Sobolev embeddings caused by the fact the problem is set in whole space $\mathbb{R}^N$. Hence, we need to perform a careful investigation of the behavior of the Palais–Smale sequences for the energy functional associated with system (S). Actually, we identify the levels in which the Palais–Smale condition can fail, giving a representation theorem for such sequences, and showing that the only obstacle to prove compactness are the solutions of the *limit problem*

$$-\Delta w + V_\infty w = a_\infty |w|^{p-1} w, \; x \in \mathbb{R}^3. \tag{P_∞}$$

In [6] the authors considered the same problem with $V \equiv 1$ and some integrability conditions on the function $a(x) - a_\infty$. By assuming that the L^2-norm of the weight K is smaller than a number related with the least energy level of two limit problems, they obtained the existence of a positive ground state solution. On the other hand, in [10] G. Vaira supposed that $V \equiv 1$, $a(x) \to a_\infty$, $K(x) \to K_\infty$ as $|x| \to +\infty$, with a_∞, $K_\infty > 0$. Under some integrability conditions on $a(x) - a_\infty$ and $K(x) - K_\infty$, and some other mild conditions on the potentials, she also obtained a positive solution. Our Theorem 1.1 complements (and is not comparable with) the existence results of [6, 10].

We finally point out that a slight modification of our approach allows us to drop condition (H_3) by the following one (see Remark 3.3):

$(\widetilde{H_3})$ there exist c_V, c_a, γ, $\theta > 0$ such that, for each $x \in \mathbb{R}^3$, there hold

$$V(x) \le V_\infty - c_V e^{-\gamma|x|}, \qquad a(x) \ge a_\infty - c_a e^{-\theta|x|},$$

with $\gamma < \min\{\theta, \alpha\} \le \max\{\theta, \alpha\} < 2\sqrt{V_\infty}$.

The paper is organized as follows: in the next section we present the variational setting of the problem and state the compactness lemma that we shall use. In Section 3 we prove the main theorem.

2. The variational setting

Throughout the paper we write $\int u$ instead of $\int_{\mathbb{R}^3} u(x)dx$. For each $u \in W^{1,2}(\mathbb{R}^3)$ we define

$$\|u\| := \left(\int (|\nabla u|^2 + V(x)u^2)\right)^{1/2}.$$

It follows from (H_2) that $\|\cdot\|$ is a norm which is equivalent to the usual one of $W^{1,2}(\mathbb{R}^3)$. For any $A \subset \mathbb{R}^3$ and $u \in L^p(A)$ we denote $\|u\|_{L^p(A)} := (\int_A |u|^p dx)^{1/p}$. If $A = \mathbb{R}^3$ we write only $\|u\|_p$. Moreover, in what follows, without any loss of generality, we assume that $a_\infty = 1$.

Since $K \in L^2(\mathbb{R}^3)$, a straightforward application of the Lax–Milgram theorem implies that, for any given $u \in W^{1,2}(\mathbb{R}^3)$, there exists a unique $\phi = \phi_u \in D^{1,2}(\mathbb{R}^3)$ such that

$$\int \nabla\phi_u \cdot \nabla v = \int K(x)u^2 v, \text{ for all } v \in D^{1,2}(\mathbb{R}^3).$$

Actually, the function ϕ_u weakly solves $-\Delta\phi = K(x)u^2$ and we can construct the application $\phi : W^{1,2}(\mathbb{R}^3) \to D^{1,2}(\mathbb{R}^3)$ which associates to each $u \in W^{1,2}(\mathbb{R}^3)$ the function $\phi(u)$ as above. From simplicity we write only ϕ_u to denote $\phi(u)$. We collect below some properties of the map ϕ (see [6, Lemma 2.1]).

Lemma 2.1. *The following hold:*

1. *ϕ is continuous and maps bounded sets into bounded sets;*
2. *$\phi_{tu} = t^2\phi_u$, for any $u \in W^{1,2}(\mathbb{R}^3)$, $t > 0$;*
3. *if $u_n \rightharpoonup u$ weakly in $W^{1,2}(\mathbb{R}^3)$ then $\phi_{u_n} \rightharpoonup \phi_u$ weakly in $D^{1,2}(\mathbb{R}^3)$.*

We shall use the following technical result.

Lemma 2.2. *If $(u_n) \subset W^{1,2}(\mathbb{R}^3)$ is such that $u_n \rightharpoonup u$ weakly in $W^{1,2}(\mathbb{R}^3)$, then*

$$\lim_{n\to\infty} \int K(x)\phi_{u_n} u_n^2 = \int K(x)\phi_u u^2$$

and

$$\lim_{n\to\infty} \int K(x)\phi_{u_n} u_n \varphi = \int K(x)\phi_u u\varphi,$$

for all $\varphi \in W^{1,2}(\mathbb{R}^3)$.

Proof. We have that

$$\int K(x)(\phi_{u_n}u_n^2 - \phi_u u^2) = \int K(x)\phi_{u_n}(u_n^2 - u^2) + \int K(x)u^2(\phi_{u_n} - \phi_u).$$

It follows from Lemma 2.1 that $\phi_{u_n} \rightharpoonup \phi_u$ weakly in $D^{1,2}(\mathbb{R}^3)$, and therefore the last term above goes to zero. Hence, in order to prove the first statement of the lemma, it suffices to check that

$$\lim_{n\to+\infty} \int K(x)\phi_{u_n}(u_n^2 - u^2) = 0. \tag{2.1}$$

By using the Hölder and Sobolev inequality we get

$$\begin{aligned} \left| \int K(x)\phi_{u_n}(u_n^2 - u^2) \right| &\le \|\phi_{u_n}\|_6 \left(\int K(x)^{\frac{6}{5}} |u_n^2 - u^2|^{\frac{6}{5}} \right)^{5/6} \\ &\le S\|u_n\|_{D^{1,2}} \left(\int K(x)^{\frac{6}{5}} |u_n^2 - u^2|^{\frac{6}{5}} \right)^{5/6}, \end{aligned} \tag{2.2}$$

where S is related with the embedding $D^{1,2}(\mathbb{R}^3) \hookrightarrow L^6(\mathbb{R}^3)$.

For any given $\rho > 0$, we can use the Hölder inequality twice to obtain

$$\int_{\mathbb{R}^3 \setminus B_\rho(0)} K(x)^{\frac{6}{5}} |u_n^2 - u^2|^{\frac{6}{5}} \mathrm{d}x \le \|K\|_{L^2(\mathbb{R}^3\setminus B_\rho(0))}^{6/5} \left(\int |u_n^2 - u^2|^3 \right)^{2/5}.$$

The Hölder inequality and the boundedness of (u_n) in $L^6(\mathbb{R}^3)$ provide $c_1 > 0$ such that

$$\left(\int |u_n^2 - u^2|^3 \right)^{2/5} \le \|u_n - u\|_6^{6/5} \|u_n + u\|_6^{6/5} \le c_1. \tag{2.3}$$

Moreover, since the condition (H_1) implies $K \in L^2(\mathbb{R}^3)$, we can choose $\rho > 0$ large in such a way that $\|K\|_{L^2(\mathbb{R}^3\setminus B_\rho(0))} < \varepsilon$. Thus, we infer from the above inequalities that

$$\int_{\mathbb{R}^3 \setminus B_\rho(0)} K(x)^{\frac{6}{5}} |u_n^2 - u^2|^{\frac{6}{5}} \mathrm{d}x \le c_1\varepsilon. \tag{2.4}$$

For any $M > 0$ we define the set $\Omega_M := \{x \in B_\rho(0) : K(x) \ge M\}$. Since $K \in L^2(\mathbb{R}^3)$, the Lebesgue measure of Ω_M goes to zero as $M \to \infty$. So, for some $M > 0$ sufficiently large, we have that

$$\left(\int_{\Omega_M} K(x)^2 \mathrm{d}x \right)^{3/5} \le \varepsilon.$$

Then we can use the Hölder inequality and (2.3) again to get

$$\begin{aligned} \int_{B_\rho(0)} K(x)^{\frac{6}{5}} |u_n^2 - u^2|^{\frac{6}{5}} \mathrm{d}x &= \int_{\Omega_M} K(x)^{\frac{6}{5}} |u_n^2 - u^2|^{\frac{6}{5}} \mathrm{d}x \\ &\quad + \int_{B_\rho(0)\setminus\Omega_M} K(x)^{\frac{6}{5}} |u_n^2 - u^2|^{\frac{6}{5}} \mathrm{d}x \\ &\le c_2\varepsilon + M^{\frac{6}{5}} \int_{B_\rho(0)\setminus\Omega_M} |u_n^2 - u^2|^{\frac{6}{5}} \mathrm{d}x. \end{aligned} \tag{2.5}$$

On the other hand

$$\int_{B_\rho(0)\setminus\Omega_M} |u_n^2 - u^2|^{\frac{6}{5}}\,\mathrm{d}x \le \|u_n + u\|^{6/5}_{L^{12/5}(B_\rho(0))}\|u_n - u\|^{6/5}_{L^{12/5}(B_\rho(0))}.$$

Since $u_n \to u$ strongly in $L^{\frac{12}{5}}(B_\rho(0))$, we obtain

$$\lim_{n\to\infty}\int_{B_\rho(0)\setminus\Omega_M} |u_n^2 - u^2|^{\frac{6}{5}}\,\mathrm{d}x = 0,$$

and therefore it follows from (2.5) that

$$\int_{B_\rho(0)} K(x)^{\frac{6}{5}}|u_n^2 - u^2|^{\frac{6}{5}}\,\mathrm{d}x \le c_2\varepsilon + o_n(1),$$

where $o_n(1)$ stands for a quantity approaching zero as $n \to \infty$. The above expression, (2.4) and (2.2) imply (2.1) and the proof of the first statement of the lemma is concluded. The second one can be proved in the same way. We omit the details. □

The main interest in function ϕ comes from the fact that it enables us dealing with system (P) as a single equation. Actually, it can be proved that $(u, \phi) \in W^{1,2}(\mathbb{R}^3) \times D^{1,2}(\mathbb{R}^3)$ is a solution of (P) if, and only if, $u \in W^{1,2}(\mathbb{R}^3)$ is a nonnegative critical point of the C^1-functional $I : W^{1,2}(\mathbb{R}^3) \to \mathbb{R}$ given by

$$I(u) := \frac{1}{2}\|u\|^2 + \int K(x)\phi_u(x)u^2 - \frac{1}{p+1}\int a(x)(u^+)^{p+1},$$

where $u^+(x) := \max\{u(x), 0\}$. Since we intend to apply critical point theory to find such critical points, we need to prove some kind of compactness properties for the functional I. In this setting, the limit problem (P_∞) plays an important role. We observe that weak solutions of (P_∞) are precisely the critical points of the functional

$$I_\infty(w) := \frac{1}{2}\int(|\nabla w|^2 + V_\infty w^2) - \frac{1}{p+1}\int(w^+)^{p+1}, \; w \in W^{1,2}(\mathbb{R}^3).$$

Let $\mathcal{N}_\infty$ be the Nehari manifold of I_∞, that is

$$\mathcal{N}_\infty := \{w \in W^{1,2}(\mathbb{R}^3) \setminus \{0\} : I'_\infty(w)w = 0\}$$

and consider the related minimization problem

$$c_\infty := \inf_{w\in\mathcal{N}_\infty} I_\infty(w).$$

The proof of the next result can be found in Berestycki–Lions [5].

Proposition 2.3. *Problem (P_∞) has a positive and radially symmetrical solution $\omega \in W^{1,2}(\mathbb{R}^3)$ such that $I_\infty(\omega) = c_\infty$. Moreover, for any $0 < \delta < \sqrt{V_\infty}$, there exists a constant $C = C(\delta) > 0$ such that*

$$\omega(x) \le Ce^{-\delta|x|}, \quad \textit{for all } x \in \mathbb{R}^3. \tag{2.6}$$

In order to prove that the functional I satisfies a local Palais–Smale condition we shall use the following version of a result due to Struwe [9] (see also [3]).

Lemma 2.4. *Let $(u_n) \subset W^{1,2}(\mathbb{R}^3)$ be such that*

$$I(u_n) \to c, \quad I'(u_n) \to 0$$

and $u_n \rightharpoonup u$ weakly in $W^{1,2}(\mathbb{R}^3)$. Then $I'(u) = 0$ and we have either

(a) *$u_n \to u$ strongly in $W^{1,2}(\mathbb{R}^3)$, or*

(b) *there exists $k \in \mathbb{N}$, $(y_n^j) \in \mathbb{R}^3$ with $|y_n^j| \to \infty$, $j = 1, \ldots, k$, and nontrivial solutions $w^1, \ldots, w^k \in W^{1,2}(\mathbb{R}^3)$ of the problem (P_∞), such that*

$$I(u_n) \to I(u) + \sum_{j=1}^{k} I_\infty(w^j) \tag{2.7}$$

and

$$\left\| u_n - u - \sum_{j=1}^{k} w^j(\cdot - y_n^j) \right\| \to 0.$$

Proof. To prove this result one can use Lemma 2.2 and similar arguments to that of [6]. Hence we omit the details. □

Corollary 2.5. *If $(u_n) \subset W^{1,2}(\mathbb{R}^3)$ is such that $I(u_n) \to c < c_\infty$ and $I'(u_n) \to 0$, then (u_n) has a convergent subsequence.*

Proof. Let $(u_n) \subset W^{1,2}(\mathbb{R}^3)$ be as in the previous statement. Since $p > 3$ by a standard argument it follows that (u_n) is bounded. Hence, up to a subsequence, $u_n \rightharpoonup u_0$ weakly in $W^{1,2}(\mathbb{R}^3)$. By Lemma 2.4 we have $I'(u_0) = 0$ and therefore

$$I(u_0) = I(u_0) - \frac{1}{2} I'(u_0)u_0 = \left(\frac{1}{2} - \frac{1}{p+1} \right) \int a(x)(u_0^+)^{p+1} \geq 0.$$

If $u_n \not\to u_0$ in $W^{1,2}(\mathbb{R}^3)$, we can invoke Lemma 2.4 again to obtain $k \in \mathbb{N}$ and nontrivial solutions $w^1, \ldots, w^k$ of (P_∞) satisfying

$$\lim_{n\to\infty} I(u_n) = c = I(u_0) + \sum_{j=1}^{k} I_\infty(w^j) \geq k c_\infty \geq c_\infty,$$

contrary to the hypothesis. Hence $u_n \to u_0$ strongly in $W^{1,2}(\mathbb{R}^3)$. □

3. The proof of Theorem 1.1

We devote this section to the proof of our main theorem. The idea is looking for critical points of the functional I by considering the following minimization problem

$$c_0 := \inf_{u \in \mathcal{N}} I(u),$$

where $\mathcal{N}$ is the Nehari manifold of I, namely

$$\mathcal{N} := \left\{ u \in W^{1,2}(\mathbb{R}^3) \setminus \{0\} : I'(u)u = 0 \right\}.$$

From now on we denote by ω a positive ground state solution of the problem (P_∞). For $x_n := (0, \dots, n)$ we also set

$$\omega_n(x) := \omega(x - x_n).$$

Since $p > 3$ we can easily check that, for each $n \in \mathbb{N}$, there exists $t_n > 0$ such that $t_n\omega_n \in \mathcal{N}$. Moreover, the following holds

Lemma 3.1. *The sequence* (t_n) *satisfies* $\lim_{n\to+\infty} t_n = 1$.

Proof. Since $I'(t_n\omega_n)(t_n\omega_n) = 0$, we can use item 2 of Lemma 2.1 to get

$$0 = t_n^2 \int (|\nabla\omega_n|^2 + V(x)\omega_n^2) + t_n^4 \int K(x)\phi_{\omega_n}\omega_n^2 - t_n^{p+1} \int a(x)\omega_n^{p+1}. \qquad (3.1)$$

By using (1.1), a change o variables and the Lebesgue Theorem we get

$$\lim_{n\to\infty} \int V(x)\omega_n^2 = \lim_{n\to\infty} \int V(x + x_n)\omega^2 = \int V_\infty\omega^2$$

and

$$\lim_{n\to\infty} \int a(x)\omega_n^{p+1} = \lim_{n\to\infty} \int a(x + x_n)\omega^{p+1} = \int \omega^{p+1}.$$

Moreover, by item 1 of Lemma 2.1, we also have that

$$\left|\int K(x)\phi_{\omega_n}(x)\omega_n^2\right| \le \|K\|_2\|\phi_{\omega_n}\|_6\|\omega\|_6 \le c_1,$$

for some $c_1 > 0$.

We claim that (t_n) is bounded. Indeed, if this is not the case, we can divide equation (3.1) by t_n^{p+1}, take the limit as $n \to \infty$ and use $p + 1 > 4$ and the above statements to conclude that $\int \omega^{p+1} = 0$, which is a contradiction. Hence (t_n) is bounded. Moreover, for some $\bar{t} > 0$, there holds $t_n \ge t > 0$. Otherwise, since $\|t_n\omega_n\|_{W^{1,2}(\mathbb{R}^3)} = t_n\|\omega\|_{W^{1,2}(\mathbb{R}^3)}$, we would have $\mathrm{dist}(\mathcal{N}, 0) = 0$, which is impossible.

The above reasoning shows that, up to a subsequence, $t_n \to t_0 > 0$. We claim that

$$\lim_{n\to\infty} \int K(x)\phi_{\omega_n}(x)\omega_n^2 = 0. \qquad (3.2)$$

Assuming the claim and taking the limit in (3.1) we obtain

$$0 = t_0^2 \int (|\nabla\omega|^2 + V_\infty\omega^2) - t_0^{p+1} \int \omega^{p+1} = I'_\infty(t_0\omega)(t_0\omega).$$

Since $\omega \in \mathcal{N}_\infty$ we conclude that $t_0 = 1$.

It remains to prove the claim. First notice that, by item 1 of Lemma 2.1, we have that $\|\phi_{\omega_n}\|_6 \le c_2$, for some $c_2 > 0$. Given $\varepsilon > 0$ we choose $\rho > 0$ such that $\|K\|_{L^2(\mathbb{R}^3\setminus B_\rho(0))} < \varepsilon$. Thus,

$$\left|\int_{\mathbb{R}^3\setminus B_\rho(0)} K(x)\phi_{\omega_n}(x)\omega_n^2 \mathrm{d}x\right| \le \|K\|^2_{L^2(\mathbb{R}^3\setminus B_\rho(0))}\|\phi_{\omega_n}\|_6\|\omega\|_6^2 \le c_2\|\omega\|_6\varepsilon. \qquad (3.3)$$

On the other hand, Hölder's inequality and a change of variables provide

$$\left|\int_{B_\rho(0)} K(x)\phi_{\omega_n}(x)\omega_n^2 \mathrm{d}x\right| \le \|K\|_2\|\phi_{\omega_n}\|_6 \left(\int_{B_\rho(x_n)} \omega^6 \mathrm{d}x\right)^{1/3} = o_n(1),$$

since $\omega \in L^6(\mathbb{R}^3)$ and $|x_n| \to \infty$, as $n \to \infty$. The above inequality and (3.3) establishes (3.2). The proof is finished. □

The following result contains the core estimate for the proof of our main theorem.

Proposition 3.2. *If* (H_1)–(H_3) *hold, then* $0 < c_0 < c_\infty$.

Proof. Let ω, ω_n and $t_n > 0$ be as in the beginning of this section. Since $t_n\omega_n \in \mathcal{N}$, a straightforward calculation provides

$$\begin{aligned} c_0 \le I(t_n\omega_n) &= I_\infty(t_n\omega) + \frac{t_n^2}{2}A_n + \frac{t_n^4}{4}D_n + \frac{t_n^{p+1}}{p+1}E_n \\ &\le c_\infty + \frac{t_n^2}{2}A_n + \frac{t_n^4}{4}D_n + \frac{t_n^{p+1}}{p+1}E_n, \end{aligned} \tag{3.4}$$

where

$$A_n := \int (V(x) - V_\infty)\omega_n^2, \; D_n := \int K(x)\phi_{\omega_n}(x)\omega_n^2$$

and

$$E_n := \int (1 - a(x))\omega_n^{p+1}.$$

Now we need to estimate the decay rate of each of the above terms. It follows from the first estimate in (1.2) that

$$A_n = \int (V(x) - V_\infty)\omega_n^2 \le c_V \int e^{-\gamma|x|}\omega_n^2 = c_V \int e^{-\gamma|x+x_n|}\omega^2.$$

Since $|x + x_n| \ge |x_n| - |x| = n - |x|$, we obtain

$$A_n \le c_V e^{-\gamma n} \int e^{\gamma|x|}\omega^2 = C_V e^{-\gamma n}, \tag{3.5}$$

with $C_V > 0$, where we have used in the last equality the exponential decay of ω given in Proposition 2.3 and that $\gamma < 2\sqrt{V_\infty}$, which implies that $\int e^{-\gamma|x|}\omega^2 < \infty$. In order to estimate D_n we use Hölder's inequality, $\alpha < 2\sqrt{V_\infty}$ and argue as above to get

$$\begin{aligned} D_n = \int K(x)\phi_{\omega_n}(x)\omega_n^2 &\le \|\phi_{\omega_n}\|_6 \left(\int K(x)^{\frac{6}{5}}\omega_n^{\frac{12}{5}}\right)^{5/6} \\ &\le c_1 \left(\int e^{-\frac{6\alpha}{5}|x+x_n|}\omega^{\frac{12}{5}}\right)^{5/6} \\ &\le C_K e^{-\alpha n}, \end{aligned} \tag{3.6}$$

with $C_K > 0$. We now use the second inequality in (1.2) to estimate E_n as follows

$$E_n = \int (1 - a(x))\omega_n^{p+1} \leq -c_a \int e^{-\theta|x|}\omega_n^{p+1} = -c_a \int e^{-\theta|x+x_n|}\omega^{p+1}.$$

Since $|x + x_n| \leq n + |x|$, we obtain $C_a > 0$ such that

$$E_n \leq -c_a e^{-\theta n} \int e^{-\theta|x|}\omega^{p+1} = -C_a e^{-\theta n}. \tag{3.7}$$

By replacing (3.5)–(3.7) in (3.4) we obtain,

$$\begin{aligned} c_0 &\leq c_\infty + e^{-\theta n}\left(\frac{t_n^2}{2}C_V e^{(\theta-\gamma)n} + \frac{t_n^4}{4}C_K e^{(\theta-\alpha)n} - \frac{t_n^{p+1}}{p+1}C_a\right) \\ &= c_\infty + e^{-\theta n}(o_n(1) - C_a), \end{aligned}$$

where we have used in the last equality that $t_n \to 1$ and $\theta < \min\{\alpha, \gamma\}$. Since $C_a > 0$ we can take n large enough to conclude that $c_0 < c_\infty$. The proposition is proved. □

We are now ready to obtain the ground state solution of (S).

Proof of Theorem 1.1. Let $(u_n) \subset \mathcal{N}$ be such that $I(u_n) \to c_0$. Since $\mathcal{N}$ is a C^1 regular manifold and is closed (see [6, Lemma 3.1]), we can use Ekeland's Variational Principle to obtain that

$$I(u_n) \to c_0 \text{ and } I'(u_n) \to 0.$$

Proposition 3.2 and Corollary 2.5 imply that the sequence (u_n) strongly converges to a function $u_0 \in W^{1,2}(\mathbb{R}^3)$ such that $I(u_0) = c_0 > 0$ and $I'(u_0) = 0$. Setting $u_0^-(x) := \max\{-u_0(x), 0\}$, we can use $0 = I'(u_0)u_0^- = -\|u_0^-\|$ to conclude that $u_0 \geq 0$ a.e. in $\mathbb{R}^3$. It follows from elliptic regularity and the strong maximum principle that $u > 0$ in $\mathbb{R}^3$. The theorem is proved. □

Remark 3.3. A simple inspection of the proof of Proposition 3.2 shows that we can drop the condition (H_3) by the hypotheses $(\widetilde{H_3})$ stated in the introduction. Indeed, with this dual condition what happens is that term A_n of the proof of the proposition becomes negative while the term E_n is positive. The choices of the numbers α, γ and θ guarantee that the desired inequality also holds in this setting.

References

[1] A. Ambrosetti, D. Ruiz, *Multiple bound states for the Schrödinger–Poisson problem.* Commun. Contemp. Math. **10** (2008), 391–404.

[2] A. Azzollini and A. Pomponio, *Ground state solutions for the nonlinear Schrödinger–Maxwell equations*, J. Math. Anal. Appl. **345** (2008), 90–108.

[3] V. Benci and G. Cerami, *Positive solutions of some nonlinear elliptic problems in exterior domains*, Arch. Rat. Mech. Anal. **99** (1987), 283–300.

[4] V. Benci and D. Fortunato, *An eigenvalue problem for the Schrödinger–Maxwell equations*, Topol. Methods Nonlinear Anal. **11** (1998), 283–293.

[5] H. Berestycki and P.L. Lions, *Nonlinear scalar field equations. I. Existence of a ground state*, Arch. Rational Mech. Anal. 82 (1983), no. 4, 313–345.

[6] G. Cerami and G. Vaira, *Positive solutions for some non-autonomous Schrödinger–Poisson systems*, J. Differential Equations **248** (2010), 521–543.

[7] M.F. Furtado, L. Maia and E.S. Medeiros, *Positive and nodal solutions for a nonlinear Schrödinger equation with indefinite potential*, Adv. Non. Studies **8** (2008), 353–373.

[8] D. Ruiz, *The Schrödinger–Poisson equation under the effect of a nonlinear local term*. J. Funct. Anal. **237** (2006), 655–674.

[9] M. Struwe, *A global compactness result for elliptic boundary value problems involving limiting nonlinearities*, Math. Z. **187** (1984), no. 4, 511–517.

[10] G. Vaira, *Ground states for Schrödinger–Poisson type systems*, Ric. Mat. **60** (2011), 263–297.

Marcelo F. Furtado and Liliane A. Maia
Universidade de Brasília
Departamento de Matemática
70910-900, Brasília – DF Brazil
e-mail: `mfurtado@unb.br`
`lilimaia@unb.br`

Everaldo S. Medeiros
Universidade Federal da Paraíba
Departamento de Matemática
58051-900, João Pessoa – PB Brazil
e-mail: `everaldo@mat.ufpb.br`

Progress in Nonlinear Differential Equations
and Their Applications, Vol. 85, 287–300

Low Energy Solutions for the Semiclassical Limit of Schrödinger–Maxwell Systems

Marco Ghimenti and Anna Maria Micheletti

Dedicated to our friend Bernhard

Abstract. We show that the number of positive solutions of Schrödinger–Maxwell system on a smooth bounded domain $\Omega \subset \mathbb{R}^3$ depends on the topological properties of the domain. In particular we consider the Lusternik–Schnirelmann category and the Poincaré polynomial of the domain.

Mathematics Subject Classification (2010). 35J60, 35J20, 58E30, 81V10.

Keywords. Schrödinger–Maxwell systems, Lusternik–Schnirelman category, Morse theory.

1. Introduction

Given real numbers $q > 0$, $\omega > 0$ we consider the following Schrödinger–Maxwell system on a smooth bounded domain $\Omega \subset \mathbb{R}^3$:

$$\begin{cases} -\varepsilon^2 \Delta u + u + \omega u v = |u|^{p-2} u & \text{in } \Omega \\ -\Delta v = q u^2 & \text{in } \Omega \\ u, v = 0 & \text{on } \partial\Omega \end{cases} \tag{1}$$

This paper deals with the semiclassical limit of the system (1), i.e., it is concerned with the problem of finding solutions of (1) when the parameter ε is sufficiently small. This problem has some relevance for the understanding of a wide class of quantum phenomena. We are interested in the relation between the number of positive solutions of (1) and the topology of the bounded set Ω. In particular we consider the Lusternik–Schnirelmann category $\operatorname{cat}\Omega$ of Ω in itself and its Poincaré polynomial $P_t(\Omega)$.

Our main results are the following.

Theorem 1. *Let $4 < p < 6$. For ε small enough there exist at least $\operatorname{cat}(\Omega)$ positive solutions of* (1).

Theorem 2. *Let $4 < p < 6$. Assume that for ε small enough all the solutions of problem* (1) *are non-degenerate. Then there are at least $2P_1(\Omega) - 1$ positive solutions.*

Schrödinger–Maxwell systems recently received considerable attention from the mathematical community. In the pioneering paper [9] Benci and Fortunato studied system (1) when $\varepsilon = 1$, $\|u\|_{L^2} = 1$ and without nonlinearity. Regarding the system in a semiclassical regime Ruiz [18] and D'Aprile–Wei [11] showed the existence of a family of radially symmetric solutions respectively for $\Omega = \mathbb{R}^3$ or a ball. D'Aprile–Wei [12] also proved the existence of clustered solutions in the case of a bounded domain Ω in $\mathbb{R}^3$.

Recently, Siciliano [19] relates the number of solution with the topology of the set Ω when $\varepsilon = 1$, and the nonlinearity is a pure power with exponent p close to the critical exponent 6. Moreover, in the case $\varepsilon = 1$, many authors proved results of existence and non existence of solution of (1) in presence of a pure power nonlinearity $|u|^{p-2}u$, $2 < p < 6$ or more general nonlinearities [1, 2, 3, 4, 10, 14, 15, 17, 20].

In a forthcoming paper [13], we aim to use our approach to give an estimate on the number of low energy solutions for Klein–Gordon–Maxwell systems on a Riemannian manifold in terms of the topology of the manifold and some information on the profile of the low energy solutions.

In the following we always assume $4 < p < 6$.

2. Notations and definitions

In the following we use the following notations.

- $B(x, r)$ is the ball in $\mathbb{R}^3$ centered in x with radius r.
- The function $U(x)$ is the unique positive spherically symmetric function in $\mathbb{R}^3$ such that
$$-\Delta U + U = U^{p-1} \text{ in } \mathbb{R}^3$$
we remark that U and its first derivative decay exponentially at infinity.
- Given $\varepsilon > 0$ we define $U_\varepsilon(x) = U\left(\frac{x}{\varepsilon}\right)$.
- We denote by supp φ the support of the function φ.
- We define
$$m_\infty = \inf_{\int_{\mathbb{R}^3} |\nabla v|^2 + v^2 dx = |v|^p_{L^p(\mathbb{R}^3)}} \frac{1}{2} \int_{\mathbb{R}^3} (|\nabla v|^2 + v^2) dx - \frac{1}{p} |v|^p_{L^p(\mathbb{R}^3)}$$
- We also use the following notation for the different norms for $u \in H^1_0(\Omega)$:
$$\|u\|^2_\varepsilon = \frac{1}{\varepsilon^3} \int_\Omega \varepsilon^2 |\nabla u|^2 + u^2 dx \quad |u|^p_{\varepsilon,p} = \frac{1}{\varepsilon^3} \int_\Omega |u|^p dx$$
$$\|u\|^2_{H^1_0} = \int_\Omega |\nabla u|^2 dx \qquad |u|^p_p = \int_\Omega |u|^p dx$$
and we denote by H_ε the Hilbert space $H^1_0(\Omega)$ endowed with the $\|\cdot\|_\varepsilon$ norm.

Definition 3. Let X a topological space and consider a closed subset $A \subset X$. We say that A has category k relative to X ($\mathrm{cat}_X A = k$) if A is covered by k closed sets A_j, $j = 1, \ldots, k$, which are contractible in X, and k is the minimum integer with this property. We simply denote $\mathrm{cat}\, X = \mathrm{cat}_X X$.

Remark 4. Let X_1 and X_2 be topological spaces. If $g_1 : X_1 \to X_2$ and $g_2 : X_2 \to X_1$ are continuous operators such that $g_2 \circ g_1$ is homotopic to the identity on X_1, then $\mathrm{cat}\, X_1 \leq \mathrm{cat}\, X_2$.

Definition 5. Let X be any topological space and let $H_k(X)$ denotes its kth homology group with coefficients in $\mathbb{Q}$. The Poincaré polynomial $P_t(X)$ of X is defined as the following power series in t

$$P_t(X) := \sum_{k \geq 0} (\mathrm{dim} H_k(X))\, t^k$$

Actually, if X is a compact space, we have that $\mathrm{dim} H_k(X) < \infty$ and this series is finite; in this case, $P_t(X)$ is a polynomial and not a formal series.

Remark 6. Let X and Y be topological spaces. If $f : X \to Y$ and $g : Y \to X$ are continuous operators such that $g \circ f$ is homotopic to the identity on X, then $P_t(Y) = P_t(X) + Z(t)$ where $Z(t)$ is a polynomial with non-negative coefficients.

These topological tools are classical and can be found, e.g., in [16] and in [5].

3. Preliminary results

Using an idea in a paper of Benci and Fortunato [9] we define the map $\psi : H_0^1(\Omega) \to H_0^1(\Omega)$ defined by the equation

$$-\Delta \psi(u) = qu^2 \text{ in } \Omega \tag{2}$$

Lemma 7. *The map $\psi : H_0^1(\Omega) \to H_0^1(\Omega)$ is of class C^2 with derivatives*

$$\psi'(u)[\varphi] = i^*(2qu\varphi) \tag{3}$$

$$\psi''(u)[\varphi_1, \varphi_2] = i^*(2q\varphi_1\varphi_2) \tag{4}$$

where the operator $i_\varepsilon^* : (L^{p'}, |\cdot|_{\varepsilon,p'}) \to H_\varepsilon$ is the adjoint operator of the immersion operator $i_\varepsilon : H_\varepsilon \to (L^p, |\cdot|_{\varepsilon,p})$.

Proof. The proof is standard. □

Lemma 8. *The map $T : H_0^1(\Omega) \to \mathbb{R}$ given by*

$$T(u) = \int_\Omega u^2 \psi(u) dx$$

is a C^2 map and its first derivative is

$$T'(u)[\varphi] = 4 \int_\Omega \varphi u \psi(u) dx.$$

Proof. The regularity is standard. The first derivative is

$$T'(u)[\varphi] = 2\int u\varphi\psi(u) + \int u^2\psi'(u)[\varphi].$$

By (3) and (2) we have

$$\begin{aligned} 2q\int u\varphi\psi(u) &= -\int \Delta(\psi'(u)[\varphi])\psi(u) = -\int \psi'(u)[\varphi]\Delta\psi(u) \\ &= \int \psi'(u)[\varphi]qu^2 \end{aligned}$$

and the claim follows. □

At this point we consider the following functional $I_\varepsilon \in C^2(H_0^1(\Omega), \mathbb{R})$

$$I_\varepsilon(u) = \frac{1}{2}\|u\|_\varepsilon^2 + \frac{\omega}{4}G_\varepsilon(u) - \frac{1}{p}|u^+|_{\varepsilon,p}^p \tag{5}$$

where

$$G_\varepsilon(u) = \frac{q}{\varepsilon^3}\int_\Omega u^2\psi(u)dx = \frac{q}{\varepsilon^3}T(u).$$

By Lemma 8 we have

$$\begin{aligned} I'_\varepsilon(u)[\varphi] &= \frac{1}{\varepsilon^3}\int_\Omega \varepsilon^2\nabla u\nabla\varphi + u\varphi + \omega u\psi(u)\varphi - (u^+)^{p-1}\varphi \\ I'_\varepsilon(u)[u] &= \|u\|_\varepsilon^2 + \omega G_\varepsilon(u) - |u^+|_{\varepsilon,p}^p \end{aligned}$$

then if u is a critical point of the functional I_ε the pair of positive functions $(u, \psi(u))$ is a solution of (1).

4. Nehari manifold

We define the following Nehari set

$$\mathcal{N}_\varepsilon = \left\{u \in H_0^1(\Omega) \setminus 0 \ : \ N_\varepsilon(u) := I'_\varepsilon(u)[u] = 0\right\}.$$

In this section we give an explicit proof of the main properties of the Nehari manifold, although standard, for the sake of completeness

Lemma 9. *$\mathcal{N}_\varepsilon$ is a C^2 manifold and* $\inf_{\mathcal{N}_\varepsilon}\|u\|_\varepsilon > 0$.

Proof. If $u \in \mathcal{N}_\varepsilon$, using that $N_\varepsilon(u) = 0$, and $p > 4$ we have

$$N'_\varepsilon(u)[u] = 2\|u\|_\varepsilon^2 + 4\omega G_\varepsilon(u) - p|u^+|_{\varepsilon,p} = (2-p)\|u\|_\varepsilon + (4-p)\omega G_\varepsilon(u) < 0$$

so $\mathcal{N}_\varepsilon$ is a C^2 manifold.

We prove the second claim by contradiction. Take a sequence $\{u_n\}_n \in \mathcal{N}_\varepsilon$ with $\|u_n\|_\varepsilon \to 0$ while $n \to +\infty$. Thus, using that $N_\varepsilon(u) = 0$,

$$\|u_n\|_\varepsilon^2 + \omega G_\varepsilon(u_n) = |u_n^+|_{p,\varepsilon}^p \le C\|u_n\|_\varepsilon^p,$$

so

$$1 < 1 + \frac{\omega G_\varepsilon(u)}{\|u_n\|_\varepsilon} \le C\|u_n\|_\varepsilon^{p-2} \to 0$$

and this is a contradiction. □

Remark 10. If $u \in \mathcal{N}_\varepsilon$, then

$$\begin{aligned} I_\varepsilon(u) &= \left(\frac{1}{2} - \frac{1}{p}\right)\|u\|_\varepsilon^2 + \omega\left(\frac{1}{4} - \frac{1}{p}\right)G_\varepsilon(u) \\ &= \left(\frac{1}{2} - \frac{1}{p}\right)|u^+|_{p,\varepsilon}^p - \frac{\omega}{4}G_\varepsilon(u). \end{aligned}$$

Lemma 11. *It holds the Palais–Smale condition for the functional I_ε on $\mathcal{N}_\varepsilon$.*

Proof. We start proving the PS condition for I_ε. Let $\{u_n\}_n \in H_0^1(\Omega)$ such that

$$I_\varepsilon(u_n) \to c \qquad |I'_\varepsilon(u_n)[\varphi]| \le \sigma_n\|\varphi\|_\varepsilon \text{ where } \sigma_n \to 0$$

We prove that $\|u_n\|_\varepsilon$ is bounded. Suppose $\|u_n\|_\varepsilon \to \infty$. Then, by PS hypothesis

$$\frac{pI_\varepsilon(u_n) - I'_\varepsilon(u_n)[u_n]}{\|u_n\|_\varepsilon} = \left(\frac{p}{2} - 1\right)\|u_n\|_\varepsilon + \left(\frac{p}{4} - 1\right)\frac{G_\varepsilon(u_n)}{\|u_n\|_\varepsilon} \to 0$$

and this is a contradiction because $p > 4$.

At this point, up to subsequence $u_n \to u$ weakly in $H_0^1(\Omega)$ and strongly in $L^t(\Omega)$ for each $2 \le t < 6$. Since u_n is a PS sequence

$$u_n + \omega i_\varepsilon^*(\psi(u_n)u_n) - i_\varepsilon^*\left((u_n^+)^{p-1}\right) \to 0 \text{ in } H_0^1(\Omega)$$

we have only to prove that $i_\varepsilon^*(\psi(u_n)u_n) \to i_\varepsilon^*(\psi(u)u)$ in $H_0^1(\Omega)$, then we have to prove that

$$\psi(u_n)u_n \to \psi(u)u \text{ in } L^{t'}$$

We have $|\psi(u_n)u_n - \psi(u)u|_{\varepsilon,t'} \le |\psi(u)(u_n - u)|_{\varepsilon,t'} + |(\psi(u_n) - \psi(u))\,u_n|_{\varepsilon,t'}$. We get

$$\int_\Omega |\psi(u_n) - \psi(u)|^{\frac{t}{t-1}}|u_n|^{\frac{t}{t-1}} \le \left(\int_\Omega |\psi(u_n) - \psi(u)|^t\right)^{\frac{1}{t-1}}\left(\int_\Omega |u_n|^{\frac{t}{t-2}}\right)^{\frac{t-2}{t-1}} \to 0,$$

thus we can conclude easily.

Now we prove the PS condition for the constrained functional. Let $\{u_n\}_n \in \mathcal{N}_\varepsilon$ such that

$$I_\varepsilon(u_n) \to c$$
$$|I'_\varepsilon(u_n)[\varphi] - \lambda_n N'(u_n)[\varphi]| \le \sigma_n\|\varphi\|_\varepsilon \text{ with } \sigma_n \to 0.$$

In particular

$$I'_\varepsilon(u_n)\left[\frac{u_n}{\|u_n\|_\varepsilon}\right] - \lambda_n N'(u_n)\left[\frac{u_n}{\|u_n\|_\varepsilon}\right] \to 0.$$

Then

$$\lambda_n\left\{(p-2)\|u_n\|_\varepsilon + (p-4)\,\omega\frac{G_\varepsilon(u_n)}{\|u_n\|_\varepsilon}\right\} \to 0$$

thus $\lambda_n \to 0$ because $p > 4$. Since $N'(u_n) = u_n - i_\varepsilon^*(4\omega\psi(u_n)u_n) - pi_\varepsilon^*(|u_n^+|^{p-1})$ is bounded we obtain that $\{u_n\}_n$ is a PS sequence for the free functional I_ε, and we get the claim. □

Lemma 12. *For all $w \in H_0^1(\Omega)$ such that $|w^+|_{\varepsilon,p} = 1$ there exists a unique positive number $t_\varepsilon = t_\varepsilon(w)$ such that $t_\varepsilon(w)w \in \mathcal{N}_\varepsilon$.*

Proof. We define, for $t > 0$

$$H(t) = I_\varepsilon(tw) = \frac{1}{2}t^2\|w\|_\varepsilon^2 + \frac{t^4}{4}\omega G_\varepsilon(w) - \frac{t^p}{p}.$$

Thus

$$H'(t) = t\left(\|w\|_\varepsilon^2 + t^2\omega G_\varepsilon(w) - t^{p-2}\right) \tag{6}$$

$$H''(t) = \|w\|_\varepsilon^2 + 3t^2\omega G_\varepsilon(w) - (p-1)t^{p-2} \tag{7}$$

By (6) there exists $t_\varepsilon > 0$ such that $H'(t_\varepsilon)$. Moreover, by (6), (7) and because $p > 4$ we have that $H''(t_\varepsilon) < 0$, so t_ε is unique. □

5. Main ingredient of the proof

We sketch the proof of Theorem 1. First of all, since the functional $I_\varepsilon \in C^2$ is bounded below and satisfies PS condition on the complete C^2 manifold $\mathcal{N}_\varepsilon$, we have, by well-known results, that I_ε has at least $\operatorname{cat} I_\varepsilon^d$ critical points in the sublevel

$$I_\varepsilon^d = \left\{u \in H^1 \ : \ I_\varepsilon(u) \le d\right\}.$$

We prove that, for ε and δ small enough, it holds

$$\operatorname{cat}\Omega \le \operatorname{cat}\left(\mathcal{N}_\varepsilon \cap I_\varepsilon^{m_\infty+\delta}\right)$$

where

$$m_\infty := \inf_{\mathcal{N}_\infty} \frac{1}{2}\int_{\mathbb{R}^3}(|\nabla v|^2 + v^2)dx - \frac{1}{p}\int_{\mathbb{R}^3}|v|^p dx$$

$$\mathcal{N}_\infty = \left\{v \in H^1(\mathbb{R}^3) \smallsetminus \{0\} \ : \ \int_{\mathbb{R}^3}(|\nabla v|^2 + v^2)dx = \int_{\mathbb{R}^3}|v|^p dx\right\}.$$

To get the inequality $\operatorname{cat}\Omega \le \operatorname{cat}\left(\mathcal{N}_\varepsilon \cap I_\varepsilon^{m_\infty+\delta}\right)$ we build two continuous operators

$$\Phi_\varepsilon \ : \ \Omega^- \to \mathcal{N}_\varepsilon \cap I_\varepsilon^{m_\infty+\delta}$$

$$\beta \ : \ \mathcal{N}_\varepsilon \cap I_\varepsilon^{m_\infty+\delta} \to \Omega^+.$$

where

$$\Omega^- = \{x \in \Omega \ : \ d(x, \partial\Omega) < r\}$$

$$\Omega^+ = \{x \in \mathbb{R}^3 \ : \ d(x, \partial\Omega) < r\}$$

with r small enough so that $\operatorname{cat}(\Omega^-) = \operatorname{cat}(\Omega^+) = \operatorname{cat}(\Omega)$.

Following an idea in [7], we build these operators Φ_ε and β such that $\beta \circ \Phi_\varepsilon : \Omega^- \to \Omega^+$ is homotopic to the immersion $i : \Omega^- \to \Omega^+$. By the properties of Lusternik–Schinerlmann category we have

$$\operatorname{cat} \Omega \le \operatorname{cat} \left(\mathcal{N}_\varepsilon \cap I_\varepsilon^{m_\infty+\delta}\right)$$

which ends the proof of Theorem 1.

Concerning Theorem 2, we can re-state classical results contained in [5, 8] in the following form.

Theorem 13. *Let I_ε be the functional (5) on $H^1(\Omega)$ and let K_ε be the set of its critical points. If all its critical points are non-degenerate then*

$$\sum_{u\in K_\varepsilon} t^{\mu(u)} = tP_t(\Omega) + t^2(P_t(\Omega) - 1) + t(1+t)Q(t) \tag{8}$$

where $Q(t)$ is a polynomial with non-negative integer coefficients and $\mu(u)$ is the Morse index of the critical point u.

By Remark 6 and by means of the maps Φ_ε and β we have that

$$P_t(\mathcal{N}_\varepsilon \cap I_\varepsilon^{m_\infty+\delta}) = P_t(\Omega) + Z(t) \tag{9}$$

where $Z(t)$ is a polynomial with non-negative coefficients. Set $m_\varepsilon = \inf_{\mathcal{N}_\varepsilon} I_\varepsilon$, we get that $\inf_\varepsilon m_\varepsilon =: \alpha > 0$, because $\lim\limits_{\varepsilon\to 0} m_\varepsilon = m_\infty$ (see (20)), and we have the following relations [5, 8]

$$P_t(I_\varepsilon^{m_\infty+\delta}, I_\varepsilon^{\alpha/2}) = tP_t(\mathcal{N}_\varepsilon \cap I_\varepsilon^{m_\infty+\delta}) \tag{10}$$

$$P_t(H_0^1(\Omega), I_\varepsilon^{m_\infty+\delta})) = t(P_t(I_\varepsilon^{m_\infty+\delta}, I_\varepsilon^{\alpha/2}) - t) \tag{11}$$

$$\sum_{u\in K_\varepsilon} t^{\mu(u)} = P_t(H_0^1(\Omega), I_\varepsilon^{m_\infty+\delta})) + P_t(I_\varepsilon^{m_\infty+\delta}, I_\varepsilon^{\alpha/2}) + (1+t)\tilde{Q}(t) \tag{12}$$

where $\tilde{Q}(t)$ is a polynomial with non-negative integer coefficients. Hence, by (9), (10), (11), (12) we obtain (8). At this point, evaluating equation (8) for $t = 1$ we obtain the claim of Theorem 2

6. The map Φ_ε

For every $\xi \in \Omega^-$ we define the function

$$W_{\xi,\varepsilon}(x) = U_\varepsilon(x-\xi)\chi(|x-\xi|)$$

where $\chi : \mathbb{R}^+ \to \mathbb{R}^+$ where $\chi \equiv 1$ for $t \in [0, r/2)$, $\chi \equiv 0$ for $t > r$ and $|\chi'(t)| \le 2/r$.

We can define a map

$$\Phi_\varepsilon : \Omega^- \to \mathcal{N}_\varepsilon$$
$$\Phi_\varepsilon(\xi) = t_\varepsilon(W_{\xi,\varepsilon})W_{\xi,\varepsilon}$$

Remark 14. We have that the following limits hold uniformly with respect to $\xi \in \Omega$

$$\begin{aligned} \|W_{\varepsilon,\xi}\|_\varepsilon &\to \|U\|_{H^1(\mathbb{R}^3)}, \\ |W_{\varepsilon,\xi}|_{\varepsilon,t} &\to \|U\|_{L^t(\mathbb{R}^3)} \text{ for all } 2 \le t \le 6. \end{aligned}$$

Lemma 15. *There exists $\bar{\varepsilon} > 0$ and a constant $c > 0$ such that*

$$G_\varepsilon(W_{\varepsilon,\xi}) = \frac{1}{\varepsilon^3} \int_\Omega q W_{\varepsilon,\xi}^2(x) \psi(W_{\varepsilon,\xi}) dx < c\varepsilon^2.$$

Proof. It holds

$$\begin{aligned} \|\psi(W_{\varepsilon,\xi})\|^2_{H_0^1(\Omega)} &= \int_\Omega q W_{\varepsilon,\xi}^2(x)\psi(W_{\varepsilon,\xi}) dx \le q \|\psi(W_{\varepsilon,\xi})\|_{L^6(\Omega)} \left(\int_\Omega W_{\varepsilon,\xi}^{12/5} dx \right)^{5/6} \\ &\le c \|\psi(W_{\varepsilon,\xi})\|_{H_0^1(\Omega)} \left(\frac{1}{\varepsilon^3} \int_\Omega W_{\varepsilon,\xi}^{12/5} dx \right)^{5/6} \varepsilon^{5/2}. \end{aligned}$$

By Remark 14 we have that $\|\psi(W_{\varepsilon,\xi})\|_{H_0^1(\Omega)} \le \varepsilon^{5/2}$ and the claim follows by applying again the Cauchy–Schwartz inequality. □

Proposition 16. *For all $\varepsilon > 0$ the map Φ_ε is continuous. Moreover for any $\delta > 0$ there exists $\varepsilon_0 = \varepsilon_0(\delta)$ such that, if $\varepsilon < \varepsilon_0$ then $I_\varepsilon(\Phi_\varepsilon(\xi)) < m_\infty + \delta$.*

Proof. It is easy to see that Φ_ε is continuous because $t_\varepsilon(w)$ depends continuously on $w \in H_0^1$.

At this point we prove that $t_\varepsilon(W_{\varepsilon,\xi}) \to 1$ uniformly with respect to $\xi \in \Omega$. In fact, by Lemma 12 $t_\varepsilon(W_{\varepsilon,\xi})$ is the unique solution of

$$\|W_{\varepsilon,\xi}\|_\varepsilon^2 + t^2 \omega G_\varepsilon(W_{\varepsilon,\xi}) - t^{p-2} |W_{\varepsilon,\xi}|_{\varepsilon,p}^p = 0.$$

By Remark 14 and Lemma 15 we have the claim.

Now, we have

$$I_\varepsilon\left(t_\varepsilon(W_{\varepsilon,\xi})W_{\varepsilon,\xi}\right) = \left(\frac{1}{2} - \frac{1}{p}\right) \|W_{\varepsilon,\xi}\|_\varepsilon^2 t_\varepsilon^2 + \omega \left(\frac{1}{4} - \frac{1}{p}\right) t_\varepsilon^4 G_\varepsilon(W_{\varepsilon,\xi}).$$

Again, by Remark 14 and Lemma 15 we have

$$I_\varepsilon\left(t_\varepsilon(W_{\varepsilon,\xi})W_{\varepsilon,\xi}\right) \to \left(\frac{1}{2} - \frac{1}{p}\right) \|U\|^2_{H^1(\mathbb{R}^3)} = m_\infty$$

that concludes the proof. □

Remark 17. We set

$$m_\varepsilon = \inf_{\mathcal{N}_\varepsilon} I_\varepsilon.$$

By Proposition 16 we have that

$$\limsup_{\varepsilon \to 0} m_\varepsilon \le m_\infty. \tag{13}$$

7. The map β

For any $u \in \mathcal{N}_\varepsilon$ we can define a point $\beta(u) \in \mathbb{R}^3$ by

$$\beta(u) = \frac{\int_\Omega x|u^+|^p dx}{\int_\Omega |u^+|^p dx}.$$

The function β is well defined in $\mathcal{N}_\varepsilon$ because, if $u \in \mathcal{N}_\varepsilon$, then $u^+ \neq 0$.

We have to prove that, if $u \in \mathcal{N}_\varepsilon \cap I_\varepsilon^{m_\infty+\delta}$ then $\beta(u) \in \Omega^+$.

Let us consider partitions of Ω. For a given $\varepsilon > 0$ we say that a finite partition $\mathcal{P}_\varepsilon = \{P_j^\varepsilon\}_{j\in\Lambda_\varepsilon}$ of Ω is a "good" partition if: for any $j \in \Lambda_\varepsilon$ the set P_j^ε is closed; $P_i^\varepsilon \cap P_j^\varepsilon \subset \partial P_i^\varepsilon \cap \partial P_j^\varepsilon$ for any $i \neq j$; there exist $r_1(\varepsilon), r_2(\varepsilon) > 0$ such that there are points $q_j^\varepsilon \in P_j^\varepsilon$ for which $B(q_j^\varepsilon, \varepsilon) \subset P_j^\varepsilon \subset B(q_j^\varepsilon, r_2(\varepsilon)) \subset B(q_j^\varepsilon, r_1(\varepsilon))$, with $r_1(\varepsilon) \geq r_2(\varepsilon) \geq C\varepsilon$ for some positive constant C; lastly, there exists a finite number $\nu \in \mathbb{N}$ such that every $x \in \Omega$ is contained in at most ν balls $B(q_j^\varepsilon, r_1(\varepsilon))$, where ν does not depends on ε.

Lemma 18. *There exists a constant $\gamma > 0$ such that, for any $\delta > 0$ and for any $\varepsilon < \varepsilon_0(\delta)$ as in Proposition 16, given any "good" partition $\mathcal{P}_\varepsilon = \{P_j^\varepsilon\}_j$ of the domain Ω and for any function $u \in \mathcal{N}_\varepsilon \cap I_\varepsilon^{m_\infty+\delta}$ there exists, for an index $\bar{j}$ a set $P_{\bar{j}}^\varepsilon$ such that*

$$\frac{1}{\varepsilon^3}\int_{P_{\bar{j}}^\varepsilon} |u^+|^p dx \geq \gamma.$$

Proof. Taking in account that $I'_\varepsilon(u)[u] = 0$ we have

$$\begin{aligned}\|u\|_\varepsilon^2 &= |u^+|_{\varepsilon,p}^p - \frac{1}{\varepsilon^3}\int_\Omega \omega u^2\psi(u) \leq |u^+|_{\varepsilon,p}^p = \sum_j \frac{1}{\varepsilon^3}\int_{P_j} |u^+|^p \\ &= \sum_j |u_j^+|_{\varepsilon,p}^p = \sum_j |u_j^+|_{\varepsilon,p}^{p-2}|u_j^+|_{\varepsilon,p}^2 \leq \max_j \left\{|u_j^+|_{\varepsilon,p}^{p-2}\right\} \sum_j |u_j^+|_{\varepsilon,p}^2\end{aligned}$$

where u_j^+ is the restriction of the function u^+ on the set P_j.

At this point, arguing as in [6, Lemma 5.3], we prove that there exists a constant $C > 0$ such that

$$\sum_j |u_j^+|_{\varepsilon,p}^2 \leq C\nu\|u^+\|_\varepsilon^2, \qquad \text{thus} \qquad \max_j \left\{|u_j^+|_{\varepsilon,p}^{p-2}\right\} \geq \frac{1}{C\nu}$$

that concludes the proof. □

Proposition 19. *For any $\eta \in (0, 1)$ there exists $\delta_0 < m_\infty$ such that for any $\delta \in (0, \delta_0)$ and any $\varepsilon \in (0, \varepsilon_0(\delta))$ as in Proposition 16, for any function $u \in \mathcal{N}_\varepsilon \cap I_\varepsilon^{m_\infty+\delta}$ we can find a point $q = q(u) \in \Omega$ such that*

$$\frac{1}{\varepsilon^3}\int_{B(q,r/2)} (u^+)^p > (1-\eta)\frac{2p}{p-2}m_\infty.$$

Proof. First, we prove the proposition for $u \in \mathcal{N}_\varepsilon \cap I_\varepsilon^{m_\varepsilon+2\delta}$.

By contradiction, we assume that there exists $\eta \in (0,1)$ such that we can find two sequences of vanishing real number δ_k and ε_k and a sequence of functions $\{u_k\}_k$ such that $u_k \in \mathcal{N}_{\varepsilon_k}$,

$$m_{\varepsilon_k} \le I_{\varepsilon_k}(u_k) = \left(\frac{1}{2}-\frac{1}{p}\right)\|u_k\|^2_{\varepsilon_k} + \omega\left(\frac{1}{4}-\frac{1}{p}\right)G_{\varepsilon_k}(u_k) \le m_{\varepsilon_k}+2\delta_k \le m_\infty + 3\delta_k \tag{14}$$

for k large enough (see Remark 17), and, for any $q \in \Omega$,

$$\frac{1}{\varepsilon_k^3}\int_{B(q,r/2)}(u_k^+)^p \le (1-\eta)\frac{2p}{p-2}m_\infty.$$

By Ekeland's principle and by definition of $\mathcal{N}_{\varepsilon_k}$ we can assume

$$\left|I'_{\varepsilon_k}(u_k)[\varphi]\right| \le \sigma_k\|\varphi\|_{\varepsilon_k} \text{ where } \sigma_k \to 0. \tag{15}$$

By Lemma 18 there exists a set $P_k^{\varepsilon_k} \in \mathcal{P}_{\varepsilon_k}$ such that

$$\frac{1}{\varepsilon_k^3}\int_{P_k^{\varepsilon_k}}|u_k^+|^p dx \ge \gamma.$$

We choose a point $q_k \in P_k^{\varepsilon_k}$ and we define, for $z \in \Omega_{\varepsilon_k} := \frac{1}{\varepsilon_k}(\Omega - q_k)$

$$w_k(z) = u_k(\varepsilon_k z + q_k) = u_k(x).$$

We have that $w_k \in H_0^1(\Omega_{\varepsilon_k}) \subset H^1(\mathbb{R}^3)$. By equation (14) we have

$$\|w_k\|^2_{H^1(\mathbb{R}^3)} = \|u_k\|^2_{\varepsilon_k} \le C.$$

So $w_k \to w$ weakly in $H^1(\mathbb{R}^3)$ and strongly in $L^t_{\text{loc}}(\mathbb{R}^3)$.

We set $\psi(u_k)(x) := \psi_k(x) = \psi_k(\varepsilon_k z + q_k) := \tilde{\psi}_k(z)$ where $x \in \Omega$ and $z \in \Omega_{\varepsilon_k}$. It is easy to verify that

$$-\Delta_z\tilde{\psi}_k(z) = \varepsilon_k^2 q w_k^2(z).$$

With abuse of language we set

$$\tilde{\psi}_k(z) = \psi(\varepsilon_k w_k).$$

Thus

$$\begin{aligned} I_{\varepsilon_k}(u_k) &= \frac{1}{2}\|u_k\|^2_{\varepsilon_k} - \frac{1}{p}|u_k^+|^p_{\varepsilon_k,p} + \frac{\omega}{4}\frac{1}{\varepsilon_k^3}\int_\Omega q u_k^2\psi(u_k) \\ &= \frac{1}{2}\|w_k\|^2_{H^1(\mathbb{R}^3)} - \frac{1}{p}\|w_k^+\|^p_{L^p(\mathbb{R}^3)} + \frac{\omega}{4}\int_{\Omega_{\varepsilon_k}} q w_k^2\psi(\varepsilon_k w_k) \\ &= \frac{1}{2}\|w_k\|^2_{H^1(\mathbb{R}^3)} - \frac{1}{p}\|w_k^+\|^p_{L^p(\mathbb{R}^3)} + \varepsilon_k^2\frac{\omega}{4}\int_{\mathbb{R}^3} q w_k^2\psi(w_k) := E_{\varepsilon_k}(w_k). \end{aligned} \tag{16}$$

By definition of $E_{\varepsilon_k} : H^1(\mathbb{R}^3) \to \mathbb{R}$, we get $E_{\varepsilon_k}(w_k) \to m_\infty$.

Given any $\varphi \in C_0^\infty(\mathbb{R}^3)$ we set $\varphi(x) = \varphi(\varepsilon_k z + q_k) := \tilde{\varphi}_k(z)$. For k large enough we have that $\text{supp}\tilde{\varphi}_k \subset \Omega$ and, by (15), that $E'_{\varepsilon_k}(w_k)[\varphi] = I'_{\varepsilon_k}(u_k)[\tilde{\varphi}_k] \to 0$.

Moreover, by definition of E_{ε_k} and by Lemma 8 we have

$$E'_{\varepsilon_k}(w_k)[\varphi] = \langle w_k, \varphi\rangle_{H^1(\mathbb{R}^3)} - \int_{\mathbb{R}^3} |w_k^+|^{p-1}\varphi + \omega\varepsilon_k^2 \int_{\mathbb{R}^3} q w_k \psi(w_k)\varphi$$
$$\to \langle w, \varphi\rangle_{H^1(\mathbb{R}^3)} - \int_{\mathbb{R}^3} |w^+|^{p-1}\varphi.$$

Thus w is a weak solution of

$$-\Delta w + w = (w^+)^{p-1} \text{ on } \mathbb{R}^3.$$

By Lemma 18 and by the choice of q_k we have that $w \neq 0$, so $w > 0$.

Arguing as in (16), and using that $u_k \in \mathcal{N}_{\varepsilon_k}$ we have

$$I_{\varepsilon_k}(u_k) = \left(\frac{1}{2} - \frac{1}{p}\right) \|u_k\|^2_{\varepsilon_k} + \omega\left(\frac{1}{4} - \frac{1}{p}\right)\frac{1}{\varepsilon_k^3}\int_\Omega q u_k^2 \psi(u_k) \tag{17}$$
$$= \left(\frac{1}{2} - \frac{1}{p}\right) \|w_k\|^2_{H^1(\mathbb{R}^3)} + \varepsilon_k^2\omega\left(\frac{1}{4} - \frac{1}{p}\right)\int_{\mathbb{R}^3} q w_k^2 \psi(w_k) \to m_\infty$$

and

$$I_{\varepsilon_k}(u_k) = \left(\frac{1}{2} - \frac{1}{p}\right) |u_k^+|^p_{p,\varepsilon_k} - \frac{\omega}{4}\frac{1}{\varepsilon_k^3}\int_\Omega q u_k^2 \psi(u_k) \tag{18}$$
$$= \left(\frac{1}{2} - \frac{1}{p}\right) |w_k^+|^p_p - \varepsilon_k^2\frac{\omega}{4}\int_{\mathbb{R}^3} q w_k^2 \psi(w_k) \to m_\infty.$$

So, by (17) we have that $\|w\|^2_{H^1(\mathbb{R}^3)} = \frac{2p}{p-2} m_\infty$ and that $\left(\frac{1}{2} - \frac{1}{p}\right)\|w_k\|^2_{H^1(\mathbb{R}^3)} \to m_\infty$ and we conclude that $w_k \to w$ strongly in $H^1(\mathbb{R}^3)$.

Given $T > 0$, by the definition of w_k we get, for k large enough

$$|w_k^+|^p_{L^p(B(0,T))} = \frac{1}{\varepsilon_k^3}\int_{B(q_k,\varepsilon_k T)} |u_k^+|^p dx \le \frac{1}{\varepsilon_k^3}\int_{B(q_k, r/2)} |u_k^+|^p dx$$
$$\le (1-\eta)\frac{2p}{p-2} m_\infty. \tag{19}$$

Then we have the contradiction. In fact, by (18) we have $\left(\frac{1}{2} - \frac{1}{p}\right)|w_k^+|^p_p \to m_\infty$ and this contradicts (19). At this point we have proved the claim for $u \in \mathcal{N}_\varepsilon \cap I_\varepsilon^{m_\varepsilon + 2\delta}$. Now, by the conclusion for $u \in \mathcal{N}_\varepsilon \cap I_\varepsilon^{m_\varepsilon+2\delta}$ and by (18) we have

$$I_{\varepsilon_k}(u_k) = \left(\frac{1}{2} - \frac{1}{p}\right)|u_k^+|^p_{p,\varepsilon_k} + O(\varepsilon^2) \ge (1-\eta)m_\infty + O(\varepsilon^2)$$

and, passing to the limit,

$$\liminf_{k\to\infty} m_{\varepsilon_k} \ge m_\infty.$$

This, combined by (13) gives us that

$$\lim_{\varepsilon\to 0} m_\varepsilon = m_\infty. \tag{20}$$

Hence, when ε, δ are small enough, $\mathcal{N}_\varepsilon \cap I_\varepsilon^{m_\infty+\delta} \subset \mathcal{N}_\varepsilon \cap I_\varepsilon^{m_\varepsilon+2\delta}$ and the general claim follows. □

Proposition 20. *There exists* $\delta_0 \in (0, m_\infty)$ *such that for any* $\delta \in (0, \delta_0)$ *and any* $\varepsilon \in (0, \varepsilon(\delta_0)$ *(see Proposition* 16*), for every function* $u \in \mathcal{N}_\varepsilon \cap I_\varepsilon^{m_\infty+\delta}$ *it holds* $\beta(u) \in \Omega^+$*. Moreover the composition*

$$\beta \circ \Phi_\varepsilon : \Omega^- \to \Omega^+$$

is homotopic to the immersion $i : \Omega^- \to \Omega^+$

Proof. By Proposition 19, for any function $u \in \mathcal{N}_\varepsilon \cap I_\varepsilon^{m_\infty+\delta}$, for any $\eta \in (0,1)$ and for ε, δ small enough, we can find a point $q = q(u) \in \Omega$ such that

$$\frac{1}{\varepsilon^3} \int_{B(q,r/2)} (u^+)^p > (1-\eta) \frac{2p}{p-2} m_\infty.$$

Moreover, since $u \in \mathcal{N}_\varepsilon \cap I_\varepsilon^{m_\infty+\delta}$ we have

$$I_\varepsilon(u) = \left(\frac{p-2}{2p}\right) |u^+|_{p,\varepsilon}^p - \frac{\omega}{4} \frac{1}{\varepsilon^3} \int_\Omega q u^2 \psi(u) \le m_\infty + \delta.$$

Now, arguing as in Lemma 15 we have that

$$\|\psi(u)\|_{H^1(\Omega)}^2 = q \int_\Omega \psi(u) u^2 \le C \|\psi(u)\|_{H^1(\Omega)} \left(\int_\Omega u^{12/5}\right)^{5/6},$$

so $\|\psi(u)\|_{H^1(\Omega)} \le \left(\int_\Omega u^{12/5}\right)^{5/6}$, then

$$\begin{aligned}\frac{1}{\varepsilon^3} \int \psi(u) u^2 &\le \frac{1}{\varepsilon^3} \|\psi\|_{H^1(\Omega)} \left(\int_\Omega u^{12/5}\right)^{5/6} \le C \frac{1}{\varepsilon^3} \left(\int_\Omega u^{12/5}\right)^{5/3} \\ &\le C\varepsilon^2 |u|_{12/5,\varepsilon}^4 \le C\varepsilon^2 \|u\|_\varepsilon^4 \le C\varepsilon^2\end{aligned}$$

because $\|u\|_\varepsilon$ is bounded since $u \in \mathcal{N}_\varepsilon \cap I_\varepsilon^{m_\infty+\delta}$.

Hence, provided we choose $\varepsilon(\delta_0)$ small enough, we have

$$\left(\frac{p-2}{2p}\right) |u^+|_{p,\varepsilon}^p \le m_\infty + 2\delta_0.$$

So,

$$\frac{\frac{1}{\varepsilon^3} \int_{B(q,r/2)} (u^+)^p}{|u^+|_{p,\varepsilon}^p} > \frac{1-\eta}{1 + 2\delta_0/m_\infty}.$$

Finally,

$$\begin{aligned}|\beta(u) - q| &\le \frac{\left|\frac{1}{\varepsilon^3} \int_\Omega (x-q)(u^+)^p\right|}{|u^+|_{p,\varepsilon}^p} \\ &\le \frac{\left|\frac{1}{\varepsilon^3} \int_{B(q,r/2)} (x-q)(u^+)^p\right|}{|u^+|_{p,\varepsilon}^p} + \frac{\left|\frac{1}{\varepsilon^3} \int_{\Omega \setminus B(q,r/2)} (x-q)(u^+)^p\right|}{|u^+|_{p,\varepsilon}^p} \\ &\le \frac{r}{2} + 2\mathrm{diam}(\Omega) \left(1 - \frac{1-\eta}{1 + 2\delta_0/m_\infty}\right),\end{aligned}$$

so, choosing η, δ_0 and $\varepsilon(\delta_0)$ small enough we proved the first claim. The second claim is standard. □

References

[1] A. Ambrosetti, D. Ruiz, *Multiple bound states for the Schrödinger–Poisson problem*, Commun. Contemp. Math. **10** (2008), 391–404.

[2] A. Azzollini, P. D'Avenia, A. Pomponio, *On the Schrödinger–Maxwell equations under the effect of a general nonlinear term*, Ann. Inst. H. Poincaré Anal. Non Linéaire **27** (2010), no. 2, 779–791.

[3] A. Azzollini, A. Pomponio, *Ground state solutions for the nonlinear Schrödinger–Maxwell equations*, J. Math. Anal. Appl. **345** (2008), no. 1, 90–108.

[4] J. Bellazzini, L. Jeanjean, T. Luo, *Existence and instability of standing waves with prescribed norm for a class of Schrödinger–Poisson equations* in press on Proc. London Math. Soc. (arXiv http://arxiv.org/abs/1111.4668).

[5] V. Benci, *Introduction to Morse theory: A new approach*, in: Topological Nonlinear Analysis, in: Progr. Nonlinear Differential Equations Appl., vol. 15, Birkhäuser Boston, Boston, MA, 1995, pp. 37–177.

[6] V. Benci, C. Bonanno, A.M. Micheletti, *On the multiplicity of solutions of a nonlinear elliptic problem on Riemannian manifolds*, Journal of Functional Analysis **252** (2007), 464–489 37–177.

[7] V. Benci, G. Cerami, *The effect of the domain topology on the number of positive solutions of nonlinear elliptic problems*, Arch. Ration. Mech. Anal. **114** (1991), 79–93.

[8] V. Benci, G. Cerami, *Multiple positive solutions of some elliptic problems via the Morse theory and the domain topology*, Calc. Var. Partial Differential Equations **2** (1994), 29–48.

[9] V. Benci, D. Fortunato, *An eigenvalue problem for the Schrödinger–Maxwell equations*, Topol. Methods Nonlinear Anal. **11** (1998), no. 2, 283–293.

[10] T. D'Aprile, D. Mugnai, *Solitary waves for nonlinear Klein–Gordon–Maxwell and Schrödinger–Maxwell equations*, Proc. Roy. Soc. Edinburgh Sect. A **134** (2004), no. 5, 893–906.

[11] T. D'Aprile, J. Wei, *Layered solutions for a semilinear elliptic system in a ball*, J. Differential Equations **226** (2006), no. 1, 269–294.

[12] T. D'Aprile, J. Wei, *Clustered solutions around harmonic centers to a coupled elliptic system*, Ann. Inst. H. Poincaré Anal. Non Linéaire **24** (2007), no. 4, 605–628.

[13] M. Ghimenti, A.M. Micheletti, *Number and profile of low energy solutions for singularly perturbed Klein–Gordon–Maxwell systems on a Riemannian manifold*, work in preparation.

[14] I. Ianni, G. Vaira, *On concentration of positive bound states for the Schrödinger–Poisson problem with potentials*, Adv. Nonlinear Stud. **8** (2008), no. 3, 573–595.

[15] Kikuchi, *On the existence of solutions for a elliptic system related to the Maxwell–Schrödinger equations*, Non-linear Anal. **67** (2007), 1445–1456.

[16] R.S. Palais, *Homotopy theory of infinite-dimensional manifolds*, Topology **5** (1966), 1–16.

[17] L. Pisani, G. Siciliano, *Note on a Schrödinger–Poisson system in a bounded domain*, Appl. Math. Lett. **21** (2008), no. 5, 521–528.

[18] D. Ruiz, *Semiclassical states for coupled Schrödinger–Maxwell equations*: Concentration around a sphere, Math. Models Methods Appl. Sci. **15** (2005), no. 1, 141–164.

[19] G. Siciliano, *Multiple positive solutions for a Schrödinger–Poisson–Slater system*, J. Math. Anal. Appl. **365** (2010), no. 1, 288–299.

[20] Z. Wang, H.S. Zhou, *Positive solution for a nonlinear stationary Schrödinger–Poisson system in* $\mathbb{R}^3$, Discrete Contin. Dyn. Syst. **18** (2007), 809–816.

Marco Ghimenti and Anna Maria Micheletti
Dipartimento di Matematica
Università di Pisa
via F. Buonarroti 1/c
I-56127 Pisa, Italy
e-mail: marco.ghimenti@dma.unipi.it
a.micheletti@dma.unipi.it

Progress in Nonlinear Differential Equations
and Their Applications, Vol. 85, 301–307

An Abstract Theorem in Nonlinear Analysis

A.C. Lazer and P.J. McKenna

Abstract. An elementary proof of the existence of multiple solutions of nonlinear operator equations is given. We show the existence, depending on a parameter in the equation, of either exactly two or at least four solutions for these equations.

Mathematics Subject Classification (2010). 35J60, (35J25, 47J30, 58E05).

Keywords. Nonlinear, boundary values, operator theory, multiplicity.

An abstract result

This paper is based on a long series of results on the nonlinear boundary value problem

$$\begin{aligned} \Delta u + bu^+ &= \phi_1(x) \quad && x \in \Omega \\ u &= 0 && x \in \partial\Omega \end{aligned} \tag{1}$$

where Ω is a bounded region in $\mathbb{R}^n$ and ϕ_1 denoted the first eigenfunction of the Laplacian with Dirichlet boundary conditions. The real number λ_1 is the first of the infinite sequence of eigenvalues $\lambda_n \to +\infty$. The nonlinearity bu^+ is a model nonlinearity for a nonlinearity of the form $f(u)$ where $f'(+\infty) = b$ and $f'(-\infty) = 0$. More generally, we can take $f'(-\infty) = a$ if we replace bu^+ by $bu^+ - au^-$.

Since $\phi_1(x) \geq 0$, it is an elementary calculation that equation (1) has two obvious solutions if $\lambda_1 < b < +\infty$. These are given by $u_1 = -\phi_1/\lambda_1 (\leq 0)$ and $u_2 = \phi_1/(b - \lambda_1)(\geq 0)$.

Generally, the literature can be summarized by the following two statements: if the interaction of the nonlinearity with the spectrum of the Laplacian is small, by which we mean $\lambda_1 < b < \lambda_2$, then the two "obvious" solutions are the only ones. One the other hand if there is more interaction with the spectrum, that is, if $b > \lambda_2$, then more "non-obvious" solutions appear. Let us make this more precise.

For the rest of this section, we will be working in a closed subspace H of the Hilbert space $L^2(\Omega)$, where Ω is a bounded region in $\mathbb{R}^n$. The unbounded selfadjoint linear operator $L : D(L) \to H$ will satisfy the following **hypotheses**

1. L has an infinite sequence of eigenvalues l_n with $0 < l_1 < l_2 < l_3 \leq l_4 \cdots \leq l_n \cdots$ with the convention that $L\phi_n + l_n\phi_n = 0$,
2. The associated eigenfunctions ϕ_n are an orthonormal basis for H with $\phi_1(x) > 0$ for all $x \in \Omega$.
3. There exists $\epsilon > 0$ such that $\phi_1(x) > \epsilon\|\phi_2(x)\|, \forall x \in \Omega$.

Typically, the space H will either be all of $L^2(\Omega)$, or a subspace defined by certain symmetries. We will also assume that the map $u \to bu^+$ leaves H invariant, which will usually be satisfied if the symmetries are *even*.

The reader will immediately notice that the Laplacian, with Dirichlet boundary conditions of equation (1) satisfies these hypotheses. Of course, they apply to a much wider class than just this operator.

The reader will also notice, after following the proof, that the operator L could be allowed to have continuous spectrum, so long as that spectrum is contained in the interval $(l_3, +\infty)$

We shall now study the operator equation in H:

$$Lu + bu^+ = s\phi_1. \tag{2}$$

Most of our results would also apply to more general equations with the nonlinearity replaced by $f(u)$ with appropriate assumptions on f (see Section 1) and the right-hand side replaced by $s\phi_1$ with s sufficiently large.

Our two main theorems are

Theorem 1. *If $l_1 < b < l_2$, equation* (3) *has exactly two solutions for $s > 0$, the zero solution for $s = 0$, and no solutions for $s < 0$.*

Theorem 2. *If $l_2 < b < l_3$, equation* (3) *has at least four solutions for $s > 0$, the zero solution for $s = 0$, and no solutions for $s < 0$.*

We start with the easy part.

Lemma 1. *If $b > l_1$, the equations*

$$Lu + bu^+ = s\phi_1 \tag{3}$$

has no solutions for $s < 0$ and only $u \equiv 0$ for $s = 0$.

Proof. Suppose $Lu + bu^+ = s\phi_1$. Re-write this as $(L - l_1 I)u + bu^+ - l_1 u = \phi_1$. Now take the inner product with ϕ_1, and observe that since $< (L - l_1 I)u, \phi_1 >= 0$ and for suitable $\alpha > 0$, we have $bu^+ - l_1 u \geq \alpha \mid u \mid$, we conclude that $\int_\Omega \alpha \mid u \mid \phi_1 dV \leq s \int_\Omega \phi_1^2 dV$. Since $\phi_1 > 0$, this immediately implies that for $s < 0$, we have a contradiction, and for $s = 0$, we have $u \equiv 0$. □

Proof of Theorem 1. We shall use the classical method of Lyapunov–Schmidt. Write $H = V \oplus W$ where V is the span of ϕ_1 and $W = V^\perp$. Let P be the orthogonal projection on V. Write $u = v + w$, where $v = Pu$ and $w = (I - P)u$. Then, equation (3) is equivalent to the system of equations

$$Lw + (I - P)b(v + w)^+ = 0, \tag{4}$$

$$Lv + P(v + w)^+ = s\phi_1, \tag{5}$$

or equivalently,

$$w = -L^{-1}(I - P)b(v + w)^+, \tag{6}$$

$$-l_1 v + Pb(v + w)^+ = s\phi_1. \tag{7}$$

Now consider equation (7). If we fix $v \in V$, then a solution of equation (7) is a fixed point of the map $T_v(w) = -L^{-1}(I - P)b(v + w)^+$, with $T_v : W \to W$. Since the operator norm $\| (L^{-1}(I - P) \| = 1/\lambda_2$, it follows that $T_v : W \to W$ is a contraction and therefore for each $v \in V$, there exists a unique $w(v) \in W$ that satisfies (7). Furthermore, $w(v)$ depends in Lipschitz-continuously on v. Thus the system (7) and (8) is equivalent to the single one-dimensional equation

$$-l_1 v + Pb(v + w(v))^+ = s\phi_1. \tag{8}$$

Finally, we remark that since v is a multiple of ϕ_1, then v is satisfies $v \geq 0$ or $v \leq 0$. In either case, we can verify that $w(v) \equiv 0$ satisfies equation (7) since if $v \leq 0$, then $v^+ = 0$ and if $v \geq 0$ then $v^+ = v$ and then $(I - P)v = 0$. Thus, letting $v = c\phi_1$, equation (9) becomes $-cl_1\phi_1 = s\phi_1$ if $c \leq 0$ or $-cl_1\phi_1 + bc\phi_1 = s\phi_1$ if $c \geq 0$. Thus equation (3) has exactly two solutions for $s > 0$; $u_1 = -s\phi_1/l_1$ and $u_2 = s\phi_1/(b - l_1)$. □

Proof of Theorem 2. First a remark. If

$$Lu + bu^+ = s\phi_1$$

has a solution for $s = s_1 > 0$, then it has a solution for $s = s_2$, for any other $s_2 > 0$. (Just multiply equation across by s_2/s_1, using homogeneity under positive multiplication of u^+.)

This proof begins in a similar fashion to that of Theorem 1. Write $H = V \oplus W$ where V is span$\{\phi_1, \phi_2\}$ and $W = V^\perp$. Let P be the orthogonal projection on V. Write $u = v + w$, where $v = Pu$ and $w = (I - P)u$. Then, equation (3) is equivalent to the system of equations

$$Lw + (I - P)b(v + w)^+ = 0, \tag{9}$$

$$Lv + Pb(v + w)^+ = s\phi_1, \tag{10}$$

or equivalently,

$$w = -L^{-1}(I - P)b(v + w)^+, \tag{11}$$

$$Lv + Pb(v + w)^+ = s\phi_1. \tag{12}$$

Now consider equation (7). If we fix $v \in V$, then a solution of equation (7) is a fixed point of the map $T_v = -L^{-1}(I - P)b(v + w)^+$, with $T_v : W \to W$. Since the operator norm $\| L^{-1}(I - P) \| = 1/\lambda_3$, it follows that $T_v : W \to W$ is a contraction and therefore for each $v \in V$, there exists a unique $w(v) \in W$ that satisfies (7). Furthermore, $w(v)$ depends in Lipschitz-continuously on v. Thus the system (11) and (12) is equivalent to the single two-dimensional equation on V;

$$Lv + Pb(v + w(v))^+ = s\phi_1. \tag{13}$$

The difference here is that since $v \in V$, the two-dimensional space spanned by ϕ_1 and ϕ_2, we cannot expect v to be of one sign. So the implicitly defined $w(v)$ will play an important role in our analysis of the two-dimensional equation (14). To analyse equation (14) more carefully, let us write $v = c_1\phi_1 + c_2\phi_2$. Then equation (14) can be written

$$-l_1c_1\phi_1 - l_2c_2\phi_2 + Pb(c_1\phi_1 + c_2\phi_2 + w(c_1\phi_1 + c_2\phi_2))^+ = s\phi_1. \qquad (14)$$

Now let us divide the plane V into four cones. Let $C_1 = \{c_1\phi_1 + c_2\phi_2, c_1 \geq 0, \mid c_2 \mid \leq \epsilon c_1\}$, $C_2 = \{c_1\phi_1 + c_2\phi_2, c_2 > 0, c_2 \geq \epsilon \mid c_1 \mid\}$, $C_3 = \{c_1\phi_1 + c_2\phi_2, c_1 \leq 0, \mid c_2 \mid \leq -\epsilon c_1\}$, and $C_4 = \{c_1\phi_1 + c_2\phi_2, c_2 \leq 0, c_2 \leq -\epsilon \mid c_1 \mid\}$.

Thus, we have divided V into four cones, the first, C_1 a narrow cone centered on the positive ϕ_1-axis, C_2, a relatively broad cone centered on the positive ϕ_2-axis, C_3, a narrow cone centered on the negative ϕ_1-axis, and C_4, a relatively broad cone centered on the negative ϕ_2-axis.

Furthermore, by hypothesis (3) at the beginning of this section, if $u \in C_1$, then $u \geq 0$ and if $u \in C_3$, then $u \leq 0$. Therefore, in C_1 and C_3, by our previous reasoning, $w(c_1\phi_1 + c_2\phi_2) \equiv 0$ and so equation (18) becomes linear. For example, in C_1, equation (15) becomes

$$-l_1c_1\phi_1 - l_2c_2\phi_2 + b(c_1\phi_1 + c_2\phi_2) = s\phi_1 \qquad (15)$$

with a linear diagonal operator on C_1 with positive entries on the diagonal on the left hand side. Thus, one can see that the cone C_1 contains the positive solution v_2 already described in Theorem 2. Similarly, the cone C_3, contains the negative solution v_1.

Now, we need to show that C_2 and C_4 also contain solutions. Let us define a line segment in C_2 by

$$L_2 = \{c_1\phi_1 + c_2\phi_2, -1 \leq c_1 \leq 1, C_2 = \epsilon\} \qquad (16)$$

and consider its image under the map $Tv = Lv + Pb(v + w(v))^+$. One endpoint of L_2 is $-\phi_1 + \epsilon\phi_2$ which is also in C_3. Thus, its image under T is $P = l_1\phi_1 - l_2\epsilon\phi_2$. Similarly, the other endpoint of L_2 is $\phi_1 + \epsilon\phi_2$ and its image under T is $Q = (b - l_1)\phi_1 + (b - l_2)\epsilon\phi_2$. Therefore the continuous curve $T(L_2)$ begins in the lower half-plane of V ($c_2 < 0$) and ends in the upper half-plane of V. Therefore it must cross the ϕ_1-axis at some point s_1, thereby giving a solution of equation (14) (and therefore (3)) for $s = s_1$. By Lemma 1, s_1 must be positive. By multiplication by s_2/s_1, equation (3) has a solution in C_2 for any $s_2 > 0$.

The proof that there is a solution in C_4 is similar. Thus we can conclude that equation(3) has at least four distinct solutions for $l_2 < b < l_3$.

Now consider equation (7). if we fix $v \in V$, then a solution of equation (7) is a fixed point of the map $T_v = -L^{-1}(I - P)b(v + w)^+$, with $T_v : W \to W$. Since the operator norm $\| (L^{-1}(I - P) \| = 1/\lambda_3$, it follows that $T_v : W \to W$ is a contraction and therefore for each $v \in V$, there exists a unique $w(v) \in W$ that satisfies (7). Furthermore, $w(v)$ depends Lipschitz-continuously on v. Thus the system (7) and (8) is equivalent to the single two-dimensional equation on V;

$$Lv + Pb(v + w(v))^+ = s\phi_1. \tag{17}$$

□

The difference here is that since $v \in V$, the two dimensional space spanned by ϕ_1 and ϕ_2, we cannot expect v to be of one sign. So the implicitly defined $w(v)$ will play an important role in our analysis of the two-dimensional equation (13). To analyse equation more carefully, let us write $v = c_1\phi_1 + c_2\phi_2$. Then equation (13) can be written

$$-l_1c_1\phi_1 - l_2c_2\phi_2 + Pb(c_1\phi_1 + c_2\phi_2 + w(c_1\phi_1 + c_2\phi_2))^+ = s\phi_1. \tag{18}$$

Remark 1. More is known in the case of L the Laplacian with Dirichlet bounday conditions. In particular, in [34], the existence of *exactly* four solutions is know. There, it is also proved that there exists an $\epsilon > 0$ such that if $l_3 < b < l_3 + \epsilon$ then at least six solutions exist. In these proofs, essential use is made of the maximum principle and eigenvalue comparison theorem. At the level, of generality of this paper, these techniques do not apply. This naturally leads to a series of questions.

Question 1. *Can one prove that in Theorem* 2*, there are* exactly *four solutions?*

Question 2. *Can one say anything along the lines of Theorem* 2 *when* $b > l_3$*?*

(In this case, the base space V would be more than two dimensional.) Can one still prove the existence of at least four solutions in this generality? (Again, this is known in the case of the Laplacian, [19], also making use of the maximum principle.)

Question 3. *What are the correct general conditions for the right hand side for equation* (3)*?*

It is easy to think of other right-hand sides ψ for which Theorem 2 might work. The main ingredients might be $\psi > 0$, and $Lu + bu = \psi > 0$ implies $u > 0$, with $\psi \perp \phi_2$.

Question 4. *Can one say more about the global structure of the natures of the map in terms of folds and cusps? Interesting results occur in* [8].

References

[1] Ambrosetti, A. and Prodi, G., On the inversion of some differentiable mappings with singularities between Banach Spaces, *Ann. Math. Pura Appl.* 93 (1973), 231–247

[2] Amann, H. and Hess, P., A multiplicity result for a class of elliptic boundary value problems, *Proc Roy. Soc. Edinb.* 84 (1979), 145–151.

[3] Berger, M. and Church, P., An application of singularity theory to nonlinear partial differential equations, *Bull. A.M.S.* 178 (1983) 119–126.

[4] Berger, M.S. and Podolak, E., On the solutions of nonlinear Dirichlet problems, *Indiana Univ. Math.Jour.* 24 (1975) 837–846.

[5] Breuer, B., McKenna, P.J., Plum, M., Multiple solutions for a semilinear boundary value problem: a computational multiplicity proof. J. Differential Equations 195 (2003), 243–269.

[6] Brezis, H. and Nirenberg, L., Characterization of the ranges of some nonlinear operators and applications to boundary value problems, *Ann. Scuola Norm. Sup. Pisa, Cl. Sci.,*, 5 (1978) 225–326.

[7] Cafagna, V. and Tarantello, G., Multiple solutions for some semilinear elliptic equations, *Math. Ann.* 276 (1987), 643–656.

[8] Cal Neto, José Teixeira and Tomei, Carlos, Numerical analysis of semilinear elliptic equations with finite spectral interaction, *J. Math. Anal. Appl.* 395 (2012) 63–77.

[9] Castro, A. and Gadam, S., The Lazer McKenna conjecture for radial solutions in the $\mathbf{R}^N$ ball.*Electron. J. Differential Equations* 1993, approx. 6 pp.

[10] Castro, A., and Shivaji, R., Multiple solutions for a Dirichlet problem with jumping nonlinearities, II *Jour. Math. Anal. Appl.,* 133 (1988), 509–528.

[11] Costa, D.G., de Figueiredo, D.G., Srikanth, P.N., The exact number of solutions for a class of ordinary differential equations through Morse index computation. *J. Differential Equations* 96 (1992), no. 1, 185–199.

[12] Dancer, E.N., On the ranges of certain weakly nonlinear elliptic partial differential equations, *J. Math. Pures et Appl.* 57 (1978), 351–366.

[13] Dalbono, Francesca and McKenna, P.J., Multiplicity results for a class of asymmetric weakly coupled systems of second-order differential equations, *Boundary Value Problems* 2 (2005) 129–151.

[14] Dancer, E.N., A counterexample to the Lazer–McKenna conjecture, *Nonlinear Analysis, T.M.A.* 13 (1989) 19–22.

[15] Dancer, E.N. and Yan, S., On the superlinear Lazer–McKenna conjecture, *J. Differential Equations* 210 (2005), 317–351.

[16] Dancer, E.N. and Yan, S.S., On the superlinear Lazer–McKenna conjecture: Part II *Comm. P.D.E.* 30 (2005) 1331–1358.

[17] Gallouet, T., and Kavian, O., Resultats d'existence et de non existence pour certain problems demilinearies a l'infini, *Ann. Fac. Sci. Toulouse,* 3 (1981), 201–246.

[18] Hammerstein, A., Nichtlineare Integralgleichungen nebst Anwendungen, *Acta Mat.* 54 (1930), 117–176.

[19] Hofer, H., Variational and topological methods in partially ordered Hilbert spaces, *Math Ann.,* (1982), 493–514.

[20] Kazdan, J.L. and Warner, F., Remarks on quasilinear elliptic equations., *Comm. Pure and Appl. Math.,* 28 (1975), 567–597.

[21] Lazer, A.C. and McKenna, P.J., On the number of solutions of a nonlinear Dirichlet problem, *Jour. Math. Anal. Appl.,* 84 (1981), 282–294.

[22] Lazer, A.C. and McKenna, P.J., Some multiplicity results for a class of semilinear elliptic and parabolic boundary value problems, *J. Math. Anal. Appl.*, 107 (1985), 371–395.

[23] Lazer, A.C. and McKenna, P.J., On a conjecture related to the number of solutions of a nonlinear Dirichlet problem, *Proc Roy. Soc. Edin.*, 95 A (1983), 275–283.

[24] Lazer, A.C. and McKenna, P.J., Critical point theory and boundary value problems with nonlinearities crossing multiple eigenvalues II, *Comm. P.D.E.*, 11 (1986), 1653–1676.

[25] Lazer, A.C., and McKenna, P.J., A symmetry theorem and applications to nonlinear partial differential equations, *J. Diff. Equations*, 71 (1988).

[26] Lazer, A.C. and McKenna, P.J., Large-amplitude periodic oscillations in suspension bridges: some new connections with nonlinear analysis, *SIAM Rev.* **32** (1990), 537–578.

[27] Lupo, D., Solimini, S., Srikanth, P.N., Multiplicity results for an ODE problem with even nonlinearity. *Nonlinear Anal. T.M.A.* 12 (1988), 657–673.

[28] Manes, A., and Micheletti, A.M., Un'estensione della teoria variazionale classica degli autovalori per operatori ellittici del secondo ordine. Boll. Un. Mat. Ital. 7 (1973), 285–301.

[29] Molle, Riccardo and Passaseo, Donato, Existence and multiplicity of solutions for elliptic equations with jumping nonlinearities, *J. Funct. Anal.*, 259, (2010), 2253–2295.

[30] Nirenberg, L., Variational and topological methods in nonlinear problems, *Bull. Amer. Math. Soc.*, 4 (1981), 267–302.

[31] Ruf, B., Remarks and generalizations related to a recent multiplicity result of A. Lazer and P. McKenna, *Nonlinear Anal. T.M.A.* 9, (1985), 1325–1330.

[32] Ruf, B., On nonlinear elliptic problems with jumping nonlinearities, *Annali di Matematica pura ed applicata* 78, (1980), 133–151.

[33] Ruf, B. and Solimini, S. On a class of superlinear Sturm–Liouville problems with arbitrarily many solutions. *SIAM J. Math. Anal.*, 17 (1986), 761–77.

[34] Solimini, S., Some remarks on the number of solutions of some nonlinear elliptic equations, *Anal. Nonlin., I.H.P.*, 2 (1985), 143–156.

A.C. Lazer
Department of Mathematics
University of Miami
Coral Gables, FL 33124, USA

P.J. McKenna
Department of Mathematics
University of Connecticut
Storrs, CT 06269, USA

and

Department of Mathematics and Statistics
King Fahd University of Petroleum and Minerals
Dhahran 31261, Saudi Arabia

Progress in Nonlinear Differential Equations
and Their Applications, Vol. 85, 309–321

Existence and Nonexistence Results for a Schrödinger Equation with Saturable Nonlinearity

Raquel Lehrer and Liliane A. Maia

Abstract. We present some recent results on the existence of positive solution for the equation

$$-\Delta u + \lambda u = a(x)\frac{u^3}{1+u^2}.$$

Using concentration compactness arguments and a general Pohozaev manifold $\mathcal{P}$, we find a bound state solution via a linking theorem. Moreover, we show that a minimizing problem, related to the existence of a ground state, has no solution.

Mathematics Subject Classification (2010). 35J20, 35J60, 35Q55.

Keywords. Asymptotically linear, Pohozaev identity, concentration compactness, Cerami sequence, barycenter.

1. Introduction and main result

In this paper we study the following Schrödinger equation:

$$-\Delta u + \lambda u = a(x)\frac{u^3}{1+u^2} \quad \text{in} \quad \mathbb{R}^N, \tag{1.1}$$

for $N \geq 3$, $\lambda > 0$ and $a(x) > 0$. We observe that the function $f(s) = \dfrac{s^3}{1+s^2}$ is asymptotically linear at infinity and is include in the class of functions considered in the work [8] of D. Costa and H. Tehrani. In their work, they used concentration compactness arguments together with comparison between energy levels in order to obtain the strong convergence of the Cerami sequence and apply the Mountain Pass Theorem to obtain a positive solution. In that case, the functional I, associated

The first author is supported by CCET/UNIOESTE, while the second author is supported by CNPq/ PQ 306388/2011-1 and PROEX/CAPES.

with equation (1.1) and the functional I_∞ associated with the limit problem, satisfy the relation $I(u) < I_\infty(u), \forall u \in H^1(\mathbb{R}^N) \setminus \{0\}$. In our study, we impose conditions on the function a such that these functionals now have the reverse relation $I_\infty < I$, and the comparison of the energy levels is not a useful tool any more.

In the asymptotically linear case at infinity it is natural to use instead of the traditional Palais–Smale condition, the Cerami condition [6] (see [7, 4] among many others):

(Ce) the functional I satisfies the Cerami condition if, for any sequence (u_n) in $H^1(\mathbb{R}^N)$ such that $I(u_n)$ is bounded and $\|I'(u_n)\|(1+\|u_n\|) \to 0$, then there exists a convergent subsequence .

We will assume the following conditions on the function a:

(A1) $a \in C^2(\mathbb{R}^N, \mathbb{R}^+)$, with $\inf_{x\in\mathbb{R}^N} a(x) > 0$;

(A2) $\lim_{|x|\to\infty} a(x) = a_\infty > \lambda$;

(A3) $\nabla a(x) \cdot x \geq 0$, for all $x \in \mathbb{R}^N$, with the strict inequality holding on a subset of positive Lebesgue measure of $\mathbb{R}^N$;

(A4) $a(x) + \dfrac{\nabla a(x) \cdot x}{N} < a_\infty$, for all $x \in \mathbb{R}^N$;

(A5) $\nabla a(x) \cdot x + \dfrac{x \cdot H(x) \cdot x}{N} \geq 0$, for all $x \in \mathbb{R}^N$, where H represents the Hessian matrix of the function a.

Later in this paper we will also assume yet another condition (A6), requiring that the supremum of $|a_\infty - a(x)|$ is not large; (see Lemma 3.11).

Remark 1.1. A model function for a is given by $a(x) = a_\infty - \dfrac{1}{|x|+k}$ with $k > \dfrac{1}{a_\infty}$.

The functional associated to (1.1) is given by

$$I(u) = \frac{1}{2}\int_{\mathbb{R}^N} |\nabla u|^2 + \lambda u^2 dx - \int_{\mathbb{R}^N} a(x)F(u)dx,$$

where

$$F(u) = \int_0^u \frac{s^3}{1+s^2} ds = \frac{u^2}{2} - \frac{1}{2}\ln(1+u^2).$$

Remark 1.2. As proved in [12], the function $F(u)$ satisfies the non quadraticity condition (NQ), i.e.,

$$\lim_{|u|\to\infty} \left(\frac{1}{2}f(u)u - F(u)\right) = +\infty \text{ and } \left(\frac{1}{2}f(u)u - F(u)\right) > 0;\ \forall\, u \in \mathbb{R}\setminus\{0\}.$$

Our main existence result is the following:

Theorem 1.3. *Assume* (A1–A6). *Then equation* (1.1) *has a positive solution* u *in* $H^1(\mathbb{R}^N)$.

Furthermore, we will also prove the following non existence theorem:

Theorem 1.4. *Assume* (A1–A4). *Then,* $p = \inf_{u \in \mathcal{P}} I(u)$ *is not a critical level for the functional* I*, with the Pohozaev manifold* $\mathcal{P}$ *defined by* (2.2). *In particular, the infimum* p *is not achieved.*

Remark 1.5. Since we are looking for positive solutions, as in [8], we take $f(s)$ defined on all $s \in \mathbb{R}$, by making $f(s) = 0$ if $s \leq 0$. Thus, since the critical points of the functional associated with equation (1.1) are weak solutions of the equation, if u is a critical point of I then

$$\begin{aligned} 0 = I'(u)u^- &= \int_{\mathbb{R}^N} \nabla u \nabla u^- + \lambda u u^- dx - \int_{\mathbb{R}^N} a(x) f(u) u^- dx \\ &= \int_{\mathbb{R}^N} |\nabla u^-|^2 + \lambda |u^-|^2 dx = \|u^-\|_\lambda^2, \end{aligned}$$

where $u^- := \min\{u, 0\}$. Hence, necessarily we have $u \geq 0$; here we use the norm $\|u\|_\lambda^2 := \int_{\mathbb{R}^N} |\nabla u|^2 + \lambda u^2 \, dx$ in $H^1(\mathbb{R}^N)$.

Moreover, as in [12], we have that given $\varepsilon > 0$, there exists $C = C(\varepsilon) > 0$ such that

$$|F(s)| \leq \frac{\varepsilon}{2}|s|^2 + C|s|^p, 2 < p < 2^* \, . \tag{1.2}$$

With this estimative and condition (A1), we can prove that the functional I satisfies the geometric conditions on the Mountain Pass Theorem.

2. Non existence result

In [10], Proposition 2.1, we proved that any solution of (1.1) satisfies the Pohozaev identity:

$$\frac{N-2}{2} \int_{\mathbb{R}^N} |\nabla u|^2 dx = N \int_{\mathbb{R}^N} G(x,u) dx + \int_{\mathbb{R}^N} \nabla a(x) \cdot x F(u) dx \tag{2.1}$$

where $G(x,u) = a(x)F(u) - \lambda \frac{u^2}{2}$. We can also define the Pohozaev manifold by

$$\mathcal{P} := \left\{ u \in H^1(\mathbb{R}^N) \setminus \{0\} \, ; J(u) = 0 \right\} \, , \tag{2.2}$$

where

$$J(u) = \frac{N-2}{2} \int_{\mathbb{R}^N} |\nabla u|^2 dx - N \int_{\mathbb{R}^N} G(x,u) dx - \int_{\mathbb{R}^N} \nabla a(x) \cdot x F(u) dx.$$

Some properties of this manifold can be found in [10], Lemma 2.2.

We now begin presenting some relations between the Pohozaev manifold $\mathcal{P}$ associated with the non-autonomous problem (1.1), and the Pohozaev manifold $\mathcal{P}_\infty$ associated with the autonomous problem at infinity

$$-\Delta u + \lambda u = a_\infty \frac{u^3}{1+u^2} \quad \text{in} \quad \mathbb{R}^N. \tag{2.3}$$

The Pohozaev identity for this equation is given by

$$\frac{N-2}{2}\int_{\mathbb{R}^N}|\nabla u|^2dx = N\int_{\mathbb{R}^N}G_\infty(u)dx \tag{2.4}$$

with $G_\infty(u) := a_\infty F(u) - \lambda\frac{u^2}{2}$, and the Pohozaev manifold in given by

$$\mathcal{P}_\infty = \left\{u \in H^1(\mathbb{R}^N)\setminus\{0\}\,;u \text{ satisfies } (2.4)\right\}.$$

We also need to consider the functional I_∞ associated with (2.3) and given by

$$I_\infty(u) = \frac{1}{2}\int_{\mathbb{R}^N}|\nabla u|^2 + \lambda u^2dx - \int_{\mathbb{R}^N}a_\infty F(u)dx$$

and the set of paths $\Gamma_\infty = \left\{\gamma \in C([0,1],H^1(\mathbb{R}^N))|\gamma(0)=0, I_\infty(\gamma(1))<0\right\}$. In this way, we can define the min-max mountain pass level (see [2])

$$c_\infty := \min_{\gamma\in\Gamma_\infty}\max_{0\le t\le 1} I_\infty(\gamma(t)).$$

Note that the hypotheses (A3) and (A4) imply that $I_\infty(u) < I(u)$ for all u in $H^1(\mathbb{R}^N)$. Inspired by the work [3] of A. Azzolini and A. Pomponio, we will show by the end of this section that

$$p := \inf_{u\in\mathcal{P}} I(u) = c_\infty,$$

and that this level is not achieved, i.e., this is not a critical level for the functional I.

Lemma 2.1. *Suppose that* $\int_{\mathbb{R}^N}G_\infty(u)dx > 0$. *Then there exist unique* $\theta_1 > 0$ *and* $\theta_2 > 0$ *such that* $u(x/\theta_1) \in \mathcal{P}$ *and* $u(x/\theta_2)\in\mathcal{P}_\infty$. *Moreover, let*

$$\mathcal{O} = \left\{u \in H^1(\mathbb{R}^N)\setminus\{0\}\,;\int_{\mathbb{R}^N}G_\infty(u)dx > 0\right\}$$

be an open subset of $H^1(\mathbb{R}^N)$. *The function* $\theta_1 : \mathcal{O}\to\mathbb{R}^+$ *defined by* $u\mapsto\theta_1(u)$, *such that* $u(x/\theta_1(u))\in\mathcal{P}$, *is continuous.*

Proof. The case of projecting on $\mathcal{P}_\infty$ can be found in Lemma 3.1 of [12]. For the case of $\mathcal{P}$ and the continuity of the function θ_1, we refer to [10], Lemmas 3.1 and 3.3. □

Lemma 2.2. *If* $u\in\mathcal{P}_\infty$, *then there exists* $\theta > 0$ *such that* $u(\cdot/\theta)\in\mathcal{P}$ *and* $\theta > 1$.

Proof. If $u \in \mathcal{P}_\infty$, then $\int_{\mathbb{R}^N}G_\infty(u)dx > 0$ and Lemma 2.1 asserts the existence of a unique θ such that $u(./\theta)\in\mathcal{P}$. Now, considering the expression $J(u(./\theta)) = 0$ and recalling that $\theta > 0$ we obtain

$$\frac{N-2}{2N}\int_{\mathbb{R}^N}|\nabla u|^2dx = \theta^2\int_{\mathbb{R}^N}\left\{a(\theta x) + \frac{\nabla a(\theta x)\cdot(\theta x)}{N}\right\}F(u) - \lambda\frac{u^2}{2}dx.$$

By condition (A4) we get $\frac{N-2}{2N}\int_{\mathbb{R}^N}|\nabla u|^2dx < \theta^2\int_{\mathbb{R}^N}G_\infty(u)dx$. Since $u\in\mathcal{P}_\infty$, the inequality above is true if and only if $\theta > 1$. □

Lemma 2.3. *If $u \in \mathcal{P}$, then there exists $\theta > 0$ such that $u(\cdot/\theta) \in \mathcal{P}_\infty$ and $\theta < 1$.*

Proof. First, by (A4) we have that if $u \in \mathcal{P}$ then

$$\frac{N-2}{2N}\int_{\mathbb{R}^N} |\nabla u|^2 dx < \int_{\mathbb{R}^N} G_\infty(u)dx,$$

and since $\int_{\mathbb{R}^N} |\nabla u|^2 dx > 0$, we have by Lemma 2.1 the existence of θ. Also, if $u(./\theta) \in \mathcal{P}_\infty$, then θ satisfies

$$\theta^2 = \frac{(N-2)\int_{\mathbb{R}^N} |\nabla u|^2 dx}{2N\int_{\mathbb{R}^N} G_\infty(u)dx} < \frac{\int_{\mathbb{R}^N} G_\infty(u)dx}{\int_{\mathbb{R}^N} G_\infty(u)dx} = 1.$$

Therefore $\theta < 1$ and the lemma is proved. □

The proofs of the following three lemmas can be found in [10], Lemmas 3.7, 3.8 and 3.9 respectively.

Lemma 2.4. *If $u \in \mathcal{P}_\infty$, then $u(\cdot - y) \in \mathcal{P}_\infty$, for all $y \in \mathbb{R}^N$. Moreover, there exists $\theta_y > 1$ such that $u\left(\frac{\cdot - y}{\theta_y}\right) \in \mathcal{P}$ and $\lim_{|y|\to\infty} \theta_y = 1$.*

Lemma 2.5. $\sup_{y\in\mathbb{R}^N} \theta_y = \bar{\theta} < \infty$ *and* $\bar{\theta} > 1$.

Lemma 2.6. *There exists a real number $\hat{\sigma} > 0$ such that $\inf_{u\in\mathcal{P}} \|\nabla u\|_2 \geq \hat{\sigma}$.*

Lemma 2.7. $p =: \inf_{u\in\mathcal{P}} I(u) > 0$.

Proof. Let $u \in \mathcal{P}$, then $I(u)$ satisfies

$$\begin{aligned} I(u) &= \frac{1}{N}\left(\int_{\mathbb{R}^N} \nabla a(x)\cdot xF(u)dx + \int_{\mathbb{R}^N} |\nabla u|^2 dx\right) \\ &\geq \frac{1}{N}\int_{\mathbb{R}^N} |\nabla u|^2 dx \geq \frac{1}{N}\hat{\sigma}^2 > 0\,, \end{aligned}$$

where we have used Lemma 2.6 and condition (A3). It follows that $p > 0$. □

Remark 2.8. We recall that L. Jeanjean and K. Tanaka have shown in [9] that

$$\inf_{u\in\mathcal{P}_\infty} I_\infty(u) = c_\infty.$$

Remark 2.9. If $u \in H^1(\mathbb{R}^N)$, with $\int_{\mathbb{R}^N} G_\infty(u)dx > 0$ and $\theta > 0$ is such that $u(\cdot/\theta) \in \mathcal{P}_\infty$, then we may write

$$I_\infty(u(x/\theta)) = \frac{\theta^{N-2}}{N}\int_{\mathbb{R}^N} |\nabla u|^2 dx\,. \tag{2.5}$$

Lemma 2.10. $p = c_\infty$.

Proof. Let $w \in H^1(\mathbb{R}^N)$ be the ground state solution (which is positive and radially symmetric) of the problem at infinity, $w \in \mathcal{P}_\infty$ and $I_\infty(w) = c_\infty$. Given any $y \in \mathbb{R}^N$, we define $w_y := w(x-y)$. From the translation invariance of the integrals, we get $w_y \in \mathcal{P}_\infty$ and $I_\infty(w_y) = c_\infty$. From Lemma 2.4, for any $y \in \mathbb{R}^N$, there exists a $\theta_y > 1$ such that $\tilde{w}_y = w_y(\cdot/\theta_y) \in \mathcal{P}$. Therefore, we have

$$|I(\tilde{w}_y) - c_\infty| \leq \frac{|\theta_y^{N-2} - 1|}{2} \int_{\mathbb{R}^N} |\nabla w|^2 dx + |\theta_y^N - 1| \int_{\mathbb{R}^N} \frac{\lambda w^2}{2} dx$$
$$+ \int_{\mathbb{R}^N} |F(w)| \left| a_\infty - \theta_y^N a(x\theta_y + y) \right| dx.$$

Since $\theta_y \to 1$, if $|y| \to \infty$, we obtain $\lim_{|y|\to\infty} I(\tilde{w}_y) = c_\infty$. Therefore, $p = \inf_{u\in\mathcal{P}} I(u) \leq c_\infty$.

On the other hand, consider $u \in \mathcal{P}$ and $0 < \theta < 1$ such that $u(\cdot/\theta) \in \mathcal{P}_\infty$. Since $u \in \mathcal{P}$, then u satisfies

$$\begin{aligned} I(u) &= \frac{1}{N} \int_{\mathbb{R}^N} |\nabla u|^2 dx + \frac{1}{N} \int_{\mathbb{R}^N} \nabla a(x) \cdot x F(u) dx \\ &> \frac{1}{N} \int_{\mathbb{R}^N} |\nabla u|^2 dx \geq \frac{\theta^{N-2}}{N} \int_{\mathbb{R}^N} |\nabla u|^2 dx = I_\infty(u(x/\theta)) \geq c_\infty \end{aligned}$$

where we have used (2.5) and (A3). Thus, for any $u \in \mathcal{P}, I(u) > c_\infty$ and hence $\inf_{u\in\mathcal{P}} I(u) \geq c_\infty$. We conclude that $p = c_\infty$. □

Now we are ready to prove Theorem 1.4, which is the main result in this section.

Proof. Suppose, by contradiction, that there exists $z \in H^1(\mathbb{R}^N)$, a critical point of the functional I at level p, i.e., $z \in \mathcal{P}$ and $I(z) = p$. Let $\theta \in (0,1)$ be such that $z(x/\theta) \in \mathcal{P}_\infty$. Then

$$\begin{aligned} p = I(z) &= \frac{1}{N} \int_{\mathbb{R}^N} |\nabla z|^2 dx + \frac{1}{N} \int_{\mathbb{R}^N} \nabla a(x) \cdot x F(z) dx \\ &> \frac{1}{N} \int_{\mathbb{R}^N} |\nabla z|^2 dx > \frac{\theta^{N-2}}{N} \int_{\mathbb{R}^N} |\nabla z|^2 dx = I_\infty(z(x/\theta)) \geq c_\infty , \end{aligned}$$

using (A3) and (2.5). Therefore $p > c_\infty$, which contradicts the previous lemma. □

As a consequence of the fact that p is not a critical level of the functional I, if $\mathcal{P}$ is a natural constraint of the functional I, the infimum p is not achieved. This is the case in our next result. Its proof can be found in [10], Lemma 3.14 and is based on the arguments found in [14], where a similar lemma is proved for an autonomous equation.

Lemma 2.11. *Assume* (A1) *and* (A5). *Then* $\mathcal{P}$ *is a natural constraint of problem* (1.1).

3. Existence of a positive solution

By the previous section, we are motivated to search for solutions in higher levels of energy of the functional I. More precisely, in this section we show that there exists a critical point of the functional I in the range of energies $(c_\infty, 2c_\infty)$, and therefore a positive solution of the equation (1.1). In order to find such a solution we will apply ideas similar to those employed by A. Ambrosetti, G. Cerami and D. Ruiz in [1]. Their argument uses linking together with the barycenter function (also used by G.S. Spradlin ([16], [17])), restricted to the Nehari manifold associated to their problem. In our case, as before, we use the Pohozaev manifold $\mathcal{P}$ instead.

We start by noting that the min-max levels of the Mountain Pass Theorem for the functionals I and I_∞ are equal, i.e., if c is the min-max mountain pass level for the functional I given by

$$c = \min_{\gamma\in\Gamma} \max_{0\leq t\leq 1} I(\gamma(t)) , \tag{3.1}$$

where $\Gamma := \{\gamma \in C([0,1], H^1(\mathbb{R}^N))|\gamma(0) = 0, I(\gamma(1)) < 0\}$, then $c_\infty = c$ and this proof can be found in [10], Lemma 4.1.

Lemma 3.1. $p = c$.

Proof. Define $p_\infty = \inf_{u\in\mathcal{P}_\infty} I_\infty(u)$. We already know that $c_\infty = p_\infty$ by [9]. Moreover, in Lemma 2.10 we have shown that $p = c_\infty$. Therefore $p = c_\infty = c$. □

We recall that a sequence (u_n) is said to be a Cerami sequence for the functional I at level d in $\mathbb{R}$, denoted by $(\mathrm{Ce})_d$, if $I(u_n) \to d$ and $\|I'(u_n)\|(1+\|u_n\|_\lambda) \to 0$. Now we show that, if $d > 0$, then any $(\mathrm{Ce})_d$ sequence for the functional I is bounded, up to a subsequence.

Lemma 3.2. *If (u_n) is a $(\mathrm{Ce})_d$ sequence with $d > 0$, then it has a bounded subsequence.*

The proof involves the Lemma of Lions [11], the non quadraticity (NQ) of the function F (as done in [19]), and Fatou's Lemma together with condition (A1).

The next step is to show the existence of a Cerami sequence for the functional I at level c.

Lemma 3.3. *Let c be as in (3.1), then there exists a $(\mathrm{Ce})_c$ sequence $(u_n) \subset H^1(\mathbb{R}^N)$.*

The proof of this lemma can be found in [10], Lemma 4.5.

Lemma 3.4. *(Splitting) Let $(u_n) \in H^1(\mathbb{R}^N)$ be a bounded sequence such that*

$$I(u_n) \to d > 0 \quad \text{and} \quad \|I'(u_n)\|(1 + \|u_n\|_\lambda) \to 0 .$$

Replacing (u_n) by a subsequence, if necessary, there exists a solution $\bar{u}$ of (1.1), a number $k \in \mathbb{N}\cup\{0\}$, k functions $u^1, u^2, \dots, u^k$ and k sequences of points $(y_n^j) \in \mathbb{R}^N, 1 \leq j \leq k$, satisfying:

a) *$u_n \to \bar{u}$ in $H^1(\mathbb{R}^N)$ or*
b) *u^j are nontrivial solutions of (2.3);*

c) $|y_n^j| \to \infty$ *and* $|y_n^j - y_n^i| \to \infty, i \neq j$;

d) $u_n - \sum_{i=1}^{k} u^i(x - y_n^i) \to \bar{u}$;

e) $I(u_n) \to I(\bar{u}) + \sum_{i=1}^{k} I_\infty(u^i)$.

Remark 3.5. Nowadays the proof of this lemma is standard and is a version of the concentration compactness of P.L. Lions [11] and found in [18]. The main ingredients are the Lions Lemma and the Brezis–Lieb Lemma [5].

Corollary 3.6. *If* $I(u_n) \to c_\infty$ *and* $\|I'(u_n)\|(1 + \|u_n\|_\lambda) \to 0$, *then either* (u_n) *is relatively compact or the splitting lemma holds with* $k = 1$ *and* $\bar{u} = 0$.

Lemma 3.7. *The functional* I *satisfies condition* (Ce) *at level* $d \in (c_\infty, 2c_\infty)$.

Proof. Consider $d \in (c_\infty, 2c_\infty)$ and a Cerami sequence $(u_n)_d \in H^1(\mathbb{R}^N)$ which, by Lemma 3.2, is bounded. Applying Lemma 3.4, up to subsequences, we have

$$u_n - \sum_{j=1}^{k} u^j(x - y_n^j) \to \bar{u}, \quad \text{in} \quad H^1(\mathbb{R}^N)\,,$$

where u^j is a weak solution of the problem at infinity, $|y_n^j| \to \infty$ and $\bar{u}$ is a weak solution of equation (1.1). Moreover,

$$I(u_n) = I(\bar{u}) + \sum_{j=1}^{k} I_\infty(u_j) + o_n(1).$$

Since $d < 2c_\infty$, then $k < 2$. If $k = 1$, we have two cases to distinguish:

1) $\bar{u} \neq 0$, which implies $I(\bar{u}) \geq c_\infty$ and therefore $I(u_n) \geq 2c_\infty$.
2) $\bar{u} = 0$, which implies $I(u_n) \to I_\infty(w)$, since w is the unique positive solution of 2.3

Since $d \in (c_\infty, 2c_\infty)$, we get a contradiction in both cases. Therefore, we must have $k = 0$ and the strong convergence $u_n \to \bar{u}$. □

Now, we will introduce the barycenter function.

Definition 3.8. *Define the barycenter function of a given function* $u \neq 0 \in H^1(\mathbb{R}^N)$ *as follows: let*

$$\mu(u)(x) = \frac{1}{|B_1|} \int_{B_1(x)} |u(y)|dy,$$

with $\mu(u) \in L^\infty(\mathbb{R}^N)$ *and is a continuous function. Subsequently, take*

$$\hat{u}(x) = \left[\mu(u)(x) - \frac{1}{2} \max \mu(u)\right]^+.$$

It follows that $\ddot{u} \in C_0(\mathbb{R}^N)$. *Now define the barycenter of* u *by*

$$\beta(u) = \frac{1}{|\hat{u}|_{L^1}} \int_{\mathbb{R}^N} x\hat{u}(x)dx \in \mathbb{R}^N.$$

Since $\hat{u}$ has compact support, by definition, $\beta(u)$ is well defined. The function β satisfies the following properties:

(a) β is a continuous function in $H^1(\mathbb{R}^N) \setminus \{0\}$.
(b) If u is radial, then $\beta(u) = 0$.
(c) Given $y \in \mathbb{R}^N$ and defining $u_y(x) := u(x - y)$, then $\beta(u_y) = \beta(u) + y$.

With this barycenter function, we define the level

$$b := \inf \{I(u);\ u \in \mathcal{P} \text{ and } \beta(u) = 0\}. \tag{3.2}$$

It is clear that $b \geq c_\infty$. Moreover, we have that

Lemma 3.9. $b > c_\infty$.

Proof. Suppose, by contradiction, that $b = c_\infty$. By the definition of b, there exists a (minimizing) sequence $\{u_n\} \in \{u \in H^1(\mathbb{R}^N)|\ u \in \mathcal{P},\ \beta(u) = 0\}$ such that

$$I(u_n) \to b > 0.$$

By Lemma 4.10 from [10], the sequence u_n is bounded. Since $b = p$ by Lemmas 4.1 from [10] and Lemma 3.1, then $\{u_n\}$ is also a minimizing sequence of I on $\mathcal{P}$. By the Ekeland Variational Principle (Theorem 8.5 in [20]) there exists another sequence $\{\tilde{u}_n\} \subset \mathcal{P}$ such that:

i) $I(\tilde{u}_n) \to p$;
ii) $I'|_{\mathcal{P}}(\tilde{u}_n) \to 0$;
iii) $\|\tilde{u}_n - u_n\| \to 0$.

The Pohozaev manifold is a natural constraint, hence in fact $I'(\tilde{u}_n) \to 0$. Indeed, suppose $I'(\tilde{u}_n)$ does not go to zero. That means there exists $\epsilon_0 > 0$ and a subsequence $\{\tilde{u}_{n_j}\}$, with $n_j \to \infty$, such that

$$\|I'(\tilde{u}_{n_j})\| > \epsilon_0 .$$

Since f is Lipschitz continuous,

$$\begin{aligned} &|(I'(\tilde{u}_{n_j}) - I'(v))\varphi| \\ &= \left| \int (\nabla \tilde{u}_{n_j} - \nabla v)\varphi dx + \lambda \int (\tilde{u}_{n_j} - v)\varphi dx - \int (f(\tilde{u}_{n_j}) - f(v))\varphi \right| \\ &\leq \|\tilde{u}_{n_j} - v\|\|\varphi\| + K\|\tilde{u}_{n_j} - v\|\|\varphi\| \\ &= (1 + K)\|\tilde{u}_{n_j} - v\|\|\varphi\| . \end{aligned}$$

Thus, if

$$\|\tilde{u}_{n_j} - v\| < \frac{\tilde{\delta}}{1+K} := 3\delta$$

then

$$\|I'(\tilde{u}_{n_j}) - I'(v)\| < \tilde{\delta}.$$

This yields,

$$\epsilon_0 - \tilde{\delta} < \|I'(\tilde{u}_{n_j})\| - \tilde{\delta} < \|I'(v)\| \,.$$

For $\tilde{\delta} > 0$ sufficiently small, we have $\lambda := \epsilon_0 - \tilde{\delta} > 0$ and for all n_j,

$$\text{for all } v \in B_{3\delta}(\tilde{u}_{n_j}), \quad \text{then} \quad \|I'(v)\| > \lambda \,.$$

Let $\epsilon := \min\{\frac{p}{2}, \frac{\lambda\delta}{8}\}$ and $S := \{\tilde{u}_{n_j}\}$. By Lemma 2.3 in [20], there is a deformation η on the level p, taking all the points of S_δ to the level $p - \epsilon$.

$$I(\eta(1, u)) \leq I(u) \quad \text{for all} \quad u \in H^1(\mathbb{R}^N) \,.$$

Moreover, for n_j sufficiently large,

$$\max_{t>0} I(\eta(1, \tilde{u}_{n_j}(\frac{\cdot}{t})) \leq p - \epsilon$$

because $\{\tilde{u}_{n_j}\}$ is a minimizing sequence, $I(\tilde{u}_{n_j}) < p + \frac{\epsilon}{2}$, for n_j sufficiently large, and since $\tilde{u}_{n_j} \in \mathcal{P}$,

$$\max_{t>0} I(\tilde{u}_{n_j}(\frac{\cdot}{t})) = I(\tilde{u}_{n_j}) \to p \,.$$

On the other hand, $\gamma_0(t) = \eta(1, \tilde{u}_{n_j}(\frac{\cdot}{t}))$ is a path in Γ and hence

$$c \leq \max_{t>0} \gamma_0(t) \leq p - \epsilon < p \,.$$

But this is a contradiction, because we have proved that $p = c$. Moreover, since f is Lipschitz continuous, $I'(\tilde{u}_n) \to 0$ and $\|\tilde{u}_n - u_n\| \to 0$ imply $I'(u_n) \to 0$, as $n \to \infty$. Therefore, the sequence $\{u_n\}$ satisfies the assumptions of Corollary 3.6 and since $p = c_\infty$ and is not attained, then the splitting lemma holds with $k = 1$. This yields

$$u_n(x) \to u^1(x - y_n) \,, \tag{3.3}$$

where $y_n \in \mathbb{R}^N$, $|y_n| \to \infty$ and u^1 is a solution of the problem at infinity. By making a translation, we obtain

$$u_n(x + y_n) = u^1(x) + o_n(1).$$

Calculating the barycenter function on both sides, we have

$$\beta(u_n(x + y_n)) = \beta(u_n) - y_n \,,$$

with $\beta(u_n) = 0$ and

$$\beta(u^1(x) + o_n(1)) \to \beta(u^1(x)),$$

since β is a continuous function. On one side, $\beta(u^1(x)) = 0$ and, on the other, $|y_n| \to \infty$ so we arrive at a contradiction. Therefore, we must have $b > c_\infty$. □

Let us consider again the positive, radially symmetric, ground state solution $w \in H^1(\mathbb{R}^N)$ of the autonomous problem at infinity. We define the operator $\Pi : \mathbb{R}^N \to \mathcal{P}$ by

$$\Pi[y](x) = w\left(\frac{x - y}{\theta_y}\right),$$

where θ_y is exactly the real number θ which projects $w(\cdot - y)$ onto the Pohozaev manifold $\mathcal{P}$. Π is a continuous function of y because θ_y is unique and $\theta_y(w(\cdot - y))$ is a continuous function of $w(\cdot - y)$.

The proofs of the following properties of this operator Π can be found in [10], Lemmas 4.13 and 4.14.

Lemma 3.10. (a) $\beta(\Pi[y](x)) = y$; (b) $I(\Pi[y]) \to c_\infty$, *if* $|y| \to \infty$;

Lemma 3.11. *Assume*

(A6) $\sup_{\mathbb{R}^N} |a_\infty - a(x)| < \dfrac{2c_\infty}{\bar{\theta}^N \|w\|_2^2}$,

where $\bar{\theta} = \sup_{y\in\mathbb{R}^N} \theta_y$. *Then* $I(\Pi[y]) < 2c_\infty$.

Proof. Since I_∞ is translation invariant, the maximum of $t \mapsto I_\infty(w(\cdot/t))$ is attained at $t = 1$. Also, recalling that $\theta_y > 1$ and using (A6), we obtain

$$I(\Pi[y]) = I_\infty(\Pi[y]) + I(\Pi[y]) - I_\infty(\Pi[y]) \le I_\infty(w) + \int_{\mathbb{R}^N} (a_\infty - a(x))F(\Pi[y])dx$$

$$< c_\infty + \frac{2c_\infty}{\bar{\theta}^N \|w\|_2^2} \int_{\mathbb{R}^N} \frac{1}{2} w^2 \left(\frac{x-y}{\theta_y}\right) dx = c_\infty + \frac{c_\infty \theta_y^N}{\bar{\theta}^N \|w\|_2^2} \|w\|_2^2 = c_\infty + c_\infty$$

where we used that $F(u) = \dfrac{1}{2}\left(u^2 - \ln(1+u^2)\right) < \dfrac{1}{2}u^2$. This yields $I(\Pi[y]) < 2c_\infty$. □

We will need a version of the Linking Theorem with the Cerami condition by P. Bartolo, V. Benci and D. Fortunato in [4] (see Theorem 2.3). We also refer to the works of E.A.B. Silva [15] and M. Schechter [13] for similar versions of the Linking Theorem with the Cerami condition.

Now we are ready to prove our main existence result, Theorem 1.3.

Proof. We have $I_\infty(\Pi[y]) < I(\Pi[y])$, for any $y \in \mathbb{R}^N$, due to condition $(A4)$. Since from Lemma 3.9 we have $b > c_\infty$ and $I(\Pi[y]) \to c_\infty$ if $|y| \to \infty$, from Lemma 3.10 (b), then there exists $\bar{\rho} > 0$ such that for every $\rho \ge \bar{\rho}$,

$$c_\infty < \max_{|y|=\rho} I(\Pi[y]) < b. \tag{3.4}$$

In order to apply the Linking Theorem, we take

$$Q := \Pi(\overline{B_{\bar{\rho}}(0)}) \quad \text{and} \quad S := \left\{u \in H^1(\mathbb{R}^N) |\ u \in \mathcal{P},\ \beta(u) = 0\right\}.$$

Since $\beta(\Pi[y]) = y$, from Lemma 3.10 (a), we have that $\partial Q \cap S = \emptyset$ (if $u \in S$, then $\beta(u) = 0$, and if $u \in \partial Q$, then $\beta(u) = y \neq 0$). Now we need to show that $h(Q) \cap S \neq \emptyset$, for any $h \in \mathcal{H}$, where

$$\mathcal{H} = \{h \in C(Q, \mathcal{P}); h|_{\partial Q} = id\}.$$

Given $h \in \mathcal{H}$, let us define $T : \overline{B_{\bar{\rho}}(0)} \to \mathbb{R}^N$ for $T(y) = \beta \circ h \circ \Pi[y]$. The function T is continuous and for any $|y| = \bar{\rho}$, we have that $\Pi[y] \in \partial Q$, thus $h \circ \Pi[y] = \Pi[y]$, because $h|_{\partial Q} = id$, and hence from Lemma 3.10 (a) $T(y) = y$. By the Fixed Point

Theorem of Brouwer, we conclude that there exists $\tilde{y} \in B_{\bar{\rho}}(0)$ such that $T(\tilde{y}) = 0$, which implies $h(\Pi[\tilde{y}]) \in S$. Therefore $h(Q) \cap S \neq \emptyset$ and S and ∂Q "link".

Furthermore, from the definitions of b and Q and the inequalities (3.4), we may write

$$b = \inf_S I > \max_{\partial Q} I .$$

Let us define

$$d = \inf_{h \in \mathcal{H}} \max_{u \in Q} I(h(u)).$$

Then we have $d \geq b$. In particular, it follows that $d > c_\infty$, because from Lemma 3.9 we know that $b > c_\infty$. Furthermore, if we take $h = id$, then

$$\inf_{h \in \mathcal{H}} \max_{u \in Q} I(h(u)) < \max_{u \in Q} I(u) < 2c_\infty,$$

by Lemma 3.11. This implies $d < 2c_\infty$. The two inequalities give $d \in (c_\infty, 2c_\infty)$, thus from Lemma 3.7 the (Ce) condition is satisfied at level d. Therefore, we can apply the Linking Theorem and conclude that d is a critical level for the functional I. This guarantees the existence of a nontrivial solution $u \in H^1(\mathbb{R}^N)$ of the equation (1.1). Reasoning as usual, because of the hypotheses on f, and using the maximum principle we may conclude that u is positive, which implies the proof of the theorem. □

References

[1] A. Ambrosetti, G. Cerami, D. Ruiz, *Solitons of linearly coupled systems of semilinear non-autonomous equations on* $\mathbb{R}^N$, J. of Func. Anal. **254**, (2008), 2816–2845.

[2] A. Ambrosetti and P.H. Rabinowitz, *Dual variational methods in critical point theory and applications*, J. Funct. Anal. **14**, (1973), 349–381.

[3] A. Azzollini, A. Pomponio, *On the Schrödinger equation in* $\mathbb{R}^N$ *under the effect of a general nonlinear term*, Indiana Univ. Math. J. **58**, no. 3, (2009), 1361–1378.

[4] P. Bartolo, V. Benci and D. Fortunato, *Abstract critical point theorems and applications to some nonlinear problems with strong resonance at infinity*, Nonlinear Anal. **7** (1983), 981–1012.

[5] H. Brezis and E.H. Lieb, *A relation between pointwise convergence of functions and convergence of functionals*, Proc. Amer. Math. Soc. **88**, no. 3, (1983), 486–490.

[6] G. Cerami, *An existence criterion for the critical points on unbounded manifolds*, (Italian) Istit. Lombardo Accad. Sci. Lett. Rend. A 112 (1978), no. 2, 332–336 (1979).

[7] D.G. Costa and C.A. Magalhães, *Variational elliptic problems which are nonquadratic at infinity*, Nonlinear Anal. **23** (1994), no. 11, 1401–1412.

[8] D.G. Costa, H. Tehrani, *On a Class of Asymptotically Linear Elliptic Problems in* $\mathbb{R}^N$, J. Dif. Equations **173**, (2001), 470–494.

[9] L. Jeanjean, K. Tanaka, *A remark on least energy solutions in* $\mathbb{R}^N$, Proc. Amer. Math. Soc. **131**, no. 8, (2002), 2399–2408.

[10] R. Lehrer, L.A. Maia, *Positive solutions to asymptotically linear equations via Pohǒzaev manifold*, J. Funct. Anal **266** (2014), 213–246.

[11] P.L. Lions, *The concentration-compactness principle in the calculus of variations. The locally compact case,* Ann. Inst. Henri Poincaré, Analyse Non Linéaire **1** (1984), 109–145 and 223–283.

[12] L.A. Maia and R. Ruviaro, *Sing-changing Solutions for a Schrödinger Equation with Saturable Nonlinearity*, Milan J. Math. vol. **79** (2011), 259–271.

[13] M. Schechter, *Linking methods in critical point theory*, Birkhäuser Boston Inc., Boston, MA, (1999).

[14] J. Shatah, *Unstable Ground State of Nonlinear Klein–Gordon Equations*, Trans. Amer. Math. Soc. **290**, no. 2, (1985), 701–710.

[15] E.A.B. Silva, *Subharmonic solutions for subquadratic Hamiltonian systems*, J. Dif. Equations **115** (1995), 120–145.

[16] G.S. Spradlin, *Existence of solutions to a hamiltonian System without convexity condition on the nonlinearity*, Elect. J. of Diff. Eq., no. 21, (2004), 1–13.

[17] G.S. Spradlin, *Interacting near-solutions of a hamiltonian system*, Calc. Var. **22**, (2005), 447–464.

[18] M. Struwe, *A global compactness result for elliptic boundary value problems involving limiting nonlinearities*, Math. Z. **187**, no. 4, (1984), 511–517.

[19] C.A. Stuart and H.S. Zhou, *Applying the Mountain Pass Theorem to an Asymptotically linear Elliptic Equation on* $\mathbb{R}^N$, Commun. Partial Diff. Eq. **9-10**, (1999), 1731–1758.

[20] M. Willem, *Minimax Theorems*, Volume 24, Birkhäuser, Boston, 1996.

Raquel Lehrer
Universidade Estadual do Oeste do Paraná – UNIOESTE
Centro de Ciências Exatas e Tecnológicas – CCET.
Rua Universitária, 2069, Jardim Universitário
85819-110 – Cascavel, PR – Brasil
e-mail: `raquel.lehrer@unioeste.br`

Liliane A. Maia
(Corresponding author)
Departamento de Matemática
Instituto de Ciências Exatas
Universidade de Brasília
Campus Darcy Ribeiro, Asa Norte
70910-900 – Brasília, DF – Brasil
e-mail: `lilimaia@unb.br`

Progress in Nonlinear Differential Equations
and Their Applications, Vol. 85, 323–340

Bubble Concentration on Spheres for Supercritical Elliptic Problems

Filomena Pacella and Angela Pistoia

Abstract. We consider the supercritical Lane–Emden problem

$$(P_\varepsilon) \qquad -\Delta v = |v|^{p_\varepsilon - 1} v \text{ in } \mathcal{A}, \quad v = 0 \text{ on } \partial\mathcal{A}$$

where $\mathcal{A}$ is an annulus in $\mathbb{R}^{2m}$, $m \geq 2$ and $p_\varepsilon = \frac{(m+1)+2}{(m+1)-2} - \varepsilon$, $\varepsilon > 0$.

We prove the existence of positive and sign changing solutions of (P_ε) concentrating and blowing-up, as $\varepsilon \to 0$, on $(m-1)$-dimensional spheres. Using a reduction method ([18, 14]) we transform problem (P_ε) into a nonhomogeneous problem in an annulus $\mathcal{D} \subset \mathbb{R}^{m+1}$ which can be solved by a Ljapunov–Schmidt finite-dimensional reduction.

Mathematics Subject Classification (2010). 35J61, 35B25, 35B40.

Keywords. Supercritical problem, concentration on manifolds.

1. Introduction

In this paper we address the question of finding solutions concentrated on manifolds of positive dimension of supercritical elliptic problems of the type

$$-\Delta v = |v|^{p-1} v \text{ in } \mathcal{A}, \quad v = 0 \text{ on } \partial\mathcal{A}, \tag{1}$$

where $\mathcal{A} := \{y \in \mathbb{R}^d \, : \, a < |y| < b\}$, $a > 0$, is an annulus in $\mathbb{R}^d$, $d > 2$ and $p > \frac{d+2}{d-2}$ is a supercritical exponent.

We remark that the critical and supercritical Lane–Emden problems are very delicate due to topological and geometrical obstruction enlightened by the Pohozaev's identity ([16]). We also point out that in the supercritical case a result of Bahri–Coron type ([2]) cannot hold in general nontrivially topological domains as shown by a nonexistence result of Passaseo ([15]), obtained exploiting critical exponents in lower dimensions. Using similar ideas, some results for exponents p which are subcritical in dimension $n < d$ and instead supercritical in dimension d have been obtained in different kind of domains in [1, 4, 6, 8, 9, 10, 11, 13].

Here we consider annuli in even dimension $d = 2m$, $m \geq 2$ and obtain results about the existence of solutions, both positive and sign changing, of different type, concentrated on $(m-1)$-dimensional spheres. More precisely, we have

Theorem 1.1 (Case of positive solutions). *Let $\mathcal{A} \subset \mathbb{R}^{2m}$, $m \geq 2$ and define $(\partial\mathcal{A})_a := \{y \in \partial\mathcal{A} \,:\, |y| = a\}$. There exists $\epsilon_0 > 0$ such that for any $\epsilon \in (0, \epsilon_0)$, the following supercritical problem*

$$-\Delta v = |v|^{p_\varepsilon - 1} v \text{ in } \mathcal{A}, \quad v = 0 \text{ on } \partial\mathcal{A}, \tag{2}$$

with $p_\varepsilon = \frac{(m+1)+2}{(m+1)-2} - \varepsilon$ has:

i) *a positive solution v_ε which concentrates and blows-up on a $(m-1)$-dimensional sphere $\Gamma \subset (\partial\mathcal{A})_a$ as $\varepsilon \to 0$,*
ii) *a positive solution v_ε which concentrates and blows-up on two $(m-1)$-dimensional orthogonal spheres $\Gamma_1 \subset (\partial\mathcal{A})_a$ and $\Gamma_2 \subset (\partial\mathcal{A})_a$ as $\varepsilon \to 0$,*

Theorem 1.2 (Case of sign changing solutions). *Let $\mathcal{A} \subset \mathbb{R}^{2m}$, $m \geq 2$ and define $(\partial\mathcal{A})_a := \{y \in \partial\mathcal{A} \,:\, |y| = a\}$. There exists $\epsilon_0 > 0$ such that for any $\epsilon \in (0, \epsilon_0)$, the supercritical problem (2) with $p_\varepsilon = \frac{(m+1)+2}{(m+1)-2} - \varepsilon$ has:*

i) *a sign changing solution v_ε such that v_ε^+ and v_ε^- concentrate and blow-up on two $(m-1)$-dimensional orthogonal spheres $\Gamma_+ \subset (\partial\mathcal{A})_a$ and $\Gamma_- \subset (\partial\mathcal{A})_a$, respectively, as $\varepsilon \to 0$,*
ii) *a sign changing solution v_ε such that v_ε^+ and v_ε^- concentrate and blow up on the same $(m-1)$-dimensional sphere $\Gamma \subset (\partial\mathcal{A})_a$, as $\varepsilon \to 0$,*
iii) *two sign changing solutions v_ε^1 and v_ε^2 each one is such that $(v_\varepsilon^i)^+$ and $(v_\varepsilon^i)^-$ concentrate and blow up on two $(m-1)$-dimensional orthogonal spheres $(\Gamma_i)_+ \subset (\partial\mathcal{A})_a$ and $(\Gamma_i)_- \subset (\partial\mathcal{A})_a$, respectively, as $\varepsilon \to 0$, $i = 1, 2$.*

We remark that the exponent $\frac{(m+1)+2}{(m+1)-2} - \varepsilon$ which is almost critical in dimension $(m+1)$ is obviously supercritical for problem (2).

To prove our results we use the reduction method introduced in [14] which allows to transform symmetric solutions to (2) into symmetric solutions of a similar nonhomogeneous problem in an annulus $\mathcal{D} \subset \mathbb{R}^{m+1}$. This method was inspired by the paper [18] where a reduction approach was used to pass from a singularly perturbed problem in an annulus in $\mathbb{R}^4$ to a singularly perturbed problem in an annulus in $\mathbb{R}^3$.

More precisely let us consider the annulus $\mathcal{D} \subset \mathbb{R}^{m+1}$ $\mathcal{D} := \{x \in \mathbb{R}^{m+1} \,:\, a^2/2 < |x| < b^2/2\}$, and, write a point $y \in \mathbb{R}^{2m}$ as $y = (y_1, y_2)$, $y_i \in \mathbb{R}^m$, $i = 1, 2$. Then we consider functions v in $\mathcal{A} \subset \mathbb{R}^{2m}$ which are radially symmetric in y_1 and y_2, i.e., $v(y) = w(|y_1|, |y_2|)$ and functions u in $\mathcal{D} \subset \mathbb{R}^{m+1}$ which are radially symmetric about the x_{m+1}-axis, i.e., $u(x) = h(|x|, \varphi)$ with $\varphi = \arccos\left(\frac{x}{|x|} \cdot \underline{e}_{m+1}\right)$ where $\underline{e}_{m+1} = (0, \ldots, 0, 1)$. We also set

$$X = \left\{v \in C^{2,\alpha}(\overline{A}) \,:\, v \text{ is radially symmetric}\right\}$$
$$Y = \left\{u \in C^{2,\alpha}(\overline{D}) \,:\, u \text{ is axially symmetric}\right\}.$$

Then, as corollary of Theorem 1.1 of [14] we have

Proposition 1.3. *There is a bijective correspondence h between solutions v of (2) in X and solutions $u = h(v)$ in Y of the following reduced problem*

$$-\Delta u = \frac{1}{2|x|}|u|^{p_\varepsilon - 1}u \quad \text{in } \mathcal{D} \subset \mathbb{R}^{m+1}, \qquad u = 0 \quad \text{on } \partial\mathcal{D}. \tag{3}$$

As a consequence of this result we can obtain solutions of problem (2) by constructing axially symmetric solutions of the lower-dimensional problem (3). This has the immediate advantage of transforming the supercritical problem (2) into the subcritical problem (3) if the exponent p_ε is taken as $\frac{(m+1)+2}{(m+1)-2} - \varepsilon$. Indeed we will prove Theorem 1.1 and Theorem 1.2 by constructing axially symmetric solutions of (3), positive or sign changing, which blow-up and concentrate in points of the annulus $\mathcal{D} \subset \mathbb{R}^{m+1}$. These solutions will give rise to solutions of (2) concentrating on $(m-1)$-dimensional spheres, because, as a consequence of the proof of Theorem 1.1 of [14] and of Remark 3.1 of [14] it holds

Proposition 1.4. *If u_ε is an axially symmetric solution of (2) concentrating, as $\varepsilon \to 0$, on a point ξ which belongs to the $x_{(m+1)}$-axis, i.e., $\xi = (0, \dots, 0, t)$ for $t \in \mathbb{R} \setminus \{0\}$, then the corresponding solution $v_\varepsilon = h^{-1}(u_\varepsilon)$ concentrates on a $(m-1)$-dimensional sphere in $\mathbb{R}^{2m}$.*

This is because, by symmetry considerations and by the change of variable performed in [14] to prove Theorem 1.1 any point ξ on the $x_{(m+1)}$-axis in $\mathcal{D} \subset \mathbb{R}^{m+1}$ is mapped into a $(m-1)$-dimensional sphere in $\mathcal{A} \subset \mathbb{R}^{2m}$. We refer to [14] for all details.

Thus let $\Omega := \{x \in \mathbb{R}^n \ : \ 1 < |x| < r\}$ be an annulus in $\mathbb{R}^n$, $n \geq 3$, and consider the problem

$$-\Delta u = \frac{1}{2|x|}|u|^{p-1-\epsilon}u \quad \text{in } \Omega, \qquad u = 0 \quad \text{on } \partial\Omega, \tag{4}$$

where $p = \frac{n+2}{n-2}$ and ε is a small positive parameter. Let $U_{\delta,\xi}(x) := \alpha_n \frac{\delta^{\frac{n-2}{2}}}{(\delta^2 + |x-\xi|^2)^{\frac{n-2}{2}}}$ with $\delta > 0$ and $x, \xi \in \mathbb{R}^n$, be the solutions to the critical problem $-\Delta u = u^p$ in $\mathbb{R}^n$. Here $\alpha_n := [n(n-2)]^{\frac{n-2}{4}}$. We have

Theorem 1.5. *There exists $\epsilon_0 > 0$ such that, for each $\epsilon \in (0, \epsilon_0)$, problem (4) has*

(i) *an axially symmetric positive solution u_ϵ with one simple positive blow-up point which converge to ξ_0 as ε goes to zero, i.e.,*

$$u_\epsilon(x) = U_{\delta_\epsilon, \xi_\epsilon}(x) + o(1) \quad \text{in } H_0^1(\Omega),$$

with

$$\epsilon^{-\frac{n-1}{n-2}}\delta_\epsilon \to d > 0, \quad \xi_\epsilon \to \xi_0;$$

(ii) *an axially symmetric positive solution* u_ϵ *with two simple positive blow-up points which converge to* ξ_0 *and* $-\xi_0$ *as* ε *goes to zero, i.e.,*

$$u_\epsilon(x) = U_{\delta_\epsilon,\xi_\epsilon}(x) + U_{\delta_\epsilon,-\xi_\epsilon}(x) + o(1),$$

with

$$\epsilon^{-\frac{n-1}{n-2}}\delta_\epsilon \to d > 0, \quad \xi_\epsilon \to \xi_0;$$

(iii) *an axially symmetric sign-changing solution* u_ϵ *with one simple positive and one simple negative blow-up points which converge to* ξ_0 *and* $-\xi_0$ *as* ε *goes to zero, i.e.,*

$$u_\epsilon(x) = U_{\delta_\epsilon,\xi_\epsilon}(x) - U_{\delta_\epsilon,-\xi_\epsilon}(x) + o(1),$$

with

$$\epsilon^{-\frac{n-1}{n-2}}\delta_\epsilon \to d > 0, \quad \xi_\epsilon \to \xi_0;$$

(iv) *an axially symmetric sign-changing solution* u_ϵ *with one double nodal blow-up point which converge to* ξ_0 *as* ε *goes to zero, i.e.,*

$$u_\epsilon(x) = U_{\delta_{1_\epsilon},\xi_{1_\epsilon}}(x) - U_{\delta_{2_\epsilon},\xi_{2_\epsilon}}(x) + o(1),$$

with

$$\epsilon^{-\frac{n-1}{n-2}}\delta_{i\epsilon} \to d_i > 0, \quad \xi_{i\epsilon} \to \xi_0$$

for $i = 1, 2$.

(v) *two axially symmetric sign-changing solutions* u_ϵ *with two double nodal blow-up points which converge to* ξ_0 *and* $-\xi_0$ *as* ε *goes to zero, i.e.,*

$$u_\epsilon(x) = \left[U_{\delta_{1_\epsilon},\xi_{1_\epsilon}}(x) - U_{\delta_{2_\epsilon},\xi_{2_\epsilon}}(x)\right] + \left[U_{\delta_{1_\epsilon},-\xi_{1_\epsilon}}(x) - U_{\delta_{2_\epsilon},-\xi_{2_\epsilon}}(x)\right] + o(1)$$

and

$$u_\epsilon(x) = \left[U_{\delta_{1_\epsilon},\xi_{1_\epsilon}}(x) - U_{\delta_{2_\epsilon},\xi_{2_\epsilon}}(x)\right] - \left[U_{\delta_{1_\epsilon},-\xi_{1_\epsilon}}(x) - U_{\delta_{2_\epsilon},-\xi_{2_\epsilon}}(x)\right] + o(1)$$

with

$$\epsilon^{-\frac{n-1}{n-2}}\delta_{i\epsilon} \to d_i > 0, \quad \xi_{i\epsilon} \to \xi_0$$

for $i = 1, 2$.

Obviously Theorem 1.1 and Theorem 1.2 derive from Theorem 1.5 for $n = m + 1$ using Proposition 1.3 and Proposition 1.4.

The proof of Theorem 1.5 relies on a very well-known Ljapunov–Schmidt finite-dimensional reduction. We omit many details on the finite-dimensional reduction because they can be found, up to some minor modifications, in the literature. In Section 2 we write the approximate solution, we sketch the proof of the Ljapunov–Schmidt procedure and we prove Theorem 1.5. In Section 3 we only compute the expansion of the reduced energy, which is crucial in this framework. In the Appendix we recall some well-known facts.

2. The Ljapunov–Schmidt procedure

We equip $\mathrm{H}_0^1(\Omega)$ with the inner product $(u,v)=\int_\Omega \nabla u \nabla v dx$ and the corresponding norm $\|u\|^2=\int_\Omega |\nabla u|^2 dx$. For $r\in[1,\infty)$ and $u\in \mathrm{L}^r(\Omega)$ we set $\|u\|_r^r=\int_\Omega |u|^r dx$.

Let us rewrite problem (4) in a different way. Let $i^*:\mathrm{L}^{\frac{2n}{n+2}}(\Omega)\to \mathrm{H}_0^1(\Omega)$ be the adjoint operator of the embedding $i:\mathrm{H}_0^1(\Omega)\hookrightarrow \mathrm{L}^{\frac{2n}{n-2}}(\Omega)$, i.e.,

$$i^*(u)=v \quad \Leftrightarrow \quad (v,\varphi)=\int_\Omega u(x)\varphi(x)dx \ \forall\ \varphi\in \mathrm{H}_0^1(\Omega).$$

It is clear that there exists a positive constant c such that

$$\|i^*(u)\|\le c\|u\|_{\frac{2n}{n+2}} \qquad \forall\ u\in \mathrm{L}^{\frac{2n}{n+2}}(\Omega).$$

Setting $f_\varepsilon(s):=|s|^{p-1-\varepsilon}s$ and using the operator i^*, problem (4) turns out to be equivalent to

$$u=i^*\left[\frac{1}{2|x|}f_\varepsilon(u)\right],\quad u\in \mathrm{H}_0^1(\Omega). \tag{5}$$

Let $U_{\delta,\xi}(x):=\alpha_n\frac{\delta^{\frac{n-2}{2}}}{(\delta^2+|x-\xi|^2)^{\frac{n-2}{2}}}$, with $\alpha_n:=[n(n-2)]^{\frac{n-2}{4}}$ be the positive solutions to the limit problem

$$-\Delta u=u^p \text{ in } \mathbb{R}^n.$$

Set

$$\psi^0_{\delta,\xi}(x):=\frac{\partial U_{\delta,\xi}}{\partial\delta}(x)=\alpha_n\frac{n-2}{2}\delta^{\frac{n-4}{2}}\frac{|x-\xi|^2-\delta^2}{(\delta^2+|x-\xi|^2)^{n/2}}$$

and for any $j=1,\dots,n$

$$\psi^j_{\delta,\xi}(x):=\frac{\partial U_{\delta,\xi}}{\partial\xi_j}(x)=\alpha_n(n-2)\delta^{\frac{n-2}{2}}\frac{x_j-\xi_j}{(\delta^2+|x-\xi|^2)^{n/2}}.$$

It is well known that the space spanned by the $(n+1)$ functions $\psi^j_{\delta,\xi}$ is the set of the solutions to the linearized problem

$$-\Delta\psi=pU_{\delta,\xi}^{p-1}\psi \text{ in } \mathbb{R}^n.$$

We also denote by PW the projection onto $\mathrm{H}_0^1(\Omega)$ of a function $W\in D^{1,2}(\mathbb{R}^n)$, i.e.,

$$\Delta PW=\Delta W \text{ in } \Omega,\quad PW=0 \text{ on } \partial\Omega.$$

Set $\xi_0:=(0,\dots,0,1)$. We look for two different types of solutions to problem (5). The solutions of the type (i), (ii) and (iii) of Theorem 1.5 will be of the form

$$u_\varepsilon=PU_{\delta,\xi}+\lambda PU_{\mu,\eta}+\phi \tag{6}$$

where $\lambda\in\{-1,0,+1\}$ ($\lambda=0$ in case (i), $\lambda=+1$ in case (ii) and $\lambda=-1$ in case (iii)) and the concentration parameters are

$$\mu=\delta \quad\text{and}\quad \delta:=\varepsilon^{\frac{n-1}{n-2}}d \text{ for some } d>0 \tag{7}$$

while the concentration points satisfy

$$\eta = -\xi \quad \text{and} \quad \xi = (1+\tau)\xi_0, \text{ with } \tau := \varepsilon t,\ t > 0. \tag{8}$$

On the other hand, the solutions of the type (iv) and (v) of Theorem 1.5 will be of the form

$$u_\varepsilon = PU_{\delta_1,\xi_1} - PU_{\delta_2,\xi_2} + \lambda\left(PU_{\mu_1,\eta_1} - PU_{\mu_2,\eta_2}\right) + \phi, \tag{9}$$

where $\lambda \in \{-1, 0, +1\}$ ($\lambda = 0$ in case (iv), $\lambda = +1$ in the first case (v) and $\lambda = -1$ in the second case (v)) and the concentration parameters are

$$\mu_i = \delta_i \quad \text{and} \quad \delta_i := \varepsilon^{\frac{n-1}{n-2}} d_i \quad \text{with} \quad d_i > 0 \tag{10}$$

while the concentration points are aligned along the x_n-axes and satisfy

$$\eta_i = -\xi_i \quad \text{and} \quad \xi_i = (1+\tau_i)\xi_0 \text{ with } \tau_i := \varepsilon t_i,\ t_i > 0. \tag{11}$$

Next, we introduce the configuration space Λ where the concentration parameters and the concentration points lie. For solutions of type (6) we set $\mathbf{d} = d \in (0, +\infty)$ and $\mathbf{t} = t \in (0, +\infty)$ and so

$$\Lambda := \left\{(\mathbf{d}, \mathbf{t}) \in (0, +\infty) \times (0, +\infty)\right\},$$

while for solutions of type (9) we set $\mathbf{d} = (d_1, d_2) \in (0, +\infty)^2$ and $\mathbf{t} = (t_1, t_2) \in (0, +\infty)^2$ and so

$$\Lambda := \{(\mathbf{d}, \mathbf{t}) \in (0, +\infty)^4 \ :\ t_1 < t_2\}.$$

In each of these cases we write

$$V_{\mathbf{d},\mathbf{t}} := PU_{\delta,\xi} + \lambda PU_{\mu,\eta} \quad \text{or} \quad V_{\mathbf{d},\mathbf{t}} := PU_{\delta_1,\xi_1} - PU_{\delta_2,\xi_2} + \lambda\left(PU_{\mu_1,\eta_1} - PU_{\mu_2,\eta_2}\right).$$

The remainder term ϕ in both cases (6) and (9) belongs to a suitable space which we now define.

We introduce the spaces

$$K_{\mathbf{d},\mathbf{t}} := \text{span}\{P\psi^j_{\delta_i,\xi_i},\ P\psi^\ell_{\mu_\kappa,\xi_\kappa} \ :\ i, \kappa = 1, 2,\ j, \ell = 0, 1, \ldots, n\}$$

(we agree that if $\lambda = 0$ then $K_{\mathbf{d},\mathbf{t}}$ is only generated by the $P\psi^j_{\delta_i,\xi_i}$'s) and

$$K^\perp_{\mathbf{d},\mathbf{t}} := \left\{\phi \in \mathcal{H}_\lambda \ :\ (\phi, \psi) = 0 \quad \forall\, \psi \in K_{\mathbf{d},\mathbf{t}}\right\},$$

where the space $\mathcal{H}_\lambda$ depends on $\lambda \in \{-1, 0, +1\}$ and is defined by

$$\begin{aligned}
\mathcal{H}_0 &:= \{\phi \in \mathrm{H}^1_0(\Omega) \ :\ \phi \text{ is axially symmetric with respect to the } x_n\text{-axes } \},\\
\mathcal{H}_{+1} &:= \{\phi \in \mathcal{H}_0 \ :\ \phi(x_1, \ldots, x_n) = \phi(x_1, \ldots, -x_n\},\\
\mathcal{H}_{-1} &:= \{\phi \in \mathcal{H}_0 \ :\ \phi(x_1, \ldots, x_n) = -\phi(x_1, \ldots, -x_n\}.
\end{aligned}$$

Then we introduce the orthogonal projection operators $\Pi_{\mathbf{d},\mathbf{t}}$ and $\Pi^\perp_{\mathbf{d},\mathbf{t}}$ in $H^1_0(\Omega)$, respectively.

As usual for this reduction method, the approach to solve problem (4) or (5) will be to find a pair $(\mathbf{d}, \mathbf{t})$ and a function $\phi \in K^\perp_{\mathbf{d},\mathbf{t}}$ such that

$$\Pi^\perp_{\mathbf{d},\mathbf{t}}\left\{V_{\mathbf{d},\mathbf{t}} + \phi - i^*\left[\frac{1}{2|x|} f_\varepsilon\left(V_{\mathbf{d},\mathbf{t}} + \phi\right)\right]\right\} = 0 \tag{12}$$

and

$$\Pi_{\mathbf{d},\mathbf{t}}\left\{V_{\mathbf{d},\mathbf{t}}+\phi-i^*\left[\frac{1}{2|x|}f_\varepsilon\left(V_{\mathbf{d},\mathbf{t}}+\phi\right)\right]\right\}=0. \tag{13}$$

First, we shall find for any $(\mathbf{d},\mathbf{t})$ and for small ε, a function $\phi\in K_{\mathbf{d},\mathbf{t}}^\perp$ such that (12) holds. To this aim we define a linear operator $L_{\mathbf{d},\mathbf{t}}:K_{\mathbf{d},\mathbf{t}}^\perp\to K_{\mathbf{d},\mathbf{t}}^\perp$ by

$$L_{\mathbf{d},\mathbf{t}}\phi:=\phi-\Pi_{\mathbf{d},\mathbf{t}}^\perp i^*\left[f_0'\left(V_{\mathbf{d},\mathbf{t}}\right)\phi\right].$$

Proposition 2.1. *For any compact sets* $\mathbf{C}$ *in* Λ *there exists* $\varepsilon_0,c>0$ *such that for any* $\varepsilon\in(0,\varepsilon_0)$ *and for any* $(\mathbf{d},\mathbf{t})\in\mathbf{C}$ *the operator* $L_{\mathbf{d},\mathbf{t}}$ *is invertible and*

$$\|L_{\mathbf{d},\mathbf{t}}\phi\|\geq c\|\phi\|\qquad\forall\,\phi\in K_{\mathbf{d},\mathbf{t}}^\perp.$$

Proof. We argue as in Lemma 1.7 of [12]. □

Now, we are in position to solve equation (12).

Proposition 2.2. *For any compact sets* $\mathbf{C}$ *in* Λ *there exists* $\varepsilon_0,c,\sigma>0$ *such that for any* $\varepsilon\in(0,\varepsilon_0)$ *and for any* $(\mathbf{d},\mathbf{t})\in\mathbf{C}$ *there exists a unique* $\phi_{\mathbf{d},\mathbf{t}}^\varepsilon\in K_{\mathbf{d},\mathbf{t}}^\perp$ *such that*

$$\Pi_{\mathbf{d},\mathbf{t}}^\perp\left\{V_{\mathbf{d},\mathbf{t}}+\phi_{\mathbf{d},\mathbf{t}}^\varepsilon-i^*\left[\frac{1}{2|x|}f_\varepsilon\left(V_{\mathbf{d},\mathbf{t}}+\phi_{\mathbf{d},\mathbf{t}}^\varepsilon\right)\right]\right\}=0.$$

Moreover

$$\left\|\phi_{\mathbf{d},\mathbf{t}}^\varepsilon\right\|\leq c\varepsilon^{\frac{1}{2}+\sigma}.$$

Proof. First, we estimate the rate of the error term

$$R_{\mathbf{d},\mathbf{t}}:=\Pi_{\mathbf{d},\mathbf{t}}^\perp\left\{V_{\mathbf{d},\mathbf{t}}-i^*\left[\frac{1}{|x|}f_\varepsilon\left(V_{\mathbf{d},\mathbf{t}}\right)\right]\right\}$$

as

$$\|R_{\mathbf{d},\mathbf{t}}\|_{\frac{2n}{n+2}}=O\left(\varepsilon^{\frac{1}{2}+\sigma}\right)$$

for some $\sigma>0$. We argue as in Appendix B of [1] using estimates of Section 3. Then we argue exactly as in Proposition 2.3 of [5]. □

Now, we introduce the energy functional $J_\varepsilon:\mathrm{H}_0^1(\Omega)\to\mathbb{R}$ defined by

$$J_\varepsilon(u):=\frac{1}{2}\int\limits_\Omega|\nabla u|^2dx-\frac{1}{p+1-\varepsilon}\int\limits_\Omega\frac{1}{2|x|}|u|^{p+1-\varepsilon}dx,$$

whose critical points are the solutions to problem (4). Let us define the reduced energy $\widetilde{J}_\varepsilon:\Lambda\to\mathbb{R}$ by

$$\widetilde{J}_\varepsilon(\mathbf{d},\mathbf{t})=J_\varepsilon\left(V_{\mathbf{d},\mathbf{t}}+\phi_{\mathbf{d},\mathbf{t}}^\varepsilon\right).$$

Next, we prove that the critical points of $\widetilde{J}_\varepsilon$ are the solution to problem (13).

Proposition 2.3. *The function* $V_{\mathbf{d},\mathbf{t}}+\phi_{\mathbf{d},\mathbf{t}}^\varepsilon$ *is a critical point of the functional* J_ε *if and only if the point* $(\mathbf{d},\mathbf{t})$ *is a critical point of the function* $\widetilde{J}_\varepsilon$.

Proof. We argue as in Proposition 1 of [3]. □

The problem is thus reduced to the search for critical points of $\widetilde{J}_\varepsilon$, so it is necessary to compute the asymptotic expansion of $\widetilde{J}_\varepsilon$.

Proposition 2.4. *It holds true that*

$$\widetilde{J}_\varepsilon(\mathbf{d},\mathbf{t}) = c_1 + c_2\varepsilon + c_3\varepsilon\log\varepsilon + \varepsilon(1+|\lambda|)\Phi(\mathbf{d},\mathbf{t}) + o(\varepsilon),$$

C^0*-uniformly on compact sets of* Λ, *where*

(i) *in case* (6)

$$\Phi(\mathbf{d},\mathbf{t}) := c_4\left(\frac{d}{2t}\right)^{n-2} + c_5 t - c_6 \ln d$$

(ii) *in case* (9)

$$\begin{aligned}\Phi(\mathbf{d},\mathbf{t}) := c_4 &\left[\left(\frac{d_1}{2t_1}\right)^{n-2} + \left(\frac{d_2}{2t_2}\right)^{n-2}\right. \\ &\left. + 2\left(d_1 d_2\right)^{\frac{n-2}{2}}\left(\frac{1}{|t_1-t_2|^{n-2}} - \frac{1}{|t_1+t_2|^{n-2}}\right)\right] \\ &+ c_5\left(t_1+t_2\right) - c_6\left(\ln d_1 + \ln d_2\right).\end{aligned}$$

Here c_i*'s are constants and* c_4, c_5 *and* c_6 *are positive.*

Proof. The proof is postponed to Section 3. □

Proof of Theorem 1.5. It is easy to verify that the function Φ of Proposition 2.4 in both cases has a minimum point which is stable under uniform perturbations. Therefore, if ε is small enough there exists a critical point $(\mathbf{d}_\varepsilon, \mathbf{t}_\varepsilon)$ of the reduced energy $\widetilde{J}_\varepsilon$. Finally, the claim follows by Proposition 2.3. □

3. Expansion of the reduced energy

It is standard to prove that

$$\widetilde{J}_\varepsilon(\mathbf{d},\mathbf{t}) = J_\varepsilon\left(V_{\mathbf{d},\mathbf{t}}\right) + o(\varepsilon)$$

(see for example [3, 5]). So the problem reduces to estimating the leading term $J_\varepsilon\left(V_{\mathbf{d},\mathbf{t}}\right)$. We will estimate it in case (9) with $|\lambda| = 1$, because in the other cases its expansion is easier and can be deduced from that. Proposition 2.4 will follow from Lemma 3.1, Lemma 3.2 and Lemma 3.3.

For future reference we define the constants

$$\gamma_1 = \alpha_n^{p+1}\int_{\mathbb{R}^n}\frac{1}{(1+|y|^2)^n}dy, \tag{14}$$

$$\gamma_2 = \alpha_n^{p+1}\int_{\mathbb{R}^n}\frac{1}{(1+|y|^2)^{\frac{n+2}{2}}}dy, \tag{15}$$

$$\gamma_3 = \alpha_n^{p+1}\int_{\mathbb{R}^n}\frac{1}{(1+|y|^2)^n}\log\frac{1}{(1+|y|^2)^{\frac{n-2}{2}}}dy. \tag{16}$$

For sake of simplicity, we set $U_i := U_{\delta_i,\xi_i}$ and $V_i := V_{\mu_i,\eta_i}$.

Lemma 3.1. *It holds true that*

$$\frac{1}{2}\int_\Omega |\nabla V_{\mathbf{d},\mathbf{t}}|^2 dx = 2\gamma_1 - \gamma_2\varepsilon\Bigg[\left(\frac{d_1}{2t_1}\right)^{n-2} + \left(\frac{d_2}{2t_2}\right)^{n-2} + (d_1 d_2)^{\frac{n-2}{2}}\left(\frac{1}{|t_1-t_2|^{n-2}} - \frac{1}{|t_1+t_2|^{n-2}}\right)\Bigg] + o(\varepsilon).$$

Proof. We have

$$\begin{aligned}\int_\Omega |\nabla V_{\mathbf{d},\mathbf{t}}|^2 dx &= \int_\Omega |\nabla PU_1|^2 dx + \int_\Omega |\nabla PU_2|^2 dx - 2\int_\Omega \nabla PU_1 \nabla PU_2 dx \\ &\quad + \int_\Omega |\nabla PV_1|^2 dx + \int_\Omega |\nabla PV_2|^2 dx - 2\int_\Omega \nabla PV_1 \nabla PV_2 dx \\ &\quad + 2\sum_{i,j=1}^{2} \lambda \int_\Omega \nabla PU_i \nabla PV_j dx \\ &= 2\left(\int_\Omega |\nabla PU_1|^2 dx + \int_\Omega |\nabla PU_2|^2 dx - 2\int_\Omega \nabla PU_1 \nabla PU_2 dx\right) + o(\varepsilon),\end{aligned} \tag{17}$$

because of the symmetry (see (10) and (11)) and the fact that

$$\int_\Omega \nabla PU_i \nabla PV_j dx = O\left(\delta_i^{\frac{n-2}{2}} \mu_j^{\frac{n-2}{2}}\right) = o(\varepsilon).$$

Let us estimate the first term in (17). The estimate of the second term is similar. We set

$$\tau := \min\left\{\mathrm{d}(\xi_1,\partial\Omega), \mathrm{d}(\xi_2,\partial\Omega), \frac{|\xi_1-\xi_2|}{2}\right\} = \min\left\{\tau_1,\tau_2,\frac{|\tau_1-\tau_2|}{2}\right\}. \tag{18}$$

We get

$$\int_\Omega |\nabla PU_1|^2 dx = \int_\Omega U_1^p PU_1 dx = \int_{B(\xi_1,\tau)} U_1^p PU_1 dx + \int_{\Omega\setminus B(\xi_1,\tau)} U_1^p PU_1 dx.$$

By Lemma A.1 we deduce

$$\int_{\Omega\setminus B(\xi_1,\tau)} U_1^p PU_1 dx = O\left(\left(\frac{\delta_1}{\tau}\right)^n\right) = o(\varepsilon)$$

$$\int_{B(\xi_1,\tau)} U_1^p PU_1 dx = \int_{B(\xi_1,\tau)} U_1^{p+1} dx + \int_{B(\xi_1,\tau)} U_1^p (PU_1 - U_1)\, dx, \tag{19}$$

with

$$\int\limits_{B(\xi_1,\tau)} U_1^{p+1} = \gamma_1 + O\left(\left(\frac{\delta_1}{\tau_1}\right)^n\right) = \gamma_1 + o(\varepsilon).$$

The second term in (19) is estimated in (i) of Lemma 3.4.

It remains only to estimate the third term in (17).

$$\int\limits_{\Omega} \nabla PU_1 \nabla PU_2 dx = \int\limits_{\Omega} U_1^p PU_2 dx = \int\limits_{B(\xi_1,\tau)} U_1^p PU_2 dx + \int\limits_{\Omega\setminus B(\xi_1,\tau)} U_1^p PU_2 dx. \quad (20)$$

We have

$$\int\limits_{\Omega\setminus B(\xi_1,\tau)} U_1^p PU_2 dx = O\left(\delta_1^{\frac{n+2}{2}}\delta_2^{\frac{n-2}{2}} \int\limits_{\Omega\setminus B(\xi_1,\tau)} \frac{1}{|x-\xi_1|^{n+2}}\frac{1}{|x-\xi_2|^{n-2}}dx\right)$$

$$= O\left(\frac{\delta_1^{\frac{n+2}{2}}\delta_2^{\frac{n-2}{2}}}{\tau^n} \int\limits_{\mathbb{R}^n\setminus B(0,1)} \frac{1}{|y|^{n+2}}\frac{1}{|y+\frac{\xi_1-\xi_2}{\tau}|^{n-2}}dy\right) = O\left(\frac{\delta_1^{\frac{n+2}{2}}\delta_2^{\frac{n-2}{2}}}{\tau^n}\right) = o(\varepsilon).$$

The first term in (20) is estimated in (ii) of Lemma 3.4.

The claim then follows collecting all the previous estimates and taking into account the choice of δ_i's and τ_i's made in (10) and (11). □

Lemma 3.2. *It holds true that*

$$\frac{1}{p+1}\int\limits_{\Omega} \frac{1}{|x|}|V_{\mathbf{d},\mathbf{t}}|^{p+1}dx = 2\left[\frac{2}{p+1}\gamma_1 - \frac{1}{p+1}\gamma_1\varepsilon\left(t_1+t_2\right)\right.$$
$$- 2\gamma_2\varepsilon\left[\left(\frac{d_1}{2t_1}\right)^{n-2} + \left(\frac{d_2}{2t_2}\right)^{n-2}\right.$$
$$\left.\left. + 2\left(d_1d_2\right)^{\frac{n-2}{2}}\left(\frac{1}{|t_1-t_2|^{n-2}} - \frac{1}{|t_1+t_2|^{n-2}}\right)\right]\right] + o(\varepsilon).$$

Proof. We have

$$\int\limits_{\Omega} \frac{1}{|x|}|V_{\mathbf{d},\mathbf{t}}|^{p+1}dx = \int\limits_{\Omega} \frac{1}{|x|}|PU_1 - PU_2 + \lambda\left(PV_1 - PV_2\right)|^{p+1}dx \quad (21)$$
$$= \int\limits_{\Omega} \frac{1}{|x|}\left(|PU_1 - PU_2 + \lambda\left(PV_1 - PV_2\right)|^{p+1}\right.$$
$$\left. - |U_1|^{p+1} - |U_2|^{p+1} - |V_1|^{p+1} - |V_2|^{p+1}\right)dx$$
$$+ \int\limits_{\Omega} \frac{1}{|x|}\left(|U_1|^{p+1} + |U_2|^{p+1} + |V_1|^{p+1} + |V_2|^{p+1}\right)dx$$

$$= \int_\Omega \frac{1}{|x|}\big(|PU_1 - PU_2 + \lambda(PV_1 - PV_2)|^{p+1}$$
$$- |U_1|^{p+1} - |U_2|^{p+1} - |V_1|^{p+1} - |V_2|^{p+1}\big)dx$$
$$+ 2\int_\Omega \frac{1}{|x|}\left(|U_1|^{p+1} + |U_2|^{p+1}\right)dx,$$

because of the symmetry (see (10) and (11)).

The last two terms in (21) are estimated in (v) of Lemma 3.4. Let τ as in (18). We split the first integral as

$$\int_\Omega \frac{1}{|x|}\big(|PU_1 - PU_2 + \lambda(PV_1 - PV_2)|^{p+1} \tag{22}$$
$$- |U_1|^{p+1} - |U_2|^{p+1} - |V_1|^{p+1} - |V_2|^{p+1}\big)dx$$
$$= \int_{B(\xi_1,\tau)} + \cdots + \int_{B(\xi_2,\tau)} + \cdots + \int_{B(-\xi_1,\tau)} + \cdots + \int_{B(-\xi_2,\tau)} + \cdots$$
$$\cdots + \int_{\Omega\setminus(B(\xi_1,\tau)\cup B(\xi_2,\tau)\cup B(-\xi_1,\tau)\cup B(-\xi_2,\tau))} \cdots$$

By Lemma A.1 we deduce

$$\int_{\Omega\setminus(B(\xi_1,\tau)\cup B(\xi_2,\tau)\cup B(-\xi_1,\tau)\cup B(-\xi_2,\tau))} \cdots$$
$$= O\left(\int_{\Omega\setminus(B(\xi_1,\tau)\cup B(\xi_2,\tau)\cup B(-\xi_1,\tau)\cup B(-\xi_2,\tau))} \left(U_1^{p+1} + U_2^{p+1} + V_1^{p+1} + V_2^{p+1}\right)dx\right)$$
$$= O\left(\frac{\delta_1^n}{\tau^n} + \frac{\delta_2^n}{\tau^n}\right) = o(\varepsilon).$$

We now estimate the integral over $B(\xi_1,\tau)$ in (22).

$$\int_{B(\xi_1,\tau)} \frac{1}{|x|}\big(|PU_1 - PU_2 + \lambda(PV_1 - PV_2)|^{p+1} \tag{23}$$
$$- |U_1|^{p+1} - |U_2|^{p+1} - |V_1|^{p+1} - |V_2|^{p+1}\big)dx$$
$$= (p+1)\int_{B(\xi_1,\tau)} \frac{1}{|x|}U_1^p\left(PU_1 - U_1 - PU_2 + \lambda(PV_1 - PV_2)\right)dx$$
$$+ \frac{p(p+1)}{2}\int_{B(\xi_1,\tau)} \frac{1}{|x|}|U_1 + \theta\rho|^{p-1}\rho^2 dx$$

$$-\int_{B(\xi_1,\tau)} \frac{1}{|x|}\left(|U_2|^{p+1}+|V_1|^{p+1}-|V_2|^{p+1}\right)dx$$

$$=(p+1)\int_{B(\xi_1,\tau)} \frac{1}{|x|}U_1^p\left(PU_1-U_1\right)dx-(p+1)\int_{B(\xi_1,\tau)} \frac{1}{|x|}U_1^pPU_2dx+o(\varepsilon),$$

where $\rho := PU_1 - U_1 - PU_2 + \lambda\,(PV_1 - PV_2)$. Indeed, by Lemma A.1 one can easily deduce that

$$\int_{B(\xi_1,\tau)} \frac{1}{|x|}U_1^p\left(PV_1-PV_2\right)dx,\ \int_{B(\xi_1,\tau)} \frac{1}{|x|}|U_2|^{p+1}dx,\ \int_{B(\xi_1,\tau)} \frac{1}{|x|}|V_i|^{p+1}dx=o(\varepsilon)$$

and also

$$\frac{p(p+1)}{2}\int_{B(\xi_1,\tau)} \frac{1}{|x|}|U_1+\theta\rho|^{p-1}\rho^2dx\le c\int_{B(\xi_1,\tau)} |U_1|^{p-1}\rho^2dx+\int_{B(\xi_1,\tau)} |\rho|^{p+1}dx$$

$$\le c\int_{B(\xi_1,\tau)} U_1^{p-1}\left(PU_1-U_1\right)^2dx+c\int_{B(\xi_1,\tau)} U_1^{p-1}\left(PU_2\right)^2dx$$

$$+c\int_{B(\xi_1,\tau)} U_1^{p-1}\left(PV_1-PV_2\right)^2dx+c\int_{B(\xi_1,\tau)} |PU_1-U_1|^{p+1}dx$$

$$+c\int_{B(\xi_1,\tau)} |U_2|^{p+1}dx+c\int_{B(\xi_1,\tau)} \left(|V_1|^{p+1}+|V_2|^{p+1}\right)dx$$

$$=o(\varepsilon).$$

The first term and the second term in (23) are estimated in (iii) and (iv) of Lemma 3.4, respectively.

Therefore, the claim follows. □

Lemma 3.3. *It holds true that*

$$\frac{1}{p+1-\varepsilon}\int_\Omega \frac{1}{|x|}|V_{\mathbf{d},\mathbf{t}}|^{p+1-\varepsilon}=\frac{1}{p+1}\int_\Omega \frac{1}{|x|}|V_{\mathbf{d},\mathbf{t}}|^{p+1}$$

$$+(1+|\lambda|)\left[\frac{\gamma_1}{(p+1)^2}-\alpha_n\frac{\gamma_1}{(p+1)}-\frac{\gamma_3}{(p+1)}+\frac{n-2}{2(p+1)}\left(\ln\delta_1+\ln\delta_2\right)\right]\varepsilon+o(\varepsilon).$$

Proof. We argue exactly as in Lemma 3.2 of [7]. □

Lemma 3.4. *Let τ as in* (18). *It holds true that*

(i)

$$\int_{B(\xi_1,\tau)} U_1^p\left(PU_1-U_1\right)dx=-\gamma_2\left(\frac{\delta_1}{2\tau_1}\right)^{n-2}+o(\varepsilon)$$

(ii)

$$\int_{B(\xi_1,\tau)} U_1^p PU_2 dx = \gamma_2 (\delta_1\delta_2)^{\frac{n-2}{2}} \left(\frac{1}{|\tau_1 - \tau_2|^{n-2}} - \frac{1}{|\tau_1 + \tau_2|^{n-2}} \right) + o(\varepsilon)$$

(iii)

$$\int_{B(\xi_1,\tau)} \frac{1}{|x|} U_1^p (PU_1 - U_1)\, dx = -\gamma_2 \left(\frac{\delta_1}{2\tau_1}\right)^{n-2} + o(\varepsilon)$$

(iv)

$$\int_{B(\xi_1,\tau)} \frac{1}{|x|} U_1^p PU_2 dx = -\gamma_2 \left(\frac{\delta_1}{2\tau_1}\right)^{n-2} + o(\varepsilon)$$

(v)

$$\int_{\Omega} \frac{1}{|x|} U_1^{p+1} dx = \gamma_1 - \gamma_1\tau_1 + o(\varepsilon).$$

Proof. Proof of (i) By Lemma A.1 we get

$$\int_{B(\xi_1,\tau)} U_1^p (PU_1 - U_1)\, dx = \int_{B(\xi_1,\tau)} U_1^p \left(-\alpha_n \delta_1^{\frac{n-2}{2}} H(x,\xi_1) + R_{\delta_1,\xi_1}\right) dx$$
$$= -\alpha_n \delta_1^{\frac{n-2}{2}} \int_{B(\xi_1,\tau)} U_1^p H(x,\xi_1) dx + \int_{B(\xi_1,\tau)} U_1^p R_{\delta_1,\xi_1} dx,$$

with

$$\int_{B(\xi_1,\tau)} U_1^p R_{\delta_1,\xi_1} dx = O\left(\left(\frac{\delta_1}{\tau_1}\right)^n\right).$$

By Lemma 3.5 we get

$$\alpha_n \delta_1^{\frac{n-2}{2}} \int_{B(\xi_1,\tau)} U_1^p H(x,\xi_1) dx = \alpha_n^{p+1} \delta_1^{n-2} \int_{B(0,\tau/\delta_1)} H(\delta_1 y + \xi_1, \xi_1) \frac{1}{(1+|y|^2)^{\frac{n+2}{2}}} dy$$
$$= \alpha_n^{p+1} \left(\frac{\delta_1}{\tau_1}\right)^{n-2} \int_{B(0,\tau/\delta_1)} \tau_1^{n-2} H(\delta_1 y + \xi_1, \xi_1) \frac{1}{(1+|y|^2)^{\frac{n+2}{2}}} dy$$
$$= \alpha_n^{p+1} \left(\frac{\delta_1}{\tau_1}\right)^{n-2} \left[\frac{1}{2^{n-2}} \int_{\mathbb{R}^n)} \frac{1}{(1+|y|^2)^{\frac{n+2}{2}}} dy + o(1) \right].$$

Proof of (ii). By Lemma A.1 and Lemma 3.5 we get

$$\int\limits_{B(\xi_1,\tau)} U_1^p PU_2 dx = \int\limits_{B(\xi_1,\tau)} U_1^p \left(U_2 - \alpha_n \delta_2^{\frac{n-2}{2}} H(x,\xi_2) + R_{\delta_2,\xi_2}\right) dx$$

$$= \alpha_n^{p+1} (\delta_1\delta_2)^{\frac{n-2}{2}} \int\limits_{B(0,\tau/\delta_1)} \frac{1}{(1+|y|^2)^{\frac{n+2}{2}}} \frac{1}{(\delta_2^2 + |\delta_1 y + \xi_1 - \xi_2|^2)^{\frac{n-2}{2}}} dy$$

$$- \alpha_n^{p+1} (\delta_1\delta_2)^{\frac{n-2}{2}} \int\limits_{B(0,\tau/\delta_1)} \frac{1}{(1+|y|^2)^{\frac{n+2}{2}}} H(\delta_1 y + \xi_1, \xi_2) dy$$

$$+ \alpha_n^{p+1} (\delta_1\delta_2)^{\frac{n-2}{2}} \int\limits_{B(0,\tau/\delta_1)} \frac{1}{(1+|y|^2)^{\frac{n+2}{2}}} R_{\delta_2,\xi_2}(\delta_1 y + \xi_1) dy$$

$$= \alpha_n^{p+1} \frac{(\delta_1\delta_2)^{\frac{n-2}{2}}}{|\tau_1 - \tau_2|^{n-2}} \int\limits_{B(0,\tau/\delta_1)} \frac{1}{(1+|y|^2)^{\frac{n+2}{2}}} \frac{|\tau_1 - \tau_2|^{n-2}}{(\delta_2^2 + |\delta_1 y + \xi_1 - \xi_2|^2)^{\frac{n-2}{2}}} dy$$

$$- \alpha_n^{p+1} \frac{(\delta_1\delta_2)^{\frac{n-2}{2}}}{|\tau_1 + \tau_2|^{n-2}} \int\limits_{B(0,\tau/\delta_1)} \frac{|\tau_1 + \tau_2|^{n-2}}{(1+|y|^2)^{\frac{n+2}{2}}} H(\delta_1 y + \xi_1, \xi_2) dy$$

$$+ O\left((\delta_1\delta_2)^{\frac{n-2}{2}} \frac{\delta_2^{\frac{n+2}{2}}}{\tau_2^n} \right)$$

$$= \alpha_n^{p+1} \frac{(\delta_1\delta_2)^{\frac{n-2}{2}}}{|\tau_1 - \tau_2|^{n-2}} \left[\int\limits_{\mathbb{R}^n} \frac{1}{(1+|y|^2)^{\frac{n+2}{2}}} dy + o(1) \right]$$

$$- \alpha_n^{p+1} \frac{(\delta_1\delta_2)^{\frac{n-2}{2}}}{|\tau_1 + \tau_2|^{n-2}} \left[\int\limits_{\mathbb{R}^n} \frac{1}{(1+|y|^2)^{\frac{n+2}{2}}} dy + o(1) \right]$$

$$+ o\left(\frac{(\delta_1\delta_2)^{\frac{n-2}{2}}}{\tau_2^{n-2}} \right).$$

Proof of (iii) *and* (iv) We argue as in the proof of (i) and (ii) using estimates (25) and (26).

Proof of (v). We have

$$\int\limits_{\Omega} \frac{1}{|x|} U_1^{p+1} dx = \int\limits_{B(\xi_1,\tau)} \frac{1}{|x|} U_1^{p+1} dx + \int\limits_{\Omega\setminus B(\xi_1,\tau)} \frac{1}{|x|} U_1^{p+1} dx, \tag{24}$$

with

$$\int\limits_{\Omega\setminus B(\xi_1,\tau)} \frac{1}{|x|} U_1^{p+1} dx = O\left(\frac{\delta_1^n}{\tau^n}\right),$$

So, we only have to estimate the first term in (24). We split it as

$$\int_{B(\xi_1,\tau)} \frac{1}{|x|} U_1^{p+1} dx = \int_{B(\xi_1,\tau)} U_1^{p+1} dx + \int_{B(\xi_1,\tau)} \left(\frac{1}{|x|} - 1\right) U_1^{p+1} dx.$$

We have

$$\int_{B(\xi_1,\tau)} U_1^{p+1} dx = \gamma_1 + O\left(\frac{\delta_1^n}{\tau^n}\right).$$

Since $\xi_1 = \xi_0(1+\tau_1)$ and $|\xi_0| = 1$, by the mean value theorem we get

$$\frac{1}{|\delta_1 y + \tau_1 \xi_0 + \xi_0|} - 1 = -\tau_1 - \delta_1 \langle y, \xi_0 \rangle + R(y), \tag{25}$$

where R satisfies the uniform estimate

$$|R(y)| \le c\left(\delta_1^2 |y|^2 + \delta_1 \tau_1 |y| + \tau_1^2\right) \text{ for any } y \in B(0, \tau/\delta_1). \tag{26}$$

Therefore we conclude

$$\int_{B(\xi_1,\tau)} \left(\frac{1}{|x|} - 1\right) U_1^{p+1} dx = \alpha_n^{p+1} \int_{B(0,\tau/\delta_1)} \left(\frac{1}{|\delta_1 y + \tau_1 \xi_0 + \xi_0|} - 1\right) \frac{1}{(1+|y|^2)^n} dy$$

$$= \alpha_n^{p+1} \int_{B(0,\tau/\delta_1)} \left(-\tau_1 - \delta_1 \tau_1 \langle y, \xi_0 \rangle + R(y)\right) \frac{1}{(1+|y|^2)^n} dy = -\gamma_1 \tau_1 + o(\tau).$$

Collecting all the previous estimates we get the claim. □

Lemma 3.5. *Let τ as in* (18). *It holds true that*

(i)

$$\int_{B(0,\tau/\delta_1)} \tau_1^{n-2} H(\delta_1 y + \xi_1, \xi_1) \frac{1}{(1+|y|^2)^{\frac{n+2}{2}}} dy = \frac{1}{2^{n-2}} \int_{\mathbb{R}^n} \frac{1}{(1+|y|^2)^{\frac{n+2}{2}}} dy + o(1),$$

(ii)

$$\int_{B(0,\tau/\delta_1)} \frac{|\tau_1 + \tau_2|^{n-2}}{(1+|y|^2)^{\frac{n+2}{2}}} H(\delta_1 y + \xi_1, \xi_2) dy = \int_{\mathbb{R}^n} \frac{1}{(1+|y|^2)^{\frac{n+2}{2}}} dy + o(1),$$

(iii)

$$\int_{B(0,\tau/\delta_1)} \frac{1}{(1+|y|^2)^{\frac{n+2}{2}}} \frac{|\tau_1 - \tau_2|^{n-2}}{(\delta_2^2 + |\delta_1 y + \xi_1 - \xi_2|^2)^{\frac{n-2}{2}}} dy = \int_{\mathbb{R}^n} \frac{1}{(1+|y|^2)^{\frac{n+2}{2}}} dy + o(1).$$

Proof. We are going to use Lebesgue's dominated convergence theorem together with Lemma A.2. First of all, taking into account that $\xi_1 = (1+\tau_1)\xi_0$ and $\bar{\xi}_1 = (1-\tau_1)\xi_0$ we deduce that

$$\tau_1^{n-2} H(\delta_1 y + \xi_1, \xi_1) \frac{1}{(1+|y|^2)^{\frac{n+2}{2}}} \to \frac{1}{2^{n-2}} \frac{1}{(1+|y|^2)^{\frac{n+2}{2}}} \text{ a.e. in } \mathbb{R}^n$$

and also that

$$H(\delta_1 y+\xi_1,\xi_1)\le C_2\frac{1}{|\delta_1 y+\xi_1-\bar{\xi}_1|^{n-2}}=C_2\frac{1}{|\delta_1 y+2\tau_1\xi_0|^{n-2}}\le C_2\frac{1}{\tau_1^{n-2}},$$

since

$$|\delta_1 y+2\tau\xi_0|\ge 2\tau_1-|\delta_1 y|\ge\tau_1 \text{ for any } y\in B(0,\tau/\delta_1).$$

That proves (i).

In a similar way, taking into account that $\xi_1=(1+\tau_1)\xi_0$ and $\bar{\xi}_2=(1-\tau_2)\xi_0$ we get

$$(\tau_2+\tau_1)^{n-2}H(\delta_1 y+\xi_1,\xi_2)\frac{1}{(1+|y|^2)^{\frac{n+2}{2}}}\to\frac{1}{(1+|y|^2)^{\frac{n+2}{2}}} \text{ a.e. in } \mathbb{R}^n$$

and also that

$$H(\delta_1 y+\xi_1,\xi_2)\le C_2\frac{1}{|\delta_1 y+\xi_1-\bar{\xi}_2|^{n-2}}=C_2\frac{1}{|\delta_1 y+(\tau_1+\tau_2)\xi_0|^{n-2}}\le C_2\frac{1}{\tau_2^{n-2}},$$

since

$$|\delta_1 y+(\tau_1+\tau_2)\xi_0|\ge\tau_1+\tau_2-|\delta_1 y|\ge\tau_2 \text{ for any } y\in B(0,\tau/\delta_1).$$

That proves (ii).

Finally, we have

$$\frac{1}{(1+|y|^2)^{\frac{n+2}{2}}}\frac{|\tau_1-\tau_2|^{n-2}}{(\delta_2^2+|\delta_1 y+\xi_1-\xi_2|^2)^{\frac{n-2}{2}}}\to\frac{1}{(1+|y|^2)^{\frac{n+2}{2}}} \text{ a.e. in } \mathbb{R}^n$$

and also that

$$\frac{1}{(\delta_2^2+|\delta_1 y+\xi_1-\xi_2|^2)^{\frac{n-2}{2}}}\le\frac{1}{|\delta_1 y+\xi_1-\xi_2|^{n-2}}\le\frac{2^{n-2}}{|\tau_1-\tau_2|^{n-2}}$$

since

$$|\delta_1 y+\xi_1-\xi_2|\ge|\xi_1-\xi_2|-|\delta_1 y|\ge\frac{|\xi_1-\xi_2|}{2} \text{ for any } y\in B(0,\tau/\delta_1).$$

That proves (iii). □

Appendix

Here we recall some well-known facts which are useful to get estimates in Section 3.

We denote by $G(x,y)$ the Green's function associated to $-\Delta$ with Dirichlet boundary condition and $H(x,y)$ its regular part, i.e.,

$$-\Delta_x G(x,y)=\delta_y(x) \quad \text{for } x\in\Omega, \quad G(x,y)=0 \quad \text{for } x\in\partial\Omega,$$

and

$$G(x,y)=\gamma_n\left(\frac{1}{|x-y|^{n-2}}-H(x,y)\right) \quad \text{where} \quad \gamma_n=\frac{1}{(n-2)|S^{n-1}|}$$

($|S^{n-1}|=(2\pi^{n/2})/\,\Gamma(n/2)$ denotes the Lebesgue measure of the $(n-1)$-dimensional unit sphere).

The following lemma was proved in [17].

Lemma A.1. *It holds true that*

$$PU_{\delta,\xi}(x) = U_{\delta,\xi}(x) - \alpha_n \delta^{\frac{n-2}{2}} H(x,\xi) + O\left(\frac{\delta^{\frac{n+2}{2}}}{\operatorname{dist}(\xi,\partial\Omega)^n}\right)$$

for any $x \in \Omega$.

Since Ω is smooth, we can choose small $\epsilon > 0$ such that, for every $x \in \Omega$ with $\operatorname{dist}(x,\partial\Omega) \le \epsilon$, there is a unique point $x_\nu \in \partial\Omega$ satisfying $\operatorname{dist}(x,\partial\Omega) = |x - x_\nu|$. For such $x \in \Omega$, we define $\bar{x} = 2x_\nu - x$ the reflection point of x with respect to $\partial\Omega$.

The following two lemmas are proved in [1].

Lemma A.2. *It holds true that*

$$\left| H(x,y) - \frac{1}{|\bar{x}-y|^{n-2}} \right| = O\left(\frac{\operatorname{dist}(x,\partial\Omega)^n}{|\bar{x}-y|^{n-2}}\right)$$

and

$$\left| \nabla_x \left(H(x,y) - \frac{1}{|\bar{x}-y|^{n-2}} \right) \right| = O\left(\frac{1}{|\bar{x}-y|^{n-2}}\right)$$

for any $x \in \Omega$ *with* $\operatorname{dist}(x,\partial\Omega) \le \epsilon$.

References

[1] N. Ackermann, M. Clápp, A. Pistoia, *Boundary clustered layers for some supercritical problems.* J. Differential Equations **254** (2013), no. 10, 4168–4193.

[2] A. Bahri, J.M. Coron, *On a nonlinear elliptic equation involving the critical Sobolev exponent: the effect of the topology of the domain.* Comm. Pure Appl. Math. **41** (1988), 253–294

[3] A. Bahri, Y. Li, O. Rey, *On a variational problem with lack of compactness: the topological effect of the critical points at infinity,* Calc. Var. Partial Differential Equations **3** (1995), 67–93

[4] T. Bartsch, M. Clápp, M. Grossi, F. Pacella, *Asymptotically radial solutions in expanding annular domains,* Math. Ann. **352** (2012), no. 2, 485–515.

[5] T. Bartsch, A. Micheletti, A. Pistoia, *On the existence and the profile of nodal solutions of elliptic equations involving critical growth,* Calc. Var. Partial Differential Equations **26** (2006), 265–282.

[6] M. Clapp, J. Faya, A. Pistoia, *Nonexistence and multiplicity of solutions to elliptic problems with supercritical exponents,* Calc. Var. PDE (2012; doi: 10.1007/s00526-012-0564-6), arXiv:1212.5137.

[7] M. Del Pino, P. Felmer, M. Musso, *Two-bubble solutions in the super-critical Bahri–Coron's problem,* Calc. Var. Partial Differential Equations **16** (2003), 113–145.

[8] M. Del Pino, M. Musso, F. Pacard, *Bubbling along boundary geodesics near the second critical exponent,* J. Eur. Math. Soc. (JEMS) **12** (2010), no. 6, 1553–1605.

[9] F. Gladial, M. Grossi, *Supercritical elliptic problem with nonautonomous nonlinearities,* J. Differential Equations **253** (2012), no. 9, 2616–2645.

[10] S. Kim, A. Pistoia, *Clustered boundary layer sign changing solutions for a supercritical problem*, Journal London Math. Soc. (2013; doi: 10.1112/jlms/jdt006).

[11] S. Kim, A. Pistoia, *Boundary towers of layers for some supercritical problems*, J. Differential Equations **255** (2013), 2302–2339.

[12] M. Musso, A. Pistoia, *Multispike solutions for a nonlinear elliptic problem involving the critical Sobolev exponent*, Indiana Univ. Math. **51** (2002), no. 3, 541–579.

[13] F. Pacard, F. Pacella, B. Sciunzi, *Solutions of semilinear elliptic equations in tubes*, Journal of Geometric Analysis **24** (2014), 445–471.

[14] F. Pacella, P.N. Srikanth, *A reduction method for semilinear elliptic equations and solutions concentrating on spheres*, J. Funct. Anal. (to appear).

[15] D. Passaseo, *Nonexistence results for elliptic problems with supercritical nonlinearity in nontrivial domains*, J. Funct. Anal. **114** (1993), 97–105.

[16] S.I. Pohozaev, *Eigenfunctions of the equation* $\Delta u + \lambda f(u) = 0$, Soviet Math. Dokl. **6** (1965), 1408–1411.

[17] O. Rey, *The role of Green's function in a nonlinear elliptic equation involving the critical Sobolev exponent*, J. Funct. Anal. **89** (1990), 1–52.

[18] B. Ruf, P.N. Srikanth, *Singularly perturbed elliptic equations with solutions concentrating on a 1-dimensional orbit*, J. Eur. Math. Soc. (JEMS) **12** (2010), no. 2, 413–427.

Filomena Pacella
Dipartimento di Matematica "G. Castelnuovo"
Università di Roma "La Sapienza"
P.le Aldo Moro 5
I-00185 Roma, Italy
e-mail: `pacella@mat.uniroma1.it`

Angela Pistoia
Dipartimento SBAI
Università di Roma "La Sapienza"
via Antonio Scarpa 16
I-00161 Roma, Italy
e-mail: `pistoia@sbai.uniroma1.it`

Progress in Nonlinear Differential Equations
and Their Applications, Vol. 85, 341–352

Normalized Solutions for a Schrödinger–Poisson System Under a Neumann Condition

Lorenzo Pisani and Gaetano Siciliano

Dedicated to Prof. Bernhard Ruf in occasion of his 60th birthday

Abstract. In this paper we study the existence of normalized standing wave solutions for a Schrödinger–Poisson system in a bounded domain of $\mathbf{R}^3$. We assign a Dirichlet boundary condition for the wave function and a Neumann boundary condition for the potential ϕ. In particular this last condition has some interesting consequences which force us to consider the case in which the interaction "constant" q is merely a constant function or not. However with very mild assumption on q we are able to find infinitely many solutions in both cases. The result presented here can be found with all the details in the papers [14, 16].

Mathematics Subject Classification (2010). 35J50, 35J57, 35Q55.

Keywords. Standing waves, electrostatic field, bounded domain, variational methods, Krasnoselskii genus.

1. Introduction

In this paper we summarize and compare two results obtained in the papers [14, 16]. They are concerned with the following Schrödinger–Poisson type system in a bounded and smooth domain $\Omega \subset \mathbf{R}^3$:

$$\begin{cases} -\Delta u + q\phi u = \omega u & \text{in } \Omega \\ -\Delta \phi = qu^2 & \text{in } \Omega \end{cases} \tag{1}$$

with the boundary conditions

$$\begin{cases} u = 0 & \text{on } \partial\Omega, \\ \frac{\partial \phi}{\partial \mathbf{n}} = h & \text{on } \partial\Omega \end{cases} \tag{2}$$

where h is a smooth function defined on $\partial\Omega$.

For a physical interpretation and a rigorous derivation of the above system we refer the reader to [6, 7] or [12]. Here we simply say that the equations (1) arise from the search of stationary solutions $\psi(x,t) = e^{-i\omega t}u(x)$ for a Schrödinger equation coupled with the Maxwell equations in the purely electrostatic case: $\mathbf{E} = -\nabla\phi(x)$, $\mathbf{A} = \mathbf{0}$. The interaction is governed by the charge density q, which is a datum of the problem and satisfies a suitable condition that will be presented in a while. The physical meaning of the boundary conditions is that the particle is constrained "to live" in Ω and that the normal component of the electric field $\mathbf{E}$ is assigned on the boundary of Ω; in other words, h prescribes the flux $\mathfrak{F} = -\int_{\partial\Omega} h ds$ of the electric field through the boundary $\partial\Omega$.

The equations above are known in the literature under the names Schrödinger–Maxwell or Schrödinger–Poisson.

The unknown of the problem are the real functions u, ϕ and the real number ω, so in our case the wave function is completely unknown: we have to find the modulus and the frequency; consequently a solution of the system is a triple (u, ω, ϕ).

We are interested in normalized wave functions, so we impose on u to satisfy the condition

$$\int_\Omega u^2 dx = 1 \tag{3}$$

which makes sense from a physical and probabilistic point of view.

Note that the Neumann datum imposes a further constraint (compatibility condition) on u: every solution has to satisfy

$$\mathfrak{F} = \int_\Omega q u^2 \, dx. \tag{4}$$

Now we distinguish two cases.

1.1. The case $q \in C(\overline{\Omega})$ and not constant

In this case we are in presence of a density of charge varying from point to point in Ω. The interested reader is referred to [1, 9] and references therein for the derivation of the system.

We set $q_{\min} := \min q(\overline{\Omega})$ and $q_{\max} := \max q(\overline{\Omega})$.

We have the two constraints

$$S := \left\{ u \in H_0^1(\Omega) : \int_\Omega u^2 dx = 1 \right\}, \tag{5}$$

$$N := \left\{ u \in H_0^1(\Omega) : \int_\Omega q u^2 = \mathfrak{F} \right\}, \tag{6}$$

different in nature: the first one is due to the fact that we have imposed the normalizing condition, the second one is imposed by the Neumann condition. We define $M := S \cap N$. Of course, any (eventual) solution u is in M and is such that

$$q_{\min} \le \mathfrak{F} \le q_{\max}.$$

So it makes sense to consider the set $q^{-1}(\mathfrak{F})$; actually a major role is played by its measure, indeed

1) if $|q^{-1}(\mathfrak{F})| = 0$, then M is not a differentiable manifold;
2) if $|q^{-1}(\mathfrak{F})| \neq 0$ and $\mathfrak{F} \in \{q_{\min}, q_{\max}\}$ then $M = \emptyset$.

Point 2) is easy to see, while point 1) requires some technicalities, on which we will return.

Our result, obtained in [16], is the following

Theorem 1.1. *Assume*

$$q_{\min} < \mathfrak{F} < q_{\max} \tag{7}$$

and

$$|q^{-1}(\mathfrak{F})| = 0. \tag{8}$$

Then there exists a solution $(u, \omega, \phi) \in H_0^1(\Omega) \times \mathbf{R} \times H^1(\Omega)$ *of system* (1)–(2)–(3) *such that* $u \geq 0$.

Moreover there exist infinitely many solutions $(u_n, \omega_n, \phi_n) \in H_0^1(\Omega) \times \mathbf{R} \times H^1(\Omega)$ *with* $\int_\Omega |\nabla u_n|^2 dx \to +\infty$.

The assumptions of Theorem 1.1 are immediately satisfied if $\mathfrak{F}$ is a regular value of q.

Note that (7) implies that q cannot be constant, and it cannot have constant value $\mathfrak{F}$ on any open subset of Ω, due to condition (8).

1.2. The case q constant and different from zero

Now we have a uniform distribution of charge in the domain Ω. Then the compatibility condition (4) becomes

$$\mathfrak{F} = q \int_\Omega u^2 dx = q \neq 0. \tag{9}$$

This implies that a necessary condition in order to have solutions simply reduces to assign h such that the relation above is satisfied.

In this case the value ω plays no role in the existence of solutions, indeed the problem enjoys some invariance: $(u, 0, \phi)$ is a solution of

$$\begin{cases} -\Delta u + q\phi u = 0 & \text{in } \Omega, \\ -\Delta \phi = qu^2 & \text{in } \Omega, \\ \int_\Omega u^2 dx = 1 & \\ u = 0 & \text{on } \partial\Omega, \\ \frac{\partial \phi}{\partial \mathbf{n}} = h & \text{on } \partial\Omega, \end{cases} \tag{10}$$

if and only if $(u, \omega, \phi - \omega/q)$ is a solution of (1)–(2). In other words, if we find $\psi(x,t) = u(x)$, a static solution of the Schrödinger–Maxwell equations, then we have also the stationary solutions $u(x)e^{-i\omega t}$ with any frequency $\omega \in \mathbf{R}$. All these solutions are associated to the same electric field, indeed the change of variable $\phi \mapsto \phi - \omega/q$ has no effect on $\mathbf{E}$.

Now our result is the following.

Theorem 1.2. *Assume condition* (9). *Then there exists a solution* $(u, \phi) \in H_0^1(\Omega) \times H^1(\Omega)$ *of system* (10) *such that* $u \geq 0$.

Moreover there exist infinitely many solutions $(u_n, \phi_n) \in H_0^1(\Omega) \times H^1(\Omega)$ *of* (10) *with* $\int_\Omega |\nabla u_n|^2 dx \to +\infty$ *and* $\int_\Omega \phi_n dx \to \infty$.

It is worth to note that the invariance of the problem by the shift $(u, 0, \phi) \to (u, \omega, \phi - \omega/q)$ can be used to reduce the problem to find solutions with $\int_\Omega \phi\, dx$ given (see the remarks at the end of the paper).

Many authors have studied this kind of systems; here we recall [1, 2, 3, 6, 8, 9, 10, 11, 13, 15, 17, 18, 19] and the reference therein. However these authors do not consider the case with given L^2-norm. The case with fixed L^2-norm on u has also been considered in the recent papers [4, 5].

The next two sections are devoted to prove, by variational methods, Theorem 1.1 and Theorem 1.2. Note that, as usual in this kinds of problems, we can also add a nonlinearity satisfying some growth condition and everything works (see [14, 16] for more details).

1.3. Notations

As a matter of notations, we endow the Sobolev space $H_0^1(\Omega)$ with the usual norm $\|\nabla u\|_2$ where $\|u\|_p$ is the L^p-norm. For an integrable function v, we will use also the notation $\bar{v} = |\Omega|^{-1} \int_\Omega v\, dx$. Finally, we will use the letter c to denote a generic positive constant (independent of u) whose value may change also from line to line.

2. Proof of Theorem 1.1

Our analysis in this case begins by studying the constraint M which is evidently symmetric with respect to the origin and weakly closed in $H_0^1(\Omega)$. The first step is to show that, under condition (7), M is not empty. To this aim, it is useful to state a general fact like the next proposition. Of course by taking $A = \Omega$ we will deduce that M is not vacuus.

Proposition 2.1. *Let A be an open subset of Ω. If $\mathfrak{F} \in (\inf q(A), \sup q(A))$ then there exists $u \in H_0^1(A)$ such that*

$$\int_A u^2\, dx = 1 \qquad \textit{and} \qquad \int_A q u^2\, dx = \mathfrak{F}.$$

For the proof we refer the reader to the paper [16].

To obtain the multiplicity result stated in Theorem 1.1, we will use the theory of *genus* of Krasnoselskii, so some topological properties of M have to be investigated. We first recall the definition of *genus*: for every C closed and symmetric subset of a topological space, the *genus* of C, denoted by $\gamma(C)$, is defined as the smallest integer $k \in \mathbf{N}$ for which there exists an odd and continuous map $h : C \to \mathbf{R}^k \setminus \{0\}$. If there is no finite such k, we set $\gamma(C) = +\infty$ and, finally, $\gamma(\emptyset) = 0$.

Theorem 2.2. *Let $u_1, \ldots, u_k \in M$ be functions with disjoint supports. Then the genus of M is at least k.*

Proof. Denoting with V_k the k-dimensional space generated by $u_1, \ldots, u_k$, it happens that $M \cap V_k = S \cap V_k$, that is the unit sphere in V_k. Indeed, it is obvious that $M \cap V_k \subset S \cap V_k$. On the other hand, if $u = \sum_{i=1}^k \alpha_i u_i \in S \cap V_k$, then $1 = \|u\|_2^2 = \sum_{i=1}^k \alpha_i^2$ which implies that

$$\int_\Omega qu^2 dx = \int_\Omega q \sum_{i=1}^k \alpha_i^2 u_i^2 dx = \sum_{i=1}^k \alpha_i^2 \int_\Omega q u_i^2 dx = \mathfrak{F} \sum_{i=1}^k \alpha_i^2 = \mathfrak{F},$$

and so $u \in M$. □

The key fact now is that we can find infinitely many functions in M with disjoint support.

Theorem 2.3. *Assume* (7). *Then, for every $k \geq 2$, there exist k functions u_1, u_2, $\ldots$, $u_k \in M$ having disjoint supports. Hence $\gamma(M) = +\infty$.*

Proof. By (7), the subsets

$$\Omega_+ = \{x \in \Omega : q(x) > \mathfrak{F}\}, \quad \Omega_- = \{x \in \Omega : q(x) < \mathfrak{F}\}$$

are open and not empty. We can choose $2k$ disjoint balls $\{Y_1, \ldots, Y_k\} \subset \Omega_-$, $\{Z_1, \ldots, Z_k\} \subset \Omega_+$, then we set $A_i = Y_i \cup Z_i$, $i = 1, \ldots, k$. It follows by construction that

$$\inf q(A_i) < \mathfrak{F} < \sup q(A_i).$$

Therefore we apply Proposition 2.1 and we find $u_i \in H_0^1(A_i)$ such that

$$\int_\Omega u_i^2\, dx = \int_{A_i} u_i^2\, dx = 1 \quad \text{and} \quad \int_\Omega q u_i^2\, dx = \int_{A_i} q u_i^2\, dx = \mathfrak{F}$$

(identifying a function in $H_0^1(A_i)$ with its trivial extension). All these functions $u_1, u_2, \ldots, u_k \in M$ have disjoint supports. □

Remark 2.4. It is easy to see that just under condition

$$|q^{-1}(\mathfrak{F})| \neq 0$$

the set M is not empty; indeed we can consider functions u whose support is contained in $q^{-1}(\mathfrak{F})$; analogously, using Theorem 2.2, the set M has infinite genus, too.

We will see now that the measure of $q^{-1}(\mathfrak{F})$ concerns with the differential structure of M. Define

$$G_1 : u \in H_0^1(\Omega) \mapsto \int_\Omega u^2 dx - 1 \in \mathbf{R}$$

$$G_2 : u \in H_0^1(\Omega) \mapsto \int_\Omega q u^2 dx - \mathfrak{F} \in \mathbf{R}$$

then we set $\mathcal{G} = (G_1, G_2)$, so that $M = \{u \in H_0^1(\Omega) : G_1(u) = G_2(u) = 0\} = \mathcal{G}^{-1}(0)$.

The next task now is to show that $\mathcal{G}$ defined above is a submersion and so M a submanifold in $H_0^1(\Omega)$ of codimension 2.

Proposition 2.5. *Assume that M is not empty. For every $u \in M$ the differentials $G_1'(u)$ and $G_2'(u)$ are linearly independent if and only if condition (8) holds, i.e.,*

$$|q^{-1}(\mathfrak{F})| = 0.$$

In this case M is a differentiable manifold of codimension 2 and for every $u \in M$ the tangent space at u is

$$T_u M = \ker \mathcal{G}'(u) = \left\{ v \in H_0^1(\Omega) : \int_\Omega uv\, dx = \int_\Omega quv\, dx = 0 \right\}.$$

We prove here just the sufficiency of condition (8) in order to be M a smooth manifold of codimension 2. For the proof of the necessity see [16].

Proof. Assume (8) and consider

$$\alpha G_1'(u) + \beta G_2'(u) = 0 \quad \text{in } H^{-1}(\Omega) \quad (\alpha, \beta \in \mathbf{R}). \tag{11}$$

Evaluating (11) on $u \in M$ we obtain $\alpha + \beta\mathfrak{F} = 0$. So it is sufficient to prove that $\beta = 0$. Now (11) becomes

$$\beta \left(-\mathfrak{F} \int_\Omega uv\, dx + \int_\Omega quv\, dx \right) = 0 \quad \forall v \in H_0^1(\Omega)$$

that is

$$\beta \int_\Omega (q - \mathfrak{F}) uv\, dx = 0 \quad \forall v \in H_0^1(\Omega).$$

If it were $\beta \neq 0$, then $(q - \mathfrak{F})u = 0$ a.e. and, by (8), it would be $u = 0$: a contradiction. □

In particular we see that the condition (8) is responsible for S and N to be "tangential" or "transversal" (see [16]).

Now we can proceed with the proof of the theorem, by describing the variational framework and defining a suitable functional whose critical points will give the solutions of our problem.

We first introduce the unique function χ which solves the auxiliary problem

$$\begin{cases} \Delta\chi = -\mathfrak{F}/\left|\Omega\right| & \text{in } \Omega, \\ \int_\Omega \chi\, dx = 0 \\ \frac{\partial\chi}{\partial\mathbf{n}} = h & \text{on } \partial\Omega \end{cases} \tag{12}$$

and then we perform the change of variables

$$\mu = \frac{1}{|\Omega|}\int_\Omega \phi\, dx \quad \text{and} \quad \varphi = \phi - \chi - \mu \tag{13}$$

so that $\overline{\varphi} = 0$. We define $\widetilde{H} = \left\{\eta \in H^1(\Omega) : \overline{\eta} = 0\right\}$ and note that $H^1(\Omega) = \widetilde{H} \oplus \mathbf{R}$.

With the new variables $(u, \omega, \varphi, \mu)$, our problem (1) with conditions (2) and (3), becomes

$$\begin{cases} -\Delta u + q(\chi + \varphi)u = \omega u - \mu q u & \text{in } \Omega \\ -\Delta\varphi = qu^2 - \mathfrak{F}/|\Omega| & \text{in } \Omega \\ \int_\Omega u^2\, dx = 1 \\ \varphi \in \tilde{H} \\ u = 0 & \text{on } \partial\Omega \\ \dfrac{\partial\varphi}{\partial\mathbf{n}} = 0 & \text{on } \partial\Omega. \end{cases} \tag{14}$$

The compatibility condition due to the Neumann condition reads again as $u \in N$, where N has been defined in (6).

Consider the C^1 functional $F : H_0^1(\Omega) \times H^1(\Omega) \longrightarrow \mathbf{R}$ defined as follows

$$F(u,\varphi) = \frac{1}{2}\int_\Omega |\nabla u|^2 dx + \frac{1}{2}\int_\Omega q(\varphi + \chi)u^2 dx - \frac{1}{4}\int_\Omega |\nabla\varphi|^2 dx - \frac{\mathfrak{F}}{2|\Omega|}\int_\Omega \varphi\, dx.$$

so that, for every $u \in H_0^1(\Omega)$ and $\varphi \in H^1(\Omega)$ we have

$$\langle F_u'(u,\varphi), v\rangle = \int_\Omega (\nabla u \nabla v + q(\varphi + \chi)uv)\, dx \qquad \forall v \in H_0^1(\Omega),$$

$$\langle F_\varphi'(u,\varphi), \xi\rangle = \frac{1}{2}\int_\Omega q\xi u^2 dx - \frac{1}{2}\int_\Omega \nabla\varphi\nabla\xi\, dx - \frac{\mathfrak{F}}{2|\Omega|}\int_\Omega \xi\, dx \quad \forall \xi \in H^1(\Omega).$$

Under our conditions (7) and (8) we know that M is not empty and a manifold of codimension 2 (Proposition 2.1 and Proposition 2.5) so that an application of the Lagrange Multipliers Theorem gives

Theorem 2.6. *Let $(u,\varphi) \in H_0^1(\Omega) \times H^1(\Omega)$. Then there exist $\omega, \mu \in \mathbf{R}$ such that (u,φ,ω,μ) is a solution of (14) if and only if (u,φ) is a critical point of F constrained on $M \times \tilde{H}$; the real constants ω and μ are the two Lagrange multipliers with respect to F_u'.*

Note that the restriction $\tilde{H}$ is a natural constraint with respect to F_φ' so no Lagrange multipliers appear with respect to this derivative.

The functional F constrained on $M \times \tilde{H}$ is unbounded from above and from below, even modulo compact perturbations.

For this kind of problem the usual strategy to proceed is the following:

1. for every fixed u, one find the unique solution $\varphi = \Phi(u)$ of the equation $F'_\varphi(u, \varphi) = 0$,
2. so a correspondence $u \mapsto \Phi(u)$ is defined and it is usually C^1,
3. the graph of Φ is a manifold and it is a natural constraint for finding critical points of F; in concrete terms we are reduced to study the functional $J(u) = F(u, \Phi(u))$, possibly constrained.

It is immediately seen, in our case, that the equation $F'_\varphi(u, \varphi) = 0$, with $\varphi \in \tilde{H}$ is just

$$\begin{cases} \Delta\varphi + qu^2 - \mathfrak{F}/|\Omega| = 0 & \text{in } \Omega, \\ \dfrac{\partial\varphi}{\partial \mathbf{n}} = 0 & \text{on } \partial\Omega, \\ \int_\Omega \varphi\, dx = 0\,. \end{cases} \tag{15}$$

When we try to apply the procedure described above, we find two obstacles. First, this problem (15) has a unique solution if and only if $u \in N$. The second difficulty concerns with the regularity of the map $\Phi : N \to \tilde{H}$; indeed, since N is not a manifold (unless we make the further assumption $\mathfrak{F} \neq 0$), we cannot require this map to be C^1 (in a classical sense).

To address these problems, we use an idea, already introduced in [14], of extending the map Φ. This is done by using the next two results, whose prove is standard.

Proposition 2.7. *For every $w \in L^{6/5}(\Omega)$ there exists a unique $L(w) \in \tilde{H}$ solution of*

$$\begin{cases} \Delta\varphi + w - \bar{w} = 0 & \textit{in } \Omega, \\ \dfrac{\partial\varphi}{\partial \mathbf{n}} = 0 & \textit{on } \partial\Omega, \\ \int_\Omega \varphi\, dx = 0. \end{cases}$$

The map $L : L^{6/5}(\Omega) \to \tilde{H}$ is linear and continuous, hence C^∞. Moreover we have

$$\|\nabla L(w)\|_2 \leq c\,\|w\|_{6/5}$$

Furthermore $\ker L = \mathbf{R}$ and if $w \in L^2(\Omega)$, then $\|L(w)\|_{H^2} \leq c\,\|w - \overline{w}\|_2$

Proposition 2.8. *The map $u \in L^6(\Omega) \longmapsto qu^2 \in L^{6/5}(\Omega)$ is of class C^1.*

As a consequence of these facts, we can define the C^1 map on all $H_0^1(\Omega)$

$$\Phi : u \in H_0^1(\Omega) \mapsto L(qu^2) \in \tilde{H}.$$

Clearly, $\Phi(u) = \Phi(-u) = \Phi(|u|)$ and for every $(u, \varphi) \in H_0^1(\Omega) \times \tilde{H}$, we have $\varphi = \Phi(u)$ if and only if, for every $\eta \in \tilde{H}$

$$\int_\Omega \nabla\varphi\nabla\eta\, dx = \int_\Omega qu^2\eta\, dx. \tag{16}$$

In particular, by taking $\eta = \Phi(u)$ we infer that

$$\int_\Omega |\nabla\Phi(u)|^2\, dx = \int_\Omega qu^2\Phi(u)\, dx, \tag{17}$$

from which we deduce $\|\nabla\Phi(u)\|_2^2 \le c\|q\|_\infty \|u\|_4^2 \|\nabla\Phi(u)\|_2$ or equivalently

$$\|\nabla\Phi(u)\|_2 \le c\|\nabla u\|_2^2 \tag{18}$$

for some positive constant c. In other words, we have proved that the map Φ is bounded on bounded sets. By standard computations we can prove more:

Lemma 2.9. *The map Φ is compact. Moreover, if $u_n \rightharpoonup u$ in $H_0^1(\Omega)$ then*

$$\int_\Omega qu_n^2\Phi(u_n)dx \to \int_\Omega qu^2\Phi(u)dx.$$

Note that $\Phi(u) \in \tilde{H}$ is the unique solution of (15) whenever $u \in N$, hence, for every $u \in N$, we have $F'_\varphi(u, \Phi(u)) = 0$. Now we can define the "reduced" functional which depends on the single variable u:

$$J : H_0^1(\Omega) \to \mathbf{R} \quad \text{such that} \quad J(u) = F(u, \Phi(u)).$$

From now on, we will use the notation $\varphi_u = \Phi(u)$, hence explicitly J is given by

$$J(u) = \frac{1}{2}\int_\Omega |\nabla u|^2 dx + \frac{1}{4}\int_\Omega |\nabla\varphi_u|^2 dx + \frac{1}{2}\int_\Omega q\chi u^2 dx \tag{19}$$

and is C^1 and even. Moreover, for every $u \in M \subset N$, the differential of J is given by

$$\begin{aligned}\langle J'(u), v\rangle &= \langle F'_u(u,\varphi_u), v\rangle + \langle F_\varphi(u,\varphi_u), \Phi'(u)v\rangle \\ &= \langle F'_u(u,\varphi_u), v\rangle \qquad \forall v \in H_0^1(\Omega)\end{aligned}$$

so we deduce, by an application of the Implicit Function Theorem, the following

Theorem 2.10. *The pair $(u, \varphi) \in M \times \tilde{H}$ is a critical point of F constrained on $M \times \tilde{H}$ if and only if u is a critical point of $J_{|M}$ and $\varphi = \Phi(u)$.*

Recalling that M is weakly closed, the first part of Theorem 1.1 is now a consequence of the following lemma.

Lemma 2.11. *The functional J on M is weakly lower semicontinuous and coercive. In particular it has a minimum u.*

Proof. Indeed by (19), $J(u) \ge \frac{1}{2}\int_\Omega |\nabla u|^2 dx - c$ so it is bounded from below and coercive on M. By Lemma 2.9 we deduce the weakly lower semicontinuity. □

Note that we can assume that the minimum is positive since $J(u) = J(|u|)$.

2.1. Multiplicity of solutions

A basic tool to prove the existence of (many) critical points is the well-known Palais–Smale condition. We recall, in general, that if J is a C^1 functional defined on a smooth manifold M, a sequence $\{u_n\} \subset M$ is a Palais–Smale sequence for J if $\{J(u_n)\}$ is bounded and $J'_{|M}(u_n) \to 0$. Moreover, the functional J is said to satisfy the Palais–Smale condition on M if every Palais–Smale sequence has a convergent subsequence to an element of M.

Proposition 2.12. *The functional J satisfies the Palais–Smale condition on M.*

We will not prove the proposition: the details can be found in [16]. The existence of infinitely many critical points $\{u_n\}$ for J on M is a consequence of Lemma 2.11, Proposition 2.12 and Theorem 2.3 (see, e.g., Corollary 4.1 of [20]). To the critical points $\{u_n\}_n$ are associated the Lagrange multipliers $\{\omega_n\}$ and $\{\mu_n\}$ and then, via (13), infinitely many solutions $(u_n, \omega_n, \phi_n) \in H_0^1(\Omega) \times \mathbf{R} \times H^1(\Omega)$ of (1)–(2)–(3).

Finally, the sublevels of the functional have finite genus (it is standard, for a general proof of this fact see, e.g., [6], Lemma 9). On the other hand, M has infinite genus hence, the critical levels have to be divergent, i.e.,

$$J(u_n) = \frac{1}{2}\int_\Omega |\nabla u_n|^2 dx + \frac{1}{4}\int_\Omega |\nabla \varphi_n|^2 dx + \frac{1}{2}\int_\Omega q\chi u_n^2 dx \to +\infty.$$

Consequently, since $\left\{\int_\Omega q\chi u_n^2 dx\right\}_n$ is bounded and $\|\nabla \varphi_n\|_2 \leq c\|\nabla u_n\|_2^2$ we infer that necessarily $\|\nabla u_n\|_2 \to +\infty$, concluding the proof of Theorem 1.1.

3. Proof of Theorem 1.2

We will just present a sketch of the proof since it follows the same lines of the proof in the previous case (the details can be found in [14]). Actually this case is simpler since we only need to work with the single constrained N defined in (6), or equivalently, with S (they differ for an insignificant constant, recall condition (9)). This constrained is, as well, a smooth manifold of codimension one, is symmetric with respect to the origin and weakly closed in $H_0^1(\Omega)$. Moreover the genus is infinite, and, as in the previous section, the main steps to prove Theorem 1.2 are:

Step 1: One introduces a function χ as in (12) and makes a change of variables to have a homogeneous boundary Neumann condition on the potential ϕ; observe that this change of variable introduces a Lagrange multiplier in the problem (10) and we arrive to system (14) with $\omega = 0$: this time we have a unique Lagrange multiplier which is μ (the average of ϕ on Ω) "acting" on S.

Step 2: The problem is reduced to find critical points of a functional of two variables F on the constraint $S \times \widetilde{H}$, which is strongly unbounded.

Step 3: Again there is a technicality to define a global map $\Phi : H_0^1(\Omega) \to \widetilde{H}$, but after that, a functional of a single variable $J(u) = F(u, \Phi(u))$ can be defined. Note that the expression of the functional J is essentially the same of the previous case.

Step 4: One is then reduced to find the critical points of J on the sphere S, and an easy computation gives $\mu_n = \langle F'_u(u_n, \Phi(u_n)), u_n\rangle = \langle J'(u_n), u_n\rangle$;

Step 5: Since J is bounded from below on S and satisfies the Palais–Smale condition, an application of the Ljusternik–Schnirelmann Theory gives the desired result. Finally it is easy to see that the divergence of the μ_k is deduced by the divergence of $\|\nabla u_k\|$.

To conclude we note two facts:

1) In both cases (q constant or not) appears a hidden Lagrange multiplier: the average of ϕ on Ω. This multiplier is not evident in the original system and appears when dealing with the Neumann condition.
2) In the case q constant the shift invariance can be also formulated in the following way: (u, ω, ϕ) is a solution of (1)–(2) if and only if $(u, \omega + q\,\mu, \phi + \mu)$ is a solution too, for every $\mu \in \mathbf{R}$. As a consequence, instead of fixing the value $\omega = 0$ as we have done, we can study the problem by fixing the value of $\int_\Omega \phi dx$, let us say zero, and allowing $\omega \neq 0$. In this case the theorem we arrive states the existence of infinitely many solutions of type (u_n, ω_n, ϕ_n) with $\int_\Omega \phi_n dx = 0$. Of course the two approaches are equivalent, but now the constraint on ϕ will appear since the beginning. The proof follows with very minor changes.

Acknowledgment

The authors are supported by M.I.U.R. "Metodi Variazionali e Topologici ed Equazioni Differenziali nonlineari" and by the Gruppo Nazionale per l'Analisi Matematica, la Probabilità e le loro Applicazioni (GNAMPA) of the Istituto Nazionale di Alta Matematica (INdAM), Italy. The second author is also supported by Projeto Temático 2011/21362-2, Pronex "Núcleo de Análise Geométrica e Aplicações", Fapesp and CNPq, Brazil.

References

[1] A. Ambrosetti, On Schrödinger–Poisson systems, *Milan J. Math.* **76** (2008), 257–274.

[2] C.O. Alves, M.A.Soares Souto and S.H.M. Soares, Schrödinger–Poisson equations without Ambrosetti–Rabinowitz condition, *J. Math. Anal. Appl.* **377** (2011), 584–592.

[3] A. Azzollini and A. Pomponio, Ground state solutions for the nonlinear Schrödinger–Maxwell equations, *J. Math. Anal. Appl.* **345** (1) (2008), 90–108.

[4] J. Bellazzini and G. Siciliano, Stable standing waves for a class of nonlinear Schrödinger–Poisson equations, *Z. Angew. Math. Phys.* **62** (2011), 267–280.

[5] J. Bellazzini and G. Siciliano, On the relation between scaling properties of functionals and existence of constrained minimizers, *J. Funct. Anal.* **261** (2011), 2486–2507.

[6] V. Benci and D. Fortunato, An eigenvalue problem for the Schrödinger–Maxwell equations, *Topol. Methods Nonlinear Anal.* **11** (2) (1998), 283–293.

[7] D. Bleecker, Gauge Theory and Variational Principles, (Dover Publications, 2005).

[8] G.M. Coclite, A multiplicity result for the nonlinear Schrödinger–Maxwell equations, *Commun. Appl. Anal.* **7** (2-3) (2003), 417–423.

[9] G. Cerami and G. Vaira, Positive solutions for some non autonomous Schrödinger–Poisson Systems, *J. Diff. Equations* **248** (3) (2010), 521–543.

[10] T. D'Aprile and D. Mugnai, Solitary waves for nonlinear Klein–Gordon–Maxwell and Schrödinger–Maxwell equations, *Proc. Roy. Soc. Edinburgh Sect. A* **134** (2004), 893–906.

[11] P. d'Avenia, Non-radially symmetric solutions of nonlinear Schrödinger equation coupled with Maxwell equations, *Adv. Nonlinear Stud.* **2** (2002), 177–192.

[12] B. Felsager, Geometry, Particles and Fields, (Springer, 1998).

[13] H. Kikuchi, On the existence of solutions for a elliptic system related to the Maxwell–Schrödinger equations, *Nonl. Anal.* **67** (2007), 1445–1456.

[14] L. Pisani and G. Siciliano, Neumann condition in the Schrödinger–Maxwell system, *Topol. Methods Nonlinear Anal.* **29** (2007), 251–264.

[15] L. Pisani and G. Siciliano, Note on a Schrödinger-Poisson system in a bounded domain, *Appl. Math. Lett.* **21** (2008), 521–528.

[16] L. Pisani and G. Siciliano, Constrained Schrödinger–Poisson system with non constant interaction, *Comm. Cont. Math.*, **15** n. 1 (2013), 1250052 (18 pages).

[17] D. Ruiz, The Schrödinger–Poisson equation under the effect of a nonlinear local term, *J. Funct. Anal.* **237** (2006), 655–674.

[18] A. Salvatore, Multiple solitary waves for a non-homogeneous Schrö-dinger-Maxwell system in $\mathbf{R}^3$, *Adv. Nonlinear Stud.* **6** (2) (2006), 157–169.

[19] G. Siciliano, Multiple positive solutions for a Schrödinger–Poisson–Slater system, *J. Math. Anal. Appl.* **365** (2010), 288–299.

[20] A. Szulkin, Ljusternik–Schnirelmann theory on C^1-manifolds, *Ann. Inst. H. Poincaré Anal. Non Linéaire* **5** (2) (1988), 119–139.

[21] M.E. Taylor, Partial Differential Equations I, (Springer, 1996).

Lorenzo Pisani
Università degli Studi di Bari
Dipartimento di Matematica
via E. Orabona 4
I-70125 Bari, Italy
e-mail: `lorenzo.pisani@uniba.it`

Gaetano Siciliano
Universidade de São Paulo
Departamento de Matemática
Rua do Matão, 1010
São Paulo, SP, 05508-090, Brazil
e-mail: `sicilian@ime.usp.br`

Progress in Nonlinear Differential Equations
and Their Applications, Vol. 85, 353–385

On Singular Liouville Systems

A. Poliakovsky and G. Tarantello

Abstract. We discuss a class of planar systems of Liouville type in presence of singular sources. When the coupling matrix admits positive entries, we provide necessary and sufficient conditions for the existence of radial solutions and corresponding uniqueness. For this purpose we point out a log HLS inequality in system's form that involves weights and holds in the radial setting.

Mathematics Subject Classification (2010). 35J47, 35J61, 34A34, 47J20.

Keywords. Liouville systems with Dirac measures, cooperative systems, radial solutions, logarithm HLS inequalities for systems.

1. Introduction

Motivated by the study of vortex configurations in several self-dual gauge field theories (see, e.g., [T], [Y], [D]), in this note we consider a class of planar 'singular' Liouville systems in presence of Dirac measures supported at a given point (say the origin). In the 'regular' situation, where the Dirac measures are neglected, a similar class of systems have emerged in various area of Physics and Applied Mathematics, as discussed for example in [CK1], [CK2], [Ki1], [Ki2], [Wo], [CSW], [CLMP1], [CLMP2], [SW1], [SW2], [JoW1], [JoW2] [W], [LZ1] see also references therein. More precisely, we are concerned with the following problem:

$$\begin{cases} -\Delta U_i = \sum_{j=1}^{m} a_{ij} e^{b_j U_j} - 4\pi n_i \delta_0 \\ \int_{\mathbb{R}^2} e^{b_j U_j} < \infty \end{cases} \tag{1.1}$$

where: $b_j > 0$, $N_j := n_j b_j > -1$, $j = 1, \dots, m$ and δ_0 denotes the Dirac measure with pole at the origin. The coupling matrix $A = (a_{i,j})$ is assumed to be symmetric and irreducible. We can allow the matrix A to be degenerate, in order to include in our analysis also a model arising from the study of self-gravitating strings (see [Y], [CGS] and [PT]).

The research of the second author was supported by Firb–Ideas Program: "Analysis and Beyond" and by PRIN Program: "Nonlinear elliptic problems in the study of vortices and related topics".

Concerning the simpler case $m = 1$, where (1.1) reduces to the well-known 'singular' Liouville equation, we refer to [PrT] for a complete characterization of the corresponding solutions (see also [ChL] for previous results), and to [T] for a discussion on related analytical aspects (e.g., blow-up analysis, Harnack-type inequalities etc). In this direction we like to mention a non-trivial classification result that covers systems. It has been established recently in [LWY] for solutions of the 'singular' Toda system, where the coupling matrix in (1.1) is given by the Cartan matrix of the group $SU(m+1)$. See also [JoW2] for an analogous classification of solutions for the Toda system in absence of Dirac measures.

We shall focus on the radial problem, and when the coupling matrix admits non-negative entries, we provide necessary and sufficient conditions for the existence and uniqueness of a radial solution for (1.1). See Theorem 3.1, Corollary 3.2 and Theorem 4.1 for the precise statements. Our results extend and complete those of Chipot–Shafrir–Wolansky [CSW] and Lin–Zhang [LZ1] concerning (1.1) in absence of Dirac measures, and with a non-degenerate coupling matrix A. Note that, while for the "regular" problem (i.e., $n_i = 0$, for every $i = 1, \dots, m$,) every solution can be shown to be radially symmetric (about some point) (see [CK1], [CSW]), this is no longer the case when $n_i > 0$ for some i= $1, \dots, m$. To this purpose, see the non-radial solutions obtained in [PrT] for the single equation, or in [PT] for a degenerate system and in [LWY] for the Toda system.

Here we exploit in a crucial way the radial framework, and identify a sharp Log HLS inequality in $\mathbb{R}^2$ for systems involving weights, see (3.66) and (3.67). Notice that our Log HLS inequality holds only for radial functions, while it fails in general. It is used to prove coerciveness of the free energy functional associated to the (radial) system (1.1). Hence it allows us to obtain a solution via minimization. A similar approach was successfully used in [CSW] to handle the "regular" system, whose free energy functional can be controlled by a convenient sharp Log-HLS inequality (without weights), that always holds. See also [W] and [SW1], [SW2] and [SW3] for further extensions of the Log-HLS inequality in system's form, also considered over compact Riemannian manifolds and in connection with the Moser–Trudinger inequality (see [Mo], [Au] and [F]). This line of work follows the 'dual' approach of Carlen–Loss [CL], which concerns the single Liouville equation in relation with the associated free energy functional and Onofri inequality in S^2 (cfr. [On]). See Beckner [B] for stronger results in this direction, including higher dimensions, and [Do] for recent developments on Log-HLS inequalities.

As already mentioned, our (radial) inequality, (see (3.66) below) involves weights and follows by using ad-hoc one-dimensional arguments. In general, it does not hold for non-radial functions, as one can check by arguing as in Lemma 2.2 in [CSW]. For the single equation similar observations are contained in [DET1], see also [Tr] in connection to surfaces with conical singularities.

According to our results, we have a rather complete understanding of radial solutions for (1.1), when the entries of the matrix A are non-negative, (the so-called 'cooperative' case, in the language of population dynamics). We hope to use such information together with a "perturbation" approach to treat more complex systems arising in the study of non-abelian Chern–Simons vortices.

On the contrary, non-radial solutions for (1.1) are far from being understood, and at this point it is not clear when to look for their existence or when to expect their non-existence.

Recent contributions for 'conflicting' systems involving coupling matrix with negative off diagonal entries can be found in [FT].

2. Preliminaries

Throughout this paper we let $I = \{1, \ldots, m\}$ and consider $A = \{a_{ij}\}_{1 \leq i,j \leq m} \in \mathbb{R}^{m \times m}$ a symmetric and irreducible matrix that is:

$$a_{ij} = a_{ji} \ \forall i, j \in I; \ \ \forall \emptyset \neq J \subsetneq I \text{ there exist } i \in J \text{ and } j \in I \setminus J \text{ such that } a_{ij} \neq 0.$$

For $b_i > 0$ and $N_i > -1$, $i \in I$, we are interested to identify the m-ple $\beta = (\beta_1, \ldots, \beta_m)$ with $\beta_i > 0$ $i \in I$, so that the following elliptic system:

$$\begin{cases} -\Delta u_i = \sum_{j=1}^m a_{ij} |x|^{2N_j} e^{b_j u_j} & \text{in } \mathbb{R}^2 \\ \frac{1}{2\pi} \int_{\mathbb{R}^2} |x|^{2N_i} e^{b_i u_i} dx = \beta_i \end{cases} \quad i \in I \tag{2.1}$$

admits a <u>radial</u> solution $u(r) = (u_1(r), \ldots, u_m(r))$. Clearly by setting: $N_i = n_i b_i$ and $U_i(x) = u_i(x) + 2n_i \ln |x|$, $i \in I$ we see that problem (2.1) is equivalent to problem (1.1) of the Introduction. Actually, by using the change of variable $r = e^t$, and the unknowns:

$$v_i(t) := u_i(e^t) \qquad i \in I, \tag{2.2}$$

we reduce to consider the following boundary value problem:

$$\begin{cases} \frac{d^2 v_i}{dt^2} + \sum_{i=1}^m a_{ij} e^{2(N_j+1)t + b_j v_j} = 0 & \text{for } t \in \mathbb{R} \\ \frac{dv_i}{dt}(-\infty) = 0, \quad v_i(-\infty) \in \mathbb{R} \\ \int_{\mathbb{R}} e^{2(N_i+1)t + b_i v_i} dt = \beta_i \end{cases} \quad i \in I. \tag{2.3}$$

For a solution $(v_1, \ldots, v_m)$ of (2.3), we define the function:

$$f_i(t) := \int_{-\infty}^t e^{2(N_i+1)s + b_i v_i(s)} ds, \quad i \in I; \tag{2.4}$$

so that,

$$\frac{dv_i}{dt}(t) + \sum_{i=1}^m a_{ij} f_j(t) = 0 \qquad \forall t \in \mathbb{R}, \quad i \in I. \tag{2.5}$$

Furthermore, for a non-empty subset $J \subseteq I$, we let,

$$\Psi_J(t) := \sum_{i \in J} \left\{ \frac{1}{b_i} e^{2(N_i+1)t + b_i v_i(t)} - \frac{2(N_i+1)}{b_i} f_i(t) + \sum_{j \in J} \frac{1}{2} a_{ij} f_i(t) f_j(t) \right\}, \tag{2.6}$$

which satisfies:

$$\Psi_J(-\infty) = 0, \qquad \Psi_J(+\infty) = \Big(\sum_{i\in J}\sum_{j\in J}\frac{1}{2}a_{ij}\beta_i\beta_j\Big) - \sum_{i\in J}\frac{2(N_i+1)}{b_i}\beta_i;$$

and

$$\frac{d\Psi_J}{dt}(t) = \sum_{i\in J}\Big\{e^{2(N_i+1)t+b_iv_i(t)}\frac{dv_i}{dt}(t) + \frac{df_i}{dt}(t)\sum_{j\in J}a_{ij}f_j(t)\Big\}$$

$$= \sum_{i\in J}\Big\{e^{2(N_i+1)t+b_iv_i(t)}\Big(\frac{dv_i}{dt}(t) + \sum_{j\in I}a_{ij}f_j(t)\Big) - e^{2(N_i+1)t+b_iv_i(t)}\sum_{j\in I\setminus J}a_{ij}f_j(t)\Big\}$$

$$= -\sum_{i\in J}\sum_{j\in I\setminus J}a_{ij}e^{2(N_i+1)t+b_iv_i(t)}\int_{-\infty}^{t}e^{2(N_j+1)s+b_jv_j(s)}ds. \tag{2.7}$$

Therefore, if for some $h \in \{-1,1\}$ we suppose that $ha_{ij} \geq 0$ for every $i \neq j$, then by using the fact that A is irreducible, we find that,

$$\frac{d\Psi_I}{dt}(t) = 0 \quad \text{and} \quad h\frac{d\Psi_J}{dt}(t) < 0 \quad \text{for } t \in \mathbb{R} \text{ and } \emptyset \neq J \subsetneqq I. \tag{2.8}$$

Thus we conclude the following:

Proposition 2.1. *Assume that the symmetric matrix A is irreducible and for fixed $h \in \{-1,1\}$ there holds: $ha_{ij} \geq 0$ for every $i \neq j$. Then the following conditions are necessary for the existence of a radial solution to (2.1):*

$$\begin{cases} \beta_i > 0 \qquad \forall i \in I \\ \Big(\sum_{i\in I}\sum_{j\in I}\frac{1}{2}a_{ij}\beta_i\beta_j\Big) - \sum_{i\in I}\frac{2(N_i+1)}{b_i}\beta_i = 0 \\ h\Big\{\Big(\sum_{i\in J}\sum_{j\in J}\frac{1}{2}a_{ij}\beta_i\beta_j\Big) - \sum_{i\in J}\frac{2(N_i+1)}{b_i}\beta_i\Big\} < 0 \quad \text{if} \quad \emptyset \neq J \subsetneqq I. \end{cases} \tag{2.9}$$

We shall focus here to the case where the matrix $A = \{a_{ij}\}_{i,j=1,\dots,m}$ admits non-negative entries, so that (2.9) holds with $h = 1$. We aim to show that actually in this case (2.9) provides also <u>sufficient</u> conditions for the solvability of (2.3). Results about the case $h = -1$, can be found in [FT].

We start to observe the following:

Lemma 2.1. *Assume $N_i > -1$, $b_i > 0$, $i \in I$ and suppose that $A = \{a_{ij}\}_{i,j=1,\dots,m}$ is a symmetric matrix with*

$$a_{ii} \geq 0 \quad \forall i \in I. \tag{2.10}$$

If (2.9) holds with $h = 1$, then for every $\emptyset \neq J \subsetneqq I$, we have:

$$\begin{cases} \Big(\sum_{i\in J}\sum_{j\in I\setminus J}a_{ij}\beta_i\beta_j\Big) + \Big(\sum_{i\in J}\sum_{j\in J}\frac{1}{2}a_{ij}\beta_i\beta_j\Big) - \sum_{i\in J}\frac{2(N_i+1)}{b_i}\beta_i > 0 \\ \Big(\sum_{i\in J}\sum_{j\in I\setminus J}a_{ij}\beta_i\beta_j\Big) > 0. \end{cases} \tag{2.11}$$

In particular A must be irreducible, and

$$\sum_{j\in I} a_{ij}\beta_j \geq \frac{1}{2}a_{ii}\beta_i + \sum_{j\in I\setminus\{i\}} a_{ij}\beta_j > \frac{2(N_i+1)}{b_i}, \qquad \textit{for every } i\in I. \tag{2.12}$$

Proof. Let $\emptyset \neq J \subsetneqq I$, then

$$\begin{aligned} 0 &= \Big(\sum_{i\in I}\sum_{j\in I}\frac{1}{2}a_{ij}\beta_i\beta_j\Big) - \sum_{i\in I}\frac{2(N_i+1)}{b_i}\beta_i \\ &= \Big(\sum_{i\in J}\Big\{\sum_{j\in J}\frac{1}{2}a_{ij}\beta_i\beta_j\Big\} - \sum_{i\in J}\frac{2(N_i+1)}{b_i}\beta_i\Big) \\ &\quad + \Big(\sum_{i\in I\setminus J}\Big\{\sum_{j\in I\setminus J}\frac{1}{2}a_{ij}\beta_i\beta_j\Big\} - \sum_{i\in I\setminus J}\frac{2(N_i+1)}{b_i}\beta_i\Big) \\ &\quad + \sum_{i\in J}\Big\{\sum_{j\in I\setminus J}\frac{1}{2}a_{ij}\beta_i\beta_j\Big\} + \sum_{i\in I\setminus J}\Big\{\sum_{j\in J}\frac{1}{2}a_{ij}\beta_i\beta_j\Big\} \\ &< \Big(\sum_{i\in J}\Big\{\sum_{j\in J}\frac{1}{2}a_{ij}\beta_i\beta_j\Big\} - \sum_{i\in J}\frac{2(N_i+1)}{b_i}\beta_i\Big) + \sum_{i\in J}\Big\{\sum_{j\in I\setminus J}a_{ij}\beta_i\beta_j\Big\} \\ &< \sum_{i\in J}\Big\{\sum_{j\in I\setminus J}a_{ij}\beta_i\beta_j\Big\}. \end{aligned} \tag{2.13}$$

and (2.11) is established. Thus from the second inequality in (2.11) we get that A is irreducible. Furthermore, by using the first inequality in (2.11) with $J = \{i\}$ we find:

$$\sum_{j\in I} a_{ij}\beta_i\beta_j \geq \frac{1}{2}a_{ii}\beta_i^2 + \sum_{j\in I\setminus\{i\}} a_{ij}\beta_i\beta_j > \frac{2(N_i+1)}{b_i}\beta_i \qquad \text{for every } i\in I;$$

and also (2.12) follows. □

3. Variational Formulation and a (radial) Log HLS inequality

Fix

$$N_i > -1 \quad \text{and} \quad b_i > 0, \qquad i\in I. \tag{3.1}$$

If the symmetric matrix $A = \{a_{ij}\}_{i,j=1,\dots,m}$ satisfies (2.10) and the m-ple $\beta = (\beta_1,\dots,\beta_m)$ satisfies (2.9) with $h = 1$, then in this section we show that problem (2.3) admits a variational formulation. Furthermore a solution of (2.3) may be obtained via a minimization procedure.

To this purpose for $\beta = (\beta_1,\beta_2,\dots,\beta_m)\in\mathbb{R}^m$ with $\beta_i > 0$ $i\in I$, we consider the set,

$$D_\beta := \Big\{r = (r_1,\dots,r_m)\in\mathbb{R}^m : 0\leq r_i\leq\beta_i\ \forall i\in I\Big\}$$

and for $\delta \geq 0$, $\emptyset \neq J \subseteq I$, we set,

$$\Phi_{\delta,\beta,J}(r) := \sum_{i\in J}\left\{ -\frac{\delta}{2}r_i(\beta_i - r_i) + \frac{2(N_i+1)}{b_i}r_i - \sum_{j\in J}\frac{1}{2}a_{ij}r_ir_j\right\}. \tag{3.2}$$

In particular, for $J = I$ we let $\Phi_{\delta,\beta} \equiv \Phi_{\delta,\beta,I}$.

Lemma 3.1. *Assume* (3.1)*, and let* $A = \{a_{ij}\}$ *be a symmetric matrix satisfying* (2.10)*. If* $\beta = (\beta_1, \dots, \beta_m) \in \mathbb{R}^m$*, satisfies:*

$$\begin{cases} \beta_i > 0 \qquad i \in I \\ \left(\sum_{i\in I}\sum_{j\in I}\frac{1}{2}a_{ij}\beta_i\beta_j\right) - \sum_{i\in I}\frac{2(N_i+1)}{b_i}\beta_i = 0 \\ \left(\sum_{i\in J}\sum_{j\in J}\frac{1}{2}a_{ij}\beta_i\beta_j\right) - \sum_{i\in J}\frac{2(N_i+1)}{b_i}\beta_i < 0 \quad \text{if} \quad \emptyset \neq J \subsetneqq I; \end{cases} \tag{3.3}$$

then, for every $\emptyset \neq J \subseteq I$*, there holds:*

$$\Phi_{0,\beta,J}(r) \geq 0, \qquad \forall r \in D_\beta; \tag{3.4}$$

and

$$\Phi_{0,\beta}(r) = 0 \quad \text{if and only if either } r = 0 \quad \text{or} \quad r = \beta. \tag{3.5}$$

Furthermore there exists $\delta_0 > 0$*, such that for* $\delta \in [0, \delta_0)$*, there holds:*

$$\Phi_{\delta,\beta}(r) \geq 0 \qquad \forall r \in D_\beta. \tag{3.6}$$

Proof. It is clear that the functions $\Phi_{0,\beta,J}(r)$ and $\Phi_{0,\beta}(r)$ correspond to (3.2) with $\delta = 0$. Hence,

$$\frac{\partial \Phi_{0,\beta,J}}{\partial r_i}(r) = \frac{2(N_i+1)}{b_i} - \sum_{j\in J} a_{ij}\, r_j \qquad \forall i \in J. \tag{3.7}$$

Denote by $\tilde{r}^{(J)} = \left(\tilde{r}_1^{(J)}, \dots, \tilde{r}_m^{(J)}\right) \in D_\beta$, a point where $\Phi_{0,\beta,J}$ attains its absolute minimum value in D_β, and set $\Phi_{0,\beta,J}\left(\tilde{r}^{(J)}\right) = H(J)$. Since $\Phi_{0,\beta,J}(0) = 0$ we see that, $H(J) \leq 0 \qquad \forall \emptyset \neq J \subseteq I$.

Claim.

$$H(J) = 0 \qquad \forall \emptyset \neq J \subseteq I, \tag{3.8}$$

and if $J \subsetneqq I$, then $\Phi_{0,\beta,J}(r) = 0$ for some $r \in D_\beta$, if and only if $r_i = 0 \quad \forall i \in J$. While if $J = I$ then $\Phi_{0,\beta}(r) = 0$ if and only if $r = 0$ or $r = \beta$.

We prove our claim by induction on $|J| = \text{card}\,(J) \in \{1, \dots, m\}$. If $|J| = 1$, say $J = \{i\}$, then the last condition in (3.3) implies that $\Phi_{0,\beta,J}(r) \geq 0$ and $\Phi_{0,\beta,J}(r) = 0$ if and only if $r_i = 0$.

Next, for $k \in \{1, \dots, m-1\}$ assume that, for $\emptyset \neq J \subset I$ such that $|J| \leq k$, we have:

$$H(J) = 0 \tag{3.9}$$

and if $\Phi_{0,\beta,J}(r) = 0$ for some $r \in D_\beta$, then $r_i = 0, \quad \forall i \in J$.

Let $J \subseteq I$ be such that $|J| = k+1$ and let $\tilde{r}^{(J)}$ be a point of absolute minimum for $\Phi_{0,\beta,J}$ in D_β. If for some index $i \in J$, we suppose that $\tilde{r}_i^{(J)} \in (0, \beta_i)$, then

$$\frac{\partial \Phi_{\beta,J}}{\partial r_i}(\tilde{r}^{(J)}) = \frac{2(N_i+1)}{b_i} - \sum_{j\in J} a_{ij}\, r_j^{(J)} = 0. \tag{3.10}$$

Observe that,

$$\begin{aligned}\Phi_{0,\beta,J}(r) :=& \left\{\frac{2(N_i+1)}{b_i} r_i - \sum_{j\in J} \frac{1}{2} a_{ij} r_i r_j\right\}\\ &+ \sum_{l\in J\setminus\{i\}} \left\{\frac{2(N_l+1)}{b_l} r_l - \sum_{j\in J\setminus\{i\}} \frac{1}{2} a_{lj} r_l r_j\right\} - \sum_{l\in J\setminus\{i\}} \frac{1}{2} a_{li} r_l r_i\\ =& \left\{\frac{2(N_i+1)}{b_i} r_i - \sum_{j\in J} a_{ij} r_i r_j\right\} + \frac{1}{2} a_{ii} r_i^2 + \Phi_{0,\beta,J\setminus\{i\}}(r).\end{aligned} \tag{3.11}$$

So, by (2.10), (3.10) and (3.11), we obtain:

$$0 \geq \Phi_{0,\beta,J}\big(\tilde{r}^{(J)}\big) \geq \frac{1}{2} a_{ii} \big(\tilde{r}_i^{(J)}\big)^2 + \Phi_{0,\beta,J\setminus\{i\}}\big(\tilde{r}^{(J)}\big) \geq H\big(J\setminus\{i\}\big) = 0, \tag{3.12}$$

that is: $a_{ii}\tilde{r}_i^{(J)} = 0$ and $\Phi_{0,\beta,J\setminus\{i\}}\big(\tilde{r}^{(J)}\big) = 0$. Therefore by the induction assumption, we get that, $\tilde{r}_j^{(J)} = 0$ for every $j \in J\setminus\{i\}$, this yields a contradiction in view of(3.11) and (3.10). Thus, there exists $J_0 \subset J$ such that $\tilde{r}_i^{(J)} = \beta_i$ if $i \in J_0$ and $\tilde{r}_i^{(J)} = 0$ if $i \in J\setminus J_0$. This implies that,

$$0 \geq \Phi_{0,\beta,J}(\tilde{r}^{(J)}) = 0 + \sum_{i\in J_0} \frac{2(N_i+1)}{b_i}\beta_i - \Big(\sum_{i\in J_0}\sum_{j\in J_0} \frac{1}{2} a_{ij}\beta_i\beta_j\Big), \tag{3.13}$$

and if $J \subsetneq I$ by (3.3), we conclude that $J_0 = \emptyset$ and $\tilde{r}_i^{(J)} = 0$ for every $i \in J$. In case $J = I$, then we have that, either $J_0 = \emptyset$ or $I \setminus J_0 = \emptyset$. Therefore we find that either $r^{(I)} \equiv 0$ or $r^{(I)} = \beta$ as claimed. In any case $H(J) = \Phi_{0,\overline{\beta},J}\big(\tilde{r}^{(J)}\big) = 0$, and (3.4), (3.5) follow.

Next, to establish (3.6) with $\delta > 0$ small, we argue by contradiction, and assume there exists $\delta_n \downarrow 0$ as $n \to +\infty$, such that, at a minimum point $\tilde{r}^{(n)} = \big(\tilde{r}_1^{(n)}, \ldots, \tilde{r}_m^{(n)}\big) \in D_\beta$, we have $\Phi_{\delta_n,\beta}\big(\tilde{r}^{(n)}\big) < 0$. We claim that, for every $n \in \mathbb{N}$ there exists $i_n \in I$ such that $\tilde{r}_{i_n}^{(n)} \in \big(0, \beta_{i_n}\big)$. As otherwise, we would find $J_n \subset I$ such that if $i \in J_n$ then $\tilde{r}_i^{(n)} = 0$ and if $i \in I \setminus J_n$ then $\tilde{r}_i^{(n)} = \beta_i$. But this would imply that $\Phi_{\delta_n,\beta}\big(\tilde{r}^{(n)}\big) \geq 0$, which is impossible. Thus, we can find a subsequence $n_k \uparrow +\infty$ as $k \to +\infty$ and a fixed $i \in I$, such that $\tilde{r}_i^{(n_k)} \in \big(0, \beta_i\big)$ and $\tilde{r}^{(n_k)} \to \tilde{r} = (\tilde{r}_1, \ldots, \tilde{r}_m) \in D_\beta$ as $k \to +\infty$. Since, $\Phi_{\delta_{n_k},\beta}\big(\tilde{r}^{(n_k)}\big) < 0$, as $k \to +\infty$ we find: $\Phi_{0,\beta}(\tilde{r}) \leq 0$. So by (3.5), we deduce that either $\tilde{r} = 0$ or $\tilde{r} = \beta$. On the other hand, since $\tilde{r}_i^{(n_k)} \in (0, \beta_i)$, we also have

$$\frac{\partial \Phi_{\delta_{n_k},\beta}}{\partial r_i}\left(\tilde{r}^{(n_k)}\right) = \delta_{n_k}\tilde{r}_i^{(n_k)} - \delta_{n_k}\beta_i/2 + \frac{2(N_i+1)}{b_i} - \sum_{j\in I} a_{ij}\,\tilde{r}_j^{(n_k)} = 0. \tag{3.14}$$

By letting $k \to +\infty$, we get $\frac{2(N_i+1)}{b_i} - \sum_{j\in I} a_{ij}\,\tilde{r}_j = 0$. Whence $\tilde{r}_j = \beta_j\ \forall j \in I$, and so, $\sum_{j\in I} a_{ij}\,\beta_j - \frac{2(N_i+1)}{b_i} = 0$, in contradiction with (2.12). □

To proceed further, for every $\beta_0 > 0$, we consider the set

$$\begin{aligned} \mathcal{W}'_{\beta_0} := \Big\{ &\varphi \in L^1(\mathbb{R}^-, \mathbb{R}),\ (\beta_0 - \varphi) \in L^1(\mathbb{R}^+, \mathbb{R}), \\ &\varphi' \in L^1(\mathbb{R}, \mathbb{R}) \text{ and } \varphi'(t) \geq 0 \text{ for a.e } t \in \mathbb{R}. \Big\}, \end{aligned} \tag{3.15}$$

and its subset,

$$\mathcal{W}_{\beta_0} := \left\{ \varphi \in \mathcal{W}'_{\beta_0} :\ \varphi'(t)\Big(\ln\left(\varphi'(t)\right)\Big)^+ \in L^1(\mathbb{R}, \mathbb{R}) \right\}, \tag{3.16}$$

where, as usual $\psi^{\pm} := \max\{\pm\psi, 0\}$ denotes the positive and negative part of ψ.

Lemma 3.2. *The sets $\mathcal{W}'_{\beta_0}$ and $\mathcal{W}_{\beta_0}$ are convex. Moreover, for every $\varphi \in \mathcal{W}'_{\beta_0}$ we have:*

(i) $\varphi \in C^0(\mathbb{R}, \mathbb{R})$ *is nondecreasing and* $\lim_{t\to-\infty} t\varphi(t) = \lim_{t\to+\infty} t\big(\beta_0 - \varphi(t)\big) = 0$. *In particular,* $\varphi(-\infty) := \lim_{t\to-\infty}\varphi(t) = 0$ *and* $\varphi(+\infty) := \lim_{t\to+\infty}\varphi(t) = \beta_0$, *and* $0 \leq \varphi(t) \leq \beta_0$.

(ii) $t\varphi'(t) \in L^1(\mathbb{R}, \mathbb{R})$ *and,*

$$\int_{\mathbb{R}} |t\varphi'(t)|dt = \int_{-\infty}^{0} \varphi(t)dt + \int_0^{+\infty} \big(\beta_0 - \varphi(t)\big)dt. \tag{3.17}$$

(iii)

$$\int_{-\infty}^{\infty} \varphi'(t)\Big(\ln\left(\varphi'(t)\right)\Big)^- dt \leq 2 + \int_{-\infty}^{0} \varphi(t)dt + \int_0^{+\infty} \big(\beta_0 - \varphi(t)\big)dt. \tag{3.18}$$

(iv) *For every $t > 0$ there holds:*

$$\varphi(-t) + \big(\beta_0 - \varphi(t)\big) \leq \frac{1}{t}\left(\int_{-\infty}^{0} \varphi(s)ds + \int_0^{+\infty} \big(\beta_0 - \varphi(s)\big)ds\right). \tag{3.19}$$

(v) *If* $\varphi(t) \in \mathcal{W}_{\beta_0}$, *then* $\left|\varphi'(t)\ln\left(\varphi'(t)\right)\right| \in L^1(\mathbb{R}, \mathbb{R})$.

Proof. Clearly, the set $\mathcal{W}'_{\beta_0}$ is convex. Moreover by the convexity of the function $h(\rho) := \rho\big(\ln(\rho)\big)^+$ in $[0, +\infty)$, we see also that $\mathcal{W}_{\beta_0}$ is convex. Next notice that $\mathcal{W}'_{\beta_0} \subset W^{1,1}_{\text{loc}}(\mathbb{R})$ so for $\varphi \in \mathcal{W}'_{\beta_0}$ and $t_1 < t_2$ there holds: $\varphi(t_2) - \varphi(t_1) = \int_{t_1}^{t_2} \varphi'(s)ds \geq 0$. Hence $\varphi(t)$ is continuous and non-decreasing. Furthermore, as $\varphi(t) \in L^1(\mathbb{R}^-, \mathbb{R})$, then we find a sequence $M_n \uparrow +\infty$ as $n \uparrow +\infty$, such that $\lim_{n\to+\infty} M_n\varphi(-M_n) = 0$. Similarly, as $\big(\beta_0 - \varphi(t)\big) \in L^1(\mathbb{R}^+, \mathbb{R})$, we find a sequence

$L_n \uparrow +\infty$ as $n \uparrow +\infty$ such that $\lim_{n\to+\infty} L_n\big(\beta_0 - \varphi(L_n)\big) = 0$. For every $M > 0$, we use integration by parts and find:

$$\int_{-M}^{0} \big|t\varphi'(t)\big| dt = -\int_{-M}^{0} t\varphi'(t)dt = -M\varphi(-M) + \int_{-M}^{0} \varphi(t)dt \leq \int_{-\infty}^{0} \varphi(t)dt. \tag{3.20}$$

Therefore, $(t)^{-}\varphi'(t) \in L^1(\mathbb{R},\mathbb{R})$, and by choosing $M := M_n$ in (3.20) and by letting $n \uparrow +\infty$, we obtain

$$\int_{-\infty}^{0} \big|t\varphi'(t)\big| dt = \int_{-\infty}^{0} \varphi(t)dt. \tag{3.21}$$

Consequently, by (3.20) and (3.21) we deduce: $\lim_{M\to+\infty} M\varphi(-M) = 0$. Similarly, for every $L > 0$, integration by parts gives:

$$\begin{aligned} \int_{0}^{L} \big|t\varphi'(t)\big| dt &= \int_{0}^{L} t\varphi'(t)dt = L\big(\varphi(L) - \beta_0\big) + \int_{0}^{L} \big(\beta_0 - \varphi(t)\big)dt \\ &\leq \int_{0}^{+\infty} \big(\beta_0 - \varphi(t)\big)dt. \end{aligned} \tag{3.22}$$

Therefore, $(t)^{+}\varphi'(t) \in L^1(\mathbb{R},\mathbb{R})$ and by choosing $L := L_n$, and letting $n \uparrow +\infty$, we find that,

$$\int_{0}^{+\infty} \big|t\varphi'(t)\big| dt = \int_{0}^{+\infty} \big(\beta_0 - \varphi(t)\big)dt, \tag{3.23}$$

and also that, $\lim_{L\to+\infty} L\big(\beta_0 - \varphi(L)\big) = 0$. Therefore, **(i)** and **(ii)** are established. Next, we recall the trivial inequality $a\big(\ln(a) - 1\big) \geq a\ln(b) - b$, valid for every $a \geq 0$ and every $b > 0$. We use it to deduce the following:

$$\begin{aligned} &\int_{\{t\in\mathbb{R}\,:\,\varphi'(t)\leq 1\}} \varphi'(t)\ln\big(\varphi'(t)\big)dt \\ &= \int_{\{t\in\mathbb{R}\,:\,\varphi'(t)\leq 1\}} \varphi'(t)dt + \int_{\{t\in\mathbb{R}\,:\,\varphi'(t)\leq 1\}} \varphi'(t)\Big(\ln\big(\varphi'(t)\big) - 1\Big)dt \\ &\geq \int_{\{t\in\mathbb{R}\,:\,\varphi'(t)\leq 1\}} \varphi'(t)dt + \int_{\{t\in\mathbb{R}\,:\,\varphi'(t)\leq 1\}} \varphi'(t)\ln\big(e^{-|t|}\big)dt - \int_{\{t\in\mathbb{R}\,:\,\varphi'(t)\leq 1\}} e^{-|t|}dt \\ &\geq -\int_{\mathbb{R}} \big|t\varphi'(t)\big|dt - \int_{\mathbb{R}} e^{-|t|}dt = -\int_{\mathbb{R}} \big|t\varphi'(t)\big|dt - 2. \end{aligned} \tag{3.24}$$

Thus (3.18) follows from (3.17) and (3.24). At this point (3.19) can be easily established, as for $t > 0$, we have:

$$\varphi(-t) + \big(\beta_0 - \varphi(t)\big) = \int_{-\infty}^{-t} \varphi'(s)ds + \int_{t}^{+\infty} \varphi'(s)ds \leq \frac{1}{t}\int_{-\infty}^{\infty} |s|\varphi'(s)ds.$$

Finally if $\varphi \in \mathcal{W}_{\beta_0}$, then by (3.18) we conclude that $\big|\varphi'(t)\ln\big(\varphi'(t)\big)\big| \in L^1(\mathbb{R},\mathbb{R})$. □

By the same argument of (3.24), we also find:

Lemma 3.3. *There exists $M_0 > 0$, such that $\forall M \geq M_0$ and $\varphi \in \mathcal{W}'_{\beta_0}$ there holds:*

$$\int_{\{t\in\mathbb{R}\,:\,|t|\geq M\}} \varphi'(t)\Big(\ln\big(\varphi'(t)\big)\Big)^{-} dt \tag{3.25}$$
$$\leq \frac{\ln(1+M^2)}{M}\left(\int_{-\infty}^{0}\varphi(t)dt + \int_{0}^{+\infty}\big(\beta_0-\varphi(t)\big)dt\right) + 2\Big(\frac{\pi}{2} - \arctan(M)\Big).$$

Proof. It suffices to use the above-mentioned inequality: $a\big(\ln(a)-1\big) \geq a\ln(b) - b$ with $a = \varphi'(t)$ and $b = \frac{1}{1+t^2}$ to derive the following:

$$\begin{aligned}
&\int_{\{t\in\mathbb{R}\,:\,|t|\geq M\}} \varphi'(t)\Big(\ln\big(\varphi'(t)\big)\Big)^{-} dt \\
&\quad = -\int_{\{t\in\mathbb{R}\,:\,\varphi'(t)\leq 1,\ |t|\geq M\}} \varphi'(t)\ln\big(\varphi'(t)\big)dt \\
&\quad \leq -\int_{\{t\in\mathbb{R}\,:\,\varphi'(t)\leq 1,\ |t|\geq M\}} \varphi'(t)dt + \int_{\{t\in\mathbb{R}\,:\,|t|\geq M\}} \varphi'(t)\ln(1+t^2)dt \\
&\qquad + \int_{\{t\in\mathbb{R}\,:\,|t|\geq M\}} \frac{dt}{1+t^2} \\
&\quad \leq \frac{\ln(1+M^2)}{M}\int_{\{t\in\mathbb{R}\,:\,|t|\geq M\}} |t|\varphi'(t)dt + 2\Big(\frac{\pi}{2} - \arctan(M)\Big),
\end{aligned} \tag{3.26}$$

for sufficiently large $M > 0$, where we have used the fact that $\dfrac{\log(1+t^2)}{t}$ is definitively monotonic decreasing. At this point the conclusion follows using (3.17). □

Next, we point out a useful "compactness" result valid in the space $\mathcal{W}_{\beta_0}$.

Lemma 3.4. *Let $\beta_0 > 0$, and $\{\varphi_n\}_{n=1}^{+\infty} \subset \mathcal{W}_{\beta_0}$ be a sequence satisfying:*

a) $\varphi_n(0) = \beta_0/2$,

b)
$$\int_{-\infty}^{+\infty} \varphi_n(t)\Big(\beta_0 - \varphi_n(t)\Big)dt + \left|\int_{-\infty}^{+\infty} \varphi'_n(t)\ln\big(\varphi'_n(t)\big)dt\right| \leq C, \tag{3.27}$$

with a suitable constant $C > 0$. Then, there exists $\varphi \in \mathcal{W}_{\beta_0}$ and a subsequence of $\{\varphi_n\}$ (denoted the same way), such that: $\varphi_n \to \varphi$ uniformly in $\mathbb{R}$,

$$\int_{\mathbb{R}} \varphi(t)\Big(\beta_0 - \varphi(t)\Big)dt \leq \varliminf_{n\to+\infty} \int_{\mathbb{R}} \varphi_n(t)\Big(\beta_0 - \varphi_n(t)\Big)dt. \tag{3.28}$$

Moreover, $\varphi'_n \rightharpoonup \varphi'$ weakly in $L^1(\mathbb{R},\mathbb{R})$ and,

$$\int_{\mathbb{R}} \varphi'(t)\Big(\ln\big(\varphi'(t)\big) - 1\Big)dt \leq \varliminf_{n\to+\infty} \int_{\mathbb{R}} \varphi'_n(t)\Big(\ln\big(\varphi'_n(t)\big) - 1\Big)dt. \tag{3.29}$$

Proof. Observe that for $\varphi_n \in \mathcal{W}_{\beta_0}$ we have:

$$\|\varphi_n'(\cdot)\|_{L^1(\mathbb{R},\mathbb{R})} = \int_{\mathbb{R}} \varphi_n'(t)dt = \beta_0, \tag{3.30}$$

and since $\varphi_n(0) = \beta_0/2$ we get,

$$\frac{\beta_0}{2}\int_{-\infty}^{0} \varphi_n(t)dt + \frac{\beta_0}{2}\int_{0}^{+\infty}\Big(\beta_0 - \varphi_n(t)\Big)dt \le \int_{-\infty}^{+\infty} \varphi_n(t)\Big(\beta_0 - \varphi_n(t)\Big)dt \le C. \tag{3.31}$$

Thus, as a consequence of (3.18), (3.19) and (3.31) we find:

$$\varphi_n(-t) + \Big(\beta_0 - \varphi_n(t)\Big) \le \frac{C}{t}, \ \forall t > 0 \quad \text{and} \quad \int_{-\infty}^{+\infty} \varphi_n'(t)\Big|\ln\left(\varphi_n'(t)\right)\Big|dt \le C. \tag{3.32}$$

Next, we use the convexity of the function $h(\rho) := \rho \cdot \ln(\rho)$ (extended by zero at $\rho = 0$), together with Jensen's inequality, to obtain that for any Borel set $\mathcal{A}$ with $0 < \mathcal{L}^1(\mathcal{A}) < +\infty$, there holds:

$$\left(\frac{1}{\mathcal{L}^1(\mathcal{A})}\int_{\mathcal{A}} \varphi_n'(t)dt\right)\ln\left(\frac{1}{\mathcal{L}^1(\mathcal{A})}\int_{\mathcal{A}} \varphi_n'(t)dt\right) \le \frac{1}{\mathcal{L}^1(\mathcal{A})}\int_{\mathcal{A}} \varphi_n'(t)\ln\left(\varphi_n'(t)\right)dt. \tag{3.33}$$

Since, $\rho\ln(\rho) \ge -\frac{1}{e}$ for every $\rho \in [0,+\infty)$, we can use (3.32) to obtain:

$$\begin{aligned}&\left(\int_{\mathcal{A}} \varphi_n'(t)dt\right)\ln\left(\frac{1}{\mathcal{L}^1(\mathcal{A})}\right)\\ &\quad\le \int_{\mathcal{A}} \varphi_n'(t)\ln\left(\varphi_n'(t)\right)dt - \left(\int_{\mathcal{A}} \varphi_n'(t)dt\right)\ln\left(\int_{\mathcal{A}} \varphi_n'(t)dt\right) \le C.\end{aligned} \tag{3.34}$$

In particular, if we take $\mathcal{A}$ a small interval $(t_1, t_2) \subset \mathbb{R}$, using (3.33), we obtain,

$$\Big|\varphi_n(t_2) - \varphi_n(t_1)\Big| \le \frac{C}{-\ln(t_2 - t_1)}. \tag{3.35}$$

Hence, the sequence $\{\varphi_n(t)\}_{n=1}^{+\infty}$ is equicontinuous, and also equibounded: $0 \le \varphi_n \le \beta_0$. Thus by the Arzelà–Ascoli theorem, up to a subsequence, we have:

$$\varphi_n(t) \to \varphi(t) \quad \text{pointwise in } \mathbb{R} \text{ and uniformly on compact sets.} \tag{3.36}$$

In particular, φ is continuous non-decreasing and satisfies: $0 \le \varphi \le \beta_0$. Furthermore, by (3.31) and Fatou's Lemma, we obtain (3.28) and we also find:

$$\int_{-\infty}^{0} \varphi(t)dt + \int_{0}^{+\infty}\Big(\beta_0 - \varphi(t)\Big)dt \le C. \tag{3.37}$$

Therefore, $\varphi(-t) + \big(\beta_0 - \varphi(t)\big) \to 0$ as $t \to +\infty$, and by the first uniform estimate in (3.32) we find that actually $\varphi_n \to \varphi$ uniformly in $\mathbb{R}$.

Next, by means of (3.17) and (3.31) we obtain that,

$$\int\limits_{\mathbb{R}\setminus[-M,M]} |\varphi_n'(t)|dt \le \frac{1}{M} \int\limits_{\mathbb{R}\setminus[-M,M]} |t|\varphi_n'(t)dt \le \frac{C}{M}, \qquad \forall M > 1. \tag{3.38}$$

Consequently, for any Borel set $B \subset \mathbb{R}$, and for every $M > 1$ we have:

$$\begin{aligned}\int_B |\varphi_n'(t)|dt &\le \int_{B\cap[-M,M]} \varphi_n'(t)dt + \frac{C}{M} \\ &\le \frac{1}{\ln(M)} \int_{\{t:\varphi_n'(t)\ge M\}} \varphi_n'(t)\ln\big(\varphi_n'(t)\big)dt \\ &\quad + M\mathcal{L}^1\Big(B\cap[-M,M]\Big) + \frac{C}{M} \\ &\le \frac{C}{\ln(M)} + M\mathcal{L}^1\Big(B\cap[-M,M]\Big) + \frac{C}{M}.\end{aligned} \tag{3.39}$$

Therefore, using the Dunford–Pettis theorem (see, e.g., [DS], p. 292), we find $\gamma(t) \in L^1(\mathbb{R},\mathbb{R})$ such that (along a subsequence) we have: $\varphi_n'(t) \rightharpoonup \gamma(t)$ weakly in $L^1(\mathbb{R},\mathbb{R})$. Since $\varphi_n(t) \to \varphi(t)$ uniformly in $\mathbb{R}$, we deduce that necessarily $\varphi \in W^{1,1}_{\mathrm{loc}}(\mathbb{R},\mathbb{R})$ and $\varphi' = \gamma \ge 0$ a.e. in $\mathbb{R}$. In other words, $\varphi \in \mathcal{W}'_{\beta_0}$. To show that actually, $\varphi \in \mathcal{W}_{\beta_0}$ we consider the functional $\Phi : L^1(\mathbb{R},\mathbb{R}) \to [0,+\infty]$ defined as follows:

$$\Phi(\psi) := \int_{-\infty}^{+\infty} |\psi(t)|\Big(\ln|\psi(t)|\Big)^+ dt.$$

Clearly Φ is convex and (by Fatou's lemma) Φ is strongly lower semicontinuous. Hence Φ is weakly lower semicontinuous, and we conclude:

$$\int\limits_{\mathbb{R}} \varphi'(t)\Big(\ln\big(\varphi'(t)\big)\Big)^+ dt \le \varliminf_{n\to+\infty} \int\limits_{\mathbb{R}} \varphi_n'(t)\Big(\ln\big(\varphi_n'(t)\big)\Big)^+ dt \le C, \tag{3.40}$$

that is $\varphi \in \mathcal{W}_{\beta_0}$ as claimed.

At this point we are left to establish (3.29). To this purpose, we use again the fact that the function: $\rho\ln(\rho)$ is convex and bounded from below. Therefore, as above, we see that, for every $M > 0$ there holds:

$$\int\limits_{-M}^{M} \varphi'(t)\Big(\ln\big(\varphi'(t)\big)\Big)dt \le \varliminf_{n\to+\infty} \int\limits_{-M}^{M} \varphi_n'(t)\Big(\ln\big(\varphi_n'(t)\big)\Big)dt. \tag{3.41}$$

At this point, for φ_n we can use (3.25) together with (3.31), and for φ we can use (3.40), in order to show that:

$$\forall\, \varepsilon > 0 \text{ there exists } M_\varepsilon > 0 \text{ sufficiently large such that,}$$

$$\int\limits_{\{t\in\mathbb{R}:\,|t|\ge M_\varepsilon\}} \varphi'(t)\Big(\ln\big(\varphi'(t)\big)\Big)^+ dt + \int\limits_{\{t\in\mathbb{R}:\,|t|\ge M_\varepsilon\}} \varphi_n'(t)\Big(\ln\big(\varphi_n'(t)\big)\Big)^- dt \le \varepsilon, \ \forall n \in \mathrm{N}.$$

Consequently,

$$\int_{-\infty}^{+\infty} \varphi'(t)\Big(\ln(\varphi'(t))\Big)dt$$

$$\le \int_{-M_\varepsilon}^{M_\varepsilon} \varphi'(t)\Big(\ln(\varphi'(t))\Big)dt + \int_{\{t\in\mathbb{R}:\,|t|\ge M_\varepsilon\}} \varphi'(t)\Big(\ln(\varphi'(t))\Big)^+dt$$

$$\le \varliminf_{n\to+\infty} \int_{-M_\varepsilon}^{M_\varepsilon} \varphi_n'(t)\Big(\ln(\varphi_n'(t))\Big)dt + \int_{\{t\in\mathbb{R}:\,|t|\ge M_\varepsilon\}} \varphi'(t)\Big(\ln(\varphi'(t))\Big)^+dt$$

$$\le \varliminf_{n\to+\infty} \left\{ \int_{-\infty}^{+\infty} \varphi_n'(t)\Big(\ln(\varphi_n'(t))\Big)dt + \int_{\{t\in\mathbb{R}:\,|t|\ge M_\varepsilon\}} \varphi_n'(t)\Big(\ln(\varphi_n'(t))\Big)^-dt \right\}$$

$$+ \int_{\{t\in\mathbb{R}:\,|t|\ge M_\varepsilon\}} \varphi'(t)\Big(\ln(\varphi'(t))\Big)^+dt$$

$$\le \varliminf_{n\to+\infty} \int_{-\infty}^{+\infty} \varphi_n'(t)\Big(\ln(\varphi_n'(t))\Big)dt + \varepsilon \qquad \forall\varepsilon>0$$

and (3.29) follows. □

Next we analyze what happens, when we remove condition **a)** in Lemma 3.4.

Lemma 3.5. *Let $\beta_0 > 0$, $\{\varphi_n\}_{n=1}^{+\infty} \subset \mathcal{W}_{\beta_0}$ be a sequence of functions satisfying (3.27) and let $\tau_n \in \mathbb{R}$ be such that $\varphi_n(\tau_n) = \beta_0/2$. Then along a suitable subsequences the following holds:*

- *if $\{\tau_n\}_{n=1}^{+\infty}$ is bounded, then the conclusion of Lemma 3.4 holds;*
- *if $\lim_{n\to+\infty} \tau_n = +\infty$, then $\varphi_n(t) \to 0$ pointwise and uniformly in sets bounded from above;*
- *if $\lim_{n\to+\infty} \tau_n = -\infty$ then $\varphi_n(t) \to \beta_0$ pointwise and uniformly in sets bounded from below.*

In the latter cases: $\varphi_n' \rightharpoonup 0$ weakly, in the sense of measures.*

Proof. Let $g_n \in \mathcal{W}_{\beta_0}$ be defined as follows:

$$g_n(t) := \varphi_n(t+\tau_n), \qquad \forall t \in \mathbb{R}. \tag{3.42}$$

Hence, Lemma 3.4 applies to g_n, and we find $g \in \mathcal{W}_{\beta_0}$ such that (up to a subsequence) $g_n \to g$ uniformly in $\mathbb{R}$ and $g_n' \rightharpoonup g'$ weakly in $L^1(\mathbb{R},\mathbb{R})$. Furthermore, (3.29) and (3.28) hold for (a subsequence of) g_n and for g. In case $\{\tau_n\}_{n=1}^{+\infty}$ is bounded, then (up to a subsequence) we can assume that $\tau_n \to \tau \in \mathbb{R}$, and for

$\varphi(t) := g(t-\tau)$ we have: $\varphi(t) \in \mathcal{W}_{\beta_0}$, $\varphi_n \to \varphi$ uniformly in $\mathbb{R}$, $\varphi_n' \rightharpoonup \varphi'$ weakly in $L^1(\mathbb{R}, \mathbb{R})$ and (3.29) and (3.28) hold.

Next assume that (along a subsequence) $\lim_{n\to+\infty} \tau_n = +\infty$. Thus, for $\varphi_n(t) = g_n(t-\tau_n)$ we find that, $\varphi_n(t) \to 0 \quad \forall t \in \mathbb{R}$. Moreover, since φ_n is nondecreasing, for arbitrary fixed $s \in \mathbb{R}$ we have

$$0 \le \varphi_n(t) \le \varphi_n(s) \to 0 \quad \forall t \le s.$$

So the convergence is uniform in sets which are bounded from above.

Similarly, if $\lim_{n\to+\infty} \tau_n = -\infty$, then $\varphi_n(t) \to \beta_0, \quad \forall t \in \mathbb{R}$. As above for fixed $s \in \mathbb{R}$, we find:

$$\beta_0 \ge \varphi_n(t) \ge \varphi_n(s) \to \beta_0 \quad \forall t \ge s,$$

So that the convergence is uniform in sets bounded from below.

Finally, using (3.38) for g_n, we easily check that in the latter cases: $\varphi_n' \rightharpoonup 0$ weakly* in the sense of measures. □

For $\beta = (\beta_1, \ldots, \beta_m) \in \mathbb{R}^m$ satisfying (3.3) we consider the set $\mathcal{U}_\beta := \mathcal{W}_{\beta_1} \times \mathcal{W}_{\beta_2} \times \cdots \times \mathcal{W}_{\beta_m}$, and we define the functional (free energy):

$$Y_\beta(g) := \int_{\mathbb{R}} \sum_{i\in I} \left\{ \frac{1}{b_i} g_i'(t)\Big(\ln\big(g_i'(t)\big) - 1\Big) + \frac{2(N_i+1)}{b_i} g_i(t) - \sum_{j\in I} \frac{1}{2} a_{ij} g_i(t) g_j(t) \right\} dt, \tag{3.43}$$

for $g = (g_1, \ldots, g_m) \in \mathcal{U}_\beta$.

Observe that, the second condition in (3.3) is necessary and sufficient to ensure that Y_β is well defined in $\mathcal{U}_\beta$. Also note that,

$$\begin{aligned} Y_\beta(g) = &\int_{\mathbb{R}} \sum_{i\in I} \frac{1}{b_i} g_i'(t)\Big(\ln\big(g_i'(t)\big) - 1\Big) dt \\ &+ \int_{\mathbb{R}} \sum_{i\in I} \frac{\delta}{2} g_i(t)\Big(\beta_i - g_i(t)\Big) dt + \int_{\mathbb{R}} \Phi_{\delta,\beta}\big(g(t)\big) dt, \end{aligned} \tag{3.44}$$

whith $\Phi_{\delta,\beta}$ defined by (3.2).

We are going to show that Y_β is bounded from below and coercive in $\mathcal{U}_\beta$.

To this end, observe that for $i \in I$ and $g_i \in \mathcal{W}_{\beta_i}$ there exist $-\infty \le s_i < \tau_i \le +\infty$ such that $0 < g_i(t) < \beta_i \ \forall t \in (s_i, \tau_i)$ and $g_i' \equiv 0$ in $\mathbb{R} \setminus [s_i, \tau_i]$. In particular, $g_i(t) = 0$ for $t \le s_i$ and $g_i(t) = \beta_i$ for $t \ge \tau_i$. By means of the already mentioned inequality: $a\big(\ln(a) - 1\big) + b \ge a \ln(b)$, with $a \ge 0$, $b > 0$; we can estimate:

$$\begin{aligned} &\int_{\mathbb{R}} \left\{ \frac{1}{b_i} g_i'(t)\Big(\ln\big(g_i'(t)\big) - 1\Big) + \frac{\delta}{2} g_i(t)\Big(\beta_i - g_i(t)\Big) \right\} dt \\ &= \int_{s_i}^{\tau_i} \left\{ \frac{1}{b_i} g_i'(t)\Big(\ln\big(g_i'(t)\big) - 1\Big) + \frac{\delta}{2} g_i(t)\Big(\beta_i - g_i(t)\Big) \right\} dt \end{aligned}$$

$$\geq \frac{1}{b_i}\int_{s_i}^{\tau_i} g_i'(t)\ln\left(\frac{b_i\delta}{2}g_i(t)\Big(\beta_i - g_i(t)\Big)\right)dt$$

$$= \frac{1}{b_i}\int_0^{\beta_i}\ln\left(\frac{b_i\delta}{2}s\Big(\beta_i - s\Big)\right)ds = \frac{\beta_i}{b_i}\ln\left(\frac{b_i\delta}{2}\right) + \frac{2\beta_i}{b_i}\Big(\ln(\beta_i) - 1\Big) := C_i(\beta_i,\delta).$$

As a consequence, by Lemma 3.1, for $0 < \delta < \delta_0$ and every $J \subseteq I$, we have:

$$\begin{aligned}
Y_\beta(g) &= \int_{\mathbb{R}}\sum_{i\in J}\frac{1}{b_i}g_i'(t)\Big(\ln(g_i'(t)) - 1\Big)dt + \int_{\mathbb{R}}\sum_{i\in J}\frac{\delta}{2}g_i(t)\Big(\beta_i - g_i(t)\Big)dt \\
&\quad + \int_{\mathbb{R}}\Phi_{\delta,\beta}\big(g(t)\big)dt \\
&\quad + \int_{\mathbb{R}}\sum_{i\in I\setminus J}\left\{\frac{1}{b_i}g_i'(t)\Big(\ln(g_i'(t)) - 1\Big) + \frac{\delta}{2}g_i(t)\Big(\beta_i - g_i(t)\Big)\right\}dt \\
&\geq \int_{\mathbb{R}}\sum_{i\in J}\frac{1}{b_i}g_i'(t)\Big(\ln(g_i'(t)) - 1\Big)dt + \int_{\mathbb{R}}\sum_{i\in J}\frac{\delta}{2}g_i(t)\Big(\beta_i - g_i(t)\Big)dt \qquad (3.45) \\
&\quad + \int_{\mathbb{R}}\Phi_{\delta,\beta}\big(g(t)\big)dt + \sum_{i\in I\setminus J}C_i(\beta_i,\delta) \\
&\geq \int_{\mathbb{R}}\sum_{i\in J}\frac{1}{b_i}g_i'(t)\Big(\ln(g_i'(t)) - 1\Big)dt + \int_{\mathbb{R}}\sum_{i\in J}\frac{\delta}{2}g_i(t)\Big(\beta_i - g_i(t)\Big)dt \\
&\quad + \sum_{i\in I\setminus J}C_i(\beta_i,\delta).
\end{aligned}$$

Using (3.45) with $J := \emptyset$, we obtain that $Y_\beta(g)$ is bounded from below on $\mathcal{U}_\beta$. Furthermore, if we take $0 \leq 2\delta < \delta_0$, and set $C_0 := \sum_{i\in I} C_i(\beta_i,\delta)$, then from (3.45) we find also that,

$$\begin{aligned}
Y_\beta(g) &\geq \int_{\mathbb{R}}\Phi_{\delta,\beta}\big(g(t)\big)dt + C_0 \\
&= \int_{\mathbb{R}}\Phi_{2\delta,\beta}\big(g(t)\big)dt + \int_{\mathbb{R}}\sum_{i\in I}\frac{\delta}{2}g_i(t)\Big(\beta_i - g_i(t)\Big)dt + C_0 \qquad (3.46) \\
&\geq \sum_{i\in I}\int_{\mathbb{R}}\frac{\delta}{2}g_i(t)\Big(\beta_i - g_i(t)\Big)dt + C_0.
\end{aligned}$$

While, by using (3.45) with $J = \{i\}$ we deduce

$$Y_\beta(g) \geq \int_{\mathbb{R}}\frac{1}{b_i}g_i'(t)\Big(\ln(g_i'(t)) - 1\Big)dt + \int_{\mathbb{R}}\frac{\delta}{2}g_i(t)\Big(\beta_i - g_i(t)\Big)dt + \bar{C}_i, \qquad (3.47)$$

where,

$$\bar{C}_i := \sum_{j\in I\setminus\{i\}} C_j(\beta_i,\delta) \qquad \forall i \in I.$$

By the above estimates, we can prove:

Proposition 3.1. *Assume* (3.1) *and let the symmetric matrix* $A = \{a_{ij}\} \in \mathbb{R}^{m\times m}$ *satisfy* (2.10)*. For any* $\beta = (\beta_1,\dots,\beta_m) \in \mathbb{R}^m$ *satisfying* (3.3)*, the functional* $Y_\beta(g)$ *is bounded from below and attains its minimum in* $\mathcal{U}_\beta$*.*

Proof. Let $g^{(n)}(t) := \big(g_1^{(n)}(t),\dots,g_m^{(n)}(t)\big) \in \mathcal{U}_\beta$ be a minimizing sequence for Y_β. Namely,

$$\gamma_0 := \inf_{g(t)\in\mathcal{U}_\beta} Y_\beta(g) = \lim_{n\to+\infty} Y_\beta(g^{(n)}), \tag{3.48}$$

and let $\tau_i^{(n)} \in \mathbb{R}$ be such that

$$g_i^{(n)}(\tau_i^{(n)}) = \beta_i/2 \qquad \forall\, i \in I,\ \forall\, n \in \mathbb{N}. \tag{3.49}$$

Without loss of generality, we may assume (after a suitable translation) that,

$$\tau_1^{(n)} = 0, \quad \text{i.e.,} \quad g_1^{(n)}(0) = \beta_1/2, \qquad \forall n \in \mathbb{N}. \tag{3.50}$$

We fix $0 < \delta < \delta_0/2$, ($\delta_0 > 0$ as given in Lemma 3.1) and use (3.45), (3.46) and (3.47) to obtain:

$$0 \le \int_{\mathbb{R}} \Phi_{\delta,\beta}(g^{(n)}(t))dt \le Y_\beta(g^{(n)}) - C_0 \le C, \tag{3.51}$$

$$0 \le \sum_{i\in I} \int_{\mathbb{R}} \frac{\delta}{2} g_i^{(n)}(t)\Big(\beta_i - g_i^{(n)}(t)\Big)dt \le Y_\beta(g^{(n)}) - C_0 \le C, \tag{3.52}$$

and

$$\begin{aligned} C_i(\beta,\delta) &\le \int_{\mathbb{R}} \frac{1}{b_i}(g_i^{(n)})'(t)\Big(\ln\Big((g_i^{(n)})'(t)\Big) - 1\Big)dt + \int_{\mathbb{R}} \frac{\delta}{2} g_i^{(n)}(t)\Big(\beta_i - g_i^{(n)}(t)\Big)dt \\ &\le Y_\beta(g^{(n)}) - \bar{C}_i \le C, \qquad \forall i \in I. \end{aligned} \tag{3.53}$$

From (3.52) and (3.53) we check that, $g_i^{(n)}(t) \in \mathcal{W}_{\beta_i}$ satisfies the assumptions of Lemma 3.5, $\forall i \in I$. So we need to see what happens to the sequence $\{\tau_i^{(n)}\} \in \mathbb{R}$, $\forall i \in I$. In general, we may claim that there exist three disjoint sets $J_1, J_2, J_3 \subset I$, such that $J_1 \cup J_2 \cup J_3 = I$ and (along a suitable subsequence) there holds:

$$\lim_{n\to+\infty} \tau_i^{(n)} = \begin{cases} \tau_i \in \mathbb{R} & \text{if } i \in J_1 \\ +\infty & \text{if } i \in J_2 \\ -\infty & \text{if } i \in J_3. \end{cases} \tag{3.54}$$

Since, by our assumptions $\tau_1^{(n)} = 0$, then $J_1 \ne \emptyset$. Moreover by Lemma 3.5, for every $i \in J_1$, there exists $g_i \in \mathcal{W}_{\beta_i}$, such that (up to a subsequence) $g_i^{(n)} \to g_i$ as

$n \to +\infty$ uniformly in $\mathbb{R}$. Moreover,

$$\int_{\mathbb{R}} g_i'(t)\Big(\ln\left(g_i'(t)\right) - 1\Big)dt \le \varliminf_{n\to+\infty} \int_{\mathbb{R}} (g_i^{(n)})'(t)\Big(\ln\left((g_i^{(n)})'(t)\right) - 1\Big)dt, \quad (3.55)$$

and

$$\int_{\mathbb{R}} g_i(t)\Big(\beta_0 - g_i(t)\Big)dt \le \varliminf_{n\to+\infty} \int_{\mathbb{R}} g_i^{(n)}(t)\Big(\beta_0 - g_i^{(n)}(t)\Big)dt, \quad (3.56)$$

while,

$$\lim_{n\to+\infty} g_i^{(n)}(t) = 0 \qquad \forall t \in \mathbb{R}, \;\; \forall\, i \in J_2, \quad (3.57)$$

and

$$\lim_{n\to+\infty} g_i^{(n)}(t) = \beta_i \qquad \forall\, t \in \mathbb{R}, \;\; \forall\, i \in J_3. \quad (3.58)$$

Recalling that,

$$0 \le \Phi_{\delta,\beta}(r) := \sum_{i\in I}\left\{ -\frac{\delta}{2} r_i(\beta_i - r_i) + \frac{2(N_i+1)}{b_i} r_i - \sum_{j\in I} \frac{1}{2} a_{ij} r_i r_j \right\}, \quad (3.59)$$

we can use (3.51) together with Fatou's Lemma, to deduce that,

$$0 \le \int_{\mathbb{R}} \Phi_{\delta,\beta}\big(\psi(t)\big)dt \le C, \quad (3.60)$$

with

$$\psi(t) := \begin{cases} g_i(t) \in \mathbb{R} & \text{if } i \in J_1 \\ 0 & \text{if } i \in J_2 \\ \beta_i & \text{if } i \in J_3. \end{cases} \quad (3.61)$$

In particular, by (3.60) we must have $\Phi_{\delta,\beta}\big(\psi(-\infty)\big) = 0$ and $\Phi_{\delta,\beta}\big(\psi(+\infty)\big) = 0$. This gives:

$$\sum_{i\in J_3}\left\{ \frac{2(N_i+1)}{b_i}\beta_i - \sum_{j\in J_3} \frac{1}{2} a_{ij}\beta_i\beta_j \right\} = 0 \quad (3.62a)$$

and

$$\sum_{i\in J_1\cup J_3}\left\{ \frac{2(N_i+1)}{b_i}\beta_i - \sum_{j\in J_1\cup J_3} \frac{1}{2} a_{ij}\beta_i\beta_j \right\} = 0. \quad (3.62b)$$

Since $J_1 \neq \emptyset$, from (3.3) we see that necessarily $J_3 = \emptyset$, $J_1 = I$ and so also $J_2 = \emptyset$. So, for every $i \in I$ we have: $\lim_{n\to+\infty} \tau_i^{(n)} = \tau_i \in \mathbb{R}, \quad \forall\, i \in I$. Consequently,

$$g_i^{(n)}(t) \to g_i(t) \quad \text{uniformly in } \mathbb{R}; \quad (3.63)$$

$$\sum_{i\in I} \int_{\mathbb{R}} g_i'(t)\Big(\ln\left(g_i'(t)\right) - 1\Big)dt \le \sum_{i\in I} \varliminf_{n\to+\infty} \int_{\mathbb{R}} (g_i^{(n)})'(t)\Big(\ln\left((g_i^{(n)})'(t)\right) - 1\Big)dt, \quad (3.64)$$

with $g(t) := \big(g_1(t), \dots, g_m(t)\big) \in \mathcal{U}_\beta$. Furthermore, by Fatou's Lemma we also get,

$$0 \leq \int_{\mathbb{R}} \Phi_{0,\beta}\big(g(t)\big)dt \leq \varliminf_{n\to+\infty} \int_{\mathbb{R}} \Phi_{0,\beta}\big(g^{(n)}(t)\big)dt. \tag{3.65}$$

Therefore,

$$\begin{aligned} Y_\beta(g) &:= \sum_{i\in I} \int_{\mathbb{R}} \frac{1}{b_i} g_i'(t)\Big(\ln\big(g_i'(t)\big) - 1\Big)dt + \int_{\mathbb{R}} \Phi_{0,\beta}\big(g(t)\big)dt \\ &\leq \varliminf_{n\to+\infty} Y_\beta\big(g^{(n)}\big) = \inf_{\sigma(t)\in\mathcal{U}_\beta} Y_\beta(\sigma). \end{aligned}$$

and we conclude that g is the desired minimizer for Y_β. □

Remark 3.1. The fact that Y_β is bounded from below in $\mathcal{U}_\beta$ can be formulated in terms of a Logarithm-HLS inequality (for systems), in presence of weights. Notice however that in general such inequality holds only for radial functions. More precisely, assuming (3.1) and (2.10), we have proved that, if $\beta = (\beta_1, \dots, \beta_m)$ satisfies (3.3), then the inequality:

$$\begin{aligned} &\sum_{i\in I} \int_{\mathbb{R}^2} \left(\frac{1}{b_i}\rho_i(x)\ln(\rho_i(x)) - \frac{2N_i}{b_i}\rho_i(x)\ln|x| \right) dx \\ &\qquad + \frac{1}{4\pi}\sum_{i,j\in I} a_{ij} \int_{\mathbb{R}^2}\int_{\mathbb{R}^2} \rho_i(x)\ln|x-y|\,\rho_j(y)dxdy \geq -C \end{aligned} \tag{3.66}$$

holds (with a suitable $C > 0$), over the radial set

$$\Gamma_{\beta,\mathrm{rad}} := \{\rho = (\rho_1, \dots, \rho_m) : \rho = \rho(|x|), \rho \in \Gamma_\beta\}$$

where,

$$\begin{aligned} \Gamma_\beta = \Big\{ &\rho = (\rho_1, \dots, \rho_m) : \rho_i \geq 0 \text{ a.e } \rho_i \ln \rho_i \in L^1(\mathbb{R}^2), \\ &\rho_i \ln(1+|x|^2) \in L^1(\mathbb{R}^2) \;\text{ and }\; \frac{1}{2\pi}\int_{\mathbb{R}^2} \rho_i = \beta_i,\ \forall i \in I \Big\}. \end{aligned} \tag{3.67}$$

Furthermore, the optimal constant in (3.66) is attained in $\Gamma_{\beta,\mathrm{rad}}$. To this respect, we refer to [CSW] and [SW2], where such inequality is established when $N_i = 0$, (and $b_i = 1$), $\forall i \in I$, which extends to systems results established for the single equation ($m = 1$) in [CL] and [B].

If we do not worry about the existence of extremals for (3.66), but aim only to find the best conditions on the m-ple $\beta = (\beta_1, \dots, \beta_m)$ so that (3.66) holds in $\Gamma_{\beta,\mathrm{rad}}$, then, as in [SW2], we can allow the equal sign in the last condition in (3.3). More precisely, for $\emptyset \neq J \subset I$, letting:

$$F_J(\beta) = \sum_{i,j\in J} \frac{1}{2} a_{ij}\beta_i\beta_j - \sum_{j\in J} \frac{2(N_i+1)}{b_i}\beta_i, \tag{3.68}$$

we can check that,

Corollary 3.1. *The inequality* (3.66) *holds in* $\Gamma_{\beta,\mathrm{rad}}$ *if and only if*

i) $F_I(\beta) = 0$
ii) *for* $\emptyset \neq J \subset I$, $F_J(\beta) \leq 0$ *and when* $F_J(\beta) = 0$ *then necessarily:*

$$F_{J\setminus\{i\}}(\beta) - \frac{1}{2}a_{ii}\beta_i^2 < 0, \quad \forall i \in J \tag{3.69}$$

To interpret (3.69), we observe that, if $F_J(\beta) = 0$ then we can argue as in Lemma 2.1, to deduce that (2.12) holds with I replaced by J, which yields to (3.69).

Geometrically the role of (3.69) is explained in Lemma 4.1 and 4.2 of [SW2].

When $N_i \leq 0$, $\forall i \in I$ then by means of Schwarz symmetrization, one sees that the right-hand side of (3.66) actually increases when tested over non-radial functions. Hence, in this case, inequality (3.66) holds over the whole set Γ_β.

On the contrary, when $N_i > 0$ for some $i \in I$, then the symmetrization argument no longer works and in fact (3.66) fails in general for non-radial functions, as one can check by arguing as in Lemma 2.2 in [CSW]. See also [DET1], where a similar issue is discussed for the single equation $m = 1$.

As in [CSW], [W], [JoW1], [SW1], [SW2], [SW3], [CL] and [B], it would be interesting to see when the inequality (3.66) could be linked (via a duality principle) to a Moser–Trudinger type inequality, within the framework of radially symmetric functions.

In this respect one could analyze, for $m = 1$, also the non-radial setting, by keeping in mind the presence of non-radial optimal minimizers for the (improved) Moser–Trudinger inequality valid for radial functions.

More precisely such optimal minimizers correspond to the regular part of the (well-known) non-radial solutions of the singular Liouville equation (1.1) with $m = 1$ and $n_1 = n \in \mathbb{N}$, as identified in [PrT] and given (in complex notation) as follows:

$$v(z) = \log \frac{\lambda}{\left(1 + \lambda\gamma_n \left|z^{n+1} + z_0\right|^2\right)^2}, \tag{3.70}$$

for every $\lambda > 0$, $z_0 \in \mathbb{C}$ and explicit $\gamma_n > 0$ depending on $n \in \mathbb{N}$ only. They have been an inspiration for the recent results of Bartolucci–Malchiodi [BM], who identified a set of suitable conditions on non-radial functions, according to which the Moser–Trudinger inequality holds with the improved constant of the radial setting.

It seems a challenging task to extend those results to system (cf. [LWY]), or interpret them in the framework of a duality principle.

Next we establish some regularity properties for the minimizer.

Lemma 3.6. *Under the assumptions of Proposition* 3.1, *a minimizer* $f \in \mathcal{U}_\beta$ *of* Y_β *satisfies* $f_i' > 0$ *a.e. in* $\mathbb{R}$, $\forall i \in I$.

More precisely for every $R > 0$, there exists a constant $C_i = C_i(R) > 0, \quad i \in I$ such that: such that

$$f_i'(t) \geq C_i e^{-b_i\left(\sum_{j\in I} a_{ij}\beta_j - \frac{2(N_i+1)}{b_i}\right)t}, \quad \textit{for a.e. } t \in [-R, 0] \tag{3.71}$$

$$f_i'(t) \geq C_i e^{-2(N_i+1)|t|}, \quad \textit{for a.e. } t \in [0, R]. \tag{3.72}$$

Remark 3.2. Recall that, by (2.12) the exponent in (3.71): $\sum\limits_{j\in I} a_{ij}\beta_j - \frac{2(N_i+1)}{b_i} > 0$.

Proof. Fix $\delta(t) = \left(\delta_1(t), \ldots, \delta_m(t)\right) \in \mathcal{U}_\beta$ and for every $s = (s_1, s_2, \ldots, s_m) \in [0,1]^m$ define $\psi_i^{(s)}(t) := (1 - s_i)f_i(t) + s_i\delta_i(t) \quad \forall t \in \mathbb{R}, \ \forall i \in I$, so that $\psi^{(s)}(t) := \left(\psi_1^{(s)}(t), \ldots, \psi_m^{(s)}(t)\right) \in \mathcal{U}_\beta$.

Thus, for fixed $i \in I$, $\tau \in (0,1)$ and $s = \tau e_i$, (e_i the ith element of the standard orthonormal basis in $\mathbb{R}^m$) we have:

$$\begin{aligned}
0 &\leq \frac{1}{\tau}\left(Y_\beta(\psi^{(s)}) - Y_\beta(f)\right)\\
&\leq \int\limits_{\mathbb{R}} \left\{\frac{1}{b_i}\left(\delta_i'(t) - f_i'(t)\right)\ln\left((\psi_i^{(s)})'(t)\right)\right.\\
&\quad \left. + \left(\delta_i(t) - f_i(t)\right)\left(\frac{2(N_i+1)}{b_i} - \sum_{j\in I}\frac{1}{2}a_{ij}\left(\psi_j^{(s)}(t) + f_j(t)\right)\right)\right\}dt.
\end{aligned} \tag{3.73}$$

Let,

$$\begin{aligned}
F_i := &\int\limits_{-\infty}^{+\infty} \frac{1}{b_i} f_i'(t)\ln\left(f_i'(t)\right)dt + \frac{2(N_i+1)}{b_i}\left(\int_{-\infty}^0 f_i(t)dt + \int_0^{+\infty}\left(f_i(t) - \beta_i\right)dt\right)\\
&- \sum_{j\in I} a_{ij}\left(\int_{-\infty}^0 f_i(t)f_j(t)dt + \int_0^{+\infty} f_i(t)\left(f_i(t) - \beta_i\right)dt\right).
\end{aligned}$$

In view of (3.3) and (3.73), by straightforward calculations we find:

$$\begin{aligned}
&\int\limits_{-\infty}^{+\infty} \frac{1}{b_i}\delta_i'(t)\ln\left(f_i'(t) + \tau\delta_i'\right)dt + \frac{2(N_i+1)}{b_i}\left(\int\limits_{-\infty}^{0}\delta_i(t)dt + \int\limits_0^{+\infty}\left(\delta_i(t) - \beta_i\right)dt\right)\\
&\quad - \sum_{j\in I} a_{ij}\left(\int\limits_{-\infty}^0 \delta_i(t)f_j(t)dt + \int\limits_0^{+\infty} f_j(t)\left(\delta_i(t) - \beta_i\right)dt\right) - \frac{\tau a_{ii}}{2}\int\limits_{-\infty}^{+\infty}\left(\delta_i(t) - f_i(t)\right)^2 dt\\
&\geq F_i - \frac{\tau a_{ii}}{2}\int_{-\infty}^{+\infty}\left(\delta_i(t) - f_i(t)\right)^2 dt > -\infty.
\end{aligned} \tag{3.74}$$

Thus, we can use the monotone convergence theorem and by letting $\tau \to 0^+$, conclude:

$$\int_{-\infty}^{+\infty} \frac{1}{b_i}\delta_i'(t) \ln\Big(f_i'(t)\Big)dt + \frac{2(N_i+1)}{b_i}\Big(\int_{-\infty}^{0} |t|\delta_i'(t)dt - \int_{0}^{+\infty} t\delta_i'(t)dt\Big)$$

$$-\sum_{j\in I} a_{ij} \int_{0}^{+\infty} \beta_j\Big(\delta_i(t) - \beta_i\Big)dt$$

$$\geq \sum_{j\in I} a_{ij}\Big(\int_{-\infty}^{0} \delta_i(t) f_j(t)dt + \int_{0}^{+\infty}\Big(\delta_i(t)-\beta_i\Big)\Big(f_j(t)-\beta_j\Big)dt\Big) + F_i$$

$$\geq -\sum_{j\in I} |a_{ij}|\beta_i\Big(\int_{-\infty}^{0} f_j(t)dt + \int_{0}^{+\infty}\Big(\beta_j - f_j(t)\Big)dt\Big) + F_i := -A_i. \tag{3.75}$$

In other words, for every $\delta_i \in \mathcal{W}_{\beta_i}$ $i \in I$, the following holds:

$$\begin{aligned}&\int_{-\infty}^{0} \delta_i'(t)\left\{\frac{1}{b_i}\ln\Big(f_i'(t)\Big) + \frac{2(N_i+1)}{b_i}|t| + \frac{A_i}{\beta_i}\right\}dt \\ &+ \int_{0}^{+\infty} \delta_i'(t)\left\{\frac{1}{b_i}\ln\Big(f_i'(t)\Big) - \Big(\frac{2(N_i+1)}{b_i} - \sum_{j\in I} a_{ij}\beta_j\Big)t + \frac{A_i}{\beta_i}\right\}dt \geq 0.\end{aligned} \tag{3.76}$$

Next, for every bounded Borel set $\mathcal{B} \subset \mathbb{R}$ with $\mathcal{L}^1(\mathcal{B}) > 0$, denote by $\chi_{\mathcal{B}}$ be the characteristic function of $\mathcal{B}$. If we take $\delta_i(t) = \frac{\beta_i}{\mathcal{L}^1(\mathcal{B})}\int_{-\infty}^{t} \chi_{\mathcal{B}}(s)\,ds \in \mathcal{W}_{\beta_i} \quad \forall i \in I$ in (3.76), then we can easily derive the estimates in (3.71) and (3.72). □

As a consequence of Lemma 3.6 and (3.73), we may conclude that every minimizer $f = (f_1, \ldots, f_m)$ of Y_β in $\mathcal{U}_\beta$ satisfies:

$$\frac{d}{dt}\left\{\frac{1}{b_i}\ln\left(f_i'(t)\right) - \frac{2(N_i+1)t}{b_i}\right\} + \sum_{j\in I} a_{ij} f_j(t) = 0 \quad \forall t \in \mathbb{R}, \quad \forall i \in \{1,\ldots,m\}. \tag{3.77}$$

Therefore, if we set

$$v_i(t) := \frac{1}{b_i}\Big(\ln\left(f_i'(t)\right) - 2(N_i+1)t\Big), \quad i \in I; \tag{3.78}$$

then,

$$f_i(t) = \int_{-\infty}^{t} e^{b_i v_i(s) + 2(N_i+1)s} ds \qquad \forall t \in \mathbb{R}, \quad \forall i \in I, \tag{3.79}$$

and $v = (v_1, \dots, v_m)$ satisfies:

$$\begin{cases} v_i''(t) + \sum_{j\in I} a_{ij} e^{b_i v_i(t)+2(N_i+1)t} = 0 \quad \forall t \in \mathbb{R}, \quad \forall i \in I, \\ v_i'(-\infty) = 0 \quad \forall i \in I, \\ \int_{\mathbb{R}} e^{b_i v_i(s)+2(N_i+1)s} ds = \beta_i \quad \forall i \in I. \end{cases} \tag{3.80}$$

In conclusion, we have established the following:

Theorem 3.1. *Let the symmetric matrix $A = \{a_{ij}\} \in \mathbb{R}^{m\times m}$ satisfy $a_{ii} \geq 0$, and let $N_i > -1$, $b_i > 0$ $\forall i \in I$. If there exists a m-ple $(\beta_1, \dots, \beta_m)$ satisfying* (3.3)*, then A must be irreducible and problem* (2.3) *admits a solution.*

By combining Theorem 3.1 with Proposition 2.1 we conclude:

Corollary 3.2. *If $A = \{a_{ij}\} \in \mathbb{R}^{m\times m}$ is symmetric, irreducible with $a_{ij} \geq 0$ for $i, j \in I$. For any fixed $N_i > -1$, $b_i > 0$ $i \in I$, the condition* (3.3) *on $(\beta_1, \dots, \beta_m)$ is necessary and sufficient for the existence of a solution for* (2.3).

In the following section we complete the statement of Corollary 3.2, by showing that actually the m-ple $\beta = (\beta_1, \dots, \beta_m)$ uniquely determines the corresponding solution of (2.3), up to the natural (translation) invariance of the system (2.3), as explicitly stated in (4.2) below.

4. Uniqueness

In this Section we show that it is possible to extend the uniqueness result of [LZ1] to the case where the coupling matrix A may fail to be invertible and under the more general assumption (2.10). In this way, we can treat a degenerate system arising in the study of selfgravitating strings, and complete the uniqueness result established in [PT], as shown in Section 5 below.

Thus, throughout this section, we assume that:

$$A = \{a_{ij}\}_{i,j\in I} \text{ is irreducible and } a_{ij} = a_{ji} \geq 0, \ \forall i, j \in I. \tag{4.1}$$

By straightforward calculations it is easy to check that problem (2.3) is invariant under the transformation:

$$v_i(t) \to v_{i,\lambda}(t) = v_i(t+\lambda) + \frac{2(N_i+1)}{b_i}\lambda, \quad \forall i \in I; \qquad \lambda \in \mathbb{R}. \tag{4.2}$$

We prove:

Theorem 4.1. *Let A satisfy* (4.1) *and assume* (2.10). *For given $\beta = (\beta_1, \dots, \beta_m)$ satisfying* (3.3)*, problem* (2.3) *admits a unique solution, modulo the transformation* (4.2).

In order to prove Theorem 4.1, we start to observe that, as a consequence of (4.2), the function: $\zeta(t) = (\zeta_1(t), \dots, \zeta_m(t))$, given by

$$\zeta_i(t) := \frac{2(N_i+1)}{b_i} + v_i'(t), \quad \forall t \in \mathbb{R}, \ i \in I, \tag{4.3}$$

satisfies the following linearized problem:

$$\begin{cases} \frac{d^2 w_i}{dt^2} + \sum_{i=1}^m a_{ij} b_j e^{2(N_j+1)t + b_j v_j} w_j = 0 & \text{for } t \in \mathbb{R} \quad \forall 1 \le i \le m \\ w_i(-\infty) \in \mathbb{R} & \forall 1 \le i \le m \\ w_i(+\infty) \in \mathbb{R} & \forall 1 \le i \le m \\ \int_{\mathbb{R}} e^{2(N_j+1)t + b_j v_j(t)} w_j(t)\, dt = 0 & \forall 1 \le i \le m. \end{cases} \tag{4.4}$$

Following [LZ1], we prove the following:

Lemma 4.1. *If $w(t)$ satisfies (4.4), then $w(t) = C\zeta(t)$, for all $t \in \mathbb{R}$, and $C \in \mathbb{R}$.*

Proof. We argue by contradiction, and assume that $w(t)$ and $\zeta(t)$ are linearly independent. Since,

$$\zeta_i(-\infty) = \frac{2(N_i+1)}{b_i}, \quad \forall i \in I,$$

we may assume, without any loss of generality, that $w_1(-\infty) = 0$ and $w_2(-\infty) < 0$. For $\alpha \in \mathbb{R}$, let,

$$w^{(\alpha)}(t) = \left(w_1^{(\alpha)}(t), \dots, w_m^{(\alpha)}(t)\right) := \alpha w(t) + \zeta(t), \quad \forall t \in \mathbb{R};$$

and consider the set,

$$\mathcal{F} := \left\{ \alpha \in \mathbb{R} : \int_{-\infty}^{t} w_i^{(\alpha)}(s) e^{2(N_i+1)s + b_i v_i(s)} ds > 0 \ \forall t \in \mathbb{R},\ \forall i \in I \right\}. \tag{4.5}$$

Clearly $0 \in \mathcal{F}$, as we have:

$$\begin{aligned} &\int_{-\infty}^{t} w_i^{(0)}(s) e^{2(N_i+1)s + b_i v_i(s)} ds \\ &\quad = \frac{1}{\beta_i} \int_{-\infty}^{t} \Big(2(N_i+1) + b_i v_i'(s)\Big) e^{2(N_i+1)s + b_i v_i(s)} ds = \frac{1}{\beta_i} e^{2(N_i+1)t + b_i v_i(t)} > 0. \end{aligned}$$

Moreover, the condition: $w_2(-\infty) < 0$, ensures that $\mathcal{F}$ is bounded from above. Set $\tilde{\alpha} := \sup \mathcal{F}$, so that by a simple limiting process there holds:

$$\begin{cases} \int_{-\infty}^{t} w_i^{(\tilde{\alpha})}(s) e^{2(N_i+1)s + b_i v_i(s)} ds \ge 0 \quad \forall t \in \mathbb{R},\ \forall i \in I \\ \int_{\mathbb{R}} w_i^{(\tilde{\alpha})}(s) e^{2(N_i+1)s + b_i v_i(s)} ds = 0 \quad \forall t \in \mathbb{R},\ \forall i \in I \\ w_i^{(\tilde{\alpha})}(-\infty) \ge 0 \quad \forall i \in I \\ (w_i^{(\tilde{\alpha})})'(t) = -\sum_{j \in I} \int_{-\infty}^{t} a_{ij} b_j w_j^{(\tilde{\alpha})}(s) e^{2(N_j+1)s + b_j v_j(s)} ds \le 0 \quad \forall t \in \mathbb{R},\ \forall i \in I. \end{cases} \tag{4.6}$$

Let,

$$\tilde{J} := \left\{ i \in I : \ w_i^{(\tilde{\alpha})}(-\infty) > 0 \right\}.$$

Since $w_1^{(\tilde{\alpha})}(-\infty) = \frac{2(N_1+1)}{b_1}$, we see that $1 \in \tilde{J}$ and so $\tilde{J} \neq \emptyset$. We claim that actually $\tilde{J} = I$. Indeed, if this was not the case, then by the irreducibility of A, we would find $j_0 \in \tilde{J}$ and $i_0 \in I \setminus \tilde{J}$ such that $a_{i_0 j_0} > 0$. Since $j_0 \in \tilde{J}$, there exists $t_0 \in \mathbb{R}$ such that

$$\int_{-\infty}^{t} a_{i_0 j_0} b_1 w_{j_0}^{(\tilde{\alpha})}(s) e^{2(N_{j_0}+1)s + b_{j_0} v_{j_0}(s)} ds > 0, \quad \forall t < t_0.$$

As a consequence,

$$\begin{aligned}(w_{i_0}^{(\tilde{\alpha})})'(t) = &- \sum_{j \in I \setminus \{j_0\}} \int_{-\infty}^{t} a_{i_0 j} b_j w_j^{(\tilde{\alpha})}(s) e^{2(N_j+1)s + b_j v_j(s)} ds \\ &- \int_{-\infty}^{t} a_{i_0 j_0} b_{j_0} w_{j_0}^{(\tilde{\alpha})}(s) e^{2(N_{j_0}+1)s + b_{j_0} v_{j_0}(s)} ds < 0, \quad \forall t < t_0.\end{aligned} \tag{4.7}$$

But since $i_0 \in I \setminus \tilde{J}$, then $w_{i_0}^{(\tilde{\alpha})}(-\infty) = 0$ and together with (4.7), we obtain a contradiction to the first condition in (4.6).

So $\tilde{J} = I$, and by our assumptions on A, we conclude that actually all inequalities in (4.6) hold with the strict sign. More precisely, we have:

$$\begin{cases} \int_{-\infty}^{t} w_i^{(\tilde{\alpha})}(s) e^{2(N_i+1)s + b_i v_i(s)} ds > 0 \quad \forall t \in \mathbb{R}, \forall i \in I \\ \int_{\mathbb{R}} w_i^{(\tilde{\alpha})}(s) e^{2(N_i+1)s + b_i v_i(s)} ds = 0 \quad \forall t \in \mathbb{R}, \forall i \in I \\ w_i^{(\tilde{\alpha})}(-\infty) > 0 \quad \forall i \in I \\ w_i^{(\tilde{\alpha})}(+\infty) < 0 \quad \forall i \in I \\ (w_i^{(\tilde{\alpha})})'(t) = -\sum_{j \in I} \int_{-\infty}^{t} a_{ij} b_j w_j^{(\tilde{\alpha})}(s) e^{2(N_j+1)s + b_j v_j(s)} ds < 0 \quad \forall t \in \mathbb{R}, \forall i \in I, \end{cases} \tag{4.8}$$

that implies $\tilde{\alpha} \in \mathcal{F}$. By means of (4.8) and the fact that $w \in L^\infty(\mathbb{R})$, we can find $\varepsilon_0 > 0$ sufficiently small and $M_0 > 0$ sufficiently large, such that, $\forall i \in I$ we have: $w_i^{(\tilde{\alpha}+\varepsilon_0)}(t) > 0$, for $t \le -M_0$; $w_i^{(\tilde{\alpha}+\varepsilon_0)}(t) < 0$ for $t \ge M_0$ and $\int_{-\infty}^{t} w_i^{(\tilde{\alpha}+\varepsilon_0)}(s) e^{2(N_i+1)s + b_i v_i(s)} ds > 0$, for $t \in [-M_0, M_0]$. But those conditions imply that $(\tilde{\alpha} + \varepsilon) \in \mathcal{F}$, in contradiction to the fact that $\tilde{\alpha} = \sup \mathcal{F}$. □

Still concerning the solutions of the linearized equation in (4.4) we have:

Lemma 4.2. *If $(w_1, \ldots, w_m)$ is a solution of the linearized equation in (4.4), then $w_i(t) = O(t)$ as $t \to +\infty$, $\forall i \in I$.*

Proof. It follows exactly as in Lemma 2.1 of [LZ1], with the obvious modifications. □

Remark 4.1. By combining Lemma 4.2 with the decay property (2.12), we easily check that the condition $w_i(+\infty) \in \mathbb{R}$ in (4.4) is actually equivalent to the condition $w_i'(+\infty) = 0$, $\forall i \in I$.

Remark 4.2. By a direct inspection of Lemma 2.1 of [LZ1], it is possible to check that the estimate in Lemma 4.2 holds uniformly, whenever the coupling matrix A varies in a compact set of the space of $m \times m$ symmetric matrices.

Going back to the proof of Theorem 4.1, we notice that in view of the invariance (4.2), it suffices to show that (2.3) admits a unique solution $(v_1,\dots,v_m)$ satisfying:

$$v_m(-\infty)=0. \tag{4.9}$$

To pursue this goal, we work first under the additional assumption that,

$$a_{ii}>0,\quad \forall\, i\in I. \tag{4.10}$$

We have:

Lemma 4.3. *Let A satisfy (4.1) and (4.10). Then for every $\tau=(\tau_1,\dots,\tau_{m-1})\in\mathbb{R}^{m-1}$, the initial value problem:*

$$\begin{cases} \frac{d^2v_i}{dt^2}+\sum_{i=1}^m a_{ij}e^{2(N_j+1)t+b_jv_j}=0, & \forall\, i\in I\\ v_i(-\infty)=\tau_i,\quad \frac{dv_i}{dt}(-\infty)=0, & i=1,\dots,m-1\\ v_m(-\infty)=0=\frac{dv_m}{dt}(-\infty) \end{cases} \tag{4.11}$$

admits a unique solution $v(t,\tau)=\big(v_1(t,\tau),\dots,v_m(t,\tau)\big)$ defined for every $t\in\mathbb{R}$ and satisfying:

$$\beta_i(\tau):=\int_{\mathbb{R}} e^{2(N_i+1)t+b_iv_i(t,\tau)}dt<+\infty,\quad \forall\, i\in I. \tag{4.12}$$

Proof. Under the given assumptions, we see that both v_i and v_i' are decreasing in their interval of existence. Moreover, using (2.8) with $J=\{i\}$ (and $h=1$) we get,

$$\frac{1}{b_i}e^{2(N_i+1)t+b_iv_i(t)}+f_i(t)\left(\frac{1}{2}a_{ii}f_i(t)-\frac{2(N_i+1)}{b_i}\right)\le 0. \tag{4.13}$$

Since, $f_i(t):=\int_{-\infty}^t e^{2(N_i+1)s+b_iv_i(s)}ds$, from (4.13) we find that the (unique) solution $v(t,\tau)$ of (4.11) is defined for all $t\in\mathbb{R}$ and (4.12) holds. ☐

In view of the uniform estimates provided by (4.13), we see that,

$$w^{(l)}(t)=\frac{\partial v}{\partial\tau_l}(t,\tau)\quad\text{and}\quad\frac{\partial\beta_i}{\partial\tau_l}(\tau)\quad l=1,\dots,m-1,\quad i\in I$$

are well defined, and there holds:

$$\begin{cases} \frac{d^2w_i}{dt^2}+\sum_{i=1}^m a_{ij}b_je^{2(N_j+1)t+b_jv_j(t,\tau)}w_j=0 & \text{for } t\in\mathbb{R}\quad \forall i\in I\\ w_i(-\infty)=\delta_{li},\quad w_i'(-\infty)=0, & \forall i\in I\\ w_i'(+\infty)=-\sum_{j\in I}a_{ij}b_j\frac{\partial}{\partial\tau_l}(\beta_j(\tau)), & \forall i\in I\\ \int_{\mathbb{R}}e^{2(N_j+1)t+b_jv_j(t,\tau)}w_j(t)\,dt=\frac{\partial}{\partial\tau_l}(\beta_i(\tau)), & \forall i\in I. \end{cases} \tag{4.14}$$

Set,

$$\frac{\partial\beta}{\partial\tau}:=\begin{pmatrix} \frac{\partial\beta_1}{\partial\tau_1} & ,\dots & ,\frac{\partial\beta_1}{\partial\tau_{m-1}}\\ \dots & ,\dots & ,\dots\\ \frac{\partial\beta_m}{\partial\tau_1} & ,\dots & ,\frac{\partial\beta_m}{\partial\tau_{m-1}} \end{pmatrix} \tag{4.15}$$

Lemma 4.4. *Under the assumption of Lemma 4.3, the matrix $\frac{\partial\beta}{\partial\tau}$ admits maximal rank $m-1$, for every $\tau \in \mathbb{R}^{m-1}$.*

Proof. By contradiction, for some $\tau_0 \in \mathbb{R}^{m-1}$, we assume that there exists

$$(s_1, \dots, s_{m-1}) \in \mathbb{R}^{m-1} \setminus \{0\} : \sum_{l=1}^{m-1} s_l \frac{\partial \beta_i}{\partial \tau_l}(\tau_0) = 0, \ \forall i \in I.$$

Setting $w(t) = \sum_{l=1}^{m-1} s_l w^{(l)}(t, \tau_0)$, then by (4.14), we see that $w(t) = \big(w_1(t), \dots, w_m(t)\big)$ gives a non trivial solution of (4.4) and also satisfies: $w_m(-\infty) = 0$, in contradiction with Lemma 4.1. □

Lemma 4.5. *Under the assumption of Lemma 4.3, the set*

$$\Pi = \Big\{\beta = (\beta_1, \dots, \beta_m) \in \mathbb{R}^m : \ (3.3) \ \textit{holds}\Big\}$$

is simply connected.

Proof. For any $\emptyset \neq J \subseteq I$ and $\beta_i > 0$, $i = 1, \dots, m$, we recall from (3.68) that

$$F_J(\beta_1, \dots, \beta_m) = \sum_{i \in J}\left(\sum_{j \in J} \frac{1}{2} a_{ij}\beta_i\beta_j - \frac{2(N_i+1)}{b_i}\beta_i\right), \qquad \text{and also we set } F_\emptyset \equiv 0.$$

We start with the following:

Claim. The set

$$\Omega_I^- := \{(\beta_1, \dots, \beta_m) : \ \beta_i > 0 \text{ for } i \in I \text{ and } F_J(\beta_1, \dots, \beta_m) < 0 \text{ for } \emptyset \neq J \subseteq I\}$$

is simply connected.

We prove the claim by induction on the cardinality $m = |I|$ of the set I.

If $m = 1$, then $I = \{1\}$ and $\Omega_I^- = \left(0, \frac{4(N_1+1)}{b_1 a_{11}}\right)$, which is clearly simply connected.

Next, for every $\emptyset \neq J \subset I$: $|J| = m-1$, we suppose that Ω_J^- is simply connected. Notice that, for $\emptyset \neq J \subseteq I$ and $l \in J$ we have:

$$F_J(\beta_1, \dots, \beta_m) = \frac{1}{2} a_{ll}\beta_l^2 - \left(\frac{2(N_l+1)}{b_l} - \sum_{j \in J\setminus\{l\}} a_{lj}\beta_j\right)\beta_l + F_{J\setminus\{l\}}\big(\beta_{J\setminus\{l\}}\big), \quad (4.16)$$

where we have set $\beta_{\tilde{J}} = (\beta_i)_{i \in \tilde{J}}$, $\forall \tilde{J} \subset I$.

Hence, by (4.16), we see that,

$$\begin{aligned} &(\beta_1, \dots, \beta_m) \in \Omega_I^- \ \text{ if and only if } \ \beta_{J\setminus\{l\}} \in \Omega^-_{J\setminus\{l\}} \\ &\text{and } \ 0 < \beta_l < \gamma_l^{(J)}\big(\beta_{J\setminus\{l\}}\big) \quad \forall J \subseteq I \ \forall l \in I, \end{aligned} \qquad (4.17)$$

with,

$$\gamma_l^{(J)}(\beta_{J\setminus\{l\}}) = \frac{1}{a_{ll}}\left\{\frac{2(N_l+1)}{b_l} - \sum_{j\in J\setminus\{l\}} a_{lj}\beta_j + \sqrt{\left(\frac{2(N_l+1)}{b_l} - \sum_{j\in J\setminus\{l\}} a_{lj}\beta_j\right)^2 - 2a_{ll}F_{J\setminus\{l\}}(\beta_{J\setminus\{l\}})}\right\}. \tag{4.18}$$

By the induction assumption, from (4.18) we easily deduce that Ω_I^- is simply connected as claimed.

Next, for any given $l \in I$, consider the set:

$$\Pi_l = \left\{(\beta_1,\dots,\beta_m):\ \beta_{I\setminus\{l\}} \in \Omega^-_{I\setminus\{l\}},\ \beta_l = \gamma_l(\beta_{J\setminus\{l\}})\right\} \tag{4.19}$$

with $\gamma_l \equiv \gamma_l^{(I)}$. Since $\Omega^-_{I\setminus\{l\}}$ is simply connected, we derive that Π_l is simply connected as well, $\forall l \in I$. At this point, we conclude easily that Π is simply connected, as $\Pi = \cap_{l\in I}\Pi_l$. □

Proof of Theorem 4.1 *under assumption* (4.10). We consider the (smooth) map:

$$\begin{aligned} \psi : \mathbb{R}^{m-1} &\to \Pi \\ \tau &\to (\beta_1(\tau),\dots,\beta_m(\tau)) := \beta(\tau) \end{aligned}$$

with $\beta_i(\tau)$ defined in (4.12). Observe that, by Corollary 3.2, the map ψ is onto.

Furthermore, ψ is a proper map, since uniform estimates on the β_i's, $i \in I$ imply uniform decay estimates for v_i and uniform estimates for $\frac{v_i'(t)}{t}$; as $t \to +\infty$.

In addition, by Lemma 4.4 we can show that the map ψ is locally invertible. Indeed, for given $\tau_0 \in \mathbb{R}^{m-1}$, by re-arranging the coordinates if necessary, we may suppose that,

$$\frac{\partial\beta}{\partial\tau} := \det\begin{pmatrix} \frac{\partial\beta_1}{\partial\tau_1} & ,\dots & ,\frac{\partial\beta_1}{\partial\tau_{m-1}} \\ \dots & ,\dots & ,\dots \\ \frac{\partial\beta_{m-1}}{\partial\tau_1} & ,\dots & ,\frac{\partial\beta_{m-1}}{\partial\tau_{m-1}} \end{pmatrix} \neq 0 \quad \text{for } \tau = \tau_0. \tag{4.20}$$

Hence, by setting $\hat\beta_0 = (\beta_1(\tau_0),\dots,\beta_{m-1}(\tau_0))$, we find a neighborhoods $V(\tau_0)$ of τ_0 and $U(\hat\beta_0) \subset \mathbb{R}^{m-1}$ of $\hat\beta_0$, such that the map:

$$\begin{aligned} V(\tau_0) &\to U(\hat\beta_0) \\ \tau &\to (\beta_1(\tau),\dots,\beta_{m-1}(\tau)) \end{aligned}$$

is invertible, with smooth inverse. On the other hand, since $\beta(\tau) \in \Pi$, then necessarily:

$$\beta_m(\tau) = \gamma_m(\beta_1(\tau),\dots,\beta_{m-1}(\tau))$$

with $\gamma_m = \gamma_m^{(I)}$ given in (4.18), and we deduce the local invertibility of ψ. Consequently, by virtue of Lemma 4.5, we can conclude the global invertibility of ψ

between $\mathbb{R}^{m-1}$ and Π, (e.g., see Theorem 1.8 of [AP]). Consequently, the uniqueness property as stated in Theorem 4.1, follows by the uniqueness of the solution for problem (4.11). □

Next we use the uniqueness property valid under the assumption (4.10), in order to complete the proof of Theorem 4.1.

Proof of Theorem 4.1 (*completed*). First of all observe that, if (4.10) is removed then (4.13) still ensures that the solution $v(t,\tau)$ of (4.11) is defined $\forall t \in \mathbb{R}$, but now we cannot guarantee any longer the integrability in $\mathbb{R}$ of the function $e^{2(N_j+1)t+b_j v_j(t,\tau)}$, $j \in \{1,\ldots,m\}$. Furthermore, we can only ensure that Π is connected. Moreover, if we let,

$$\Pi_1 = \Big\{\tau \in \mathbb{R}^{m-1} : e^{2(N_j+1)t+b_j v_j(t,\tau)} \in L^1(\mathbb{R},\mathbb{R}), \ \ \forall j \in I\Big\},$$

then by Corollary 3.2, we see that Π_1 is not empty. Clearly Π_1 is open, and the map $\psi(\tau) := \beta(\tau)$ is well defined proper and locally invertible over Π_1, since the matrix $\frac{\partial \beta}{\partial \tau}$ in (4.20) admits maximal rank when if $\tau \in \Pi_1$. By Corollary 3.2 we also know that ψ is onto over the set Π. Thus, to conclude the proof of Theorem 4.1 it suffices to show that $\psi : \Pi_1 \to \Pi$ is (globally) one to one.

To this purpose, we argue by contradiction and assume that there exist $\tau^{(1)} \neq \tau^{(2)} \in \Pi_1$ such that $\psi(\tau^{(1)}) = \psi(\tau^{(2)})$.

For ε close to zero, let $A_\varepsilon = A + \varepsilon^2 Id$, and denote by $v^{(\varepsilon)}(t,\tau) = \big(v_1^{(\varepsilon)}(t,\tau), \ldots, v_m^{(\varepsilon)}(t,\tau)\big)$ the unique solution of (4.11) with the matrix A replaced by A_ε, and let

$$\beta_j(\varepsilon,\tau) := \int_{\mathbb{R}} e^{2(N_j+1)t+b_j v_j^{(\varepsilon)}(t,\tau)} dt < +\infty, \quad j \in I. \tag{4.21}$$

By the uniform decay estimates of v_ε, as $t \to +\infty$ and those of the solutions of the linearized problem (see Remark 4.1), we can show the smooth dependence of $\beta_i(\varepsilon,\tau)$ and $\frac{\partial \beta_i}{\partial \tau_l}(\varepsilon,\tau)$ with respect to $\varepsilon \in \mathbb{R}$ and $\tau \in \mathbb{R}^{m-1}$. Furthermore for $\tau \in \Pi_1$, we have:

$$\beta_j(\varepsilon,\tau) \to \beta_j(\tau), \qquad \text{as } \varepsilon \to 0; \ j \in I$$

$$\frac{\partial \beta_j}{\partial \tau_l}(\varepsilon,\tau) \to \frac{\partial \beta_j}{\partial \tau_l}(\varepsilon,\tau), \quad \text{as } \varepsilon \to 0; \ j \in I, \ l \in \{1,\ldots,m-1\}.$$

Without loss of generality, we can assume that (4.20) holds with $\tau = \tau^{(1)}$.

Therefore, we can apply the Implicit Function Theorem (e.g., see [AP]) to the function:

$$F(\tau,\varepsilon) = \Big(\beta_1(\varepsilon,\tau) - \beta_1(\varepsilon,\tau^{(2)}), \beta_2(\varepsilon,\tau) - \beta_2(\varepsilon,\tau^{(2)}), \ldots, \beta_{m-1}(\varepsilon,\tau) - \beta_{m-1}(\varepsilon,\tau^{(2)})\Big)$$

at the point $\tau = \tau^{(1)}$ and $\varepsilon = 0$, and obtain a function $\tau = \tau(\varepsilon) : (-\varepsilon_0, \varepsilon_0) \to B_{r_0}(\tau^{(1)})$, with $r_0 > 0$, $\varepsilon_0 > 0$ sufficiently small, such that,

$$\tau(0) = \tau^{(1)}, \qquad \beta_j(\varepsilon,\tau(\varepsilon)) = \beta_j(\varepsilon,\tau^{(2)}) \ \ \forall j = 1,\ldots,m-1. \tag{4.22}$$

Then necesarily:

$$\beta_m(\varepsilon,\tau(\varepsilon)) = \gamma_m\Big(\beta_1(\varepsilon,\tau(\varepsilon)),\dots,\beta_{m-1}(\varepsilon,\tau(\varepsilon))\Big)$$
$$= \gamma_m\Big(\beta_1(\varepsilon,\tau^{(2)}),\dots,\beta_{m-1}(\varepsilon,\tau^{(2)})\Big) = \beta_m(\varepsilon,\tau^{(2)}).$$

(with $\gamma_m = \gamma_m^{(I)}$ defined in (4.18))

In conclusion,

$$\beta_j(\varepsilon,\tau(\varepsilon)) = \beta_j(\varepsilon,\tau^{(2)}), \quad \forall j \in \{1,\dots,m\}.$$

Moreover, by virtue of (4.22), for ε sufficiently close to zero, we see that, $\tau(\varepsilon) \neq \tau^{(2)}$, in contradiction with our previous uniqueness result that applies to the matrix $A_\varepsilon = A + \varepsilon^2 Id$ for $\varepsilon \neq 0$. This completes the proof of Theorem 4.1. □

5. Application to a degenerate system

Motivated by the study of selfgravitating strings, in [PT] we have analyzed the range of $\beta > 0$ such that, the problem:

$$\begin{cases} -\Delta u = e^{au} + |x|^{2N} e^u \quad \text{in } \mathbb{R}^2 \\ \frac{1}{2\pi}\int_{\mathbb{R}^2} \left(e^{au} + |x|^{2N} e^u\right) dx = \beta\,, \end{cases} \tag{5.1}$$

admits a radial solution, provided $N > -1$ and $a > 0$. Clearly, (5.1) can be thought as a degenerate system of the type (2.1), with $u_1 = u = u_2$ and $m = 2$, $a_{ij} = 1$ $\forall i,j \in \{1,2\}$, $b_1 = a$, $b_2 = 1$, $N_1 = 0$, $N_2 = N$. By setting,

$$\beta_j = \frac{1}{2\pi}\int_{\mathbb{R}^2} |x|^{N_j} e^{b_j u_j} dx \qquad j = 1,2,$$

then the conditions in (3.3) can be stated as follows:

$$0 < \beta_1 < \frac{4}{a}, \quad 0 < \beta_2 < 4(N+1) \tag{5.2}$$

$$\frac{1}{2}(\beta_1+\beta_2)^2 - \frac{2}{a}\beta_1 - 2(N+1)\beta_2 = 0. \tag{5.3}$$

In terms of $\beta = \beta_1 + \beta_2$, condition (5.2) can be stated as as follows:

$$\beta\Big(\beta - 4(N+1)\Big) + 4\Big((N+1) - \frac{1}{a}\Big)\beta_1 = 0 = \beta\Big(\beta - \frac{4}{a}\Big) - 4\Big((N+1) - \frac{1}{a}\Big)\beta_2. \tag{5.4}$$

Thus if $a = \frac{1}{N+1}$ then $\beta = 4(N+1)$ and in this case the equation (5.1) becomes invariant under the scaling: $u(x) \to u_\lambda(x) := u(\lambda x) + 2(N+1)\ln\lambda$, and under the Kelvin transform. We refer to [CGS] for a discussion of this situation. Here, we focus to the case $a \neq \frac{1}{N+1}$. If $a > \frac{1}{N+1}$, then necessarily $\frac{4}{a} < \beta < 4(N+1)$. In addition, from (5.4) it follows that:

$$\begin{aligned}&\Big(\beta-2(N+1)\Big)^2-\Big(2(N+1)-\frac{4}{a}\Big)^2\\&=\beta\Big(\beta-4(N+1)\Big)+4\Big((N+1)-\frac{1}{a}\Big)\cdot\frac{4}{a}\\&>\beta\Big(\beta-4(N+1)\Big)+4\Big((N+1)-\frac{1}{a}\Big)\beta_1=0.\end{aligned}\tag{5.5}$$

Thus (5.5) implies also that: $\beta > 4(N+1)-\frac{4}{a}$. In other words,

$$\text{if } a>\frac{1}{N+1} \text{ then } \max\left\{\frac{4}{a}\,,\,4(N+1)-\frac{4}{a}\right\}<\beta<4(N+1). \tag{5.6}$$

Similarly one finds that,

$$\text{if } 0<a<\frac{1}{N+1} \text{ then } \max\left\{4(N+1)\,,\,\frac{4}{a}-4(N+1)\right\}<\frac{4}{a}. \tag{5.7}$$

Clearly (5.6) and (5.7) provide necessary conditions for the existence of a radial solution for (5.1). As a matter of fact, in [PT] was shown that actually (5.6) and (5.7) are also sufficient to guarantee the existence of a radial solution for (5.1), consistently with the results obtained in the previous sections for more general systems.

We can also extend our uniqueness result to hold for problem (5.1), when $a \neq \frac{1}{N+1}$.

Indeed for $\tau \in \mathbb{R}$, let $v(t,\tau)$ be the unique solution of the Cauchy problem:

$$\begin{cases} v''(t)+e^{2t+av(t)}+e^{2(N+1)t+v(t)}=0 & \text{for } t\in\mathbb{R}\\ v(-\infty)=\tau, \quad v'(-\infty)=0. \end{cases}\tag{5.8}$$

We know that, $v(t,\tau)$ is well defined for all $t \in \mathbb{R}$, depends smoothly on τ and satisfies:

$$\beta_1(\tau)=\int_{-\infty}^{+\infty} e^{2t+av(t,\tau)}dt<+\infty \quad\text{and}\quad \beta_2(\tau)=\int_{-\infty}^{+\infty} e^{2(N+1)t+v(t,\tau)}dt<+\infty$$

$$\beta(\tau)=\beta_1(\tau)+\beta_2(\tau)=\int_{-\infty}^{+\infty}\Big(e^{2t+av(t,\tau)}+e^{2(N+1)t+v(t,\tau)}\Big)dt<+\infty.$$

Furthermore, $w(t,\tau) := \frac{\partial v}{\partial \tau}(t,\tau)$ is a well-defined solution of the following linearized problem, which satisfies:

$$\begin{cases} w''(t)+ae^{2t+av(t,\tau)}w(t)+e^{2(N+1)t+v(t,\tau)}w(t)=0 & \text{for } t\in\mathbb{R}\\ w(-\infty)=1, \quad w'(-\infty)=0 \quad\text{and}\quad w'(+\infty)=\beta'(\tau)\\ \int_{-\infty}^{+\infty} e^{2t+av(t,\tau)}w(t)dt=\beta_1'(\tau) \quad\text{and}\quad \int_{-\infty}^{+\infty} e^{2(N+1)t+v(t,\tau)}w(t)dt=\beta_2'(\tau). \end{cases}$$

All those properties were also checked in [PT]. We claim that,

$$\beta'(\tau)\neq 0 \quad \forall \tau\in\mathbb{R}.$$

Indeed, if by contradiction, we suppose there exists τ_0 such that $\beta'(\tau_0) = 0$, then by (5.4) (since $a \neq \frac{1}{N+1}$), we see that also $\beta_1'(\tau_0) = 0 = \beta_2'(\tau_0)$. Thus, we would find a solution $v(t) = v(t, \tau_0)$ of the equation in (5.8), such that its linearized problem admits a bounded solution $w(t) = w(t, \tau_0)$. By viewing the equation as a degenerate 2×2 system, then from Lemma 4.1 we find a constant $C \neq 0$ such that, simultaneously must hold

$$w(t) = C\Big(2(N+1) + v'(t)\Big) \quad \text{and} \quad w(t) = C\Big(\frac{2}{a} + v'(t)\Big).$$

Clearly, this is possible if and only if $a = 1/(N+1)$. Hence, for $a \neq 1/(N+1)$, we see that $\beta'(\tau) \neq 0 \; \forall \tau \in \mathbb{R}$. So $\beta(\tau)$ is strictly monotone and onto, and it implies the uniqueness for the radial solution of (5.1). In conclusion, we can complete the result of [PT] as follows:

Theorem 5.1. *Let $N > -1$ and $0 < a \neq 1/(N+1)$. Then* (5.6) *and* (5.7) *are necessary and sufficient conditions for the existence of a radial solution for* (5.1). *Furthermore, for such β's the corresponding radial solution is unique.*

It is interesting to compare the uniqueness result of Theorem (5.1) with the multiplicity result in [DET2].

Acknowledgment

After this work was completed we learnt from the referee that results comparable to ours were obtained in [LZ2] under stronger assumptions on the coupling matrix A.

References

[AP] A. Ambrosetti and G. Prodi "A Premier of Nonlinear Analysis", Cambridge University Press, **34** Cambridge, UK 1993.

[Au] (MR1636569) T. Aubin, "Some Nonlinear Problems in Riemannian Geometry", Springer-Verlag, Berlin Heidelberg New York, 1998.

[BM] D. Bartolucci and A. Malchiodi, *An improved Geometric Inequality via Vanishing Moments, with Applications to Singular Liouville Equations*, Comm. Math. Phys., in press.

[B] W. Beckner, *Sharp Sobolev inequalities on the sphere and the Moser–Trudinger inequality*, Ann. Math. **128** (1993), 213–242.

[CLMP1] (MR1145596) E. Caglioti, P.L. Lions, C. Marchioro and M. Pulvirenti, *A special class of stationary flows for two-dimensional Euler equations, a statistical mechanics description*, part I, Comm. Math. Phys., **143** (1992), 501–525.

[CLMP2] (MR1362165) E. Caglioti, P.L. Lions, C. Marchioro and M. Pulvirenti, *A special class of stationary flows for two-dimensional Euler equations, a statistical mechanics description*, part II, Comm. Math. Phys., **174** (1995), 229–260.

[CL] E. Carlen and M. Loss, *Competing Symmetries, the logarithmic HLS inequality and Onofri inequality on S^n*, Geom. Funct. Analysis, **1** (1992), 90–104.

[CK1] (MR1262195) S. Chanillo and M. Kiessling, *Rotational symmetry of solutions of some nonlinear problems in statistical mechanics and in geometry*, Comm. Math. Phys., **160** (1994), 217–238.

[CK2] (MR1361515) S. Chanillo and M. Kiessling, *Conformally invariant systems of nonlinear PDE of Liouville type*, Geom. Funct. Analysis, **5** (1995), 924–947.

[CGS] R. Chen, Y. Guo and D. Spirn, *Asymptotic behavior and symmetry of condensate solutions in electroweak theory*, preprint 2010.

[ChL] (MR1233443) W. Chen and C. Li, *Qualitative properties of solutions of some nonlinear elliptic equations in R^2*, Duke Math. J., **71** (1993), 427–439.

[CSW] (MR1473855) M. Chipot, I. Shafrir and G. Wolansky, *On the solutions of Liouville systems*, Jour. Diff. Eq., **140** (1997), 59–105, Erratum J.D.E., **178** (2002), 630.

[Do] J. Dolbeault, *Sobolev and Hardy–Littlewood–Sobolev inequalities: duality and fast diffusion*, Math. Res. Lett. **18** (2011), 1037–1050.

[DET1] (MR2437030) J. Dolbeault, M.J. Esteban and G. Tarantello, *The role of Onofri type inequalities in the symmetry properties of extremals for Caffarelli–Kohn–Nirenberg inequalities, in two space dimensions*, Ann. Sc. Norm. Super. Pisa Cl. Sci. (5), **7** (2008), 313–341.

[DET2] (MR2509375) J. Dolbeault, M.J. Esteban and G. Tarantello, *Multiplicity results for the assigned Gauss curvature problem in $\mathbb{R}^2$*, Nonlinear Anal. T.M.A., **70** (2009), 2870–2881.

[DS] N. Dunford, J.T. Schwartz, "Linear Operator. Part I: General Theory", Wiley Interscience New York, 1958.

[D] G. Dunne, "Self-Dual Chern–Simons Theories", Lect. Notes in Phys. vol. m **36** New Series, Springer, 1995.

[F] L. Fontana, *Sharp borderline Sobolev inequalities on compact Riemannian manifolds*, Comment. Math. Helv. **68** (1993), 415–454.

[FT] R. Fortini and G. Tarantello, *The role of Liouville systems in the study of non-abelian Chern–Simons vortices*, in preparation.

[JoW1] (MR1846799) J. Jost and G. Wang, *Analytic aspects of the Toda system: I. A Moser–Trudinger inequality*, Comm. Pure Appl. Math., **54** (2001), 1289–1319.

[JoW2] (MR1877003) J. Jost and G. Wang, *Classification of solutions of a Toda system in R^2*, Int. Math. Res. Not., **6** (2002), 277–290.

[Ki1] (MR1193342) M.K.H. Kiessling, *Statistical mechanics of classical particles with logarithmic interaction*, Comm. Pure Appl. Math., **46** (1991), 27–56.

[Ki2] (MR1797146) M.K.H. Kiessling, *Statistical mechanics approach to some problems in conformal geometry*, Phys. A, **79** (2000), 353–368.

[LWY] C.S. Lin, J. Wei and D. Ye *Classification and nondegeneracy of* $\mathrm{SU}(n+1)$ *Toda Systems with singular sources*, Invent. Math. To appear. 117–143.

[LZ1] C.S. Lin and L. Zhang *Profile of bubbling solutions to a Liouville system*, Ann. I.H.P. Anal. Non Linéaire, **27** (2010), 117–143.

[LZ2] C.S. Lin and L. Zhang *Classification of radial solutions to Liouville systems with singularities*, Preprint 2013, arXiv: 1302.3866v1.

[Mo] (MR0301504) J. Moser, *A sharp form of an inequality by N. Trudinger*, Indiana Univ. Math. J., **20** (1971), 1077–1092.

[On] (MR0677001) E. Onofri, *On the positivity of the effective action in a theory of random surfaces*, Comm. Math. Phys., **86** (1982), 321–326.

[PT] A. Poliakovsky and G. Tarantello, *On a planar Liouville-type equation in the study of selfgravitating strings*, J. Diff. Eq., **252** (2012), 3668–3693.

[PrT] (MR1855007) J. Prajapat and G. Tarantello, *On a class of elliptic problems in R^2: Symmetry and Uniqueness results*, Proc. Roy. Soc. Edinburgh A, **131** (2001), 967–985.

[SW1] (MR1859233) I. Shafrir and G. Wolansky, *Moser–Trudinger type inequalities for systems in two dimensions*, C.R.A.S. Paris série I, **333** (2001), 439–443.

[SW2] (MR2159222) I. Shafrir and G. Wolansky, *Moser–Trudinger and logarithmic HLS inequalities for systems*, J. Euro. Math. J. **7** (2005), 413–448.

[SW3] I. Shafrir and G. Wolansky, *The logarithmic HLS inequalities for systems on compact manifolds*, J. Funct. Anal. **227** (2005), 200–226.

[Su] (MR1186683) T. Suzuki, *Global analysis for a two dimensional elliptic eigenvalue problem with exponential nonlinearity*, Ann. I.H.P. Anal. Non Linéaire, **9** (1992), 367–398.

[T] (MR2403854) G. Tarantello, *Selfdual gauge field vortices: An analytical approach*, P.N.L.D.E., **72**, Birkäuser, Boston MA, 2008.

[Tr] (MR1005085) M. Troyanov, *Prescribing curvature on compact surfaces with conical singularities*, Trans. Amer. Math. Soc., **324** (1991), 793–821.

[W] (MR1689869) G. Wang, *Moser–Trudinger inequality and Liouville systems*, C.R. A.S. Paris serie I, **328** (1999), 895–900.

[Wo] (MR1226964) G. Wolansky, *On the evolution of self-interacting clusters and applications to semilinear equations with exponential nonlinearity*, J. Anal. Math., **59** (1992), 251–272.

[Y] (MR1838682) Y. Yang, "Solitons in Field Theory and Nonlinear Analysis," Springer Monographs in Mathematics, Springer-Verlag New York, 2001.

A. Poliakovsky
Department of Mathematics
Ben Gurion University of the Negev
P.O.B. 653
Be'er Sheva, 84105 Israel
e-mail: `poliakov@math.bgu.ac.il`

G. Tarantello
Università degli Studi di Roma "Tor Vergata"
Dipartimento di Matematica
via della Ricerca Scientifica
I-00133, Rome, Italy
e-mail: `tarantel@mat.uniroma2.it`

Progress in Nonlinear Differential Equations
and Their Applications, Vol. 85, 387–403

Multiple Positive Solutions for a Nonsymmetric Elliptic Problem with Concave Convex Nonlinearity

C. Saccon

Abstract. We consider a semilinear elliptic problem, with principal part possibly non symmetric, having a singular nonlinear term which is convex near zero and concave at infinity. We prove the existence of two positive solutions when a suitable parameter is small and a nonexistence result when the parameter is large. These results are closely related to a well known paper by Ambrosetti, Brezis, Cerami, where the principal part is the Laplace operator and the non linearity has no singularity. We use monotonicity arguments to get rid of the singular term. Since the problem has no variational structure, we use degree arguments to exploit the topological features of the problem; in particular, to use continuation arguments, we prove a global bound for all positive solutions.

Mathematics Subject Classification (2010). 35J75, 35J87, 47H05, 47H11.

Keywords. Elliptic semilinear singular problems, multiple positive solutions, variational inequalities, degree theory.

1. Introduction

We study the multiplicity of positive solutions to problem

$$Lu = u^q + \lambda u^p \qquad \text{in } W_0^{1,2}(\Omega) \tag{P}$$

where $q < 0$ and where Lu is a second-order differential operator in divergence form, not necessarily symmetric. In the case L is the Laplace operator and $0 < q < 1$ a well-known result of Ambrosetti, Brezis, Cerami's (see [1]) shows that (P) has (at least) two solutions for $\lambda > 0$ small and no solutions for λ large. The singular version, i.e., the case with a negative q, has been treated in several papers (see the references of [9] for an exhaustive list); to the author's knowledge the most complete result is proved in [9], where it is shown that, in the case

$L = -\Delta$, the above-mentioned result holds for $-3 < q < 0$ and also for $q \le -3$ if solutions are searched for in a wider space than $W_0^{1,2}(\Omega)$. The proof relies on a variational approach which leads to a study of the topological properties of the energy functional

$$I(u) = \frac{1}{2}\int_\Omega |\nabla u|^2\,dx - \frac{1}{q+1}u^{q+1}\,dx - \frac{\lambda}{p+1}\int_\Omega u^{p+1}\,dx$$

(which in the singular case poses some problems) combined with a clever usage of the order properties of the Laplace operator. Such an approach can be easily extended to allow more general linear operators in the principal part, provided they are symmetric.

There seem to be not so many results concerning semilinear singular problems in the nonsymmetric case, where the variational approach cannot be used. We refer the reader to the pionering papers [4, 13], to [7] and to [2] (for the first solution). In this work we extend to a non variational framework the "monotonicity techniques" used in [9] (where subdifferentials of convex functions were used) and use degree arguments to prove a two solutions theorem for $\lambda >$ small. To be precise we shall prove the following theorem. For the meaning of solution we refer to the next section, notice however that solutions are taken in $W_0^{1,2}(\Omega)$.

1.1. Theorem. *Let Ω be a bounded open subset of $\mathbb{R}^N$ with boundary of class C^2. Assume that $a_{i,j} \in C^1(\bar\Omega)$ and $b_i \in L^\infty(\Omega)$, $i, j = 1, \ldots, N$, are such that the differential operator $Lu := -\mathrm{div}[a(x)\cdot\nabla u] + b(x)\cdot\nabla u$ is uniformly elliptic and coercive. Let $-3 < q < 0$ and $1 < p < 2^* - 1$. Then there exists $\bar\lambda > 0$ such that Problem* (P)

- *has at least two solutions $u_1 \le u_2$ for $0 < \lambda < \bar\lambda$;*
- *has at least one solution for $\lambda = \bar\lambda$;*
- *has no solutions for $\lambda > \bar\lambda$.*

2. Assumptions for the general setting

Let $N \ge 3$ (for simplicity) and Ω be a bounded open subset of $\mathbb{R}^N$ with C^2 boundary. We denote by $\|\cdot\|$ and $\langle\cdot,\cdot\rangle$ the norm and the inner product, respectively, in the Sobolev space $W_0^{1,2}(\Omega)$, while $\|\cdot\|_p$ will indicate the norm in $L^p(\Omega)$. In the case $p = 2$ $\langle\cdot,\cdot\rangle_2$ will denote the $L^2(\Omega)$ inner product. We shall denote by $L_c^\infty(\Omega)$ the set of bounded functions with compact support.

We remind that $2^* = \dfrac{2N}{N-2}$. We shall also denote by $i^* : L^2(\Omega) \to W_0^{1,2}(\Omega)$ the adjoint of the embedding of $W_0^{1,2}(\Omega)$ into $L^2(\Omega)$, that is the map such that

$$\int_\Omega uv\,dx = \langle i^*(u), v\rangle \qquad \forall v \in W_0^{1,2}(\Omega). \tag{1}$$

As is well known $i^*(= (-\Delta)^{-1})$ is a compact linear map.

We consider $a_{ij} \in L^\infty(\Omega), b_i, c_i \in \mathrm{L}^N(\Omega), d \in \mathrm{L}^{N/2}(\Omega)$ such that (ellipticity):

$$\sum_{ij} a_{ij}\xi_i\xi_j \geq \nu\|\xi\|^2 \quad \forall \xi \in \mathbb{R}^N, \tag{A.0}$$

with $\nu > 0$. We also define the bilinear form $\alpha : W_0^{1,2}(\Omega) \times W^{1,2}(\Omega) \to \mathbb{R}$ by:

$$\alpha(u,v) := \sum_{ij} \int_\Omega [(a_{ij}(x)D_ju + c_i(x)u)\, D_iv + (b_i(x)D_iu + d(x)u)\, v]\; dx, \tag{2}$$

for $u, v \in W^{1,2}(\Omega)$, and the linear operator $\mathcal{A} : W_0^{1,2}(\Omega) \to W_0^{1,2}(\Omega)$ by:

$$\alpha(u,v) = \langle \mathcal{A}u, v\rangle \qquad \forall v \in W_0^{1,2}(\Omega).$$

We can also think as $\alpha(u,v) = \int_\Omega Luv\, dx$, where the differential operator L is defined by:

$$Lu = -\mathrm{div}[a \cdot \nabla u + cu] + b \cdot \nabla u + du.$$

Actually in the application 1.1 we consider much stronger regularity assumptions on the coefficients of L, which make the presence of c, d unnecessary. However, since such regularity is not needed in the framework we are going to introduce in Section 3, we maintain the general setting for as long as possible.

As usual we say that α (or $\mathcal{A}/L$) is *coercive* if there exists $\nu_1 > 0$ such that

$$\alpha(u,u) \geq \nu_1\|u\|^2 \qquad \forall u \in W_0^{1,2}(\Omega). \tag{A.1}$$

2.1. Remark. It is well known (see, e.g., [12]) that (A.0) implies:

$$\alpha(u,u) \geq \nu_1\|u\|^2 - \bar{C}\|u\|_2^2 \qquad \forall u \in W_0^{1,2}(\Omega) \tag{3}$$

for suitable $\nu_1 > 0$ and $\bar{C} \in \mathbb{R}$ (so (A.1) corresponds to $\bar{C} = 0$). It is also clear that (A.1) corresponds to requiring that the first eigenvalue of the symmetrized operator $L^*u = -\mathrm{div}\left[a \cdot \nabla u + \frac{b+c}{2}u\right] + \frac{b+c}{2} \cdot \nabla u + du$ be positive.

We also consider two functions $G : \Omega \times \mathbb{R} \to]-\infty, \infty]$ and $f : \Omega \times \mathbb{R} \to \mathbb{R}$ and the following conditions on G, f.

$$\begin{cases} \text{for all } s \in \mathbb{R}\ G(\cdot, s) \text{ is measurable, for almost all } x \in \Omega \\ G(x,\cdot) \text{ is convex and } C^1(]\alpha(x), \beta(x)[) \text{ where} \\ ]\alpha(x), \beta(x)[:= \mathrm{int}\,(\{s : G(x,s) < +\infty\})\,; \\ \text{there exists } u_0 \in W_0^{1,2}(\Omega) \text{ such that } G(\cdot, u_0) \in L^1(\Omega)); \end{cases} \tag{G.0}$$

$$\begin{cases} \text{for all } s \in \mathbb{R}\ f(\cdot, s) \text{ is measurable,} \\ \text{for almost all } x \in \Omega\ f(x,\cdot) \text{ is } C^1(\Omega), \exists\, a, b \geq 0,\ p < 2^* - 1 \text{ such that:} \\ \qquad |f'(x,s)| \leq a + b|s|^{p-1}. \end{cases} \tag{f.0}$$

In the following we denote (for a.e. $x \in \Omega$):

$$g(x,s) := \frac{\partial}{\partial s} G(x,s) \text{ if } s \in]\alpha(x), \beta(x)[, \qquad F(x,s) := \int_0^s f(x,\sigma)\, d\sigma$$

and we agree that $g(x,s) = -\infty$, if $z < \alpha(x)$, $g(x,s) = +\infty$, if $z > \beta(x)$; we also agree that $g(\alpha(x)) = \lim_{s \to \alpha(x)^+} g(x,s)$ and $g(\beta(x)) = \lim_{s \to \beta(x)^-} g(x,s)$.

2.2. Definition. Let $\lambda \in \mathbb{R}$. We say that u is a **weak subsolution** (resp. supersolution) to the problem:

$$\begin{cases} Lu + g(x,u) = f(x,u) & \text{on } \Omega \\ u = 0 & \text{on } \partial\Omega \end{cases} \tag{P}$$

if $u \in \mathrm{W}^{1,2}_{\mathrm{loc}}(\Omega)$, $u^+ \in W^{1,2}_0(\Omega)$ (resp. $u^- \in W^{1,2}_0(\Omega)$), $g(\cdot,u) \in \mathrm{L}^1_{\mathrm{loc}}(\Omega)$, and for all $\psi \in W^{1,2}(\Omega) \cap L^\infty_c(\Omega)$ with $\psi \geq 0$ one has:

$$\alpha(u,\psi) + \int_\Omega g(x,u)\psi\,dx \leq \int_\Omega f(x,u)\psi\,dx \qquad (\text{resp. } \geq 0). \tag{4}$$

We say that u is a **strict** weak subsolution/supersolution if the above inequalities are strict whenever $\psi \neq 0$.

We say that u is a **weak solution** of (P), if u is both a sub and a supersolution to (P).

2.3. Remark. u is a subsolution (a supersolution) to (P) if and only if (4) holds with for all ψ in $\mathcal{C}^\infty_0(\Omega)$ (with $\psi \geq 0$); this can be easily proved via a density argument. If in addition $u \in W^{1,2}(\Omega)$ and we have either $g(x,u) \geq 0$ or $g(x,u) \leq 0$, then:

$$\begin{aligned} &\forall v \in W^{1,2}_0(\Omega) \text{ with } v \geq 0 \quad g(\cdot,u)^+ v \in L^1(\Omega)\ (g(\cdot,u)^- v \in L^1(\Omega)), \\ &\text{and } \ \alpha(u,v) + \int_\Omega (g(x,u) - f(x,u))v\,dx \leq 0\ (\geq 0). \end{aligned} \tag{5}$$

To prove this it suffices to approximate v in $W^{1,2}_0(\Omega)$ with a sequence (ψ_n) in $W^{1,2}_0(\Omega) \cap \mathrm{L}^\infty_c(\Omega)$, such that $0 \leq \psi_n \leq \psi_{n+1} \leq v\ \forall n$ (see, e.g., [8, Lemma A.1]), plug ψ_n in (4), and pass to the limit using the Monotone Convergence Theorem in the integral containing $g(x,u)\psi_n$. With the same monotonicity argument one can show that, if u is a strict subsolution (supersolution), then (5) holds with the strict inequality for all $v \in W^{1,2}_0(\Omega) \setminus \{0\}$.

Notice that the integral $\int_\Omega g(x,u)v\,dx$ in (5) is allowed to be $+\infty$ (resp. $-\infty$). Also notice that in the previous argument we can replace $g(x,s)$ by $g(x,s) - f_0(x,s)$ where $f_0(x,s)$ is a function verifying (f.0).

As a consequence of the above, if either $g(x,s) \geq 0$ or $g(x,s) \leq 0$, then u is a solution of (P) if and only if $u \in W^{1,2}_0(\Omega)$ and for all $v \in W^{1,2}_0(\Omega)$:

$$g(\cdot,u)v \in L^1(\Omega), \quad \alpha(u,v) + \int_\Omega g(x,u)v\,dx = \int_\Omega f(x,u)v\,dx. \tag{6}$$

3. A variational inequalities approach

In this section we provide a framework for problem (P) in terms of *variational inequalities*. This will allow to derive existence theorems as well as to deal with sub and super solutions as *natural constraints* (see Theorem 3.11). In this section we always assume the validity of (A.0), (G.0), and (f.0); unless explicitely mentioned (in the existence theorems) (A.1) will not be assumed. We omit most of the proofs in this section, since they can be easily deduced by imitating the analogous ones of [11, Sections 3,4].

For the proof of the following lemma see [9, Lemma 2].

3.1. Lemma. *For any measurable function u the compositions $G(u)$ and $g(u)$ are measurable.*

Let $\mathcal{G} : W_0^{1,2}(\Omega) \to [-\infty, +\infty]$ be defined by $\mathcal{G}(u) := \int_\Omega G(x,u)\,dx$. We denote by $\mathbb{K}_g$ the *domain* of $\mathcal{G}$, that is $\mathbb{K}_g := \left\{u \in W_0^{1,2}(\Omega) : G(u) \in \mathrm{L}^1(\Omega)\right\}$.

The following result is simple to prove (see also [11, Lemma 3.3]).

3.2. Lemma. *$\mathcal{G}$ is convex, lower semicontinuous and proper (i.e., $\mathbb{K}_g \neq \emptyset$).*

We remind that the *subdifferential* of $\mathcal{G}$, in the sense of convex analysis, is the (multivalued) operator $\partial\mathcal{G}$ defined by:

$$w \in \partial\mathcal{G}(u) \Leftrightarrow u \in \mathbb{K}_g,\ \mathcal{G}(v) \geq \mathcal{G}(u) + \langle w, v-u\rangle \quad \forall v \in \mathbb{K}_g. \tag{7}$$

As is well known (see [3, Exemple 2.3.4]) $\partial\mathcal{G}$ is a maximal monotone operator. Let $\mathcal{A}_g : W_0^{1,2}(\Omega) \to 2^{W_0^{1,2}(\Omega)}$ be the (multivalued) operator:

$$\mathcal{A}_g := \mathcal{A} + \partial\mathcal{G}.$$

3.3. Theorem. *Assume that* (A.0), (A.1), *and* (G.0) *hold. Then $\mathcal{A}_g$ is a maximal monotone operator. Moreover $\mathcal{A}_g$ admits a Lipschitz continuous inverse with Lipschitz constant equal to $1/\nu_1$. This means that for any $w \in W_0^{1,2}(\Omega)$ there exists a unique $u \in W_0^{1,2}(\Omega)$ such that $w \in \mathcal{A}_g(u)$, and that:*

$$w_i \in \mathcal{A}_g(u_i)\ i = 1,2 \quad \Rightarrow \|u_1 - u_2\| \leq \frac{1}{\nu_1}\|w_1 - w_2\|.$$

Sketch of proof. (For more details see [11, Theorem 3.7].) Since $\mathcal{A}$ is monotone and Lipschitz continuous and $\partial\mathcal{G}$ is maximal monotone, then $\mathcal{A}_g$ is maximal monotone by [3, Lemme 2.4]. Using (7) and the definition of $\mathcal{A}_g$ (taking $v = u_0$ as test) we can find a constant k such that:

$$\langle w, u - u_0\rangle \geq \frac{\nu_1}{2}\|u\|^2 - k \qquad \forall u \in \mathbb{K}_g,\ \forall w \in \mathcal{A}_g(u).$$

By [3, Corollaire 2.4] this implies that $\mathcal{A}_g$ is surjective, that is $w \in \mathcal{A}_g(u)$ is solvable in u for every $w \in W_0^{1,2}(\Omega)$. Lipschitz continuity is a standard fact. □

We now estabilish a connection between $\mathcal{A}_g$ and the solutions of (P) in terms of a suitable variational inequality. This is done in the following lemmas whose proofs can be carried on by adaptating those in [11, Section 3].

3.4. Lemma. *Let $u, w \in W_0^{1,2}(\Omega)$. The following facts are equivalent*

(a) *$u \in \mathbb{K}_g$, $w \in \mathcal{A}_g(u)$;*

(b) *the following variational inequality holds*

$$\begin{cases} u \in \mathbb{K}_g, \quad g(\cdot,u)(v-u) \in L^1(\Omega) \quad \forall v \in \mathbb{K}_g, \\ a(u,v-u) + \displaystyle\int_\Omega g(u)(v-u)\,dx \ge \langle w, v-u\rangle \ \ \forall v \in \mathbb{K}_g. \end{cases} \tag{8}$$

3.5. Remark. If $u, v \in \mathbb{K}_g$ then

$$g(x,u)(v-u) \le G(x,v) - G(x,u) \quad \in \mathrm{L}^1(\Omega) \tag{9}$$

so the integral $\int_\Omega g(u)(v-u)\,dx$ in (8) always makes sense, possibly being $-\infty$.

3.6. Lemma.

(a) *If u is a solution of* (P), *then $i^*(f(\cdot,u)) \in \mathcal{A}_g(u)$, that is:*

$$\begin{cases} u \in \mathbb{K}_g,\ g(\cdot,u)(v-u) \in L^1(\Omega)\ \forall v \in \mathbb{K}_g, \\ a(u,v-u) + \displaystyle\int_\Omega g(x,u)(v-u)\,dx \ge \int_\Omega f(x,u)(v-u)\,dx\ \forall v \in \mathbb{K}_g. \end{cases} \tag{10}$$

(b) *Let $u \in W_0^{1,2}(\Omega)$ verify the variational inequality* (10). *Suppose that there exist a subsolution φ_1 (a supersolution φ_2) for* (P) *such that $\varphi_1 \le u$ ($u \le \varphi_2$). Then u is a weak subsolution (supersolution) to* (P).

From now on we consider two mesurable functions $\varphi_1, \varphi_2 : \Omega \to [-\infty, +\infty]$ such that $\varphi_1 \le \varphi_2$ a.e. in Ω.

3.7. Definition. We set

$$\mathbb{K}_{\varphi_1}^{\varphi_2} := \{u \in W_0^{1,2}(\Omega) : \varphi_1 \le u \le \varphi_2 \text{ a.e. in } \Omega\}.$$

For the sake of brevity we write $\mathbb{K}_\varphi := \mathbb{K}_\varphi^{+\infty}$ and $\mathbb{K}^\varphi := \mathbb{K}_{-\infty}^\varphi$.

In the following we denote by $\mathcal{I}_E$ the "indicator function" relative to a subset E of $W_0^{1,2}(\Omega)$, defined by $\mathcal{I}_E(u) = 0$ whenever $u \in E$ and $\mathcal{I}_E(u) = +\infty$ otherwise. We then define:

$$\mathcal{G}_{\varphi_1}^{\varphi_2} := \mathcal{G} + \mathcal{I}_{\mathbb{K}_{\varphi_1}^{\varphi_2}}, \qquad \mathcal{A}_{g,\varphi_1}^{\varphi_2} := \mathcal{A} + \partial \mathcal{G}_{\varphi_1}^{\varphi_2}.$$

Again for brevity we write $\mathcal{A}_{g,\varphi} := \mathcal{A}_{g,\varphi}^{+\infty}$ and $\mathcal{A}_g^\varphi := \mathcal{A}_{g,-\infty}^\varphi$.

The following results are analogous to the previously considered ones.

3.8. Lemma. *$\mathcal{G}_{\varphi_1}^{\varphi_2}$ is convex and lower semicontinuous.*

3.9. Theorem. *If* (A.1) *holds and $\mathbb{K}_g \cap \mathbb{K}_{\varphi_1}^{\varphi_2} \neq \emptyset$, then $\mathcal{A}_{g,\varphi_1}^{\varphi_2}$ is maximal monotone and coercive, so $\mathcal{A}_{g,\varphi_1}^{\varphi_2}$ is surjective and its inverse is Lipschitz continuous with constant $\dfrac{1}{\nu_1}$ (irrespective of φ_1, φ_2).*

3.10. Lemma. *Let $u, w \in W_0^{1,2}(\Omega)$. The following facts are equivalent:*

(a) $u \in \mathbb{K}_g \cap \mathbb{K}_{\varphi_1}^{\varphi_2}$, $w \in \mathcal{A}_{g,\varphi_1}^{\varphi_2}(u)$;

(b) *the following variational inequality holds:*

$$\begin{cases} u \in \mathbb{K}_g \cap \mathbb{K}_{\varphi_1}^{\varphi_2}, \quad g(\cdot,u)(v-u) \in L^1(\Omega) \ \forall v \in \mathbb{K}_g \cap \mathbb{K}_{\varphi_1}^{\varphi_2}, \\ \alpha(u, v-u) + \int_\Omega g(x,u)(v-u)\, dx \geq \langle w, v-u \rangle \ \forall v \in \mathbb{K}_g \cap \mathbb{K}_{\varphi_1}^{\varphi_2}. \end{cases} \qquad (V.I.)_{\varphi_1}^{\varphi_2}$$

Again for convenience we shall use $(V.I)_{\varphi_1}$ (resp. $(V.I)^{\varphi_2}$) to refer to the variational inequality $(V.I.)_{\varphi_1}^{\varphi_2}$ when $\varphi_1 = -\infty$ (resp. $\varphi_2 = +\infty$) and simply $(V.I.)$ in the case both the obstacles are infinite (in this case $\mathcal{A}_{g,\varphi_1}^{\varphi_2} = \mathcal{A}_g$).

The following theorem shows that sub and super solutions can be used as "natural contraints".

3.11. Theorem. *The following facts are true:*

- *if φ_1 is a subsolution for* (P) *and $u \in \mathbb{K}_g \cap \mathbb{K}_{\varphi_1}^{\varphi_2}$, then:*

$$i^*(f(\cdot,u)) \in \mathcal{A}_{g,\varphi_1}^{\varphi_2}(u) \Leftrightarrow i^*(f(\cdot,u)) \in \mathcal{A}_g^{\varphi_2}(u);$$

- *if φ_2 is a supersolution for* (P) *and $u \in \mathbb{K}_g \cap \mathbb{K}_{\varphi_1}^{\varphi_2}$, then:*

$$i^*(f(\cdot,u)) \in \mathcal{A}_{g,\varphi_1}^{\varphi_2}(u) \Leftrightarrow i^*(f(\cdot,u)) \in \mathcal{A}_{g,\varphi_1}(u);$$

Sketch of the proof. We consider the "$\Rightarrow$" implication in the first claim. Let u verify $(V.I.)_{\varphi_1}^{\varphi_2}$ with $w = i^*(h(u))$ and let $v \in \mathbb{K}_g \cap \mathbb{K}^{\varphi_2}$. Given $t > 0$ we set $v_t := (u + t(v-u)) \vee \varphi_1 = u + t(v-u) + w_t$ with $w_t = (\varphi_1 - u - t(v-u))^+$. Then $v_t \in \mathbb{K}_g \cap \mathbb{K}_{\varphi_1}^{\varphi_2}$ and can be used as a test in $(V.I.)_{\varphi_1}^{\varphi_2}$. This yieds:

$$\begin{aligned} t&\left(\alpha(u, v-u) + \int_\Omega (g(x,u) - f(x,u))(v-u)\, dx\right) \\ &\geq \alpha(w_t, w_t) + t\alpha(v-u, w_t) - \alpha(\varphi_1, w_t) - \int_\Omega (g(x,u) - f(x,u)) w_t\, dt \\ &\geq \alpha(w_t, w_t) + t\alpha(v-u, w_t) + \int_\Omega (g(x,\varphi_1) - g(x,u) - f(x,\varphi_1) + f(x,u)) w_t\, dt \\ &\geq \nu_1 \|w_t\|^2 - \bar{C} t^2 \|v-u\|_2^2 - t\|\alpha\| \|v-u\| \|w_t\| \\ &\quad + t \int_{\{t(v-u) < \varphi_1 - u\}} (g(x,u) - g(x,\varphi_1))(v-u)\, dx \\ &\quad - t \int_\Omega |f(x,\varphi_1) - f(x,u)|\, |v-u|\, dt \end{aligned}$$

(we have used the fact that φ_1 is a subsolution). Letting $t \to 0$ yields $\|w_t\| \to 0$. Dividing by t and letting $t \to 0^+$ gives the conclusion:

$$\alpha(u, v-u) + \int_\Omega (g(x,u) - f(x,u))(v-u)\, dx \geq 0. \qquad \square$$

3.12. Lemma. *Let $\varphi \in W_{\mathrm{loc}}^{1,2}(\Omega)$ with $\varphi^+ \in W_0^{1,2}(\Omega)$ ($\varphi^- \in W_0^{1,2}(\Omega)$) and $g(\cdot,\varphi) \in L_{\mathrm{loc}}^1(\Omega)$. Then φ is a subsolution (a supersolution) for* (P) *if and only if for every*

$M \in \mathbb{R}$ there exists $\varepsilon_M > 0$ such that:

$$\alpha(\varphi, v) + \int_\Omega (g(x,\varphi) - f(x,\varphi))\, v\, dx \le \varepsilon_M \|v\|_2 \; (\ge \varepsilon_M \|v\|_2) \tag{11}$$

for all $v \in W^{1,2}(\Omega) \cap L^\infty_c(\Omega)$ with $v \ge 0$ and $\|v\| \le M\|v\|_2$.

Proof. We prove the "only if" part, which is the nontrivial one. By contradiction suppose that there exists $M \in \mathbb{R}$ and a sequence (v_n) in $W^{1,2}_0(\Omega) \cap L^\infty_c(\Omega)$ with $v_n \ge 0$, $\|v_n\|_2 = 1$, $\|v_n\| \le M$, and

$$0 > \Lambda(v_n) := \alpha(\varphi, v_n) + \int_\Omega (g(x,\varphi) - f(x,\varphi))\, v_n\, dx \to 0.$$

Up to a subsequence we can suppose that $v_n \rightharpoonup v$ for $v \in W^{1,2}_0(\Omega)$, $v \ge 0$, and $\|v\|_2 = 1$. We can take $\eta \in C^\infty_0(\Omega)$ and $K > 0$ such that $0 \le \eta \le 1$ and $\|(\eta v) \wedge K\|_2 \ge 1/2$. Setting $w_n := (\eta v_n) \wedge K$ and $w := (\eta v) \wedge K$ it is clear that $w_n \rightharpoonup w$, $w \ne 0$, and $0 \le w_n \le v_n$, so that $0 \le \Lambda(w_n) \le \Lambda(v_n) \to 0$. Since $g(\cdot, \varphi)\eta \in L^1(\Omega)$ it follows from weak convergence that:

$$\alpha(\varphi, w) + \int_\Omega (g(x,\varphi) - f(x,\varphi))\, w\, dx = 0$$

which contradicts that fact that φ is a strict subsolution. □

3.13. Proposition. *Let φ_1 be a strict subsolution (φ_2 be a strict supersolution) for* (P)*. Let $K > 0$. There exist $\delta > 0$ such that for any convex and closed set if $\mathbb{K}$ in $W^{1,2}_0(\Omega)$ with $\mathbb{K}^{\varphi_2}_{\varphi_1} \subset \mathbb{K} \subset \mathbb{K}^{\varphi_2}$ ($\mathbb{K}^{\varphi_2}_{\varphi_1} \subset \mathbb{K} \subset \mathbb{K}_{\varphi_1}$), if*

$$w \in \mathbb{K} \cap \mathbb{K}_g,\; \alpha(w, v-w) + \int_\Omega (g(x,w) - f(x,w))(v-w)\, dx \ge 0 \;\forall v \in \mathbb{K} \cap \mathbb{K}_g, \tag{12}$$

(remind Remark 3.5*) $\|w\| \le K$, and $\operatorname{dist}_{L_2}(w, \mathbb{K}^{\varphi_2}_{\varphi_1}) < \delta$, then $w \in \mathbb{K}^{\varphi_2}_{\varphi_1}$.*

Sketch of proof. We prove the supersolution case. Let $\mathbb{K}$ an w be as above. Let $\pi(w) := w \wedge \varphi_2$: since $\pi(w) \in \mathbb{K}^{\varphi_2}_{\varphi_1} \subset \mathbb{K}$, we can use $v = \pi(w)$ in (12). Let $h(x,u) := g(x,u) - f(x,u)$. Doing some manipulations this yields:

$$\begin{aligned} 0 \ge \alpha(w - \pi(w), w - \pi(w)) &+ \int_\Omega (h(x,w) - h(x,\pi(w)))\,(w - \pi(w))\, dx \\ &+ \alpha(\pi(w), w - \pi(w)) + \int_\Omega h(x,\pi(w))(w - \pi(w)\, dx \\ &\ge \nu\|w - \pi(w)\|^2 - \bar{C}\|w - \pi(w)\|_2^2 \end{aligned}$$

for a suitable $\bar{C}$ (g is nondecreasing; f verifies (f.0),$\|w\| \le K$); moreover if $w(x) > \pi(w)(x)$ then $\pi(w)(x) = \varphi_2(x)$, so we can replace $\pi(w)$ with φ_2. Since φ_2 is a strict supersolution we can take $\varepsilon := \varepsilon_M$ with $M = \bar{C}/\nu$, as in Lemma 3.12, so:

$$2\bar{C}\|w - \pi(w)\|_2^2 \ge \nu\|w - \pi(w)\|^2 + \bar{\varepsilon}\|w - \pi(w)\|_2$$

which implies that $\|w - \pi(w)\|_2 = 0$ if $\|w - \pi(w)\|_2 < \delta$ is small enough. □

We introduce now the map $\Phi^{\varphi_2}_{\varphi_1} : W^{1,2}_0(\Omega) \to W^{1,2}_0(\Omega)$:

$$\Phi^{\varphi_2}_{\varphi_1} := (\mathcal{A}^{\varphi_2}_{g,\varphi_1})^{-1} \circ i^* \circ \mathcal{N}_f \circ i \tag{13}$$

where $i : W^{1,2}_0(\Omega) \to L^2(\Omega)$ is the embedding and $\mathcal{N}_f$ is the Nemytskii operator associated with f. We have:

$$i^*(f(\cdot, u)) \in \mathcal{A}^{\varphi_2}_{g,\varphi_1}(u) \Leftrightarrow u = \Phi^{\varphi_2}_{\varphi_1}(u). \tag{14}$$

As before we also consider Φ_{φ_1}, Φ^{φ_2}, and Φ replacing $\mathcal{A}^{\varphi_2}_{g,\varphi_1}$ by $\mathcal{A}_{g,\varphi_1}$, $\mathcal{A}^{\varphi_2}_g$, and $\mathcal{A}_g$ in the definition. It is clear that $\Phi^{\varphi_2}_{\varphi_1}$ is compact, therefore it makes sense to consider the Leray–Schauder degree of $Id - \Phi^{\varphi_2}_{\varphi_1}$, relative to 0 in any bounded open subset $U \subset W^{1,2}_0(\Omega)$ such that $0 \notin (Id - \Phi^{\varphi_2}_{\varphi_1})^{-1}(\partial U)$. Such a degree will be denoted by:

$$\deg(Id - \Phi^{\varphi_2}_{\varphi_1}, U, 0).$$

3.14. Theorem (degree inheritance). *Assume that u is an isolated solution of $u = \Phi^{\varphi_2}_{\varphi_1}(u)$.*

1. *If φ_1 is a strict subsolution, then u is an isolated solution of $u = \Phi^{\varphi_2}(u)$, and for $\delta > 0$ small:*

   $$\deg(Id - \Phi^{\varphi_2}_{\varphi_1}, B(u,\delta), 0) = \deg(Id - \Phi^{\varphi_2}, B(u,\delta), 0),$$

   where $B(u,\delta)$ is the ball centered at u with radius δ;
2. *if φ_2 is a strict supersolution, then u is an isolated solution of $u = \Phi_{\varphi_1}(u)$, and for $\delta > 0$ small:*

   $$\deg(Id - \Phi^{\varphi_2}_{\varphi_1}, B(u,\delta), 0) = \deg(Id - \Phi_{\varphi_1}, B(u,\delta), 0).$$

Sketch of proof. We prove the supersolution case. For $t \in [0,1]$ let: $\varphi_{2,t}(x) := \varphi_2(x) + \dfrac{t}{1-t}$ so $\varphi_{2,0} = \varphi_2$ and $\varphi_{2,1} = +\infty$. It can be proved (for the details see [11, Section 5]) that the map $\Phi_t := (\mathcal{A}^{\varphi_{2,t}}_{g,\varphi_1})^{-1} \circ i^* \circ \mathcal{N}_f \circ i$ is compact for all $t \in [0,1]$, $(u,t) \mapsto \Phi_t(u)$ is continuous, $\Phi_0 = \Phi^{\varphi_2}_{\varphi_1}$, and $\Phi_1 = \Phi_{\varphi_1}$. It is therefore possible to use the continuation property of the Leray–Schauder degree.

Let $B(u,\rho)$ be a ball such that u is the only fixed point of $\Phi^{\varphi_2}_{\varphi_1}$ in $\overline{B(u,\rho)}$; up to shrinking $\rho > 0$ we have $B(u,\rho) \subset B_2(u,\delta)$, where $\delta > 0$ is as from Proposition (3.13). Using Proposition (3.13) (with $\mathbb{K} = \mathbb{K}^{\varphi_{2,t}}_{\varphi_1}$) we get that for any $t \in [0,1]$ u is the unique fixed point for $\Phi_t(v)$ in $\overline{B(u,\rho)}$. By the continuation property of the degree we get the conclusion. □

4. Problems with concave-convex nonlinearities

4.1. Theorem. *Let $a_{ij} \in L^\infty(\Omega), b_i, c_i \in \mathrm{L}^s(\Omega), d \in \mathrm{L}^{s/2}(\Omega)$ with $s > N$; assume that* (A.0) *and* (A.1) *hold, and that*

$$d - \mathrm{div}(c) \geq 0 \tag{A.2}$$

in the sense of distributions. Let $q < 0$ and let $a_0 : \Omega \to \mathbb{R}$ be such that $a_0 \geq 0$, $a_0 \neq 0$, and there exists $u_0 \in W^{1,2}_0(\Omega)$ with $a_0 u_0^{q+1} \in L^1(\Omega)$.

Then there exists a unique weak solution $\tilde{u}$ in $W_0^{1,2}(\Omega)$ of

$$(Lu :=) - \operatorname{div}(a \cdot \nabla u + bu) + c \cdot \nabla u + du = a_0(x)u^q. \tag{15}$$

Such a $\tilde{u}$ is continuous and positive in Ω, and $a_0\tilde{u}^q v \in L^1(\Omega)\, \forall v \in W_0^{1,2}(\Omega)$ (in particular $\tilde{u}^{q+1} \in L^1(\Omega)$). Moreover $\tilde{u}(x) \geq \epsilon \underline{u}(x)$, for $\epsilon > 0$ small, where $\underline{u} \in W_0^{1,2}(\Omega)$ is a solution of $-\operatorname{div}(a \cdot \nabla u + bu) + c \cdot \nabla u + du = a_0(x) \wedge 1$.

If in addition $a_0 \in L^s(\Omega)$ with $s > N/2$, then we can take as $\underline{u}$ the solution of $-\operatorname{div}(a \cdot \nabla u + bu) + c \cdot \nabla u + du = a_0(x)$ and we also have $\tilde{u} \leq \bar{u}^{\frac{1}{1-q}}$ where $\bar{u}$ is a solution of $Lu = (1-q)a_0$.

Proof. Notice that the functions $\underline{u}$ and $\bar{u}$ (if defined) are positive in Ω by the strong maximum principle (see, e.g., [14, Corollary 5.1] or [6, Theorem 8.19] in the case of bounded coefficients) and are continuous in $\bar{\Omega}$ (see, e.g., [12, Theorem 7.3] or [6, Theorem 8.31]). Let:

$$G(x,s) := \begin{cases} -\dfrac{a_0(x)s^{q+1}}{q+1} & \text{if } q \neq -1 \text{ and } s > 0, \\ -a_0(x)\ln(s) & \text{if } q = -1 \text{ and } s > 0, \\ +\infty & \text{if } s < 0; \end{cases}$$

it is clear that (g.0) holds and that $g(x,s) = -a_0(x)s^q$ for $s > 0$. By Theorem 3.3 there exists a unique solution $\tilde{u}$ of $0 \in \mathcal{A}_g(u)$. Notice that $a_0(x)\tilde{u}^{q+1} \in L^1(\Omega)$, hence for every $\psi \in C_0^\infty(\Omega)$ with $\psi \geq 0$ we have $a_0(x)(\tilde{u}+\psi)^{q+1} \in L^1(\Omega)$. Using $v = \tilde{u} + \psi$ in (8) shows that $\tilde{u}$ is a supersolution. Now let $\varphi_1 := \varepsilon\underline{u}$. Simple computations show that (in the sense of distributions)

$$L\varphi_1 - a_0(x)\varphi_1^q = \varepsilon a_0(x) \wedge 1 - \varepsilon_0^q a_0(x)\underline{u}^q \leq \varepsilon a_0(x) \wedge 1(1 - \varepsilon^{q-1}\underline{u}^q) < 0$$

for $0 < \varepsilon < \|\underline{u}\|_\infty^{\frac{q}{1-q}}$, which implies that φ_1 is a subsolution. By Theorem 3.9 we can find $\tilde{u}_1$ such that $0 \in \mathcal{A}_{g,\varphi_1}(\tilde{u}_1)$; by Theorem 3.11 we have $0 \in \mathcal{A}_g(\tilde{u}_1)$ and by uniqueness of $\tilde{u}$ we have $\tilde{u} = \tilde{u}_1$. We have thus proved that $\tilde{u} \geq \varphi_1 > 0$. By (b) of Lemma 3.6 it follows that $\tilde{u}$ is a subsolution, so it is a solution to (P). Since φ_1 is continuous in Ω it follows that $\tilde{u}$ is locally bounded away from zero in Ω and by regularity theory $\tilde{u}$ is continuous in Ω. Using Remark 2.3 we have that $a_0 u^q v \in L^1(\Omega)\, \forall v \in W_0^{1,2}(\Omega)$.

It is clear that, if $a_0 \in \mathrm{L}^s(\Omega)$ with $s > N/2$, then we can replace $a_0 \wedge 1$ by a_0. In this case it is also possible to define $\bar{u}$ as above. Taking $\varphi_2 := \bar{u}^\beta$, with $\beta := \dfrac{1}{1-q}$, then $\varphi_2 \in \mathrm{W}^{1,2}_{\mathrm{loc}}(\Omega)$ and:

$$\begin{aligned} L\varphi_2 - a_0\varphi_2^q = &- \beta(\beta-1)\bar{u}^{\beta-2}(a \cdot \nabla\bar{u}) \cdot \nabla\bar{u} + \beta u^{\beta-1}L\bar{u} \\ &+ (1-\beta)u^\beta(d - \operatorname{div} c) - a_0\bar{u}^{\beta q} \geq a_0\bar{u}^{\frac{q}{1-q}} + a_0\bar{u}^{\frac{q}{q-1}} = 0. \end{aligned}$$

(again in the sense of distributions – notice that $\beta - 1 = \beta q = \frac{q}{1-q}$). So φ_2 is a supersolution. Repeating the previous argument shows that $\tilde{u} \leq \varphi_2$. □

4.2. Remark. If further regularity is assumed on the coefficients one could improve the regularity of the solution $\tilde{u}$, by using the property $\tilde{u} \geq \varphi_1$.

4.3. Theorem. *Suppose that* (A.0), (A.1), (A.2), (f.0) *hold, that* $q < 0$, $a_0 \in \mathrm{L}^r(\Omega)$ *with* $r > N/2$, $\inf_\Omega a_0 > 0$, *that there exists* $u_0 \in W_0^{1,2}(\Omega)$ *such that* $a_0 u_0^{q+1} \in L^1(\Omega)$, *and that* $f(x,s) > 0$, *whenever* $x \in \Omega$ *and* $s > 0$. *There exists* $\bar{\lambda}$ *with* $0 < \bar{\lambda} \leq +\infty$ *such that the problem*

$$-\mathrm{div}\,[a \cdot \nabla u + cu] + b \cdot \nabla u + du - a_0 u^q = \lambda f(\cdot, u) \qquad (P_\lambda)$$

has at least one solution for $0 < \lambda < \bar{\lambda}$ *and has no solutions for* $\lambda > \bar{\lambda}$.

Proof. Let φ_1 be the solution $\tilde{u}$ found in Theorem 4.1. It is clear that φ_2 is a subsolution for the problem (P_λ). Let $\bar{u}$ be constructed as in Theorem 4.1 taking $2a_0$ instead of a_0 and let $\varphi_2 := \bar{u}^{\frac{1}{1-q}}$. Arguing as in the proof of 4.1:

$$L\varphi_2 - a_0\varphi_2^q - \lambda f(\cdot, \varphi_2) \geq a_0\, \bar{u}^{\frac{q}{1+q}} - \lambda\left(a + b\bar{u}^{\frac{p}{1-q}}\right) > 0$$

for $\lambda > 0$ small enough. Let $\Phi_{\varphi_1}^{\varphi_2}$ be defined as in (13). Since $\Phi_{\varphi_1}^{\varphi_2} \subset \mathbb{K}_{\varphi_1}^{\varphi_2}$ is bounded it follows that, for $R > 0$ large enough, there are no fixed points for $\Phi_{\varphi_1}^{\varphi_2}$ lying on ∂B_R, where $B_R = \left\{u \in W_0^{1,2}(\Omega) : \|u\| < R\right\}$, and that

$$\deg(Id - \Phi_{\varphi_1}^{\varphi_2}, B_R, 0) = 1.$$

It follows that there exists $u \in W_0^{1,2}(\Omega)$ such that $u = \Phi_{\varphi_1}^{\varphi_2}$, that is, by (14) $i^*(f(\cdot,u)) \in \mathcal{A}_{g,\varphi_1}^{\varphi_2}(u)$. Since φ_1 is a subsolution and φ_2 is a supersolution, we deduce from Theorem 3.11 that $i^*(f(\cdot,u)) \in \mathcal{A}_g(u)$ and from Lemma 3.6 that u is a solution for (P_λ).

Finally we prove that, if (P_λ) has a solution, then (P_μ) has a solution for all $\mu < \lambda$; if this is true then the last conclusion follows taking $\bar{\lambda} := \sup\{\mu > 0 : (P_\mu)$ has a solution$\}$. This is easily seen since, if u solves (P_λ) and $0 < \mu < \lambda$, then u is a supersolution for (P_μ) □

4.4. Remark. Looking at the previous proof it is clear that $\inf_\Omega a_0 > 0$ is only needed to construct the supersolution φ_2. This could be obtained under weaker condition, specifying a suitable rate with which the function a_0 grows away from zero (provided some regularity on the coefficients of L is added).

From now on we consider a more regular setting: we assume

$$a \in C^1(\bar{\Omega}), \quad b \in C^0(\bar{\Omega}), \quad c = d = 0, \quad a_0 \in \mathbb{R},\ a_0 > 0. \qquad (16)$$

In particular the regularity of a implies that the weak solutions of the linear problem $Lu = h \in L^2(\Omega)$ are actually strong solutions, which in turn implies the validity of the Hopf maximum principle (see, e.g., [10]). Such a principle will be intensively used in the following; moreover the regularity of the coefficients is also needed in Theorem 4.9. A first consequence of the above assumptions is a more accurate estimate of the rate of growth away from zero of the solution $\tilde{u}$. In the following we set: $\Omega_\delta := \{x \in \Omega : \mathrm{dist}(x, \partial\Omega) < \delta\}$, where $\delta > 0$.

4.5. Lemma. *Assume that* (16) *holds, let* $-3 < q < -1$*, and let* $\tilde{u}$ *a solution of* $Lu - a_0 u^q = 0$*, as from Theorem* 4.1*. Then there exists* $\bar{\varepsilon} > 0$ *such that*

$$\tilde{u}(x) \geq \bar{\varepsilon}\,\mathrm{dist}(x, \partial\Omega)^{\frac{2}{1-q}} \qquad \forall x \in \Omega. \tag{17}$$

Proof. Let u_0 be the solution of $Lu = 1$. By regularity theory (see [6, Theorem 8.34]) and the Hopf maximum principle there exist $k_0 \geq \varepsilon_0 > 0$, $\delta_0 > 0$ such that $|\nabla u_0| \leq k_0$ in Ω and $|\nabla u_0| \geq \varepsilon_0$ in Ω_δ. Let $\varphi_1 := \varepsilon u_0^{\frac{2}{1-q}}$, then

$$L\varphi_1 - a_0\varphi_1^q = -\frac{2\varepsilon(1+q)}{(1-q)^2} u_0^{\frac{2q}{1-q}} a(x)\nabla u_0 \nabla u_0 + \frac{2\varepsilon}{1-q} u_0^{\frac{1+q}{1-q}} - \varepsilon^q u_0^{\frac{2q}{1-q}} < 0$$

if $\varepsilon > 0$ is small enough. Arguing as in the proof of Theorem 4.1 we get $\tilde{u} \geq \varphi_1$, which implies (17). □

The proof of the following lemma is just a matter of computations.

4.6. Lemma. *Let* $A > 0, B > 0$ *in* $\mathbb{R}$ *and consider the o.d.e. problem*

$$-\chi''(s) = A\chi(s)^q + B\chi(s)^p \qquad \chi(0) = 0.$$

Such a problem has a family of increasing, positive solutions $\chi_M : [0, s_M] \to [0, M]$ *defined by* $\chi_M(s) = \Gamma_M^{-1}(s)$*, where* $\Gamma_M : [0, M] \to [0, s_M]$ *is given by*

$$\Gamma_M(\xi) = \int_0^\xi \frac{d\zeta}{\sqrt{H(\zeta) - H(M)}} \qquad H(\zeta) = \begin{cases} -A\dfrac{\zeta^{q+1}}{q+1} - B\dfrac{\zeta^{p+1}}{p+1} & \text{if } q \neq 1 \\ -A\ln(\zeta) - B\dfrac{\zeta^{p+1}}{p+1} & \text{if } q = 1, \end{cases}$$

as M *varies in* $]0, +\infty[$*. In particular*

$$\chi_M'(s_M) = 0 \quad \text{and} \quad s_M = \Gamma_M(M) \Leftrightarrow \chi_M(s_M) = M.$$

Moreover (by the change of variable $\xi = M\theta$*):*

$$\Gamma_M(\xi) = \int_0^{\xi/M} \frac{M\,d\theta}{\sqrt{AM^{q+1}H_q(\theta) + BM^{p+1}H_p(\theta)}},$$

where

$$H_r(\theta) = \begin{cases} \dfrac{1-\theta^{r+1}}{r+1} & \text{if } r \neq 1, \\ -\ln(\theta) & \text{if } r = 1. \end{cases}$$

As a consequence, if $0 \leq \xi \leq M$*:*

$$\frac{M\underline{\Gamma}(\xi/M)}{\sqrt{AM^{q+1} + BM^{p+1}}} \leq \Gamma_M(\xi) \leq \frac{M\overline{\Gamma}(\xi/M)}{\sqrt{AM^{q+1} + BM^{p+1}}} \tag{18}$$

where

$$\underline{\Gamma}(s) := \int_0^s \frac{d\theta}{\sqrt{H_q(\theta) \vee H_p(\theta)}}, \quad \overline{\Gamma}(s) := \int_0^s \frac{d\theta}{\sqrt{H_q(\theta) \wedge H_p(\theta)}}. \tag{19}$$

The following lemma is a standard consequence of the regularity of $\partial\Omega$.

4.7. Lemma. *There exist $\delta_0 > 0$ and a constant $\tilde{k}$ such that*

$$\int_{\Omega_\delta} |\nabla w|^2 \, dx \geq \frac{\tilde{k}^2}{\delta^2} \int_{\Omega_\delta} w^2 \, dx \qquad \forall w \in \mathrm{W}_0^{1,2}(\Omega_\delta), \forall \delta \leq \delta_0.$$

4.8. Lemma. *Assume that (16), (A.0), and (A.1) hold. Let $-3 < q < 0$ and $\lambda > 0$, $1 < p < 2^* - 1$, and let $\tilde{u}$ be the solution of $\tilde{L}u - a_0 \tilde{u}^q = 0$ as in Theorem 4.1. If u is a (positive) weak solution of $Lu - a_0 u^q - \lambda u^p = 0$ such that $u \geq \tilde{u}$, then u is bounded. Morever there exist constants M_0, C_0 and C_1 such that:*

$$u(x) \leq C_0(1 + \|u\|_\infty) \underline{\Gamma}^{-1} \left(C_1 \left(1 + \|u\|^{q-1} + (1+\lambda)\|u\|_\infty^{p-1}\right)^{\frac{1}{2}} \mathrm{dist}(x, \partial\Omega) \right) \quad (20)$$

where $\underline{\Gamma}$ is as in (19) (we can agree that $\underline{\Gamma}^{-1}(s) = 1$, for $s > \underline{\Gamma}(1)$).

Proof. Since $u \geq \tilde{u}$, then u is continuous on Ω. To see that u is bounded one can use the same arguments of [9, Lemmata 13, 14]. To prove (20) let u_0 be the solution of $Lu_0 = 1$: as already seen there exist $k_0 \geq \varepsilon_0 > 0$, $\delta_0 > 0$ such that $\varepsilon_0 \leq |\nabla u_0(x)| \leq k_0$ for $x \in \Omega_{\delta_0}$. Let $A := \dfrac{a_0}{\varepsilon_0^2 \nu_1}$, $B := \dfrac{\lambda}{\varepsilon_0^2 \nu_1}$ and let χ_M, Γ_M be defined as in Lemma 4.6. By (18), since $p > 1$, we have $s_M < \varepsilon_0 \delta_0$ if $M > M_0 := \left(\frac{\nu_1 \overline{\Gamma}(1)}{\lambda \delta_0^2}\right)^{\frac{1}{p-1}}$. Let $\bar{C} := \left(\frac{\varepsilon_0 \overline{\Gamma}(1)}{k_0 \tilde{k}}\right)^{\frac{2}{p-1}}$ and $M := p^{\frac{1}{p-1}} \bar{C} \|u\|_\infty \vee M_0$. Then $\bar{u}(x) := \chi_M(u_0(x))$ is well defined in Ω_{δ_M}, where $\delta_M := s_M / k_0$, and

$$\begin{aligned} L\bar{u} - a_0 \bar{u}^q - \lambda \bar{u}^p &= -\chi_M''(u_0) a(x) \nabla u_0 \nabla u_0 + \chi_M'(u_0) L u_0 - a_0 \bar{u}^q - \lambda \bar{u}^p \\ &\geq \nu_1 \varepsilon_0^2 (-\chi_m''(u_0) - A \chi_M(u_0)^q - B \chi_M(u_0)^p) = 0, \end{aligned}$$

in Ω_{δ_M}. We can also set $\bar{u} = M$ on $\Omega \setminus \Omega_M$; in this way $\bar{u} \in W_0^{1,2}(\Omega)$. Using $w :- (u - \bar{u})^+ (w \geq 0,\ w = 0$ outside $\Omega_{\delta_M})$ as test function (cf. Remark 2.3):

$$\begin{aligned} 0 &= \alpha(u, w) - \int_\Omega (a_0 u^q + \lambda u^p) w \, dx \\ &= \alpha(\bar{u}, w) - \int_{\Omega_{\delta_M}} (a_0 \bar{u}^q + \lambda \bar{u}^p) w \, dx + \alpha(u - \bar{u}, w) \\ &\quad - a_0 \int_{\Omega_{\delta_M}} (u^q - \bar{u}^q) w \, dx - \lambda \int_{\Omega_{\delta_M}} (u^p - \bar{u}^p) w \, dx \\ &\geq \nu_1 \int_{\Omega_{\delta_M}} |\nabla w|^2 \, dx - \lambda p \|u\|_\infty^{p-1} \int_{\Omega_M} w^2 \, dx \\ &\geq \left(\frac{\nu_1 \tilde{k}^2}{\delta_M^2} - \frac{\lambda p M^{p-1}}{p \bar{C}^{p-1}} \right) \int_{\Omega_M} w^2 \, dx \geq \left(\frac{k_0^2 \tilde{k}^2}{\varepsilon_0^2 \overline{\Gamma}(1)^2} - \frac{1}{\bar{C}^{p-1}} \right) \lambda M^{p-1} \int_{\Omega_M} w^2 \, dx \end{aligned}$$

since, by Lemma 4.6, $\delta_M = k_0^{-1} s_M = k_0^{-1} \Gamma_M(M) \leq \frac{\varepsilon_0 \overline{\Gamma}(1) \sqrt{\nu_1}}{k_0 \sqrt{\lambda}} M^{-\frac{p-1}{2}}$. By the way we have choosen $\bar{C}$ we have $w = (u - \bar{u})^+ = 0$, that is $u \leq \bar{u}$. By the definition of $\bar{u}$ (18) follows. □

4.9. Theorem. *Assume that* (16) *holds,* $-3 < q < 0$, $1 < p < 2^* - 1$. *Let* λ_n *be a sequence of positive numbers with* $+\infty > \overline{\lambda} := \sup_n \lambda_n \geq \underline{\lambda} := \inf_n \lambda_n > 0$. *Let* u_n *be a sequence of weak solutions of* $Lu - a_0 u^q - \lambda_n u^p = 0$ *with* $u_n \geq \tilde{u}$, *where* $\tilde{u}$ *is the unique solution of* $Lu_n - a_0 u_b^q = 0$, *as provided by Theorem 4.1. Then* $\sup_n \|u_n\|_\infty < +\infty$.

Proof. We follow the idea of [5]. Let x_n be such that $u_n(x_n) = \|u_n\|_\infty$ and suppose by contradiction that $\|u_n\|_\infty \to +\infty$. Using (20) we have, for n large

$$\|u_n\|_\infty = u_n(x_n) \leq 2C_0 \|u_n\|_\infty \underline{\Gamma}^{-1}\left(\left((C_1(2+\lambda))\|u_n\|_\infty^{p-1}\right)^{\frac{1}{2}} \mathrm{dist}(x_n, \partial\Omega)\right)$$

which implies that, for a suitable $\varepsilon_1 > 0$, we have:

$$\mathrm{dist}(x_n, \partial\Omega) \geq \varepsilon_1 \|u_n\|_\infty^{\frac{2}{p-1}}. \tag{21}$$

Let $\mu_n := u_n(x_n)^{\frac{2}{p-1}}$ and set $v_n(y) := \mu_n^{-\frac{2}{p-1}} u_n\left(x_n + \mu_n^{-1} y\right)$, so $u_n(x) = \mu_n^{\frac{2}{p-1}} v_n(\mu_n(x - x_n))$ and $v_n(0) = 1$. It follows that v_n is a weak solution of

$$L_n v_n = a_0 \mu_n^{\frac{2(q-p)}{p-1}} v_n^q + \lambda_n v_n^p \tag{22}$$

in $\mathrm{W}_0^{1,2}(\Omega_n^*)$, where $L_n v = -\mathrm{div}[a_n(y)\nabla v] + b_n(y)\nabla v$ with

$$a_n(y) := a(x_n + \mu_n^{-1} y),\ b_n(y) := \mu_n^{-1} b(x_n + \mu_n^{-1} y),$$

and where $\Omega_n^* := \left\{y \in \mathbb{R}^N : x_n + \mu_n^{-1} y \in \Omega\right\} = \mu_n\left(\Omega - x_n\right)$. From the property $u_n \geq \tilde{u}$ and (17), we get:

$$v_n(y) \geq \mu_n^{-\frac{2}{p-1}} \tilde{u}(x_n + \mu_n^{-1} y) \geq \bar{\varepsilon} \mu_n^{2\frac{q-p}{(p-1)(1-q)}} \mathrm{dist}(y, \Omega_n^*)^{\frac{2}{1-q}} \tag{23}$$

Let $x_n' \in \partial\Omega$ be such that $|x_n - x_n'| = \mathrm{dist}(x_n, \partial\Omega)$. Up to subsequences we can suppose that $\delta_n := \mu_n |x_n - x_n'| \to \delta \in [0, +\infty]$ and $\lambda_n \to \lambda \in]0, +\infty[$. By (21), we have $\delta > \varepsilon_1 > 0$. We distinguish the cases $\delta < +\infty$ and $\delta = +\infty$.

Case $0 < \delta < +\infty$. In this case both x_n and x_n' converge to a point $x_0 \in \partial\Omega$. Moreover $y_n' := \mu_n(x_n' - x_n) \to -\delta \mathrm{n}(x_0) =: y'$. Performing a translation and a rotation we can suppose that $y_n' = (0, -\delta_n)$ (we are writing $y = (\hat{y}, y_N)$ with $\hat{y} \in \mathbb{R}^{N-1}$).

Given $R > 0$ let $\eta_R : \mathbb{R}^N \to [0,1]$ be a smooth function such that $\eta_R(x) = 1$ if $|x| \leq R$, $\eta(x) = 0$ for $|x| \geq R + 1$ and $|\nabla \eta_R| \leq 2$. Multiplying the equation (22) by $\eta_R v_n$ the equation for v_n and integrating on Ω_n^* yields:

$$\alpha_n(v_n, \eta_R v_n) = \mu_n^{2\frac{q-p}{p-1}} \int_{\Omega_n^*} a_0 v_n^{q+1} \eta_R(y)\, dy + \lambda_n \int_{\Omega_n^*} v_n^{p+1} \eta_R(y)\, dy. \tag{24}$$

The second term on the right is clearly bounded (since $v_n \leq 1$). By (23):

$$\mu_n^{2\frac{q-p}{p-1}} \int_{\Omega_n^*} v_n^{q+1} \eta_R(y)\, dy \leq \bar{\varepsilon}^{q+1} \mu_n^{\frac{4(q-p)}{(p-1)(1-q)}} \int_{\Omega_n^* \cap B(0,R+1)} \mathrm{dist}(y, \partial\Omega_n^*)^{\frac{2(q+1)}{1-q}}\, dy.$$

Since $q > -3$, we have $\dfrac{2(q+1)}{1-q} > -1$ so the integrals above are finite; it is then clear that the integrals converge and the above expression tends to zero as $n \to \infty$. Therefore the right-hand side of (24) is bounded. Consider $v_n = 0$ outside Ω_n^*; it is simple to check that, for n large:

$$a_n(v_n, \eta_R v_n) \geq \frac{\nu}{2} \int_{B(0,R)} |\nabla v_n|^2 \, dy$$

so (24) implies that v_n are bounded in $\mathrm{W}^{1,2}(B(0,R))$. We can therefore suppose that $v_n \rightharpoonup v$ in $\mathrm{W}^{1,2}(B(0,R))$; moreover considering a sequence of R_k with $R_k \to +\infty$ and using a diagonalization argument we see that v is defined on the whole $\mathbb{R}^N$ and that $v_n \rightharpoonup v$ in $\mathrm{W}^{1,2}(B(0,R))$ for every $R > 0$.

Now let y_0, R be such that $\overline{B(y_0,R)} \subset H^- := \{y = (\hat{y}, y_N) : y_N < -\delta\}$. It is clear that for n large $B(y_0,R) \cap \Omega_n^* = \emptyset$ so $_n = 0$ on $B(y_0,R)$. This implies that $v = 0$ (a.e.) in H^-. Due to the regularity of ∂H^- this implies that $v = 0$ on ∂H^- in the sense of H^1: indeed v is the $\mathrm{W}^{1,2}_{\mathrm{loc}}$ limit of suitable mollifications of $v(\hat{y}, y_n - 1/n)$, which are zero on ∂H^-.

Let $H^+ := \{y = (\hat{y}, y_N) : y_N > -\delta\}$ and let $\varphi \in C_0^\infty(H^+)$. It is clear that there exists $\rho > 0$ such that for n large we have $\mathrm{dist}(x, \mathbb{R}^N \setminus \Omega_n^*) \geq \rho$ for any x with $\varphi(x) \neq 0$; in particular $\varphi \in C_0^\infty(\Omega_n^*)$. By (23):

$$\int_{\Omega_n^*} \mu_n^{\frac{2(q-p)}{p-1}} v_n^q(y)\varphi(y)\, dy \leq \mu_n^{\frac{2(q-p)}{(p-1)(1-q)}} \bar{\varepsilon}^q \rho^{\frac{2q}{1-q}} \int_{\mathbb{R}^N} |\varphi(y)|\, dy \to 0$$

as $n \to \infty$. Hence, multiplying (22) by φ, integrating on Ω_n^*, and letting $n \to \infty$:

$$\int_{\mathbb{R}^N} a(x_0)\nabla v \nabla \varphi \, dy = \lambda \int_\Omega v^p \varphi \, dy \qquad \forall \varphi \in C_0^\infty(H^+),$$

$v \in \mathrm{W}1,2_{\mathrm{loc}}(H^+)$, and $v = 0$ on ∂H^+ (in the sense of H^1). By [5] v must be zero. But if $\delta > \delta_1 > 0$ we have that the right-hand side of (22) is uniformly bounded in $B(0,\delta_1)$: use again (23) and the fact that $\mathrm{dist}(B(0,\delta_1), \mathbb{R}^N \setminus \Omega_n^*) \geq (\delta - \delta_1)/2 > 0$ for n large. Then v_n verifies an equation of the form $L_n v_n = h_n$ with L_n uniformly elliptic and with equibounded coefficients and with equibouned data h_n. This implies that v_n are equi-Lipschitz continuous in $B(0,\delta_1/2)$ so they converge uniformly to v in $B(0,\delta_1/2)$. In particular $v(0) = 1$ which contradicts $v = 0$.

Case $0 < \delta = +\infty$. This case can be treated with analogous arguments, following the idea of [5], the difference being that the limit v would be a nontrivial solution of $L_\infty v - \lambda v^p = 0$ in the whole $\mathbb{R}^N$: also this is impossible so $u_n(x_n)$ has to be bounded. □

Proof of Theorem 1.1. Let $\varphi_1 := \tilde{u}$ (the solution of $Lu = a_0 u^q$): it is clear that φ_1 is a subsolution for the problem $Lu - a_o u^q - \lambda u^p$, for all $\lambda > 0$. Let:

$$\bar{\lambda} := \sup\{\lambda > 0 : \text{ there exists a supersolution } \varphi_2 \text{ with } \varphi_2 \geq \varphi_1\}.$$

Since $\varphi_1 > 0$ it is clear that a supersolution $\varphi_2 \geq \varphi_1$ for $Lu - a_o u^q - \lambda u^p$ is a strict supersolution for $Lu - a_o u^q - \mu u^p$ whenever $0 < \mu < \lambda$. So for every $\lambda \in]0, \bar{\lambda}[$ there exists a strict supersolution $\varphi_{2,\lambda} \geq \varphi_1$. By Theorem 4.3 we know that $\bar{\lambda} > 0$ and that for every $\lambda \in]0, \bar{\lambda}[$ there exists a solution $u_{1,\lambda}$ with $\varphi_1 \leq u_{1,\lambda} \leq \varphi_{2,\lambda}$. Moreover, looking at the proof of 4.3, we see that, if $u_{1,\lambda}$ is unique among all $u \in \mathbb{K}_{\varphi_1}^{\varphi_{2,\lambda}}$ which solve $Lu = a_0 u^q + \lambda u^p$, then

$$\deg\left(Id - \Phi_{\varphi_1}^{\varphi_{2,\lambda}}, B(u_{1,\lambda}, \rho), 0\right) = 1.$$

for all $\rho > 0$ (by the excision property of the degree). By Theorem 3.14 there exists $\rho_\lambda > 0$ such that $u_{1,\lambda}$ is the unique solution in $\mathbb{K}_{\varphi_1} \cap B(u_{1,\lambda}, \rho_\lambda)$ and

$$\deg\left(Id - \Phi_{\varphi_1}, B(u_{1,\lambda}, \rho_\lambda), 0\right) = 1. \tag{25}$$

We claim that $Lu = a_0 u^q + \lambda u^p$ has no solutions for λ large; in particular $\bar{\lambda} < +\infty$. Indeed by coerciveness L has a positive eigenvalue λ_1 with positive eigenfunction e_1. If u is a solution of $Lu = a_0 u^q + \lambda u^p$, then:

$$\int_\Omega (\lambda_1 u - a_0 u^q - \lambda u^p)\, e_1\, dx = 0.$$

Since the function $s \mapsto \lambda_1 - a_0 s^{q-1} - \lambda s^{p-1}$ is negative in $]0, +\infty[$ for λ large the above equality is impossible for such λ's. In particular we have:

$$\deg\left(Id - \Phi_{\varphi_1}, B(0, R), 0\right) = 0 \qquad \forall \lambda > \bar{\lambda},\ \forall R > 0.$$

Let $0 < \lambda' < \bar{\lambda} < \lambda'' < +\infty$. By Theorem 4.9 there exist $\bar{R} > 0$ such that for all $\lambda \in [\lambda', \lambda'']$ the problem $Lu = a_0 u^q + \lambda u^p$ has no solutions u outside the ball $B(0, \bar{R})$. By continuation:

$$\deg\left(Id - \Phi_{\varphi_1}, B(0, \bar{R}), 0\right) = 0 \qquad \forall \lambda \in [\lambda', \lambda''].$$

Along with (25), this implies that for $\lambda' \leq \lambda < \bar{\lambda}$ $u_{1,\lambda}$ cannot be the unique solution, so there exists another solution $u_{2,\lambda}$. Replacing φ_1 with $u_{1,\lambda}$ in the previous argument, it is clear that $u_{2,\lambda} \geq u_{1,\lambda}$. Since $\lambda' > 0$ is arbitrary, then we can find $u_{2,\lambda}$ for all $\lambda \in]0, \bar{\lambda}[$.

Finally, taking $\lambda_n < \bar{\lambda}$ with $\lambda_n \to \bar{\lambda}$, it follows from Theorem 4.9 that u_{1,λ_n} is bounded, thus providing, at the limit, a solution for the case $\lambda = \bar{\lambda}$. □

References

[1] A. Ambrosetti, H. Brezis, and G. Cerami. Combined effects of concave and convex nonlinearities in some elliptic problems. *J. Funct. Anal.*, 122:519–543, 1994.

[2] L. Boccardo. A Dirichlet problem with singular and supercritical nonlinearities. *Nonlinear Anal. Th. Meth. Appl.*, 75:4436–4440, 2012.

[3] H. Brezis. *Opérateurs maximaux monotones et semi-groupes de contractions dans les espaces de Hilbert.* 1973.

[4] M.G. Crandall, P.H. Rabinowitz, and L. Tartar. On a Dirichlet problem with a singular nonlinearity. *Comm. Part. Diff. Eq.*, 2:193–222, 1977.

[5] B. Gidas and J. Spruck. A priori bounds for positive solutions of nonlinear elliptic equations. *Comm. Partial Differ. Equations*, 6 (8):883–901, 1981.

[6] D. Gilbarg and N.S. Trudinger. *Elliptic Partial Differential Equations of Second Order*. 1998.

[7] J. Hernandez, F. Mancebo, and J. Vega. Positive solutions for singular nonlinear elliptic equations. *Proc. Royal Soc. Edinburgh*, 137A:41–62, 2007.

[8] N. Hirano, C. Saccon, and N. Shioji. Existence of multiple positive solutions for singular elliptic problems with concave and convex nonlinearities. *Adv. Diff. Eq.*, 9(1-2):197–220, 2004.

[9] N. Hirano, C. Saccon, and N. Shioji. Brezis–Nirenberg type problems and multiplicity of positive solutions for a singular elliptic problem. *J. Diff. Eq.*, 245:1997–2037, 2008.

[10] A.I. Nazarov. A centennial of the Zaremba–Hopf–Oleinik lemma. *SIAM J. Math. Anal.*, 44:437–453, 2012.

[11] C. Saccon. A variational approach to a class of singular semilinear elliptic equations. *EJDE*, to appear, 2013. Proceedings of the Conference "Variational and Topological Methods", Flagstaff, June 6–9, 2012.

[12] G. Stampacchia. "Le problème de Dirichlet pour les équations elliptiques du second ordre à coefficients discontinus". *Ann. Inst. Fourier*, 15(1):189–258, 1965.

[13] C.A. Stuart. Existence and approximation of solutions of non-linear elliptic equations. *Math. Z.*, 137A:53–63, 1976.

[14] N.S. Trudinger. Linear elliptic operators with measurable coefficients. *Ann. Scuola Norm. Sup. Pisa*, 27(2):265–308, 1973.

C. Saccon
Dipartimento di Matematica
Largo Bruno Pontecorvo 5
I-56127 Pisa, Italy
e-mail: saccon@dm.unipi.it

Progress in Nonlinear Differential Equations
and Their Applications, Vol. 85, 405–421

Nonuniqueness for the Dirichlet Problem for Fully Nonlinear Elliptic Operators and the Ambrosetti–Prodi Phenomenon

Boyan Sirakov

Abstract. We study the uniformly elliptic fully nonlinear PDE

$$F(D^2u, Du, u, x) = f(x) \quad \text{in } \Omega,$$

where F is a convex positively 1-homogeneous operator and $\Omega \subset \mathbb{R}^N$ is a regular bounded domain. We prove non-existence and multiplicity results for the Dirichlet problem, when the two principal eigenvalues of F are of different sign. Our results extend to more general cases, for instance, when F is not convex, and explain in a new light the classical results of Ambrosetti–Prodi type in elliptic PDE.

Mathematics Subject Classification (2010). 35J25, 35B34, 35P30.

Keywords. Bellman operator, Dirichlet problem, fully nonlinear, principal eigenvalue, nonuniqueness, Ambrosetti–Prodi.

1. Introduction and main results

This paper is devoted to the study of the existence and the uniqueness of solutions of the Dirichlet boundary value problem

$$\begin{cases} H(D^2u, Du, u, x) = f(x) & \text{in } \Omega \\ u = 0 & \text{on } \partial\Omega, \end{cases} \tag{1}$$

where $\Omega \subset \mathbb{R}^N$ is a bounded domain with $C^{1,1}$-boundary, $f \in L^\infty(\Omega)$, and $H(M, p, u, x)$ is an uniformly elliptic fully nonlinear operator, globally Lipschitz in (M, p) and locally Lipschitz in u. A particular type of operators to which our results apply are Isaacs and Hamilton–Jacobi–Bellman operators. Boundary value problems of this type have been very extensively studied in the framework of classical, strong and viscosity solutions, see for example [35], [24], [28], [19], [14], [17]. Most work on fully nonlinear problems concerns *proper* operators, that is, the case when H is nonincreasing in u. Recently nonproper problems of type (1) have been studied in [36] and [37], see also the references in these papers. The present work continues a study started in [37].

For all $M \in \mathcal{S}_N(\mathbb{R}), p \in \mathbb{R}^N$, define the extremal operators $\mathcal{L}^-, \mathcal{L}^+$ by

$$\mathcal{L}^-(M,p) = \mathcal{M}^-_{\lambda,\Lambda}(M) - \gamma\,|p|, \qquad \mathcal{L}^+(M,p) = \mathcal{M}^+_{\lambda,\Lambda}(M) + \gamma\,|p|,$$

for some positive constants λ, Λ, γ. Here $\mathcal{M}^+, \mathcal{M}^-$ denote the Pucci operators $\mathcal{M}^+_{\lambda,\Lambda}(M) = \sup_{A\in\mathcal{A}} \mathrm{tr}(AM)$, $\mathcal{M}^-_{\lambda,\Lambda}(M) = \inf_{A\in\mathcal{A}} \mathrm{tr}(AM)$, where $\mathcal{A} \subset \mathcal{S}_N$ denotes the set of matrices whose eigenvalues lie in the interval $[\lambda, \Lambda]$.

We suppose that H in (1) satisfies the following hypothesis: for all $M \in \mathcal{S}_N(\mathbb{R}), p \in \mathbb{R}^N, u \in \mathbb{R}, x \in \Omega$, and for some constants A_0, c, δ,

$$F(M,p,u,x) - A_0 \le H(M,p,u,x) \le \mathcal{L}^+(M,p) + c|u| + A_0, \tag{2}$$

where $F(M,p,u,x)$ is some nonlinear operator, such that

$$\begin{cases} \mathcal{L}^-(M,p) - \delta|u| \le F(M,p,u,x) \le \mathcal{L}^+(M,p) + \delta|u| \\ F(tM,tp,tu,x) = tF(M,p,u,x) \ \text{ for } t \ge 0 \\ F \text{ is convex in } (M,p,u), \qquad F(M,0,0,x) \in C(\mathcal{S}_N(\mathbb{R}) \times \overline{\Omega}, \mathbb{R}). \end{cases} \tag{3}$$

We assume that H is Lipschitz continuous and uniformly elliptic, in the following sense: for each $R \in \mathbb{R}$ there exists $c_R \in \mathbb{R}$ such that for all $M, N \in \mathcal{S}_N(\mathbb{R})$, $p, q \in \mathbb{R}^N$, $x \in \Omega$, $u, v \in [-R, R]$,

$$\begin{cases} H(M,p,u,x) - H(N,q,v,x) & \ge & \mathcal{L}^-(M-N,p-q) - c_R|u-v| \\ H(M,p,u,x) - H(N,q,v,x) & \le & \mathcal{L}^+(M-N,p-q) + c_R|u-v|. \end{cases} \tag{4}$$

Note that (3) implies (4) with $H = F$ and $c_R = \delta$, see [37] (or inequalities (7) below). Our final standing assumption is that the proper operator

$$H_v[u] := H(D^2u, Du, v(x), x) - u \tag{5}$$

satisfies the comparison principle for each $v \in C(\overline{\Omega})$, in the sense that if $H_v[u_1] \ge f \ge H_v[u_2]$ in Ω, and $v_1 = v_2 = 0$ on $\partial\Omega$ then $v_1 \le v_2$ in Ω. This is satisfied for instance when H is Hölder continuous in x with a sufficiently large Hölder constant or when H is convex in M and $H(M,0,0,x)$ is uniformly continuous. Many other conditions which ensure uniqueness for proper equations can be found in [19], [31], [14], [32].

For instance, F can be a Hamilton–Jacobi–Bellman (HJB) operator, that is, a supremum of linear second-order operators with bounded coefficients and continuous second-order coefficients – see [37] for examples and discussions. HJB operators are basic in control theory. On the other hand, H can be an Isaacs operator, that is, a sup-inf of linear operators (these operators are essential in game theory). The Dirichlet problem for such operators has been widely studied in the proper case, and still many open question subsist, see the references above. Of course H can be a semilinear or quasilinear operator satisfying the hypotheses we made.

It was shown in [34], [9], [37] (see also [11], [5] for related results) that under hypothesis (3) F has two principal eigenvalues $\lambda_1^+(F,\Omega) \le \lambda_1^-(F,\Omega)$, which correspond to a positive and a negative eigenfunction, such that (1) with $H = F$ has a unique solution for all f if $\lambda_1^+ > 0$, while if $\lambda_1^- > 0 \ge \lambda_1^+$ then (1) has a solution for $f \ge 0$ but (1) does not have solutions for $f \le 0$, $f \not\equiv 0$. The question of uniqueness in the last case was left open, since $\lambda_1^- > 0$ alone does not imply a comparison principle. It is this question that we address in the present article. We will show that uniqueness fails when only one of the two eigenvalues is positive.

We will use the following decomposition of the right-hand side $f(x)$ in (1)

$$f(x) = -t\phi(x) + h(x),$$

where $t \in \mathbb{R}$, $\phi = \varphi_1^+(F_0,\Omega)$ is the first positive eigenfunction of the operator $F_0(M,p,x) = F(M,p,0,x)$, normalized so that $\max_\Omega \phi = 1$. The existence of $\phi \in W^{2,p}_{\text{loc}}(\Omega) \cap C(\overline{\Omega})$, $p < \infty$, $\phi > 0$ in Ω, $F_0(D^2\phi, D\phi, x) = -\lambda_0^+\phi$ in Ω was shown in [37]. Since F_0 is proper, we have $\lambda_0^+ = \lambda_1^+(F_0,\Omega) > 0$, see [37].

Whenever we speak of a solution of (1) we shall mean a function in $C(\overline{\Omega})$ which satisfies (1) in the L^N-viscosity sense. See [14] for definitions and properties of these solutions. Note that $u \in W^{2,N}_{\text{loc}}(\Omega) \cap C(\overline{\Omega})$ satisfies (1) almost everywhere in Ω if and only if it is a L^N-viscosity solution of (1).

Here is our main result.

Theorem 1. *Suppose F and H verify* (2)–(5), *and*

$$\lambda_1^+(F,\Omega) < 0 < \lambda_1^-(F,\Omega). \tag{6}$$

Then for each $h \in L^\infty(\Omega)$ there exists a number $t^*(h) \in \mathbb{R}$ such that:

(1) *if $t < t^*(h)$ then* (1) *has at least two solutions;*
(2) *if $t = t^*(h)$ then* (1) *has at least one solution;*
(3) *if $t > t^*(h)$ then* (1) *has no solutions.*

The map $h \to t^*(h)$ is continuous from $L^\infty(\Omega)$ to $\mathbb{R}$.

Remark 1. If $H(M,p,u,x)$ is convex in M then the solutions obtained in Theorem 1 belong to $W^{2,p}_{\text{loc}}(\Omega) \cap C(\overline{\Omega})$, for all $p < \infty$.

The acknowledged reader may have noticed that the conclusion in Theorem 1 is similar to results obtained in the framework of the so-called Ambrosetti–Prodi problem, classical in the theory of semilinear elliptic PDE's. We shall quote here the original work [3], as well as the subsequent developments [10], [33], [20], [29], [23], [38], [16], [22]. Quasilinear operators were recently considered in [4], [7]. Here is the most typical Ambrosetti–Prodi type result: given the operator $H_L(M,p,u,x) = \text{tr}(A(x)M) + b(x).p + g(x,u)$, if g is a Lipschitz function such that $g(x,u) \ge c_1u^+ - c_2u^- - c_0$, and if $c_1 > \lambda_1 > c_2$, where λ_1 is the usual first eigenvalue of the linear operator $L(M,p,x) = \text{tr}(A(x)M) + b(x).p$, then the same conclusion as in Theorem 1 holds for (1) with $H = H_L$. Actually, this statement is nothing but Theorem 1 applied to $H = H_L$ and $F = F_L$, where

$$F_L(M,p,u,x) = \text{tr}(A(x)M) + b(x).p + c_1u^+ - c_2u^-.$$

Here $u^+ = \max\{u, 0\}, u^- = -\min\{u, 0\}$, $A \in C(\overline{\Omega})$ is a positive definite matrix, b is a bounded vector, and $c_1 > c_2$. Then $\lambda_1^+(F_L, \Omega)$ (resp. $\lambda_1^-(F_L, \Omega)$) is obviously equal to $\lambda_1 - c_1$ (resp. $\lambda_1 - c_2$).

In other words, and this is the second main conclusion of the paper, the Ambrosetti–Prodi phenomenon turns out to be due to nonuniqueness of solutions of the Dirichlet problem for a convex nonlinear operator with one positive and one negative principal eigenvalue.

Remark 2. Many of the quoted papers on the Ambrosetti–Prodi problem contain results also for systems of equations or for the case when $g(x, u)$ in H_L does not have a linear but rather a power growth in u. Such extensions for fully nonlinear equations and systems of type (1) might be the subject of a future work.

Remark 3. It is only a matter of technicalities to show the results extend to the case when $h(x)$ in (1) and A_0 in (2) belong to $L^p(\Omega)$, $p > N$.

The next section contains the proof of Theorem 1. Its overall scheme (that is, the statements of the steps of the proof) is similar to the classical one used to prove the Ambrosetti–Prodi type results quoted above. It combines Perron's method with a priori bounds and degree theory, see the next section for more details. Of course, the proofs of some steps are rather different, and require a specific nonlinear approach. We find it quite remarkable how naturally the theory of viscosity solutions and eigenvalues for fully nonlinear operators permit to carry out these proofs. We begin the next section by an overview.

Remark. The results of this paper first appeared as a preprint in 2008. This preprint was later used a basis for the more detailed study of resonance problems in [26], as well as for the result of the same type in [21] on more general fully nonlinear operators modeled on $|\nabla u|^\gamma \mathcal{M}^+_{\lambda,\Lambda}(D^2u)$. Other related results can be found in [6], [27].

2. Proof of Theorem 1

From now on $h \in L^\infty(\Omega)$ will be fixed and we shall refer to (1) as problem $(\mathcal{P}_t)$ or $(\mathcal{P}_{t,h})$, when we need to stress the dependence on t or h.

We first give the plan of the proof of Theorem 1.

1. prove an a priori upper bound on t, such that $(\mathcal{P}_t)$ has a solution;
2. prove an a priori bound on u, for $t \geq -C$;
3. prove subsolutions of $(\mathcal{P}_t)$ exist for all t, supersolutions exist for sufficiently small t, deduce by Perron's method that solutions of $(\mathcal{P}_t)$ exist for $t \in (-\infty, t^*)$;
4. prove for each $t \in (-\infty, t^*)$ there exists a subsolution of $(\mathcal{P}_t)$ which is smaller than all solutions of $(\mathcal{P}_t)$;
5. use fixed point theorems and degree theory to conclude.

Let us review the main points and the difficulties in the proofs. Steps 1 and 2 above are rather classical for operators in divergence form, that is, for cases when (1) has an equivalent formulation in terms of integrals. Then one can prove Step 1 by testing the equation with the first eigenfunction of F_0 and after that carry out a contradiction (blow-up) argument to obtain the statement in Step 2. This is not possible for operators in non-divergence form. Recently a different method was developed in [22], for the semilinear operators F_L, H_L, which gives a simultaneous proof of Steps 1 and 2, and which applies to operators with power growth in u. The proof in [22] depends on the linearity of $L = F_0$. We will show here that it is actually the nonlinear structure of F and H, as described in our hypotheses, which provides for such a method to be applicable.

Further, Step 3 above is proved with the help of an one-sided Alexandrov–Bakelman–Pucci (ABP) inequality combined with an existence result, both obtained in [37], for operators with only one positive principal eigenvalue, which we recall below.

Another important difference with the semilinear case appears in proving Step 4. If $F = F_L$ then it is automatic that the restriction of F_L to the cone $\{(M,p,u,x)\ :\ u \le 0\}$ satisfies a comparison principle in this cone (since F_L is linear and coercive there). In the nonlinear case this is not clear; however we manage to prove that subsolutions can be chosen to satisfy properties which permit to us to use a more restrictive comparison result, which we establish, based on the fraction rather than the difference between the two functions that we compare – see Lemma 2.5 and the comments there.

Finally, the multiplicity result (Step 5) relies on an argument which uses the properties of the Leray–Schauder degree of compact maps.

We next list several preliminary results, mostly from [37]. It was shown in [37] that hypothesis (3) implies

$$\begin{cases} F(M-N,p-q,u-v,x) \ge F(M,p,u,x) - F(N,q,v,x) \\ F(M+N,p+q,u+v,x) \le F(M,p,u,x) + F(N,q,v,x), \end{cases} \tag{7}$$

for all $M, N \in \mathcal{S}_N(\mathbb{R}), p, q \in \mathbb{R}^N, u, v \in \mathbb{R}, x \in \Omega$.

We recall that the principal eigenvalues of F are defined by

$$\lambda_1^+(F,\Omega) = \sup\{\lambda \in \mathbb{R} \mid \exists \psi > 0 \text{ in } \Omega,\ F(D^2\psi, D\psi, \psi, x) + \lambda\psi \le 0 \text{ in } \Omega\},$$
$$\lambda_1^-(F,\Omega) = \sup\{\lambda \in \mathbb{R} \mid \exists \psi < 0 \text{ in } \Omega,\ F(D^2\psi, D\psi, \psi, x) + \lambda\psi \ge 0 \text{ in } \Omega\}.$$

In the sequel we shall need the following *one-sided* ABP estimate, obtained in [37]. A complete version of the Alexandrov–Bakelman–Pucci inequality for proper operators can be found in [14] (an ABP inequality for the Pucci operator was first proved in [13]). We recall that λ_1^+, λ_1^- are bounded above and below by constants which depend only on $N, \lambda, \Lambda, \gamma, \delta, \Omega$, and that both principal eigenvalues of any proper operator are positive, see [37].

Theorem 2 ([37]). *Suppose the operator F satisfies* (3).

I. *If $\lambda_1^-(F,\Omega) > 0$ then for any $u \in C(\overline{\Omega})$, $f \in L^N(\Omega)$, the inequality $F(D^2u, Du, u, x) \leq f$ implies*

$$\sup_{\Omega} u^- \leq C(\sup_{\partial\Omega} u^- + \|f^+\|_{L^N(\Omega)}),$$

where C depends on Ω, $N, \lambda, \Lambda, \gamma, \delta$.

II. *In addition, if $\lambda_1^+(F,\Omega) > 0$ then $F(D^2u, Du, u, x) \geq f$ implies*

$$\sup_{\Omega} u \leq C(\sup_{\partial\Omega} u^+ + \|f^-\|_{L^N(\Omega)}).$$

Set $E_p = W^{2,p}_{\mathrm{loc}}(\Omega) \cap C(\overline{\Omega})$, $p \geq N$. We shall use the following existence result.

Theorem 3 ([37]). *Suppose the operator F satisfies* (3).

I. *If $\lambda_1^-(F,\Omega) > 0$ then for any $f \in L^p(\Omega), p \geq N$, such that $f \geq 0$ in Ω, there exists a solution $u \in E_p$ of $F(D^2u, Du, u, x) = f$ in Ω, $u = 0$ on $\partial\Omega$, such that $u \leq 0$ in Ω.*

II. *In addition, if $\lambda_1^+(F,\Omega) > 0$ then for any $f \in L^p(\Omega), p \geq N$, there exists a unique solution $u \in E_p$ of $F(D^2u, Du, u, x) = f$ in Ω, $u = 0$ on $\partial\Omega$.*

We will actually need to apply parts II in the above two theorems only to the proper operator $F_0(M,p,x) = F(M,p,0,x)$.

We now move to the proof of Theorem 1. First we will show that solutions of $(\mathcal{P}_t)$ admit an a priori bound, which is uniform in $t \in (m,\infty)$, for each $m \in \mathbb{R}$. In the sequel C will denote a constant which may change from line to line and which depends on $N, \lambda, \Lambda, \gamma, \delta$, A_0, c, Ω, and $\|h\|_{L^\infty(\Omega)}$.

The next proposition realizes Steps 1 and 2 (see the beginning of this section) of the proof of Theorem 1.

Proposition 2.1. *For each $m_0 \in \mathbb{R}_+$ there exists a constant C such that for any $t \geq -m_0$ and any solution u of $(\mathcal{P}_t)$ we have*

$$\|u\|_{L^\infty(\Omega)} \leq C \qquad \text{and} \qquad t \leq C.$$

In particular, there do not exist solutions of $(\mathcal{P}_t)$ for large t.

Proof. We divide the proof in three steps.

Claim 1. For each $m_0 \in \mathbb{R}_+$ there exists a constant $C = C(m_0)$ such that for any $t \geq -m_0$ and any solution u of $(\mathcal{P}_t)$ with this t we have

$$\|u^-\|_{L^\infty(\Omega)} \leq C.$$

Proof. This is an immediate consequence of (2), (6), and Theorem 2 I with f replaced by $m_0\phi + h$. □

Claim 2. There exists a constant C such that for solution u of $(\mathcal{P}_t)$ we have

$$t \leq C(1 + \|u\|_{L^\infty(\Omega)}).$$

Proof. By (2) and the definition of ϕ we have

$$F(D^2u, Du, u, x) - \frac{t}{\lambda_0^+} F(D^2\phi, D\phi, 0, x) \leq h(x) + A_0. \tag{8}$$

By (7) and (3) we have (recall we have set $F_0(M, p, x) = F(M, p, 0, x)$)

$$\begin{aligned} F(M, p, u, x) &\geq F(M, p, 0, x) - F(0, 0, -u, x) \\ &\geq F_0(M, p, x) - \delta|u|. \end{aligned}$$

Hence, by (7), (8), and the homogeneity of F

$$-F_0\left(D^2\left(-u + \frac{t}{\lambda_0^+}\phi\right), D\left(-u + \frac{t}{\lambda_0^+}\phi\right), x\right) \leq h(x) + A_0 + \delta|u|.$$

Then the second part of Theorem 2 implies that for all $x \in \Omega$

$$-u(x) + \frac{t}{\lambda_0^+}\phi(x) \leq C\|h(x) + A_0 + \delta|u|\|_{L^\infty(\Omega)}.$$

Taking x such that $\phi(x) = \max_\Omega \phi = 1$ finishes the proof of Claim 2. □

Conclusion. Suppose the a priori bound on u in the statement of Proposition 2.1 is false, that is, there exist sequences $\{t_n\}, \{u_n\}$ such that $t_n \geq -m_0$, $\|u_n\|_{L^\infty(\Omega)} \to \infty$, and

$$H(D^2u_n, Du_n, u_n, x) = -t_n\phi + h.$$

By (2), (3) and Claim 2 we have

$$\begin{aligned} \mathcal{L}^-(D^2u_n, Du_n) &\leq \delta\|u_n\|_{L^\infty(\Omega)} + m_0 + A_0 + h \\ \mathcal{L}^+(D^2u_n, Du_n) &\geq -C(1 + \|u_n\|_{L^\infty(\Omega)}) + h. \end{aligned}$$

Hence, setting $v_n = u_n/\|u_n\|$ (so that $\|v_n\|_{L^\infty(\Omega)} = 1$),

$$\mathcal{L}^-(D^2v_n, Dv_n) \leq C \quad \text{and} \quad \mathcal{L}^+(D^2v_n, Dv_n) \geq -C.$$

We now use the following result from the general theory of viscosity solutions of fully nonlinear PDE (it is a particular case, for instance, of Proposition 4.2 in [17]).

Proposition 2.2. *For any given $M \in \mathbb{R}$ the set of functions $u \in C(\overline{\Omega})$ such that*

$$\|u\|_{L^\infty(\Omega)} \leq M, \qquad \mathcal{L}^-(D^2u, Du) \leq M, \qquad \text{and} \quad \mathcal{L}^+(D^2u, Du) \geq -M$$

is precompact in $C(\overline{\Omega})$.

Hence a subsequence of $\{v_n\}$ converges uniformly to a function v in $\overline{\Omega}$. Note that $v \geq 0$ in Ω, by Claim 1, and $\|v\|_{L^\infty(\Omega)} = 1$.

Again by (2) $F(D^2u_n, Du_n, u_n, x) \leq m_0 + A_0 + h$, so, by the homogeneity of F

$$F(D^2v_n, Dv_n, v_n, x) \leq o(1).$$

By viscosity solutions theory (see Theorem 3.8 in [14]) we can pass to the limit in this inequality, obtaining

$$F(D^2v, Dv, v, x) \leq 0. \tag{9}$$

We recall the following strong maximum principle (Hopf lemma), a consequence from the results in [8].

Proposition 2.3 ([8]). *Let $\mathcal{O} \subset \mathbb{R}^N$ be a regular domain and let $\gamma \in \mathbb{R}$, $\delta \leq 0$. Suppose $w \in C(\overline{\mathcal{O}})$ is a viscosity solution of*

$$\mathcal{M}^-_{\lambda,\Lambda}(D^2 w) - \gamma|Dw| - \delta w \leq 0 \text{ in } \mathcal{O},$$

and $w \geq 0$ in $\mathcal{O}$. Then either $w \equiv 0$ in $\mathcal{O}$ or $w > 0$ in $\mathcal{O}$ and at any point $x_0 \in \partial\mathcal{O}$ at which $w(x_0) = 0$ we have

$$\liminf_{t \searrow 0} \frac{w(x_0 + t\nu) - w(x_0)}{t} > 0,$$

where ν is the interior normal to $\partial\mathcal{O}$ at x_0.

Therefore $v > 0$ in Ω. Using (9), the existence of such function contradicts the definition of $\lambda_1^+(F, \Omega)$ and the hypothesis $\lambda_1^+(F, \Omega) < 0$.

Hence $\|u\|_{L^\infty(\Omega)}$ is bounded, and, by Claim 2, t is bounded as well. □

We turn to existence of subsolutions and supersolutions of $(\mathcal{P}_t)$. We shall need the following boundary Lipschitz estimate for fully nonlinear equations (for a proof see Proposition 4.9 in [37]).

Proposition 2.4. *Suppose H satisfies (4) and Ω satisfies a uniform exterior sphere condition. Suppose $u \in C(\overline{\Omega})$ satisfies $H(D^2u, Du, u, x) = h$, $u = 0$ on $\partial\Omega$, where $h \in L^\infty(\Omega)$. Then there exists a constant k depending on $N, \lambda, \Lambda, \gamma, \delta$,* diam (Ω), $\|u\|_{L^\infty(\Omega)}$, $\|h\|_{L^\infty(\Omega)}$, *and the radius of the exterior spheres, such that for each $x_0 \in \partial\Omega$*

$$|u(x)| \leq k|x - x_0| \qquad \text{for each } x \in \Omega.$$

First we deal with the existence of supersolutions.

Lemma 2.1. *There exists $t_0 \in \mathbb{R}$, depending on the constants in (2)–(4) and on $\|h\|_{L^\infty(\Omega)}$, such that for each $t \leq t_0$ there exists a supersolution $\overline{u}$ of $(\mathcal{P}_t)$, such that $\overline{u} \geq 0$ in Ω, $\overline{u} \in E_p$, $p < \infty$.*

Proof. Let $\overline{u}$ be the unique solution of the Dirichlet problem (see Theorem 3 above, or Corollary 3.10 in [14])

$$\begin{cases} \mathcal{L}^+(D^2\overline{u}, D\overline{u}) = -h^-(x) & \text{in } \Omega \\ u = 0 & \text{on } \partial\Omega, \end{cases} \tag{10}$$

The ABP inequality shows that $\overline{u} \geq 0$ in Ω, $\|\overline{u}\|_{L^\infty(\Omega)} \leq C$ and $\overline{u}$ satisfies the boundary inequality in Proposition 2.4. On the other hand, the Hopf lemma and the inequality $F_0(D^2\phi, D\phi, x) \leq 0$ imply that there exists a constant $\alpha > 0$ such that for all $x_0 \in \partial\Omega$

$$\liminf_{t \searrow 0} \frac{\phi(x_0 + t\nu) - \phi(x_0)}{t} \geq \alpha,$$

where ν is the inner normal to $\partial\Omega$. Therefore there exists $t_0 < 0$ such that $-t_0\phi \geq \delta\overline{u}$ in Ω. Hence by (2) we have $H(D^2\overline{u}, D\overline{u}, \overline{u}, x) \leq -t\phi + h$, for all $t \leq t_0$, which is the required result. □

The next lemma concerns the existence of subsolutions.

Lemma 2.2. *For any $t \in \mathbb{R}$ there exists a subsolution $\underline{u} \le 0$ in Ω, $\underline{u} \in E_p$, $p < \infty$, of $(\mathcal{P}_t)$. In addition, given a compact interval $I \subset \mathbb{R}$, $\underline{u}$ can be chosen so that $\underline{u} \le u$ in Ω, for all solutions u of $(\mathcal{P}_t)$, $t \in I$.*

The difficulty in Lemma 2.2 is in the second statement. As a step in its proof, we will obtain the following uniform boundary Hopf lemma, which is of independent interest.

Lemma 2.3. *Assume Ω satisfies an uniform interior sphere condition. Suppose F satisfies (3), $\lambda_1^-(F,\Omega) > 0$, and $f \not\equiv 0, 0 \le f \le M$ in Ω. Then there exists $\alpha_0 > 0$ depending only on $\lambda, \Lambda, \nu, \delta$, Ω, and M, such that for any solution of $F(D^2u, Du, u, x) = f$ in Ω, $u \le 0$ in Ω, $u = 0$ on $\partial\Omega$, and all $x_0 \in \partial\Omega$ we have*

$$V_{x_0}(u) := \liminf_{t \searrow 0} \frac{u(x_0) - u(x_0 + t\nu)}{t} \ge \alpha_0.$$

Proof. Suppose the lemma is false, that is, there is a sequence of solutions $u_n \le 0$ in Ω and points $x_n \in \partial\Omega$ (we can suppose $x_n \to x \in \partial\Omega$) such that $V_{x_n}(u_n) \to 0$. Note that $\|u_n\|_{L^\infty(\Omega)} \le C$, by Theorem 2. From (3) we have

$$\mathcal{L}^-(D^2u_n, Du_n) \le C \quad \text{and} \quad \mathcal{L}^+(D^2u_n, Du_n) \ge -C.$$

By Proposition 2.2 a subsequence of $\{u_n\}$ converges uniformly to a function u in $\overline{\Omega}$, and $F(D^2u, Du, u, x) = f$ in Ω. Note that, by the strong maximum principle, $u_n < 0$ and $u < 0$ in Ω (since $f \not\equiv 0$ excludes $u_n \equiv 0$ or $u \equiv 0$).

By (3) and properties of Pucci operators ($\mathcal{M}^-(M) = -\mathcal{M}^+(-M)$), the positive functions $v_n = -u_n$ satisfy

$$\mathcal{M}^-_{\lambda,\Lambda}(D^2v_n) - \nu|Dv_n| - \delta v_n \le 0 \tag{11}$$

in Ω. Let ρ be the radius of the interior spheres. Fix $p \in \partial\Omega$ and let $B_\rho \subset \Omega$ be a ball tangent to $\partial\Omega$ at p. Introduce the (standard) barrier function, defined in B_ρ,

$$z(r) = e^{-\beta r^2} - e^{-\beta\rho^2},$$

where r is the distance to the center of B_ρ and β is a positive constant yet to be chosen. We recall the following fact.

Lemma 2.4. *Suppose $u \in C^2(B)$ is a radial function, defined on a ball B, say $u(x) = g(|x|)$. Then the matrix $D^2u(x)$ has $g''(|x|)$ as a simple eigenvalue, and $|x|^{-1}g'(|x|)$ as an eigenvalue of multiplicity $N-1$.*

Using this lemma and the fact that

$$\mathcal{M}^-_{\lambda,\Lambda}(M) = \lambda \sum_{\{e_i>0\}} e_i + \Lambda \sum_{\{e_i<0\}} e_i, \quad \mathcal{M}^+_{\lambda,\Lambda}(M) = \Lambda \sum_{\{e_i>0\}} e_i + \lambda \sum_{\{e_i<0\}} e_i,$$

where e_i denote the eigenvalues of M, an elementary computation shows that

$$\mathcal{M}^-_{\lambda,\Lambda}(D^2z) - \nu|Dz| - \delta z \ge 0 \tag{12}$$

in the annulus $B_\rho \setminus B_{\rho/2}$, if $\beta = \beta(\rho)$ is chosen sufficiently large. Let the point $q_n \in \partial B_{\rho/2}$ be such that $v_n(q_n) = \min_{\partial B_{\rho/2}} v_n$ and set

$$\sigma_n = \frac{v_n(q_n)}{e^{-\beta(\rho/2)^2} - e^{-\beta\rho^2}}.$$

Then $\sigma_n z \leq v_n$ on $\partial(B_\rho \setminus B_{\rho/2})$ and, by the comparison principle for proper operators (see [14] or [37], note that the operator which appears in (11),(12) is proper), $\sigma_n z \leq v_n$ in $B_\rho \setminus B_{\rho/2}$. Hence

$$\sigma_n \frac{\partial z}{\partial \nu}(p) \leq -V_p(v_n) = V_p(u_n),$$

which implies

$$\min_{\partial B_{\rho/2}} v_n \leq a_0 V_p(u_n)$$

for some $a_0 > 0$, which depends on the appropriate quantities, and for all $p \in \partial\Omega$. Therefore, there exists a sequence of points $y_n \in \Omega$ such that $\mathrm{dist}(y_n, \partial\Omega) \geq \rho/2$ and $v_n(y_n) \to 0$. Hence there exists a point $y \in \Omega$ such that $v(y) = 0$, a contradiction. □

Proof of Lemma 2.2. Set $M = A_0 + \sup_{t\in I} \|-t\phi + h\|_{L^\infty(\Omega)}$. By Theorem 3, there exists a solution $\underline{u} < 0$ in Ω of $F(D^2\underline{u}, D\underline{u}, \underline{u}, x) = M$ in Ω, $\underline{u} = 0$ on $\partial\Omega$. Hence $\underline{u}$ is a subsolution of $(\mathcal{P}_t)$ for $t \in I$, by (2).

Next, note that if u is a solution of $(\mathcal{P}_t)$ for some $t \in I$, then both functions $\psi = u$ and $\psi = 0$ are solutions of the inequality

$$F(D^2\psi, D\psi, \psi, x) \leq F(D^2\underline{u}, D\underline{u}, \underline{u}, x).$$

Since $-u^- = \min\{u, 0\}$ and the minimum of two viscosity supersolutions is a viscosity supersolution, we have

$$F(D^2(-u^-), D(-u^-), -u^-, x) \leq F(D^2\underline{u}, D\underline{u}, \underline{u}, x).$$

Observe we cannot directly infer from this inequality that $\underline{u} \leq -u^- \leq u$ since F does not satisfy a comparison principle ($\lambda_1^+(F, \Omega) < 0$). However, as we will show now, we can gain enough information on these functions in order to prove the inequality by considering their quotient instead of their difference.

By Proposition 2.4 and Lemma 2.3 we can fix k sufficiently large so that for any solution u of $(\mathcal{P}_t)$, $t \in I$, and any $x_0 \in \partial\Omega$ we have

$$\limsup_{t\searrow 0} \frac{-u^-(x_0 + t\nu)}{k\underline{u}(x_0 + t\nu)} \leq \frac{1}{4}$$

Note that $k\underline{u}$ is a subsolution of $(\mathcal{P}_t)$ for $k \geq 1$ and $t \in I$, by (2) and (3).

Fix a solution u of $(\mathcal{P}_t)$, $t \in I$. Then there exists $d > 0$ sufficiently small, so that, setting $\Omega_d = \{x \in \Omega : \mathrm{dist}(x, \partial\Omega) > d\}$, we have

$$0 < w := \frac{-u^-}{k\underline{u}} \leq \frac{1}{2} \qquad \text{in } \Omega \setminus \Omega_d.$$

The proof of Lemma 2.2 is finished with the help of the following comparison result.

Lemma 2.5. *Suppose v_1, v_2 are such that $v_1 \le 0, v_2 < 0$ in Ω, $v_2 \in E_p$, $p < \infty$,*

$$F(D^2v_1, Dv_1, v_1, x) \le F(D^2v_2, Dv_2, v_2, x) \qquad \text{in } \Omega, \tag{13}$$

$$0 < F(D^2v_2, Dv_2, v_2, x) \qquad \text{in } \Omega, \tag{14}$$

and, for some $d > 0$, $w := \dfrac{v_1}{v_2} < \dfrac{1}{2}$ in $\Omega \setminus \Omega_d$. Then $v_1 > v_2$ in Ω.

Remark. Considering the quotient rather than the difference of two functions can often be a successful technique when proving comparison results for a nonlinear operator. For fully nonlinear equations this has been used, in a different setting, for instance in [11].

Proof of Lemma 2.5. For any two vectors $p, q \in \mathbb{R}^N$ we denote the symmetric tensorial product by $p \otimes q = \frac{1}{2}(p_iq_j + p_jq_i)_{i,j=1}^N \in \mathcal{S}_N$. By replacing v_1 by wv_2 in (13) and by using (7) and the homogeneity of F we get

$$\begin{aligned} &wF(Dv_2, Dv_2, v_2, x) + v_2F(D^2w + 2\frac{Dv_2}{v_2} \otimes Dw, Dw, 0, x) \\ &= F(wDv_2, wDv_2, wv_2, x) - F(-v_2D^2w - 2Dv_2 \otimes Dw, -v_2Dw, 0, x) \\ &\le F(D^2v_1, Dv_1, v_1, x) \le F(D^2v_2, Dv_2, v_2, x), \end{aligned} \tag{15}$$

where we have used the equality

$$D^2(u_1u_2) = u_1Du_2 + 2Du_1 \otimes Du_2 + u_2Du_1,$$

valid for $u_1, u_2 \in E_p$. In case u_1 is only continuous, we use test functions in E_p to prove (15) – this is very standard, so we shall omit it.

We obtain from (15)

$$\tilde{F}(D^2(w-1), D(w-1), x) + c(x)(w-1) \ge 0 \quad \text{in } \Omega_{d/2}, \tag{16}$$

where we have set

$$\tilde{F}(M, p, x) = F(M + 2b(x) \otimes p, p, 0, x),$$

$$b(x) = \frac{Dv_2(x)}{v_2(x)} \in L^\infty(\Omega_{d/2}),$$

$$c(x) = \frac{F(D^2v_2(x), Dv_2(x), v_2(x), x)}{v_2(x)} < 0,$$

by (14). Note that $w - 1 < 0$ in a neighbourhood of $\partial\Omega_{d/2}$. Then the existence of a point in $\Omega_{d/2}$ at which $w - 1$ attains a positive maximum would contradict (16) – just test (16) with a constant function. So $w - 1 \le 0$. Finally, $w - 1 < 0$ is a consequence of the strong maximum principle. □

The following existence result is an easy consequence from the previous lemmas.

Proposition 2.5. *There exists a number t^* such that problem $(\mathcal{P}_t)$ has a solution for $t \le t^*$ and does not have a solution for $t > t^*$.*

Proof. We use the following lemma which is based on Perron's method. This type of result in the viscosity setting goes back at least to Ishii [30] – recall we assumed in the introduction that the operator $H_v[u]$ satisfies the comparison principle.

Lemma 2.6. *Suppose $u_0 \in E_N$ is a subsolution and $v_0 \in E_N$ is a supersolution of $H(D^2u, Du, u, x) = f$, where $f \in L^\infty(\Omega)$, H satisfies (4). Suppose in addition that $u_0 \le v_0$ in Ω, $u_0 \le 0$ on $\partial\Omega$, and $v_0 \ge 0$ on $\partial\Omega$. Then there exists a solution u of*

$$\begin{cases} H(D^2u, Du, u, x) = f & \text{in } \Omega \\ \qquad\qquad\qquad\;\; u = 0 & \text{on } \partial\Omega. \end{cases}$$

For a proof of this lemma see for example Lemma 4.3 in [37].

Next, set

$$t^* = \sup\{t \in \mathbb{R} : (\mathcal{P}_t) \text{ has a supersolution}\}.$$

It follows from Lemmas 2.6 and 2.2 that if for some t problem $(\mathcal{P}_t)$ has supersolution then it has a solution. It is obvious that if u is a supersolution for $(\mathcal{P}_{t_0})$ then it is also a supersolution for all $(\mathcal{P}_t)$, $t < t_0$. By Lemma 2.1 t^* is well defined and by Proposition 2.1 t^* is finite. The existence of solution for $t = t^*$ follows from a passage to the limit $t_n \to t^*$, thanks to Proposition 2.2 and Theorem 3.8 in [14]. □

Now we can move to the realization of Step 5 of the proof of Theorem 1. The argument which follows is inspired by a classical reasoning of Amann [1], [2]. We refer for instance to [15] for a systematic treatment of existence results based on degree theory.

In what follows we shall use the following global $C^{1,\alpha}$-estimate, proved in [40], [39], [41].

Theorem 4. *Suppose H satisfies (4), Ω is a $C^{1,1}$-domain and u is a solution of (1). Then there exists $\alpha, C_0 > 0$ depending on $N, \lambda, \Lambda, \gamma, \delta, \Omega$, such that $u \in C^{1,\alpha}(\Omega)$, and*

$$\|u\|_{C^{1,\alpha}(\Omega)} \le C_0 \left(\|u\|_{L^\infty(\Omega)} + \|f\|_{L^\infty(\Omega)}\right).$$

Let t_1 be such that there exists a solution $\overline{u}$ for $(\mathcal{P}_{t_1})$. Fix $t < t_1$. Then $\overline{u}$ is a strict supersolution of $(\mathcal{P}_t)$. By Lemma 2.2 there is a subsolution $\underline{u}$ of $(\mathcal{P}_t)$ such that $\underline{u} < \overline{u}$ in Ω. By the choice of $\underline{u}, \overline{u}$ and Hopf's lemma, we can also ensure that $\frac{\partial \underline{u}}{\partial \nu} < \frac{\partial \overline{u}}{\partial \nu}$ on $\partial\Omega$.

Let c_{R_0} is the constant from hypothesis (4), with

$$R_0 = \max\{\|\underline{u}\|_{L^\infty(\Omega)}, \|\overline{u}\|_{L^\infty(\Omega)}\}.$$

For any $v \in C(\overline{\Omega})$ we define $H_v(M,p,x) = H(M,p,v(x),x)$. For each $v \in C(\overline{\Omega})$ we denote with $u = K_t(v)$ the solution of the Dirichlet problem

$$\begin{cases} H_v(D^2u, Du, x) - c_{R_0}u = f(x) - c_{R_0}v & \text{in } \Omega \\ u = 0 & \text{on } \partial\Omega. \end{cases} \tag{17}$$

This problem has a unique solution, by hypothesis (5), Theorem 3, and Perron's method. By the ABP inequality K_t maps bounded sets in $C(\overline{\Omega})$ into bounded sets in $C(\overline{\Omega})$. Hence, by Proposition 2.2 and Theorem 4 (recall $C^{1,\alpha}(\Omega) \hookrightarrow C^1(\overline{\Omega})$ is compact) the map K_t sends bounded sets in $C^1(\overline{\Omega})$ into precompact sets in $C^1(\overline{\Omega})$, that is, $K_t : C^1(\overline{\Omega}) \to C^1(\overline{\Omega})$ is a compact map. Note that solutions of (1) are fixed points of K_t and vice versa.

Define

$$\mathcal{O} = \left\{ v \in C^1(\overline{\Omega}) \ : \ \underline{u} < v < \overline{u} \text{ in } \Omega \ \text{ and } \ \frac{\partial \underline{u}}{\partial \nu} < \frac{\partial v}{\partial \nu} < \frac{\partial \overline{u}}{\partial \nu} \text{ on } \partial\Omega \right\}.$$

Note the set $\mathcal{O}$ is open in $C^1(\overline{\Omega})$.

Claim. $K_t(\overline{\mathcal{O}}) \subset \mathcal{O}$. In particular, $K_t(\overline{\mathcal{O}}) \cap \partial\mathcal{O} = \emptyset$.

To prove this claim it is sufficient to show that if $\underline{u} \le v \le \overline{u}$ in Ω then $\underline{u} < K_t(v) < \overline{u}$ in Ω and $\frac{\partial \underline{u}}{\partial \nu} < \frac{\partial K_t(v)}{\partial \nu} < \frac{\partial \overline{u}}{\partial \nu}$ on $\partial\Omega$.

So let $v \in C(\overline{\Omega})$ be such that $\underline{u} \le v \le \overline{u}$ and set $u = K_t(v)$. Then we have, by (4),

$$\begin{aligned} H(D^2u, Du, \overline{u}(x), x) &= H(D^2u, Du, \overline{u}(x), x) + c_R\overline{u} - c_R\overline{u} \\ &\ge H(D^2u, Du, v(x), x) + c_R v - c_R\overline{u} \\ &= f(x) + c_R u - c_R\overline{u} \\ &> H(D^2\overline{u}, D\overline{u}, \overline{u}(x), x) + c_R(u - \overline{u}). \end{aligned}$$

This implies, again by (4),

$$\mathcal{L}^+(D^2(u-\overline{u}), D(u-\overline{u})) - c_R(u-\overline{u}) > 0$$

in Ω, and $u - \overline{u} = 0$ on $\partial\Omega$. It follows from the maximum principle for proper operators (or from Theorem 2) and from the strong maximum principle that $u < \overline{u}$ in Ω and $\frac{\partial u}{\partial \nu} < \frac{\partial \overline{u}}{\partial \nu}$ on $\partial\Omega$. In the same way we obtain the inequality for $\underline{u}$. □

To finish the proof of our main theorem we shall use the following lemma, concerning the Leray–Schauder degree of the compact map $I - K_t$. It is well known how to prove this type of result, we give a proof for completeness.

Lemma 2.7. *For any $t_0 \in (-\infty, t^*)$ there exist $R_1, R_2 \in \mathbb{R}$ such that $R_1 < R_2$ and*

$$\deg(I - K_{t_0}, \mathcal{O} \cap \mathcal{B}_{R_1}, 0) = 1 \quad \textit{and} \quad \deg(I - K_{t_0}, \mathcal{B}_{R_2}, 0) = 0, \tag{18}$$

where $\mathcal{B}_R = \{u \in C^1(\overline{\Omega}) \ : \ \|u\|_{C^1(\overline{\Omega})} < R\}$.

Proof. Let $\bar{R}$ be an upper bound (given by Theorem 4) for $C^1(\overline{\Omega})$-norms of solutions of (17) with $\|v\|_{L^\infty(\Omega)} \le R_0$. Set $R_1 = \max\{\bar{R}, \|\underline{u}\|_{C^1(\overline{\Omega})}, \|\overline{u}\|_{C^1(\overline{\Omega})}\} + 1$. To prove the first equality in (18), fix $w \in \mathcal{O} \cap \mathcal{B}_{R_1}$ and consider the compact homotopy $H(s,v) = H_s(v) = sK_{t_0}(v) + (1-s)w$, for $s \in [0,1]$, $v \in C(\overline{\Omega})$. By the choice of R_1 and the claim above we have $(I - H_s)(u) \neq 0$ for all $u \in \partial(\mathcal{O} \cap \mathcal{B}_{R_1})$ and all $s \in [0,1]$. Hence

$$\deg(I - H_1, \mathcal{O} \cap \mathcal{B}_{R_1}, 0) = \deg(I - H_0, \mathcal{O} \cap \mathcal{B}_{R_1}, 0) = 1,$$

since H_0 is a constant mapping.

By combining Proposition 2.1 with Theorem 4 we see that for each m_0 there exists an uniform bound $\tilde{C}(m_0)$ for the $C^1(\overline{\Omega})$-norms of the solutions of $(\mathcal{P}_t)$ with $t \ge m_0$. Then we take $R_2 = \max\{\tilde{C}+1, R_1+1\}$, where $\tilde{C} = \tilde{C}(t_0)$. Set $t_1 = t^* + 1$. Clearly the mapping $K(t,u) = K_t(u), t \in [t_0, t_1]$, is a compact homotopy linking K_{t_0} to K_{t_1}. Further, we have $(I - K_t)(u) \neq 0$ for all $u \in \partial\mathcal{B}_{R_2}$ and all $t \in [t_0, t_1]$, by Proposition 2.1 and the choice of R_2. Hence

$$\deg(I - K_{t_0}, \mathcal{B}_{R_2}, 0) = \deg(I - K_{t_1}, \mathcal{B}_{R_2}, 0).$$

But the last degree is zero, since K_{t_1} has no fixed points at all, by Proposition 2.5. This proves the second equality in (18). □

So, to complete the proof of the multiplicity result in Theorem 1 we can use the excision property of the degree together with Lemma 2.7, which leads to $\deg(I - K_{t_0}, \mathcal{B}_{R_2} \setminus (\overline{\mathcal{O} \cap \mathcal{B}_{R_1}}), 0) = -1$, hence problem (1) (i.e., problem $(\mathcal{P}_{t_0})$) has a second solution in $\mathcal{B}_{R_2} \setminus (\mathcal{O} \cap \mathcal{B}_{R_1})$, apart from the solution in $\mathcal{O} \cap \mathcal{B}_{R_1}$, given by Proposition 2.5.

Finally, let us show the mapping $h \to t^*(h)$ is continuous. Suppose that $h_n \rightrightarrows h$ in $\overline{\Omega}$. Set $t_n^* = t^*(h_n), t^* = t^*(h)$. Note t_n^* is bounded above, by Proposition 2.1. Furthermore, we have $t_n^* \ge t^*(-\|h\|_{L^\infty(\Omega)} - 1)$ for large n, since any solution of (1) with h replaced by $-\|h\|_{L^\infty(\Omega)} - 1$ is a supersolution of $(\mathcal{P}_{t_n^*, h_n})$. So t_n^* is bounded. Take a subsequence of t_n^* and let a be the limit of some subsequence of this subsequence (which we denote by t_n^* again). Let u_n be a solution of $(\mathcal{P}_{t_n^*, h_n})$ (we already know such a solution exists). By Proposition 2.1 $\{u_n\}$ is bounded in $L^\infty(\Omega)$. Hence, by the equation satisfied by u_n, (4) and Proposition 2.2, some subsequence of u_n converges to a solution of $(\mathcal{P}_{a,h})$. Hence $a \le t^*$.

Suppose $a < a + 3\varepsilon < t^*$, for some $\varepsilon > 0$. Let $\overline{u}$ be a positive supersolution of $(\mathcal{P}_{a+3\varepsilon, h})$ – we already know such supersolutions exist. Let w_n be the solution of the Dirichlet problem

$$\begin{cases} \mathcal{L}^+(D^2 w_n, Dw_n) = h_n - h & \text{in } \Omega \\ w_n = 0 & \text{on } \partial\Omega. \end{cases}$$

By the ABP inequality and the boundary estimate (Theorem 2 and Proposition 2.4), we have $w_n \rightrightarrows 0$ and $c_R|w_n| \le \varepsilon\phi$ in Ω for large n, where c_R is the constant from (4), with $R = \|\overline{u}\|_{L^\infty(\Omega)} + 1$.

Set $v_n = \overline{u} + w_n$. Then, by (4), if n is sufficiently large,

$$\begin{aligned} H(D^2 v_n, Dv_n, v_n, x) &\leq H(D^2 v_n, Dv_n, v_n, x) - H(D^2\overline{u}, D\overline{u}, \overline{u}, x) \\ &\quad - (a + 3\varepsilon)\phi + h \\ &\leq \mathcal{L}^+(D^2 w_n, Dw_n) + c_R w_n - (t_n^* + 2\varepsilon)\phi + h \\ &\leq -(t_n^* + \varepsilon)\phi + h_n, \end{aligned}$$

Hence v_n is a positive supersolution of $(\mathcal{P}_{t_n^*+\varepsilon, h_n})$ which implies that this problem has a solution as well (we know subsolutions always exist). This is a contradiction with the definition of t_n^*.

Acknowledgement

The author is indebted to F. Demengel for a number of remarks which lead to a substantial improvement of the paper.

References

[1] H. Amann, On the number of solutions of nonlinear equations in ordered Banach spaces, J. Funct. Anal. 14 (1973), 346–384.

[2] H. Amann, Fixed point equations and elliptic eigenvalue problems in ordered Banach spaces, SIAM Rev. 18 (1976), 620–709.

[3] A. Ambrosetti, G. Prodi, On the inversion of some differentiable mappings with singularities between Banach spaces. Ann. Mat. Pura Appl. 93 (1972), 231–246.

[4] D. Arcoya, J. Carmona, Quasilinear elliptic problems interacting with its asymptotic spectrum. Nonl. Anal. 52(6) (2003), 1591–1616.

[5] S.N. Armstrong, Principal eigenvalues and an anti-maximum principle for homogeneous fully nonlinear elliptic equations. J. Diff. Eq. 246(7) (2009), 2958–2987.

[6] S.N. Armstrong, The Dirichlet problem for the Bellman equation at resonance. J. Diff. Eq. 247(3) (2009), 931–955.

[7] D. Arcoya, D. Ruiz, The Ambrosetti–Prodi problem for the p-Laplacian, Comm. Part. Diff. Eq. 31 (2006), 841–865.

[8] M. Bardi, F. Da Lio, On the strong maximum principle for fully nonlinear degenerate elliptic equations. Arch. Math. 73 (1999), 276–285.

[9] H. Berestycki, L. Nirenberg, S.R.S. Varadhan, The principal eigenvalue and maximum principle for second order elliptic operators in general domains, Comm. Pure Appl. Math. 47(1) (1994), 47–92.

[10] M.S. Berger, E. Podolak, On the solutions of a nonlinear Dirichlet problem. Indiana Univ. Math. J. 24 (1974/75), 837–846.

[11] I. Birindelli, F. Demengel, Eigenvalue, maximum principle and regularity for fully non linear homogeneous operators. Comm. Pure Appl. Anal. 6(2) (2007), 335–366.

[12] J. Busca, B. Sirakov, Harnack type estimates for nonlinear elliptic systems and applications, Ann. Inst. H. Poincaré, Anal. Nonl. 21(5) (2004), 543–590.

[13] L.A. Caffarelli, Interior estimates for fully nonlinear elliptic equations, Ann. Math. 130 (1989), 189–213.

[14] L.A. Caffarelli, M.G. Crandall, M. Kocan, A. Świech, On viscosity solutions of fully nonlinear equations with measurable ingredients, Comm. Pure Appl. Math. 49 (1996), 365–397.

[15] K.C. Chang, Methods in nonlinear analysis. Springer Monographs in Mathematics (2005).

[16] Chang, K.C. Ambrosetti–Prodi type results in elliptic systems. Nonl. Anal. 51(4) (2002), 553–566.

[17] M.G. Crandall, M. Kocan, P.L. Lions, A. Świech, Existence results for boundary problems for uniformly elliptic and parabolic fully nonlinear equations, Elec. J. Diff. Eq. 24 (1999), 1–20.

[18] M.G. Crandall, M. Kocan, A. Świech, L^p theory for fully nonlinear uniformly parabolic equations, Comm. Part. Diff. Eq. 25(11&12) (2000), 1997–2053.

[19] M.G. Crandall, H. Ishii, P.-L. Lions, User's guide to viscosity solutions of second-order partial differential equations, Bull. Amer. Math. Soc. 27(1) (1992), 1–67.

[20] E.N. Dancer, On the ranges of certain weakly nonlinear elliptic partial differential equations. J. Math. Pures Appl. 57(4) (1978), 351–366.

[21] F. Demengel, Non uniqueness of solutions for Dirichlet problem related to fully nonlinear singular or degenerate operator, Advances in Differential equations, 14(11-12), (2009), 1107–1126.

[22] D.G. de Figueiredo, B. Sirakov, On the Ambrosetti–Prodi problem for non-variational elliptic systems, J. Diff. Eq. 240(2) (2007), 35–374

[23] D.G.de Figueiredo, S. Solimini, A variational approach to superlinear elliptic problems. Comm. Part. Diff. Eq. 9(7) (1984), 699–717.

[24] L.C. Evans, Classical solutions of fully nonlinear, convex, second-order elliptic equations, Comm. Pure Appl. Math. 25 (1982), 333–363.

[25] P. Felmer, A. Quaas, Positive solutions to "semilinear" equation involving the Pucci's operator, J. Diff. Eq. 199(2) (2004), 376–393.

[26] P. Felmer, A. Quaas, B. Sirakov, Resonance phenomena for stochastic control equations, SIAM J. Math. Anal. 42(3) (2010), 997–1024.

[27] P. Felmer, A. Quaas, B. Sirakov, Landesman–Lazer type results for second order Hamilton–Jacobi–Bellman equations. J. Funct. Anal. 258(12) (2010), 4154–4182.

[28] D. Gilbarg, N.S. Trudinger, Elliptic Partial Differential Equations of Second Order, 2nd edition, Springer Verlag.

[29] P. Hess, On a nonlinear elliptic boundary value problem of the Ambrosetti–Prodi type. Boll. Un. Mat. Ital. 17(1) (1980), 187–192.

[30] H. Ishii On uniqueness and existence of viscosity solutions of fully nonlinear second-order elliptic PDEs. Comm. Pure Appl. Math. 42(1) (1989), 15–45.

[31] H. Ishii, P.-L. Lions, Viscosity solutions of fully nonlinear second-order elliptic partial differential equations, J. Diff. Eq. 83 (1990), 26–78.

[32] R. Jensen, A. Swiech, Uniqueness and existence of maximal and minimal solutions of fully nonlinear PDE, Comm. Pure Appl. Anal. 4(1) (2005), 187–195.

[33] J. Kazdan, F.W. Warner, Remarks on some quasilinear elliptic equations. Comm. Pure Appl. Math. 28(5) (1975), 567–597.

[34] P.L. Lions, Bifurcation and optimal stochastic control, Nonl. Anal. Th. and Appl. 7(2) (1983), 177–207.

[35] P.L. Lions, Résolution analytique des problèmes de Bellman–Dirichlet, Acta Math. 146 (1981), 151–166.

[36] A. Quaas, B. Sirakov, Existence results for nonproper elliptic equations involving the Pucci operator, Comm. Part. Diff. Eq. 31 (2006), 987–1003.

[37] A. Quaas, B. Sirakov, Principal eigenvalues and the Dirichlet problem for fully nonlinear elliptic operators, to appear in Advances in Mathematics; results announced in Comptes Rendus Math. 342(2) (2006), 115–118.

[38] B. Rynne, Half-eigenvalues of elliptic operators, Proc. Royal Soc. Edinburgh 132A (2002), 1439–1451.

[39] A. Święch, $W^{1,p}$-estimates for solutions of fully nonlinear uniformly elliptic equations, Adv. Diff. Eq. 2(6) (1997), 1005–1027.

[40] N. Trudinger, On regularity and existence of viscosity solutions of nonlinear second order, elliptic equations, "PDE and the calculus of variations", Vol. II (Progr. Nonl. Diff. Eq. Appl. 2, Birkhäuser Boston) (1989), 939–957.

[41] N. Winter, $W^{2,p}$ and $W^{1,p}$-estimates at the boundary for solutions of fully nonlinear, uniformly elliptic equations, to appear in Z. Anal. Adwend. (Journal for Analysis and its Applications) 28(2) (2009), 129–164.

Boyan Sirakov
Pontificia Universidade Católica do Rio de Janeiro (PUC-Rio)
Departamento de Matemática
Rua Marques de São Vicente, 225, Gávea
Rio de Janeiro – RJ, CEP 22451-900, Brasil
e-mail: `bsirakov@mat.puc-rio.br`

Progress in Nonlinear Differential Equations
and Their Applications, Vol. 85, 423–443

Bifurcation at Isolated Eigenvalues for Some Elliptic Equations on $\mathbb{R}^N$

C.A. Stuart

Abstract. This paper concerns the bifurcation of bound states $u \in L^2(\mathbb{R}^N)$ for a class of second-order nonlinear elliptic eigenvalue problems that includes cases which are already known to exhibit some surprising behaviour. By treating a larger class of nonlinearities we cover new cases such as a situation where there is no bifurcation at a simple isolated eigenvalue lying at the bottom of the spectrum of the linearization. As an application of recent work on bifurcation for problems that are only Hadamard differentiable, we also establish bifurcation at all isolated eigenvalues of odd multiplicity which are sufficiently far from the essential spectrum.

Mathematics Subject Classification (2010). 35J61, 35P30, 47J15.

Keywords. Bifurcation, bound state, nonlinear elliptic equation, Hadamard derivative.

1. Introduction

As has already been shown in several earlier contributions [3, 15, 17], the study of bifurcation for bound states $u \in L^2(\mathbb{R}^N)$ of simple looking elliptic equations such as

$$-\Delta u + Vu + \frac{u^3}{\xi^2 + u^2} = \lambda u, \tag{1.1}$$

where $V \in L^\infty(\mathbb{R}^N)$ and $\xi \in L^2(\mathbb{R}^N) \cap C(\mathbb{R}^N)$ with $\xi > 0$ on $\mathbb{R}^N$, reveals a number of surprising phenomena. For example, there are potentials V for which bifurcation can occur at points not belonging to the spectrum of the linearized problem

$$-\Delta u + Vu = \lambda u.$$

On the other hand, as one might expect there is bifurcation at all eigenvalues of $-\Delta + V$ lying below the essential spectrum. However, it is shown in Section 5 below that this is no longer the case for the equation

$$-\Delta u + Vu - \frac{u^3}{\xi^2 + u^2} = \lambda u, \tag{1.2}$$

for some choices of V and ξ and the results in [3, 15, 17] do not apply to (1.2). The occurence of the unusual phenomena mentioned above has nothing to do with a lack of smoothness of the functions V and ξ since the conclusions are the same even if the assumption that V and ξ are infinitely differentiable is added.

The purpose of the present paper is to study bifurcation at isolated eigenvalues of the linearization for a class of equation that includes both (1.1) and (1.2), namely

$$-\Delta u + V(x)u + g(x,u) + h(x,\nabla u) + \xi(x)f(\eta(x)u) = \lambda u \tag{1.3}$$

where V, ξ and $\eta : \mathbb{R}^N \to \mathbb{R}$ and the nonlinear functions $g : \mathbb{R}^N \times \mathbb{R} \to \mathbb{R}$, $h : \mathbb{R}^N \times \mathbb{R}^N \to \mathbb{R}$ and $f : \mathbb{R} \to \mathbb{R}$ are such that $g(x,0) = h(x,0) = f(0) = 0$ and define terms of order higher than linear near $u \equiv 0$. The precise hypotheses are formulated in Section 3 and the main result is Theorem 4.1. Taking $g \equiv 0, h \equiv 0, \eta = 1/\xi$ and $f(s) = \pm s^3/(1+s^2)$, we recover (1.1), respectively (1.2). We seek solutions (λ, u) where $\lambda \in \mathbb{R}$ and $u \not\equiv 0$ lies in the usual Sobolev space $H^2(\mathbb{R}^N)$ since any distributional solution $u \in L^2(\mathbb{R}^N)$ of equations (1.1) or (1.2) lies in this space.

Under our hypotheses, the equations (1.1) to (1.3) can all be written in the form $M(u) = \lambda u$ where $M : H^2(\mathbb{R}^N) \to L^2(\mathbb{R}^N)$ is a continuous mapping such that $M(0) = 0$. However, M is not Fréchet differentiable at 0 and consequently the classical results about bifurcation cannot be applied in the cases of interest here. (See parts (2) and (3) of Theorem 3.4.) Nonetheless, M is Gâteaux differentiable at 0 and it is also Lipschitz continuous in an open neighbourhood of 0. These properties imply that M is actually Hadamard differentiable at 0. By exploiting this, new conclusions about bifurcation of bound states for (1.3) are obtained in Theorem 4.1 by using a recent abstract result about bifurcation for such problems proved in [19]. The relevant parts of the abstract theory are set out in Section 2. These results provide information about bifurcation at points which are not too close to the essential spectrum of the linearised operator $-\Delta + V$ and, for such points, the conclusions resemble those for smooth situations.

The other main contribution of this paper is to show that this restriction cannot be avoided without introducing new restrictions on the behaviour of the term $\xi f(\eta u)$ in (1.3). A situation of this kind is treated in Section 5 where we show that there may be no bifurcation at a simple eigenvalue Λ lying at the bottom of the spectrum of $-\Delta + V$ and below its essential spectrum, if Λ is too near the essential spectrum. It is important to note that all the other hypotheses of Theorem 4.1 are satisfied and yet the conclusions (ii) and (iii) fail. Thus this situation serves to show that the restriction involving the distance from the essential spectrum in the abstract result is also necessary since all the other hypotheses are satisfied there too.

Exploiting bifurcation theory for Hadamard differentiable mappings is not the only way to deal with problems like (1.1) to (1.3). Rabier [10, 11] has shown that, for an appropriate class of weights ξ and η, the equations can be treated in weighted Sobolev spaces where Fréchet differentiability of the relevant operators holds. It then follows that bifurcation occurs at every isolated eigenvalue of odd multiplicity of $-\Delta + V$. The situation discussed in Section 5 shows that there are choices of ξ and η for which this method cannot be used.

2. Bifurcation without Fréchet differentiability

For real Banach spaces X and Y,

- $B(X,Y) = \{L : X \to Y : L \text{ is linear and bounded}\}$
- $Iso(X,Y) = \{L \in B(X,Y) : L : X \to Y \text{ is an isomorphism}\}$
- $\Phi_0(X,Y) = \{L \in B(X,Y) : L \text{ is a Fredholm operator of index zero}\}$.

Let X and Y be real Banach spaces and consider the equation $F(\lambda, u) = 0$ where $F : \mathbb{R} \times X \to Y$ with $F(\lambda, 0) = 0$ for all $\lambda \in \mathbb{R}$. Setting $\mathcal{S} = \{(\lambda, u) \in \mathbb{R} \times X : F(\lambda, u) = 0 \text{ and } u \neq 0\}$, λ_0 is called a bifurcation point for the equation $F(\lambda, u) = 0$ if there exists a sequence $\{(\lambda_n, u_n)\} \subset \mathcal{S}$ such that $\lambda_n \to \lambda_0$ and $\|u_n\| \to 0$ as $n \to \infty$. There is continuous bifurcation at λ_0 if there exists a bounded connected subset $\mathcal{C}$ of $\mathcal{S}$ such that $\overline{\mathcal{C}} \cap [\mathbb{R} \times \{0\}] = \{(\lambda_0, 0)\}$. In these statements, $\mathcal{S}$ is treated as a metric space with the metric inherited from $\mathbb{R} \times X$.

In this paper, we only deal with the situation where X and Y are Hilbert spaces with $X \subset Y$ and $F(\lambda, u) = M(u) - \lambda u$ for a mapping $M : X \to Y$.

Let $(H, \langle\cdot,\cdot\rangle, \|\cdot\|)$ be a real Hilbert space. For a self-adjoint operator $S : D(S) \subset H \to H$ acting in H, the graph norm of S on $D(S)$ is defined by

$$\|u\|_S = \{\|u\|^2 + \|Su\|^2\}^{1/2} \text{ for } u \in D(S).$$

Recall that since S is closed, the graph space $(D(S), \|\cdot\|_S)$ is a Hilbert space. The following result, which is an easy consequence of the closed graph theorem (see Section 5 of [18]), provides a useful way of identifying the associated topology in concrete situations.

Proposition 2.1. *Let $S : D(S) \subset H \to H$ and $T : D(T) \subset H \to H$ be two self-adjoint operators having the same domain $X = D(S) = D(T)$. Then $\|\cdot\|_S$ and $\|\cdot\|_T$ are equivalent norms on the subspace X and $S, T \in B(X, H)$ for any of these norms.*

For a self-adjoint operator $S : D(S) \subset H \to H$, the spectrum and essential spectrum are denoted by $\sigma(S)$ and $\sigma_e(S)$, respectively. If X denotes the graph space of S then (see, for example, [2])

- $\sigma(S) = \{\lambda \in \mathbb{R} : S - \lambda I \notin Iso(X, H)\}$ and $\Lambda = \inf \sigma(S)$
- $\sigma_e(S) = \{\lambda \in \sigma(S) : S - \lambda I \notin \Phi_0(X, H)\}$ and $\Lambda_e = \inf \sigma_e(S)$
- $S - \lambda I \in \Phi_0(X, H) \Leftrightarrow \lambda \notin \sigma_e(S)$
- $\sigma_d(S) = \sigma(S) \backslash \sigma_e(S)$ consists of isolated eigenvalues of finite multiplicity.

The following result concerning bifurcation for an equation of the form $M(u) = \lambda u$ appears as Corollary 6.6 in [19]. Most of [19] is devoted to the more general equation $F(\lambda, u) = 0$ in the setting of Banach spaces.

Proposition 2.2. *Let $(Y, \langle\cdot,\cdot\rangle, \|\cdot\|)$ be a real Hilbert space and let $(X, \|\cdot\|_X)$ be the graph space of some self-adjoint operator acting in Y. For $\delta > 0$, $B_X(0,\delta) = \{u \in X : \|u\|_X < \delta\}$. Consider the equation $M(u) = \lambda u$ where the function $M : X \to Y$ has the following properties.*

(H1) $M(0) = 0$.
(H2) *M is Gâteaux differentiable at 0 and $M'(0) \in B(X, Y)$ is also a self-adjoint operator acting in Y with domain X.*
(H3) *For some $\delta > 0$, $M = M_1 + M_2$ where $M_1 \in C^1(B_X(0,\delta), Y)$ with $M_1'(0) = M'(0)$ and there exists a constant L such that $\|M_2(u) - M_2(v)\|_Y \le L\|u - v\|_Y$ for all $u, v \in B_X(0,\delta)$. Let*

$$L^Y(M_2) = \lim_{\delta\to 0} \sup_{\substack{u,v\in B_X(0,\delta)\\ u\neq v}} \frac{\|M_2(u) - M_2(v)\|_Y}{\|u - v\|_Y} < \infty.$$

Then, for λ_0 such that $d(\lambda_0, \sigma_e(M'(0))) > L^Y(M_2)$ we have the following conclusions.

(i) *If $\ker\{M'(0) - \lambda_0 I\} = \{0\}$, λ_0 is not a bifurcation point.*
(ii) *If $\dim\ker\{M'(0) - \lambda_0 I\}$ is odd, λ_0 is a bifurcation point and there is continuous bifurcation at λ_0.*
(iii) *If $\ker\{M'(0) - \lambda_0 I\} = \operatorname{span}\{\phi\}$ where $\|\phi\|_Y = 1$, λ_0 is a bifurcation point and, for any sequence $\{(\lambda_n, u_n)\} \subset S$ such that $\lambda_n \to \lambda_0$ and $\|u_n\|_X \to 0$, we have that $u_n = \langle u_n, \phi\rangle\{\phi + w_n\}$ where $\langle w_n, \phi\rangle = 0$ and $\|w_n\|_X \to 0$.*

Remark. There is an example at the end of Section 6 in [19] in which $X = Y = L^2(0,1)$ and (H1) to (H3) are satisfied with $M_1 = 0$ and $L^Y(M_2) = 1$. The mapping M is the Nemytskii operator defined by $M(u)(x) = u(x)^2/(1+|u(x)|)$ for $u \in Y$ and it is shown that the set of bifurcation points for the equation $Mu = \lambda u$ is $[-1, 1]$. Since $M'(0) = 0$, $\sigma(M'(0)) = \sigma_e(M'(0)) = 0$ and so for $\lambda_0 = 1$, we have bifurcation at a point where $\ker(M'(0) - \lambda_0 I) = \{0\}$ and $d(\lambda_0, \sigma_e(M'(0))) = L^Y(M_2)$, showing that the conclusion (i) can fail if $d(\lambda_0, \sigma_e(M'(0))) \not> L^Y(M_2)$. In Corollary 5.2 we see that parts (ii) and (iii) can also fail when (H1) to (H3) are satisfied but $d(\lambda_0, \sigma_e(M'(0))) \not> L^Y(M_2)$.

3. An elliptic equation on $\mathbb{R}^N$

In this section we present and prove our main results concerning bound states $u \in L^2(\mathbb{R}^N)$ of the equation (1.3). In the following subsections we introduce our hypotheses term by term and discuss their main consequences.

3.1. The linear term $-\Delta u + Vu$

Instead of restricting attention to bounded potentials, we deal with a larger class which allows for singularities. We suppose that the potential V belongs to the Kato-Rellich class $T_N(q)$ for some $q \geq 2$ with $q > N/2$. This means that

(V) $V = P + Q$ where $P \in L^\infty(\mathbb{R}^N)$ and $Q \in L^r(\mathbb{R}^N)$ for all $r \in [1, q]$ for some $q \geq 2$ with $q > N/2$.

Clearly (V) is satisfied when $V \in L^\infty(\mathbb{R}^N)$, but $V(x) = |x|^{-\alpha}$ is also allowed provided that $0 \leq \alpha < \min\{2, N/2\}$.

An important consequence of condition (V) is that $S = -\Delta + V : D(S) \subset L^2(\mathbb{R}^N) \to L^2(\mathbb{R}^N)$ is a self-adjoint operator with domain $D(S) = H^2(\mathbb{R}^N)$, see [1, 14, 12] for example. Furthermore, elementary Fourier analysis shows that the graph norm of $S = -\Delta$ is equivalent to the usual Sobolev norm on $H^2(\mathbb{R}^N)$, [16] for example. Then by Proposition 2.1 this is also true for $S = -\Delta + V$ whenever V satisfies the condition (V). In particular, $S \in B(H^2(\mathbb{R}^N), L^2(\mathbb{R}^N))$.

3.2. The term $g(x, u)$

The first nonlinear term in (1.3) is required to satisfy the following condition.

(G) $g : \mathbb{R}^N \times \mathbb{R} \to \mathbb{R}$ is a Carathéodory function such that, for all $x \in \mathbb{R}^N$, $g(x, 0) = 0$ and $g(x, \cdot) \in C^1(\mathbb{R})$ with

$$|\partial_s g(x, s)| \leq A\{|s|^\alpha + |s|^\beta\} \text{ for all } (x, s) \in \mathbb{R} \times \mathbb{R}^N$$

for some constant A and exponents α, β satisfying $0 < \alpha \leq \beta < \infty$ for $N \leq 4$ and $0 < \alpha \leq \beta \leq \frac{4}{N-4}$ for $N > 4$.

Theorem 3.1. *Let g satisfy* (G) *and set $G(u)(x) = g(x, u(x))$ for $u \in H^2(\mathbb{R}^N)$. Then $G \in C^1(H^2(\mathbb{R}^N), L^2(\mathbb{R}^N))$ with $DG(u)v = \partial_s g(x, u)v$ for all $u, v \in H^2(\mathbb{R}^N)$. In particular, $G(0) = 0$ and $DG(0) = 0$.*

Proof. The restrictions on α and β in condition (G) ensure that the following intervals A_N and B_N are non-empty:

for $N \leq 4$,

$$A_N = \left(0, \tfrac{\alpha}{2}\right] \cap \left(0, \tfrac{2}{N}\right] \cap \left(0, \tfrac{1}{2}\right) \quad \text{and} \quad B_N = \left(0, \tfrac{\beta}{2}\right] \cap \left(0, \tfrac{2}{N}\right] \cap \left(0, \tfrac{1}{2}\right)$$

and for $N > 4$,

$$A_N = \left[\tfrac{\alpha(N-4)}{2N}, \tfrac{\alpha}{2}\right] \cap \left(0, \tfrac{2}{N}\right] \cap \left(0, \tfrac{1}{2}\right) \quad \text{and} \quad B_N = \left[\tfrac{\beta(N-4)}{2N}, \tfrac{\beta}{2}\right] \cap \left(0, \tfrac{2}{N}\right] \cap \left(0, \tfrac{1}{2}\right).$$

Note that for $N > 4$, $A_N \cap B_N = \emptyset$ if $\alpha < \frac{\beta(N-4)}{N}$. For this reason, we decompose $\partial_s g$ in the following way.

Let $\psi \in C^\infty(\mathbb{R})$ be such that $0 \leq \psi(s) \leq 1$ for all s with $\psi(s) \equiv 1$ for $|s| \leq 1$ and $\psi(s) \equiv 0$ for $|s| \geq 2$.

Set $\gamma_1(x, s) = \psi(s)\partial_s g(x, s)$ and $\gamma_2(x, s) = \{1 - \psi(s)\}\partial_s g(x, s)$ so that $\partial_s g(x, s) = \gamma_1(x, s) + \gamma_2(x, s)$ where

$$|\gamma_1(x, s)| \leq C_1|s|^\alpha \text{ and } |\gamma_2(x, s)| \leq C_2|s|^\beta \text{ for all } (x, s) \in \mathbb{R}^N \times \mathbb{R}.$$

Noting that (G) implies that γ_1 and γ_2 are Carathéodory functions, set $\Gamma_i(u)(x) = \gamma_i(x, u(x))$ for $i = 1, 2$.

Choosing p such that $1/p \in A_N$, we have that $p > 2$, $\alpha p \geq 2$ and, for $N > 4$, αp and $2p/(p-2) \leq 2N/(N-4)$.

By the fundamental result concerning Nemytskii operators, we have that $\Gamma_1 : L^{\alpha p}(\mathbb{R}^N) \to L^p(\mathbb{R}^N)$ is a bounded continuous mapping. For $u \in L^{\alpha p}(\mathbb{R}^N)$, Hölder's inequality then shows that $T_1(u)v = \Gamma_1(u)v$ defines a bounded linear operator $T_1(u) : L^{r_1}(\mathbb{R}^N) \to L^2(\mathbb{R}^N)$ where $r_1 = \frac{2p}{p-2}$ and that

$$T_1 \in C(L^{\alpha p}(\mathbb{R}^N), B(L^{r_1}(\mathbb{R}^N), L^2(\mathbb{R}^N))).$$

Recalling that $H^2(\mathbb{R}^N)$ is continuously embedded in $L^t(\mathbb{R}^N)$ for $2 \leq t < \infty$ if $N \leq 4$ and for $2 \leq t \leq 2N/(N-4)$ for $N > 4$, this implies that $T_1 \in C(H^2(\mathbb{R}^N), B(H^2(\mathbb{R}^N), L^2(\mathbb{R}^N)))$.

Choosing q such that $1/q \in B_N$, the same arguments show that $\Gamma_2 \in C(L^{\beta q}(\mathbb{R}^N), L^q(\mathbb{R}^N))$ and $T_2 \in C(L^{\beta q}(\mathbb{R}^N), B(L^{r_2}(\mathbb{R}^N), L^2(\mathbb{R}^N)))$ where $T_2(u) = \Gamma_2(u)v$ and $r_2 = \frac{2q}{q-2}$. Hence $T_2 \in C(H^2(\mathbb{R}^N), B(H^2(\mathbb{R}^N), L^2(\mathbb{R}^N)))$.

We now have that $T = T_1 + T_2 \in C(H^2(\mathbb{R}^N), B(H^2(\mathbb{R}^N), L^2(\mathbb{R}^N)))$ where $T(u)v(x) = \partial_s g(x, u(x))v(x)$ for $u, v \in H^2(\mathbb{R}^N)$.

From condition (G), it follows that $|g(x,s)| \leq \frac{A}{\alpha+1}\{|s|^{\alpha+1} + |s|^{\beta+1}\}$ where $1 < \alpha + 1 \leq \beta + 1 < \infty$ for $N \leq 4$ and $1 < \alpha + 1 \leq \beta + 1 \leq \frac{N}{N-4}$ for $N > 4$. Also $g = g_1 + g_2$ where

$$|g_1(x,s)| = |\psi(s)g(x,s)| \leq K_1|s|^{\alpha+1}$$

and

$$|g_2(x,s)| = |\{1 - \psi(s)\}g(x,s)| \leq K_2|s|^{\beta+1},$$

so we have that $G_1 \in C(L^{(\alpha+1)2}(\mathbb{R}^N), L^2(\mathbb{R}^N))$ and $G_2 \in C(L^{(\beta+1)2}(\mathbb{R}^N), L^2(\mathbb{R}^N))$ where $G_1(u)(x) = g_1(x, u(x))$ and $G_2(u)(x) = g_2(x, u(x))$. Hence G_1 and $G_2 \in C(H^2(\mathbb{R}^N), L^2(\mathbb{R}^N))$ and therefore $G = G_1 + G_2 \in C(H^2(\mathbb{R}^N), L^2(\mathbb{R}^N))$.

Now we show that $G : H^2(\mathbb{R}^N) \to L^2(\mathbb{R}^N)$ is Fréchet differentiable at u with $DG(u) = T(u)$ where $T = T_1 + T_2$, so that $DG(u)v = \partial_s g(x,u)v$. For $u, v \in H^2(\mathbb{R}^N)$,

$$\begin{aligned}
&\int_{\mathbb{R}^N} \{G(u+v) - G(u) - T(u)v\}^2 dx \\
&= \int_{\mathbb{R}^N} \left\{\int_0^1 \frac{d}{dt} g(x, u+tv)\, dt - \partial_s g(x,u)v\right\}^2 dx \\
&= \int_{\mathbb{R}^N} \left\{\int_0^1 \partial_s g(x, u+tv) - \partial_s g(x,u) dt\, v\right\}^2 dx \\
&\leq \int_{\mathbb{R}^N} \int_0^1 \{\partial_s g(x, u+tv) - \partial_s g(x,u)\}^2 dt\, v^2 dx \\
&= \int_0^1 \|[T(u+tv) - T(u)]v\|_{L^2}^2 dt \leq \int_0^1 \|T(u+tv) - T(u)\|_{B(H^2,L^2)}^2 dt \|v\|_{H^2}^2.
\end{aligned}$$

Since $T \in C(H^2(\mathbb{R}^N), B(H^2(\mathbb{R}^N), L^2(\mathbb{R}^N)))$, for all $\varepsilon > 0$, there exists $\delta > 0$ such that $\|T(u+w) - T(u)\|^2_{B(H^2,L^2)} < \varepsilon$ whenever $\|w\|_{H^2} < \delta$. Hence

$$\int_0^1 \|T(u+tv) - T(u)\|^2_{B(H^2,L^2)} dt \le \varepsilon \text{ when } \|v\|_{H^2} < \delta,$$

proving the Fréchet differentiability of $G : H^2(\mathbb{R}^N) \to L^2(\mathbb{R}^N)$ at u with $DG(u) = T(u)$. Hence $G \in C^1(H^2(\mathbb{R}^N), L^2(\mathbb{R}^N))$. □

3.3. The term $h(x, \nabla u)$

The nonlinear function of the gradient in (1.3) is required to satisfy the following conditions.

(H) $h : \mathbb{R}^N \times \mathbb{R}^N \to \mathbb{R}$ is a Carathéodory function such that, for all $x \in \mathbb{R}^N$, $h(x, 0) = 0$ and $h(x, \cdot) \in C^1(\mathbb{R}^N)$ with

$$|\nabla_\xi h(x, \xi)| \le A\{|\xi|^\alpha + |\xi|^\beta\} \text{ for all } x, \xi \in \mathbb{R}^N$$

for some constant A and exponents α, β satisfying $0 < \alpha \le \beta < \infty$ for $N \le 2$ and $0 < \alpha \le \beta \le \frac{2}{N-2}$ for $N > 2$.

Theorem 3.2. *Let h satisfy* (H) *and set $H(u)(x) = h(x, \nabla u(x))$ for $u \in H^2(\mathbb{R}^N)$. Then $H \in C^1(H^2(\mathbb{R}^N), L^2(\mathbb{R}^N))$ with $DH(u)v = \nabla_\xi h(x, \nabla u) \cdot \nabla v$ for all $u, v \in H^2(\mathbb{R}^N)$. In particular, $H(0) = 0$ and $DH(0) = 0$.*

Proof. Setting $Ju = \nabla u$, we have that $J \in B(H^2(\mathbb{R}^N), [H^1(\mathbb{R}^N)]^N)$ and $H(u) = N(Ju)$ where $N : [H^1(\mathbb{R}^N)]^N \to L^2(\mathbb{R}^N)$ is defined by $N(w)(x) = h(x, w(x))$ for $w \in W = [H^1(\mathbb{R}^N)]^N$. Hence it is enough to prove that $N \in C^1(W, L^2(\mathbb{R}^N))$ with $DN(w)z = \nabla_\xi h(x, w) \cdot z$ for all $w, z \in W$. This can be done by following the same approach as was used to prove Theorem 3.1 so we need only mention a few crucial points. First of all, $\nabla_\xi h$ is decomposed using a radial cut-off function ψ. For the continuity of the resulting Nemytskii operators from $[L^{\alpha p}(\mathbb{R}^N)]^N$ into $L^p(\mathbb{R}^N)$, see [8] for example. Recall that $H^1(\mathbb{R}^N)$ is continuously embedded in $L^t(\mathbb{R}^N)$ for $2 \le t < \infty$ if $N \le 2$ and for $2 \le t \le 2N/(N-2)$ for $N > 2$. In the present case, the intervals A_N and B_N are given by

$$A_N = (0, \tfrac{\alpha}{2}] \cap (0, \tfrac{1}{N}] \cap (0, \tfrac{1}{2}) \quad \text{and} \quad B_N = \left(0, \tfrac{\beta}{2}\right] \cap (0, \tfrac{1}{N}] \cap (0, \tfrac{1}{2})$$

for $N \le 2$ and

$$A_N = \left[\tfrac{\alpha(N-2)}{2N}, \tfrac{\alpha}{2}\right] \cap (0, \tfrac{1}{N}] \cap (0, \tfrac{1}{2}) \quad \text{and} \quad B_N = \left[\tfrac{\beta(N-2)}{2N}, \tfrac{\beta}{2}\right] \cap (0, \tfrac{1}{N}] \cap (0, \tfrac{1}{2})$$

for $N > 2$. The restrictions on α and β in (H) ensure that these intervals are non-empty. □

3.4. The term $\xi(x)f(\eta(x)u)$

We now come to the term in (1.3) which is not Fréchet differentiable in many interesting cases. The following basic assumption is assumed to hold throughout the discussion and is sufficient for our main result about bifurcation

(F) (i) $f \in C(\mathbb{R})$ with $\lim_{s\to 0} \frac{f(s)}{s} = 0$ and $|f(s) - f(t)| \leq \ell|s-t|$ for all $s, t \in \mathbb{R}$.
(ii) ξ and η are real-valued measurable functions on $\mathbb{R}^N$ such that $\xi\eta \in L^\infty(\mathbb{R}^N)$.

Under the hypothesis (F), we have that $|\xi(x)f(\eta(x)s)| \leq |\xi(x)|\ell|\eta(x)s| \leq \ell\|\xi\eta\|_{L^\infty}|s|$ for all $x \in \mathbb{R}^N$ and $s \in \mathbb{R}$. Setting $F(u)(x) = \xi(x)f(\eta(x)u(x))$ for $u \in L^2(\mathbb{R}^N)$, it follows that $F(u) \in L^2(\mathbb{R}^N)$ and then in the same way, that

$$F(0) = 0 \text{ and } \|F(u) - F(v)\|_{L^2} \leq \ell\|\xi\eta\|_{L^\infty}\|u - v\|_{L^2} \text{ for all } u, v \in L^2(\mathbb{R}^N).$$

Theorem 3.3. *Let the condition* (F) *be satisfied.*

Then $F : L^2(\mathbb{R}^N) \to L^2(\mathbb{R}^N)$ is Gâteaux differentiable at 0 *with $F'(0) = 0$ and it is also Lipschitz continuous with Lipschitz constant $\ell\|\xi\eta\|_{L^\infty}$. Hence $F : L^2(\mathbb{R}^N) \to L^2(\mathbb{R}^N)$ is also Hadamard differentiable at* 0.

'A fortiori', the same conclusions hold for $F : H^2(\mathbb{R}^N) \to L^2(\mathbb{R}^N)$.

Proof. For $v \in L^2(\mathbb{R}^N)$ and $t \in \mathbb{R}$, we have that $|F(tv)(x)| \leq \ell\|\xi\eta\|_{L^\infty}|tv(x)|$ and the dominated convergence theorem shows that

$$\left\|\frac{F(tv)}{t}\right\|_{L^2} \to 0 \text{ as } t \to 0.$$

This proves that $F : L^2(\mathbb{R}^N) \to L^2(\mathbb{R}^N)$ is Gâteaux differentiable at 0 with $F'(0) = 0$.

The Lipschitz continuity of F is already established in the remarks following (F). □

The main result about bifurcation only requires F to satisfy the condition (F). As we now show, additional restrictions are required in order to obtain properties of $F : H^2(\mathbb{R}^N) \to L^2(\mathbb{R}^N)$ such as Fréchet differentiability at 0 or compactness. To facilitate the discussion of these results we formulate some extra properties of the weights ξ and η.

(W1) For some $R > 0$, $\eta \in C^2(|x| > R)$ with

$$\eta(x) > 0 \text{ and } \partial^\alpha\eta/\eta \in L^\infty(|x| > R) \text{ for all multi-indices with } |\alpha| \leq 2.$$

Furthermore

$$\frac{1}{\eta} \in L^2(|x| > R) \quad \text{and} \quad \liminf_{n\to\infty} \int_{|x|>n+1} \xi(x)^2 dx \Big/ \int_{|x|>n} \frac{1}{\eta(x)^2} dx > 0.$$

Here are some typical examples of weights satisfying (F) and (W1)

Examples

(i) For some $R, K > 0$, $\eta(x) = |x|^t$ where $t > N/2$ and $|\xi(x)| \geq K|x|^{-t}$ for $|x| > R$. Note that by (F) we must also have that $|\xi(x)| \leq C|x|^{-t}$ for $|x| > R$.

(ii) For some $R, K > 0$, $\eta(x) = e^{c|x|}$ where $c > 0$ and $|\xi(x)| \geq Ke^{-c|x|}$ for $|x| > R$. By (F) we also have that $|\xi(x)| \leq Ce^{-c|x|}$ for $|x| > R$.

(W2) For some $R > 0$, $\xi \in L^2(|x| > R)$ and there exists $\delta > 0$ such that $|\xi(x)\eta(x)| \geq \delta$ a.e. on $\{x \in \mathbb{R}^N : |x| > R\}$.

Theorem 3.4. *Let the condition* (F) *be satisfied.*

(1) *If* (W1) *is also satisfied, then* $F : H^2(\mathbb{R}^N) \to L^2(\mathbb{R}^N)$ *is Fréchet differentiable at* 0 *if and only if* $f \equiv 0$.

(2) *If* (W2) *is satisfied and* $\lim_{|s|\to\infty} \frac{f(s)}{s} = A \in \mathbb{R}$ *exists, then* $F : H^2(\mathbb{R}^N) \to L^2(\mathbb{R}^N)$ *is compact if and only if* $A = 0$. *If* $A \neq 0$, *then* $F : H^2(\mathbb{R}^N) \to L^2(\mathbb{R}^N)$ *is not Fréchet differentiable at* 0.

(3) *If* $\lim_{|s|\to\infty} \frac{f(s)}{s} = B \in \mathbb{R}$ *and* $|x|^{-N/2}\eta(x) \to \infty$ *as* $|x| \to \infty$, *then* $L(F) \geq |B| \liminf_{|x|\to\infty} |\xi\eta(x)|$, *where* $L(F)$ *is the Lipschitz modulus of* $F : H^2(\mathbb{R}^N) \to L^2(\mathbb{R}^N)$ *defined by*

$$L(F) = \lim_{\delta\to 0} \sup_{\substack{u,v\in B_{H^2}(0,\delta)\\ u\neq v}} \frac{\|F(u) - F(v)\|_{L^2}}{\|u - v\|_{H^2}} < \infty$$

and 'a fortiori', we have the same lower bound for the L^2*-Lipschitz modulus*

$$L^{L^2}(F) = \lim_{\delta\to 0} \sup_{\substack{u,v\in B_{H^2}(0,\delta)\\ u\neq v}} \frac{\|F(u) - F(v)\|_{L^2}}{\|u - v\|_{L^2}} < \infty,$$

which will be used in applying Proposition 2.2 *to* (1.3).

Remarks. The hypotheses (F)(ii) and (W1) imply that $\xi \in L^2(|x| > R)$ for some $R > 0$. As the proof shows, in part (2) the property $\lim_{s\to 0} \frac{f(s)}{s} = 0$ in (F)(i) can be weakened to $f(0) = 0$.

Combining Theorem 3.3 and part (3) we see that, if (F) holds with

$$\lim_{|s|\to\infty} \frac{f(s)}{s} = \pm\ell, \quad \lim_{|x|\to\infty} |x|^{-N/2}\eta(x) = \infty$$

and

$$\liminf_{|x|\to\infty} |\xi\eta(x)| = \|\xi\eta\|_{L^\infty}, \quad \text{then} \quad L(F) = L^{L^2}(F) = \ell\|\xi\eta\|_{L^\infty}$$

for $F : H^2(\mathbb{R}^N) \to L^2(\mathbb{R}^N)$.

Proof. (1) It follows from Theorem 3.3 that, if F is Fréchet differentiable at 0, then $F'(0) = 0$. Suppose that there exists $T \neq 0$ such that $f(T) \neq 0$. It suffices to show that there exists a sequence $\{u_n\} \subset H^2(\mathbb{R}^N)\backslash\{0\}$ such that $\|u_n\|_{H^2} \to 0$ and $\|F(u_n)\|_{L^2}/\|u_n\|_{H^2} \not\to 0$. We construct such a sequence as follows.

Let $\varphi \in C^\infty(\mathbb{R})$ have the following properties:

$$\varphi(s) = 0 \text{ for } s \leq 0, 0 \leq \varphi(s) \leq 1 \text{ for } 0 < s < 1, \varphi(s) = 1 \text{ for } s \geq 1.$$

For $n > R$, where R is the radius given in (W1), set $u_n(x) = \frac{T\varphi(|x|-n)}{\eta(x)}$ for $|x| \geq n$ and $u_n(x) = 0$ for $|x| \leq n$. Then $u_n \in C^2(\mathbb{R}^N)$ and, for $|x| = r > n$,

$$\partial_i u_n(x) = \frac{T}{\eta(x)} \left\{ \frac{\varphi'(r-n)x_i}{r} - \frac{\varphi(r-n)\partial_i \eta(x)}{\eta(x)} \right\}$$

and

$$\begin{aligned}\partial^2_{ij} u_n(x) = &- \frac{T\partial_i \eta(x)}{\eta(x)^2} \left\{ \frac{\varphi'(r-n)x_i}{r} - \frac{\varphi(r-n)\partial_i \eta(x)}{\eta(x)} \right\} \\ &+ \frac{T}{\eta(x)} \left\{ \frac{\varphi''(r-n)x_i x_j}{r^2} + \frac{\varphi'(r-n)\delta_{ij}}{r} - \frac{\varphi'(r-n)x_i x_j}{r^3} \right. \\ &\left. - \frac{\varphi'(r-n)x_i \partial_i \eta(x)}{r\eta(x)} - \frac{\varphi(r-n)\partial^2_{ij}\eta(x)}{\eta(x)} + \frac{\varphi(r-n)\partial_i \eta(x)\partial_j \eta(x)}{\eta(x)^2} \right\}.\end{aligned}$$

Using (W1), these formulae show that there is a constant C such that, for $|\alpha| \leq 2$, $|\partial^\alpha u_n(x)| \leq C|\frac{1}{\eta(x)}|$ for $|x| > R$ and all $n > R$. Therefore it follows from (W1) that $u_n \in H^2(\mathbb{R}^N)$ and there is a constant C such that $\|u_n\|_{H^2} \leq C\|\frac{1}{\eta}\|_{L^2(|x|>n)}$. Hence $\|u_n\|_{H^2} \to 0$ as $n \to \infty$. Furthermore,

$$\|F(u_n)\|^2_{L^2} = \int_{\mathbb{R}^N} \xi^2 f(\eta u_n)^2 dx \geq \int_{|x|>n+1} \xi(x)^2 f(T)^2 dx$$

and so

$$\frac{\|F(u_n)\|^2_{L^2}}{\|u_n\|^2_{H^2}} \geq f(T)^2 \frac{\int_{|x|>n+1} \xi(x)^2 dx}{C^2 \|\frac{1}{\eta}\|^2_{L^2(|x|>n)}}.$$

Thus $\liminf_{n\to\infty} \frac{\|F(u_n)\|_{L^2}}{\|u_n\|_{H^2}} > 0$ by (W1) and $F : H^2(\mathbb{R}^N) \to L^2(\mathbb{R}^N)$ is not Fréchet differentiable at 0.

(2) Suppose first that $A = 0$. Then, for every $\varepsilon > 0$, there exists $A_\varepsilon > 0$ such that $|f(s)| \leq A_\varepsilon + \varepsilon|s|$ for all $s \in \mathbb{R}$. Let $\{u_n\}$ be a bounded sequence in $H^2(\mathbb{R}^N)$. Passing to subsequence, we can suppose that $u_n \rightharpoonup u$ weakly in $H^2(\mathbb{R}^N)$ and we now show that $\|F(u_n) - F(u)\|_{L^2} \to 0$, which establishes the compactness of $F : H^2(\mathbb{R}^N) \to L^2(\mathbb{R}^N)$.

For any $r > R$,

$$\int_{|x|>r} F(u_n)^2 dx \leq 2\int_{|x|>r} \xi^2\{A^2_\varepsilon + \varepsilon^2\eta^2 u_n^2\}dx \leq 2A^2_\varepsilon \int_{|x|>r} \xi^2 dx + 2\varepsilon^2\|\xi\eta\|^2_{L^\infty} M^2$$

where $\|u_n\|_{L^2} \leq M$ for all n, and the same estimate hold for $\int_{|x|>r} F(u)^2 dx$. On the other hand,

$$\int_{|x|\leq r} \{F(u_n)-F(u)\}^2 dx \leq \int_{|x|\leq r} \xi^2\ell^2\eta^2(u_n-u)^2 dx \leq \ell^2\|\xi\eta\|^2_{L^\infty} \int_{|x|\leq r} (u_n-u)^2 dx$$

and hence

$$\int_{\mathbb{R}^N} \{F(u_n) - F(u)\}^2 dx$$
$$\le 2\int_{|x|>r} F(u_n)^2 dx + 2\int_{|x|>r} F(u)^2 dx + \int_{|x|\le r} \{F(u_n) - F(u)\}^2 dx$$
$$\le 8A_\varepsilon^2 \int_{|x|>r} \xi^2 dx + 8\varepsilon^2 \|\xi\eta\|_{L^\infty}^2 M^2 + \ell^2 \|\xi\eta\|_{L^\infty}^2 \int_{|x|\le r} (u_n - u)^2 dx.$$

Since $H^2(|x| < r)$ is compactly embedded in $L^2(|x| < r)$, this shows that

$$\limsup_{n\to\infty} \|F(u_n) - F(u)\|_{L^2}^2 \le 8A_\varepsilon^2 \int_{|x|>r} \xi^2 dx + 8\varepsilon^2 \|\xi\eta\|_{L^\infty}^2 M^2.$$

But $\int_{|x|>r} \xi^2 dx \to 0$ as $r \to \infty$, so

$$\limsup_{n\to\infty} \|F(u_n) - F(u)\|_{L^2}^2 \le 8\varepsilon^2 \|\xi\eta\|_{L^\infty}^2 M^2 \text{ for all } \varepsilon > 0,$$

proving that $\|F(u_n) - F(u)\|_{L^2} \to 0$ as required.

Suppose now that $A \ne 0$. Setting $g(s) = f(s) - As$ and then $G(u) = \xi g(\eta u)$, the preceding argument shows that $G : H^2(\mathbb{R}^N) \to L^2(\mathbb{R}^N)$ is compact. Choose $w \in C_0^\infty(\mathbb{R}^N)$ such that $w \not\equiv 0$ and $\operatorname{supp} w \subset B(0, 1/2)$. Consider the sequence defined by $u_n(x) = w(x - ne_1)$ where $e_1 = (1, 0, \ldots, 0) \in \mathbb{R}^N$. It is easily seen that $u_n \rightharpoonup 0$ weakly in $H^2(\mathbb{R}^N)$ and hence, by the argument just used to prove the compactness of F when $A = 0$, we have $\|G(u_n)\|_{L^2} \to 0$ since $G(0) = 0$. But, for $m, n \ge R + 1$ and $m \ne n$,

$$\|F(u_n) - F(u_m)\|_{L^2} \ge \|A\xi\eta(u_n - u_m)\|_{L^2} - \|G(u_n) - G(u_m)\|_{L^2}$$

where

$$\|A\xi\eta(u_n - u_m)\|_{L^2}^2 \ge A^2\delta^2 \int_{\mathbb{R}^N} (u_n - u_m)^2 dx = 2A^2\delta^2 \int_{\mathbb{R}^N} w^2 dx$$

since $\operatorname{supp} u_n \cap \operatorname{supp} u_m = \emptyset$ and

$$\|G(u_n) - G(u_m)\|_{L^2} \to 0 \text{ as } n, m \to \infty.$$

Thus $\{F(u_n)\}$ has no convergent subsequence and consequently, $F : H^2(\mathbb{R}^N) \to L^2(\mathbb{R}^N)$ is not compact.

Furthermore, since $G(u) = \xi g(\eta u) = F(u) - A\xi\eta u$, it follows from Theorem 3.3 that $G : H^2(\mathbb{R}^N) \to L^2(\mathbb{R}^N)$ is Hadamard differentiable at 0 with $G'(0)u = -A\xi\eta u$. But we have just shown that $\|G'(0)(u_n - u_m)\|_{L^2} \ge \sqrt{2}A\delta\|w\|_{L^2} > 0$ for all $m, n \ge R + 1$ and $m \ne n$, from which it follows that $G'(0) : H^2(\mathbb{R}^N) \to L^2(\mathbb{R}^N)$ is not a compact linear operator. Since the Fréchet derivative of a compact operator is always compact (see [9], for example), this implies that $G : H^2(\mathbb{R}^N) \to L^2(\mathbb{R}^N)$ is not Fréchet differentiable at 0. But the bounded linear operator $u \mapsto A\xi\eta u$ is Fréchet differentiable from $H^2(\mathbb{R}^N)$ to $L^2(\mathbb{R}^N)$. Hence $F : H^2(\mathbb{R}^N) \to L^2(\mathbb{R}^N)$ cannot be Fréchet differentiable at 0.

(3) Choose some $u \in C_0^\infty(\mathbb{R}^N)$ such that $u \geq 0$ on $\mathbb{R}^N$ and $\|u\|_{L^2} = 1$. Then, for $\delta > 0$ and $n \in \mathbb{N}$, define u_n^δ by

$$u_n^\delta(x) = \frac{\delta}{2n^{N/2}} u\left(\frac{x - ne_1}{n}\right) \text{ where } e_1 = (1, 0, \ldots, 0).$$

Then, using the change of variable $z = \frac{x-ne_1}{n}$ in the integrals, we have that

$$\|u_n^\delta\|_{L^2} = \frac{\delta}{2}\|u\|_{L^2}, \|\partial_i u_n^\delta\|_{L^2} = \frac{\delta}{2n}\|\partial_i u\|_{L^2}, \|\partial_i\partial_j u_n^\delta\|_{L^2} = \frac{\delta}{2n^2}\|\partial_i\partial_j u\|_{L^2}$$

for $1 \leq i, j \leq N$. Hence $u_n^\delta \in B_{L^2}(0, \delta)$ for all n and there exists n_0 such that $u_n^\delta \in B_{H^2}(0, \delta)$ for all $n \geq n_0$. Also

$$\begin{aligned}\|F(u_n^\delta)\|_{L^2}^2 &= \int_{\mathbb{R}^N} \xi(x)^2 f\left(\eta(x)\frac{\delta}{2n^{N/2}} u\left(\frac{x - ne_1}{n}\right)\right)^2 dx \\ &= \int_{\mathbb{R}^N} \xi(n[z + e_1])^2 f\left(\eta(n[z + e_1])\frac{\delta}{2n^{N/2}} u(z)\right)^2 n^N dz \\ &= \int_{\{z:u(z)>0\}} \xi(n[z + e_1])^2 \left\{\frac{f(w_n(z))}{w_n(z)}\right\}^2 \eta(n[z + e_1])^2 \left\{\frac{\delta u(z)}{2}\right\}^2 dz\end{aligned}$$

where

$$w_n(z) \equiv \eta(n[z + e_1])\frac{\delta}{2n^{N/2}} u(z) \to \infty \text{ as } n \to \infty$$

for all $z \neq -e_1$ such that $u(z) > 0$. Hence

$$\begin{aligned}&\liminf_{n\to\infty} \xi(n[z + e_1])^2 \left\{\frac{f(w_n(z))}{w_n(z)}\right\}^2 \eta(n[z + e_1])^2 \left\{\frac{\delta u(z)}{2}\right\}^2 \\ &\geq B^2 \left\{\frac{\delta u(z)}{2}\right\}^2 \liminf_{|x|\to\infty} (\xi\eta)^2(x)\end{aligned}$$

for almost all $z \in \mathbb{R}^N$. By Fatou's Lemma,

$$\liminf_{n\to\infty} \|F(u_n^\delta)\|_{L^2}^2 \geq \left[\frac{B\delta}{2}\right]^2 \liminf_{|x|\to\infty} [\xi\eta(x)]^2 \int_{\mathbb{R}^N} u(z)^2 dz$$

and hence

$$\liminf_{n\to\infty} \|F(u_n^\delta)\|_{L^2} \geq \frac{|B|\delta}{2}\|u\|_{L^2} \liminf_{|x|\to\infty} |\xi\eta(x)|.$$

For all $\delta > 0$, this implies that

$$\begin{aligned}\sup_{\substack{u,v\in B_{H^2}(0,\delta)\\ u\neq v}} \frac{\|F(u) - F(v)\|_{L^2}}{\|u - v\|_{H^2}} &\geq \sup_{n\geq n_0} \frac{\|F(u_n^\delta)\|_{L^2}}{\|u_n^\delta\|_{H^2}} \geq \liminf_{n\to\infty} \frac{\|F(u_n^\delta)\|_{L^2}}{\|u_n^\delta\|_{H^2}} \\ &= \liminf_{n\to\infty} \frac{\|F(u_n^\delta)\|_{L^2}}{\|u_n^\delta\|_{L^2}} \frac{\|u_n^\delta\|_{L^2}}{\|u_n^\delta\|_{H^2}} \geq |B| \liminf_{|x|\to\infty} |\xi\eta(x)|,\end{aligned}$$

since $\|u_n^\delta\|_{L^2} = \frac{\delta}{2}\|u\|_{L^2}$ and $\frac{\|u_n^\delta\|_{L^2}}{\|u_n^\delta\|_{H^2}} \to 1$ as $n \to \infty$.

Thus $L(F) \geq |B| \liminf_{|x|\to\infty} |\xi\eta(x)|$ for $F : H^2(\mathbb{R}^N) \to L^2(\mathbb{R}^N)$. □

4. Results about bifurcation for (1.3)

Under the hypotheses (V), (G), (H) and (F), we can now treat (1.3) as a special case of Proposition 2.2 and hence of the more general Theorem 6.3 in [19]. For this we choose

$$X = H^2(\mathbb{R}^N), Y = L^2(\mathbb{R}^N) \text{ and } M = S + G + H + F : X \to Y$$

where S, G, H and F are defined in Sections 3.1 to 3.4.

In Subsection 3.1, we have already noted that $S = -\Delta + V : X \subset Y \to Y$ is self-adjoint and that its graph norm on X is equivalent to the usual Sobolev norm for $H^2(\mathbb{R}^N)$.

It follows from Theorems 3.1 to 3.3 that $M \in C(X, Y)$ with $M(0) = 0$ and that $M : X \to Y$ is Gâteaux differentiable at 0 with $M'(0)u = Su$ for all $u \in X$. Setting $M_1 = S + G + H$ and $M_2 = F$, these results also show that $M_1 \in C^1(X, Y)$ with $M_1'(0) = S$ and that $L^Y(M_2) \leq \ell\|\xi\eta\|_{L^\infty}$. Thus the conditions (H1) to (H3) of Proposition 2.2 are satisfied. Furthermore, setting

$$F(\lambda, u) = M(u) - \lambda u \text{ for } (\lambda, u) \in \mathbb{R} \times X,$$

we have that conditions (B1) to (B5) of [19] are satisfied for all $\lambda_0 \notin \sigma_e(S)$. In particular, for $\lambda_0 \notin \sigma_e(S)$, Proposition 2.2 can be applied to (1.3) provided that $d(\lambda_0, \sigma_e(S)) > \ell\|\xi\eta\|_{L^\infty}$. In Section 5 we shall provide examples, see Corollary 5.2 in particular, with $G = H = 0$, where $0 < d(\lambda_0, \sigma(S)) < L^Y(M_2) = \ell\|\xi\eta\|_{L^\infty}$ and the conclusions of Proposition 2.2 fail. These examples also show that the hypothesis (6.1) in Theorem 6.3 of [19] plays an essential role since the other assumptions are satisfied yet the conclusion (6.1) fails.

In the present context, λ_0 is a bifurcation point for the equation $M(u) = \lambda u$ if and only if there exists a sequence $\{(\lambda_n, u_n)\} \subset \mathbb{R} \times H^2(\mathbb{R}^N)$ of solutions of (1.3) with $u_n \not\equiv 0$ such that $\lambda_n \to \lambda_0$ and $\|u_n\|_{H^2} \to 0$.

From the preceding remarks, as an immediate consequence of Proposition 2.2 we obtain the following result. By the methods used in Section 3.2 and 3.3 a nonlinearity of the form $k(x, u(x), \nabla u(x))$ could be treated instead of the separated case $g(x, u) + h(x, \nabla u)$ adopted here.

Theorem 4.1. *Consider the equation* (1.3) *under the hypotheses* (V), (G), (H) *and* (F) *and* λ_0 *such that* $d(\lambda_0, \sigma_e(S)) > \ell\|\xi\eta\|_{L^\infty}$ *where* $S = -\Delta + V$.

(i) *If* $\ker\{S - \lambda_0 I\} = \{0\}$, *then* λ_0 *is not a bifurcation point.*
(ii) *If* $\dim\ker\{S - \lambda_0 I\}$ *is odd, then there is continuous bifurcation at* λ_0.
(iii) *If* $\ker\{S - \lambda_0 I\} = \operatorname{span}\{\phi\}$ *where* $\|\phi\| = 1$ *there is continuous bifurcation at* λ_0 *and, for any sequence* $\{(\lambda_n, u_n)\} \subset \mathbb{R} \times H^2(\mathbb{R}^N)$ *of solutions of* (1.3) *with* $u_n \not\equiv 0$ *such that* $\lambda_n \to \lambda_0$ *and* $\|u_n\|_{H^2} \to 0$, *we have that* $u_n = \langle u_n, \phi\rangle_{L^2}\{\phi + w_n\}$ *where* $\langle w_n, \phi\rangle_{L^2} = 0$ *and* $\|w_n\|_{H^2} \to 0$.

Remarks. If $f \equiv 0$, the result applies to all points $\lambda_0 \notin \sigma_e(S)$ and the conclusions follow from standard bifurcation theory since $M \in C^1(X, Y)$. For $f \not\equiv 0$, previous work deals with the case $g \equiv h \equiv 0$ under much more restrictive assumptions the term F. The following proposition summarises most of the earlier contributions. Its hypotheses imply that (V), (G), (H) and (F) are all satisfied with $G = H = 0$ and $\eta = 1/\xi$. Hence in Proposition 4.2 we are discussing bifurcation for a special case of (1.3), which includes (1.1) but not (1.2). As pointed out in [17], any distributional solution $u \in L^2(\mathbb{R}^N)$ lies in $W^{2,p}(\mathbb{R}^N)$ for all $p \in [2,\infty)$ and bifurcation for (4.1) with respect to the H^2-norm, as is discussed in Theorem 4.1, is equivalent to bifurcation with respect to the L^2-norm.

Proposition 4.2. *Consider the equation*

$$-\Delta u + Vu + \xi f\left(\frac{u}{\xi}\right) = \lambda u \text{ for } u \in H^2\left(\mathbb{R}^N\right) \tag{4.1}$$

under the following hypotheses: $V \in L^\infty(\mathbb{R}^N), \xi \in L^2(\mathbb{R}^N)$ *with* $\xi > 0$ *a.e. and* $f \in C^1(\mathbb{R}^N)$ *is an odd function such that*

(i) $\ell = \sup_{s\in\mathbb{R}} |f'(s)| < \infty$, $f'(0) = 0$ *and* $\left(\frac{f(s)}{s}\right)' > 0$ *for all* $s > 0$,
(ii) *there exists* $A > 0$ *such that* $\sup_{s>0} |As - f(s)| < \infty$.

We have the following conclusions about bifurcation for (4.1), *where* $S = -\Delta + V$. *Recall that* $\Lambda = \inf \sigma(S)$ *and* $\Lambda_e = \inf \sigma_e(S)$.

(a) *If* $\ker(S - \lambda_0 I) = \{0\}$ *and either* $d(\lambda_0, \sigma_e(S)) > A$ *or* $\lambda_0 < \Lambda_e$, *then* λ_0 *is not a bifurcation point.*
(b) *If* $\ker(S - \lambda_0 I) \neq \{0\}$ *and* $\lambda_0 < \Lambda_e$, *then* λ_0 *is a bifurcation point.*
(c) *Suppose that* $N \leq 3$ *and that* $\eta = 1/\xi$ *has the following properties:*

$$\eta \in W^{2,\infty}_{loc}(\mathbb{R}^N),\ \inf \eta > 0 \text{ and for}$$
$$\text{some } t > 0,\ \partial^\alpha \eta^t \in L^\infty(\mathbb{R}^N) \text{ for } 1 \leq |\alpha| \leq 2.$$

If $\lambda_0 \notin \sigma_e(S)$ *and* $\dim\ker(S - \lambda_0 I)$ *is odd, then* λ_0 *is a bifurcation point.*

(d) *If* $A > \Lambda_e - \Lambda$ *and* $\lambda_0 \in [\Lambda_e, \Lambda + A]$, *then* λ_0 *is a bifurcation point.*
(e) *Suppose that* $V \equiv 0$ *and* $\xi(x) = (1 + |x|^2)^t$ *for some* $t > N/4$. *If* $\lambda_0 > A[1 + \frac{4t-N}{2}]$, *then* λ_0 *is not a bifurcation point and* $\lambda_0 > A = \Lambda + A$ *since* $\Lambda = \Lambda_e = 0$.

Remark 1. Since $f(0) = 0$, it follows from (i) and (ii) that

$$\lim_{s\to 0} \frac{f(s)}{s} = 0 < \frac{f(s)}{s} < A = \lim_{s\to\infty} \frac{f(s)}{s} \text{ for all } s > 0.$$

Also $f'(s) > \frac{f(s)}{s}$ for all $s > 0$, so $\ell \geq A$ and in some case the inequality is strict. For example, $f(s) = |s|^{2\sigma} s/(1 + s^2)^\sigma$ satisfies (i) and (ii) for all $\sigma > 0$, but $\ell = \sup_{s>0} f'(s) = f'(\sqrt{2\sigma+1}) > 1 = A$. On the other hand, $f(s) = s - \tanh s$ also satisfies (i) and (ii) and in this case $\ell = A = 1$.

Remark 2. With $\eta = 1/\xi$ and $g = h = 0$, we see that (4.1) is a special case of (1.3) satisfying the conditions (V), (G), (H), (F) and (W2). Therefore Theorem 4.1 applies and yields information not contained in the conclusions (a) to (e). Notice however that if f satisfies the hypotheses of Proposition 4.2, $-f$ does not. Of course, $-f$ still satisfies (F) and so Theorem 4.1 can treat this case too, but as is shown in Section 5, the conclusions (b) and (d) of the proposition fail in this case, as does part (c) for $\xi(x) = e^{-\alpha|x|}$ with large positive α. Theorem 4.1 places much weaker restrictions on the weights ξ and η.

Proof. Defining h by $h(0) = 0$ and $h(s) = f(s)/(As)$ for $s \neq 0$, our hypotheses on f imply that h satisfies the conditions (H3) to (H5) of [17] and our equation (4.1) is just (1.1) of [17] with $q = V$. Noting Proposition 2.1 of [17], the conclusions (a), (b) and (d) are consequences of statements (R1) to (R3) in Section 3 of [17], whereas (e) is just a restatement of the example following Theorem 3.1 in that paper.

The hypotheses on η made in part (c) mean that η is a transference weight of order 2 in the sense introduced by P.J. Rabier in [10, 11]. They ensure that $\xi = 1/\eta$ satisfies the condition (H2)* of Section 4 of [17], where the results of [10] are applied to (4.1). In particular, the weighted Sobolev space $W_\eta^{2,2}$ is continuously embedded in $H^2(\mathbb{R}^N)$ so bifurcation at λ_0 in $W_\eta^{2,2}$ implies that λ_0 is a bifurcation point for (4.1) in the sense of the present paper. Thus part (c) follows from statement (C2) in Section 4.2 of [17]. In fact, as (C2) shows, Rabier's work provides a stronger statement about bifurcation at such points. □

Commentaries on the conclusions

(1) Since $A \leq \ell$, the conclusion (a) is sharper than (i) of Theorem 4.1 for the equation (4.1).
(2) Under the hypotheses of the proposition, consider a potential V such that $\Lambda = \Lambda_e$ and such that there exist $b > a > \Lambda_e$ such that $(a, b) \cap \sigma(S) = \emptyset$. Then choose f with $A > b - \Lambda_e$. We now have that $(a, b) \subset [\Lambda_e, \Lambda + A]$ and hence every $\lambda_0 \in (a, b)$ is a bifurcation point by part (d) despite the fact that $\lambda_0 \notin \sigma(S)$. Note that at these points, $d(\lambda_0, \sigma_e(S)) \leq \lambda_0 - \Lambda_e < b - \Lambda_e < A \leq \ell\|\xi\eta\|_{L^\infty}$ since $\xi\eta \equiv 1$.
(3) In the next section we show that there are functions f satisfying (F) and weights ξ for which statement (b) of the proposition fails even when $\lambda_0 = \Lambda$ is a simple eigenvalue.
(4) The approach devised by Rabier can be used to establish bifurcation for (4.1) at eigenvalues of odd multiplicity of S under much weaker hypotheses on f provided that $\eta = 1/\xi$ is a transference weight. See Section 5 of [10].

5. A case where there is no bifurcation at a simple eigenvalue

In this section we consider a special case of (1.3) in which the hypotheses (V),(G), (H) and (F) are satisfied and $\Lambda = \inf \sigma(S) < \inf \sigma_e(S)$ is a simple eigenvalue of S. Consider the equation

$$-u'' + Vu + e^{-\alpha|x|} f(e^{\alpha|x|}u) = \lambda u \text{ on } \mathbb{R}, \tag{5.1}$$

where α is a positive constant,

(V_0) $V \in C_0(\mathbb{R})$ with $V \leq 0$ but $V \not\equiv 0$ on $\mathbb{R}$,

and

(K) $f \in C^1(\mathbb{R})$ is an odd function with $f'(0) = 0$, $(\frac{f(s)}{s})' \le 0$ for $s > 0$ and $\ell \equiv \sup_{s>0} |f'(s)| < \infty$.

Clearly $V \in L^\infty(\mathbb{R})$ and so $V \in T_1(q)$ for all $q \ge 2$. Thus $Su = -u'' + Vu$ defines a self-adjoint operator $S : H^2(\mathbb{R}) \subset L^2(\mathbb{R}) \to L^2(\mathbb{R})$. It follows from ($V_0$) that $\Lambda \equiv \inf \sigma(S) < 0 = \inf \sigma_e(S)$ and that Λ is a simple eigenvalue of S with an eigenfunction $\phi \in C^2(\mathbb{R})$ which is strictly positive on $\mathbb{R}$ with $\|\phi\|_{L^2} = 1$. Note that $d(\Lambda, \sigma_e(S)) = |\Lambda|$.

For the ensuing calculations it is convenient to write f in the form $f(s) = k(s)s$ with $k(0) = 0$, where (K) ensures that $k \in C(\mathbb{R})$ is an even function having the properties

$$k \in C^1((0,\infty)) \text{ with } k' \le 0 \text{ on } (0,\infty) \text{ and } -L \le k \le 0 \text{ on } \mathbb{R},$$

where $L \equiv -\lim_{s\to\infty} k(s) \in [0,\ell]$. Note that if a function f satisfies the hypotheses of Proposition 4.2, then $-f$ satisfies the condition (K). In particular, $f(s) = -|s|^{2\sigma}s/(1+s^2)^\sigma$ satisfies (K) for all $\sigma > 0$.

Recall that $H^2(\mathbb{R})$ is continuously embedded in $C^1(\mathbb{R})$. Since V and k are continuous, $u \in C^2(\mathbb{R})$ for any solution $(\lambda, u) \in \mathbb{R}\times H^2(\mathbb{R}^N)$ of (5.1). This equation can then be written as

$$-u'' + \{V + k(e^{\alpha|x|}u)\}u = \lambda u. \tag{5.2}$$

Let $Z > 0$ be such that $V(x) = 0$ for $|x| \ge Z$. Note that on $\mathbb{R}\backslash(-Z, Z)$, for $\lambda < 0$, the equation can be written as

$$u'' = \{k(e^{\alpha|x|}u) + |\lambda|\}u. \tag{5.3}$$

Theorem 5.1. *Suppose that the conditions* (V_0) *and* (K) *are satisfied and set* $L \equiv -\lim_{s\to\infty} \frac{f(s)}{s}$. *Note that* $0 \le L \le \ell$.

(i) *If* $|\Lambda| > \ell$, *there is continuous bifurcation at* Λ. *Furthermore, for any sequence* $\{(\lambda_n, u_n)\} \subset \mathbb{R} \times H^2(\mathbb{R})$ *of solutions of* (5.1) *with* $u_n \not\equiv 0$ *such that* $\lambda_n \to \Lambda$ *and* $\|u_n\|_{H^2} \to 0$, *we have that* $u_n = \langle u_n, \phi\rangle_{L^2}\{\phi + w_n\}$ *where* $\langle w_n, \phi\rangle_{L^2} = 0$ *and* $\|w_n\|_{H^2} \to 0$. *Also, for* n *large enough,* $u_n \in C^2(\mathbb{R})$ *has no zeros and so there is a sequence* $\{(\lambda_n, u_n)\} \subset \mathbb{R} \times H^2(\mathbb{R})$ *of solutions of* (5.1) *with* $u_n > 0$ *on* $\mathbb{R}$ *and* $\lambda_n \le \Lambda$ *such that* $\lambda_n \to \Lambda$ *and* $\|u_n\|_{H^2} \to 0$.

(ii) *If* $|\Lambda| < L$ *and* $\alpha > |\Lambda|^{1/2}$, *then* Λ *is not a bifurcation point for* (5.1). *Indeed, setting* $\varepsilon = \min\{(L - |\Lambda|)/2, \alpha^2 - |\Lambda|, |\Lambda|\}$, $u \equiv 0$ *is the only solution of* (5.1) *in* $H^2(\mathbb{R})$ *for* $\lambda \in (\Lambda - \varepsilon, \Lambda + \varepsilon)$.

Remark 1. Inspecting the proof of part (i), we observe that $\lambda_n < \Lambda$ provided that $f(s) < 0$ for all $s > 0$, since

$$\Lambda = \int_{\mathbb{R}} (\phi')^2 + V\phi^2 dx > \int_{\mathbb{R}} (\phi')^2 + W_n\phi^2 dx \ge \inf \sigma(S_n) = \lambda_n,$$

in this case.

Remark 2. In part (ii), the proof in fact shows that $u \equiv 0$ is the only solution with $u(x) \to 0$ as $x \to \infty$ for $\lambda \in (\Lambda - \varepsilon, \Lambda + \varepsilon)$. Note that in Step 1, the monotonicity of J implies that $\lim_{x\to\infty} u'(x)^2$ exists, and hence $u'(x) \to 0$ if $u(x) \to 0$ as $x \to \infty$. The rest of the proof is the same.

Proof. (i) The first part of the conclusion is a special case of Theorem 4.1(iii), so we only need to justify the claims about the signs of u_n and $\lambda_n - \Lambda$. Using the oddness of f, we can suppose that there is a sequence of solutions converging to $(\Lambda, 0)$ in $\mathbb{R} \times H^2(\mathbb{R})$ and, in addition, that $\langle u_n, \phi\rangle_{L^2} > 0$ for all n. Let Z be such that $\operatorname{supp} V \subset [-Z, Z]$. Since $m \equiv \inf_{|x|\le Z} \phi(x) > 0$ and $\|w_n\|_{L^\infty} \to 0$, there exists n_0 such that $\phi + w_n \ge m/2$ on $[-Z, Z]$ for all $n \ge n_0$. By increasing n_0 if necessary, we can also suppose that $\lambda_n < 0$ and $|\lambda_n - \Lambda| < |\Lambda| - \ell$ for all $n \ge n_0$. But, for $|x| \ge Z$, by (5.3) we have that

$$u_n'' = \{k(e^{\alpha|x|}u_n) - \lambda_n\}u_n \le \{-\ell + |\Lambda| - |\lambda_n - \Lambda|\}u_n < 0$$

at points where $u_n < 0$ since $-\ell \le -L \le k(s) \le 0$ for all $s \in \mathbb{R}$. Hence u_n cannot have a negative minimum in the set $(-\infty, -Z] \cup [Z, \infty)$. Since $u_n(-Z) > 0, u_n(Z) > 0$ and $\lim_{|x|\to\infty} u_n(x) = 0$, it follows that $u_n \ge 0$ on $(-\infty, -Z] \cup [Z, \infty)$ and hence on $\mathbb{R}$. Thus any zero of u_n is at least a double zero and the existence of such a value implies that $u \equiv 0$, by the uniqueness of the solution of (5.1) with the conditions $u_n(x_0) = u_n'(x_0) = 0$. Hence we have that $u_n > 0$ on $\mathbb{R}$ for all $n \ge n_0$.

Setting $W_n(x) = V(x) + k(e^{\alpha|x|}u_n(x))$, we see from (5.2) that $u_n \in H^2(\mathbb{R})$ is a positive eigenfunction with eigenvalue λ_n of the operator $S_n u = -u'' + W_n u$. Since $k(s) \ge -L$ for all $s \in \mathbb{R}$, we have that $W_n(x) \ge -L$ for $|x| \ge Z$ and so $\inf \sigma_e(S_n) \ge -L$. On the other hand, $k \le 0$ on $\mathbb{R}$ and hence $\inf \sigma(S_n) \le \inf \sigma(S) = \Lambda < -\ell \le -L$. This implies that $\inf \sigma(S_n)$ is a simple eigenvalue of S_n with a positive eigenfunction and consequently $\lambda_n = \inf \sigma(S_n)$, showing that $\lambda_n \le \Lambda$.

(ii) Let (λ, u) be a non-trivial solution with $\lambda \in (\Lambda - \varepsilon, \Lambda + \varepsilon)$ and $u \in H^2(\mathbb{R})$. We show that this leads to a contradiction.

Step 1, in which we show that u cannot change sign in (Z, ∞).

For $x, s \in \mathbb{R}$, let

$$L(x, s) = e^{-2\alpha x}\Phi(e^{\alpha x}s) \text{ where } \Phi(t) = \int_0^t f(s)ds = \int_0^t k(s)s\,ds.$$

Then $L(\cdot,\cdot) \in C^2(\mathbb{R}^2)$ with

$$\partial_x L(x, s) = \alpha e^{-2\alpha x}\psi(e^{\alpha x}s) \text{ where } \psi(t) = f(t)t - 2\Phi(t)$$

and

$$\partial_s L(x, s) = e^{-\alpha x} f(e^{\alpha x}s) = k(e^{\alpha x}s)s.$$

We observe that

$$\psi(t) = k(t)t^2 - 2\int_0^t k(s)sds = \int_0^t k'(s)s^2 ds \le 0 \text{ for all } t \in \mathbb{R}$$

by (K).

Consider now the function $J : \mathbb{R} \to \mathbb{R}$ defined by

$$J(x) = \frac{1}{2}\{u'(x)^2 + \lambda u(x)^2\} - L(x, u(x)).$$

Clearly, $J \in C^1(\mathbb{R})$ and, for $x > Z$,

$$\begin{aligned}\frac{d}{dx}J(x) &= u'(x)\{u''(x) + \lambda u(x)\} - \partial_x L(x, u(x)) - \partial_s L(x, u(x))u'(x) \\ &= u'(x)\{u''(x) + \lambda u(x) - k(e^{\alpha x}u(x))u(x)\} - \partial_x L(x, u(x)) \\ &= -\partial_x L(x, u(x) = -\alpha e^{-2\alpha x}\psi(e^{\alpha x}u(x)) \geq 0.\end{aligned}$$

We also have that $|f(s)| \leq \ell|s|$ for all $s \in \mathbb{R}$ and so $|\Phi(s)| \leq \frac{1}{2}\ell s^2$ and then $|L(x, s)| \leq \frac{1}{2}\ell s^2$, too.

Since $u \in H^2(\mathbb{R})$ implies that $\lim_{x\to\infty} u(x) = \lim_{x\to\infty} u'(x) = 0$, it follows that $J(x) \to 0$ as $x \to \infty$ and then from the monotonicity of J that $J(x) \leq 0$ for all $x > Z$.

Suppose that $u(x_0) = 0$ for some $x_0 > Z$. Then $0 \geq J(x_0) = \frac{1}{2}u'(x_0)^2$ since $L(x_0, 0) = 0$. By the uniqueness of the solution of (5.1) satisfying the conditions $u(x_0) = u'(x_0) = 0$, this implies that $u \equiv 0$ on $\mathbb{R}$, whereas we have supposed that u is a non-trivial solution. Hence u has no zeros in the interval (Z, ∞).

Step 2, in which we prove that $\lim_{x\to\infty} e^{\alpha x}u(x) = \infty$ or $-\infty$.

In view of step 1 and the oddness of f, we can suppose that $u > 0$ on (Z, ∞). Since $|\lambda| - \alpha^2 = -\lambda - \alpha^2 < (-\Lambda + \varepsilon) - \alpha^2 = |\Lambda| - \alpha^2 + \varepsilon \leq 0$ we can choose $\beta \in (|\lambda|^{1/2}, \alpha)$ and then set $w(x) = ce^{-\beta x}$ where $c = \frac{1}{2}u(R)e^{\beta R}$ and $R = Z + 1$. Then $c > 0$ and we consider the function $z = u - w$. Since $\beta > 0$ and $u \in H^2(\mathbb{R})$, $z(x) \to 0$ as $x \to \infty$. By the choice of c, $z(R) = \frac{1}{2}u(R) > 0$. Let $\Omega = \{x > R : z(x) < 0\}$ and suppose that $\Omega \neq \emptyset$. Then $z \in C^2(\mathbb{R})$ and there exists a point $x_0 \in \Omega$ such that $z(x_0) = \min\{z(x) : x \in \Omega\} < 0$ and $z''(x_0) \geq 0$. But, on Ω,

$$z'' = u'' - w'' = \{k(e^{\alpha x}u) + |\lambda|\}u - \beta^2 w \leq |\lambda|u - \beta^2 u < 0$$

since $k \leq 0$ on $\mathbb{R}$, $w > u > 0$ on Ω and $|\lambda| < \beta^2$. In particular, $z''(x_0) < 0$ contradicting the fact that z attains its minimum at z_0. Hence $\Omega = \emptyset$ and we have proved that $u(x) \geq ce^{-\beta x}$ for all $x > R = Z+1$. But then, $e^{\alpha x}u(x) \geq ce^{(\alpha-\beta)x}$ for all $x > R$, where $c > 0$ and $\alpha - \beta > 0$. Thus $\lim_{x\to\infty} e^{\alpha x}u(x) = \infty$, as required.

Step 3, in which we obtain a contradiction to the conclusion of Step 1.

As in Step 2, we can assume without loss of generality that $e^{\alpha x}u(x) \to \infty$ as $x \to \infty$ and hence $k(e^{\alpha x}u(x)) \to -L$ as $x \to \infty$. But

$$\begin{aligned}k(e^{\alpha x}u(x)) + |\lambda| &= \{k(e^{\alpha x}u(x)) + L\} - L + |\lambda| \\ &\leq \{k(e^{\alpha x}u(x)) + L\} - L + |\Lambda| + |\lambda - \Lambda| < \{k(e^{\alpha x}u(x)) + L\} - \varepsilon\end{aligned}$$

since $L - |\Lambda| \geq 2\varepsilon$ and $|\lambda - \Lambda| < \varepsilon$. Hence there exists $R_1 > Z + 1$ such that $k(e^{\alpha x}u(x)) + |\lambda| < -\varepsilon/2$ for all $x > R_1$.

Setting $v(x) = \sin\sqrt{\frac{\varepsilon}{2}}x$, we have that $v'' = -\frac{\varepsilon}{2}v$ and the zeros of v are $x_n = \sqrt{\frac{2}{\varepsilon}}n\pi$ for $n \in \mathbb{Z}$. For n even, $v > 0$ on (x_n, x_{n+1}). Now consider an even integer n such that $x_n > R_1$. Then

$$\int_{x_n}^{x_{n+1}} uv'' - u''v dx = uv'|_{x_n}^{x_{n+1}} = -\sqrt{\frac{\varepsilon}{2}}\{u(x_n) + u(x_{n+1})\} < 0.$$

On the other hand,

$$\int_{x_n}^{x_{n+1}} uv'' - u''v\, dx = \int_{x_n}^{x_{n+1}} -u\frac{\varepsilon}{2}v + \{\lambda - k(e^{\alpha x}u)\}uv\, dx$$

$$= -\int_{x_n}^{x_{n+1}} \{\frac{\varepsilon}{2} + |\lambda| + k(e^{\alpha x})\}uv\, dx > 0,$$

since $uv > 0$ on (x_n, x_{n+1}) by step 1 and $k(e^{\alpha x}u(x)) + |\lambda| < -\varepsilon/2$ by the choice of R_1. □

It is natural to look for a result similar to Theorem 5.1 for $N \geq 2$. This can easily be done for part (i), but for part (ii) which is the main point of Theorem 5.1 it is not so clear how to proceed. For the approach used here, the obstacle at present is generalizing Step 1. Steps 2 and 3 can be extended to higher dimensions so one could obtain the conclusion that there is no bifurcation of positive solutions at Λ for potentials V having compact support and for which $\Lambda < \inf \sigma_e(-\Delta + V)$.

Returning to the case $N = 1$, minor modifications of the proof of part (ii) yield the same conclusion for other types of potential. For example (V_0) could be replaced by

(V_1) $V \in L^\infty(\mathbb{R})$ with $V \leq 0$ a.e. on $\mathbb{R}$ and there exists $a < b < Z$ such that $V \in C((a, b))$ with $V(x) < 0$ for $x \in (a, b)$ and $V(x) = 0$ for $|x| > Z$.

or

(V_2) $V \in L^\infty(\mathbb{R})$ with $\lim_{|x|\to\infty} V(x) = 0$ and there exists $Z > 0$ such that $V \in C^1((Z, \infty))$ and $V'(x) \leq 0$ for all $x > Z$.

Unlike (V_0) and (V_1), (V_2) does not ensure that $\inf \sigma(S) < 0$ so this condition has to be added in that case.

Finally, to draw some important information from Theorem 5.1 we specify a class of nonlinearities f which satisfy (K) and for which $L = \ell$.

(Q) $f \in C^1(\mathbb{R})$ is an odd function with $f'(0) = 0$ which is concave on $[0, \infty)$ with $f'(\infty) \equiv \lim_{s\to\infty} f'(s) > -\infty$.

Examples of functions satisfying (Q) are given by $f(s) = \ell\{\arctan s - s\}$ and $f(s) = \ell\{\tanh s - s\}$ for any $\ell > 0$.

As already noted, functions of the form $f(s) = -|s|^{2\sigma}s/(1 + s^2)^\sigma$ satisfy (K) for all $\sigma > 0$, but they do not satisfy (Q) since $L = -\lim_{s\to\infty} f'(s) = 1$ and $\ell = \sup_{s\in\mathbb{R}} |f'(s)| = -f'(\sqrt{2\sigma + 1}) > 1$ for all $\sigma > 0$. On the other hand, for functions of the form $f(s) = -|s|^\gamma s/(1 + |s|^\gamma)$, we find that (K) is satisfied for all $\gamma > 0$ whereas (Q) is satisfied if and only if $0 < \gamma \leq 1$.

The hypothesis (Q) implies that $f' \leq 0$ on $\mathbb{R}$ and that $\sup_{s\in\mathbb{R}} |f'(s)| = \ell$ where $\ell = -f'(\infty) = -\lim_{s\to\infty} \frac{f(s)}{s}$. Setting

$$\xi(x) = e^{-\alpha|x|} \text{ and } \eta(x) = e^{\alpha|x|} \text{ and then } F(u) = \xi f(\eta u),$$

it follows that the hypotheses (F), (W1) and (W2) of Section 4 are satisfied and hence from Theorem 3.3 that $F : H^2(\mathbb{R}) \to L^2(\mathbb{R})$ is Hadamard differentiable at 0 with $F'(0) = 0$. However, except in the trivial case $f \equiv 0$, Theorem 3.4 shows that this mapping is not Fréchet differentiable at 0 and it is not compact. Furthermore, $L(F) = L^{L^2}(F) = \ell$.

Corollary 5.2. *Consider the equation* (5.1) *under the hypotheses* (V_0) *and* (Q).

(i) *If* $|\Lambda| > \ell$, *there is continuous bifurcation at* Λ.

(ii) *If* $|\Lambda| < \ell$ *and* $\alpha > |\Lambda|^{1/2}$, *then* Λ *is not a bifurcation point. Indeed, setting* $\varepsilon = \min\{(\ell - |\Lambda|)/2, \alpha^2 - |\Lambda|, |\Lambda|\}$, $u \equiv 0$ *is the only solution of* (5.1) *in* $H^2(\mathbb{R})$ *for* $\lambda \in (\Lambda - \varepsilon, \Lambda + \varepsilon)$.

Remark 1. Noting that $d(\Lambda, \sigma_e(S)) = |\Lambda|$ and $L(F) = L^{L^2}(F) = \ell$, we see that $d(\Lambda, \sigma_e(S)) < L(F)$ in part (ii) and that in this case, Λ is not a bifurcation point for (5.1). The other hypotheses of Theorem 4.1 are satisfied in both parts (i) and (ii). Hence, under the assumptions (V_0) and (Q), (5.1) is a special case of (1.3) which satisfies the hypotheses (H1) to (H3) of Proposition 2.2 and so also the conditions (B1) to (B5) of Theorem [19], as discussed at the beginning of Section 4.

Remark 2. Equation (5.1) is just (4.1) with $\xi(x) = e^{-\alpha|x|}$ but the assumption (K) means that f does not satisfy the hypotheses of Proposition 4.2 and (ii) shows that the statement (b) of that proposition does not hold for $\lambda_0 = \Lambda$ in the present situation. Of course, statement (c) also fails for (5.1) but it should be realized that this happens solely because $\eta(x) = 1/\xi(x) = e^{\alpha|x|}$ is not a transference weight. See commentary 4 in Section 4.

Acknowledgment

The work for this paper was started in the spring of 2012 at the WIPM, C.A.S., Wuhan while I was suppored by the Chinese Academy of Sciences Visiting Professorship for Senior International Scientists, Grant No. Z010T1J10. It is with great pleasure that I record my sincere thanks to the C.A.S. for this generous support.

References

[1] H.L. Cycon, R.G. Froese, W. Kirsch and B. Simon, Schrödinger Operators, Springer-Verlag, Berlin 1987.

[2] D.E. Edmunds and W.D. Evans, *Spectral Theory and differential operators*, Oxford University Press, Oxford 1987.

[3] G. Evéquoz and C.A. Stuart, Hadamard differentiability and bifurcation, Proc. Royal Soc. Edinburgh, 137A (2007), 1249–1285.

[4] G. Evéquoz and C.A. Stuart, On differentiability and bifurcation, Adv. Math. Ecom., 8 (2006), 155–184.

[5] G.Evéquoz and C.A. Stuart, Bifurcation points of a degenerate elliptic boundary-value problem, Rend. Lincei Mat. Appl., 17 (2006), 309–334.

[6] G.Evéquoz and C.A.Stuart, Bifurcation and concentration of radial solutions of a nonlinear degenerate elliptic eigenvalue problem, Adv. Nonlinear Studies, 6 (2006), 215–232.

[7] D. Gilbarg and N.S. Trudinger, *Elliptic Partial Differential Equations of Second Order*, second edition, Springer, Berlin 1983.

[8] D. Idczak and A. Rogowski, On a generalization of Krasnoselskii's theorem, J.Austral.Math.Soc., 72 (2002), 389–394.

[9] M.A. Krasnoselskii, *Topological Methods in the Theory of Nonlinear Integral Equations*, Pergamon, New York, 1964.

[10] P.J. Rabier, Bifurcation in weighted spaces, Nonlinearity, 21 (2008), 841–856.

[11] P.J. Rabier, Decay transference and Fredholmness of differential operators in weighted Sobolev spaces, Differential Integral Equations, 21 (2008), 1001–1018.

[12] P.J. Rabier and C.A. Stuart, Fredholm properties of Schrödinger operators in $L^p(\mathbb{R}^N)$, Diff. Integral Eqns., 13 (2000), 1429–144.

[13] P.J. Rabier and C.A. Stuart, Global bifurcation for quasilinear elliptic equations on $\mathbb{R}^N$, Math. Z., 237 (2001), 85–124.

[14] B. Simon, Schrödinger semigroups, Bull. AMS, 7 (1982), 447–526.

[15] C.A. Stuart, Bifurcation for some non-Fréchet differentiable problems, Nonlinear Anal., TMA, 69 (2008), 1011–1024.

[16] C.A. Stuart, An introduction to elliptic equations on $\mathbb{R}^N$, in Nonlinear functional analysis and applications to differential equations, ed., A. Ambrosetti, K.C. Chang and I. Ekeland, World Scientific, Singapore 1998.

[17] C.A. Stuart, Bifurcation and decay of solutions for a class of elliptic equations on $\mathbb{R}^N$, Cont. Math., 540 (2011), 203–230.

[18] C.A. Stuart, Asymptotic linearity and Hadamard differentiability, Nonlinear Analysis, 75 (2012), 4699–4710.

[19] C.A. Stuart, Bifurcation at isolated singular points of the Hadamard derivative, preprint 2012.

[20] C.A. Stuart, Asymptotic bifurcation and second order elliptic equations on $\mathbb{R}^N$, preprint 2012.

C.A. Stuart
Section de mathématiques
Station 8, EPFL
CH-1015 Lausanne, Switzerland
e-mail: `charles.stuart@epfl.ch`

Progress in Nonlinear Differential Equations
and Their Applications, Vol. 85, 445–456

Geometric Aspects of Ambrosetti–Prodi Operators with Lipschitz Nonlinearities

Carlos Tomei and André Zaccur

Dedicated to Bernhard, with affection and admiration

Abstract. Let the function u satisfy Dirichlet boundary conditions on a bounded domain Ω. What happens to the critical set of the Ambrosetti–Prodi operator $F(u) = -\Delta u - f(u)$ if the nonlinearity is only a Lipschitz map? It turns out that many properties which hold in the smooth case are preserved, despite of the fact that F is not even differentiable at some points. In particular, a global Lyapunov–Schmidt decomposition of great convenience for numerical solution of $F(u) = g$ is still available.

Mathematics Subject Classification (2010). 35B32, 35J91, 65N30.

Keywords. Semilinear elliptic equations, Ambrosetti–Prodi theorem.

1. Introduction

A familiar set of hypotheses for the celebrated Ambrosetti–Prodi theorem is the following. Let $\Omega \subset \mathbb{R}^n$ be an open, bounded, connected domain with smooth boundary $\partial\Omega$ and denote by $0 < \lambda_1 < \lambda_2 \leq \cdots$ the eigenvalues of the free Dirichlet Laplacian $-\Delta$ on Ω. Let $f : \mathbb{R} \to \mathbb{R}$ be a smooth, strictly convex function, with asymptotically linear derivative so that

$$\operatorname{Ran} f' = (a, b)\,, \quad a < \lambda_1 < b < \lambda_2\,.$$

Under such hypotheses, the theorem states that the equation

$$F(u) = -\Delta u - f(u) = g, \quad u|_{\partial\Omega} = 0 \qquad (1)$$

for, say, $g \in C^{0,\alpha}(\Omega)$, has (exactly) zero, one or two solutions in $C^{2,\alpha}(\Omega)$.

The authors gratefully acknowledge support from CAPES, CNPq and FAPERJ..

1.1. A first approach – locating C and $F(C)$

In the original arguments ([1], [11]), a fundamental role is played by the critical set C of $F : X = C^{2,\alpha}_D(\Omega) \to Y = C^{0,\alpha}(\Omega)$. Here $C^{2,\alpha}_D(\Omega)$ is the subspace of functions of $C^{2,\alpha}(\Omega)$ satisfying Dirichlet boundary conditions. The proof follows a few steps, which follow from the inverse function theorem and the characterization of fold points.

1. The critical set $C \subset X$ is a hypersurface; every critical point is a fold.
2. F is proper, its restriction to C is injective and $F^{-1}(F(C)) = C$.
3. The spaces $X - C$ and $Y - F(C)$ have two components. Each component of $X - C$ is taken injectively to the same component of $Y - F(C)$.

1.2. A global Lyapunov–Schmidt decomposition

Berger and Podolak came up with a different approach [2], which is easier to phrase for $F : X \to Y$ between Sobolev spaces, $X = H^1_0(\Omega)$ and $Y = H^{-1}(\Omega) \simeq H^1_0(\Omega)$. Their main result is the construction of a global Lyapunov–Schmidt decomposition for F. Let φ_1 be the (positively normalized) eigenfunction associated to the ground state λ_1, set $V_X = V_Y = \langle \varphi_1 \rangle$ and consider the orthogonal decompositions

$$X = V_X \oplus W_X \quad \text{and} \quad Y = V_Y \oplus W_Y$$

into *vertical* and *horizontal* subspaces. With a different vocabulary, their proof essentially goes through the verification of the following properties, by making use of spectral estimates on the Jacobians $DF(u) : X \to Y$.

1. Horizontal affine subspaces of X are taken by F to *sheets.*
2. The inverse under F of vertical affine subspaces of Y are *fibers.*
3. Sheets are *essentially flat*, fibers are *essentially steep.*

An affine horizontal subspace of X is a set of the form $x + W_X$, for a fixed $x \in X$; an affine vertical subspace of Y is of the form $y + V_Y$, for $y \in Y$. Sheets are graphs of smooth functions from W_Y to V_Y and fibers are graphs of smooth functions from V_X to W_X. The third property states that the inclination of the tangent spaces to sheets, with respect to horizontal subspaces, and to fibers, with respect to vertical subspaces, is uniformly bounded from above.

In the words of [4], F is a *flat* map. In the Ambrosetti–Prodi case, vertical spaces and fibers are one-dimensional. More generally, the dimension equals the number k of eigenvalues of $-\Delta_D$ in the set $\overline{f'(\mathbb{R})}$ (we suppose non-resonance, i.e., the extreme values of $\operatorname{Ran} f'$ are not eigenvalues) and similar results still hold.

1.3. Sheets and fibers allow for weaker hypotheses

There is an interesting bonus obtained from considering this global Lyapunov–Schmidt decomposition. For many nonlinearities f, the related nonlinear map

$$F : X \to Y, \quad F(u) = -\Delta u - f(u)$$

is not proper, and part of the technology related to degree theory simply breaks down. The Ambrosetti–Prodi hypotheses yield properness of F, but it is not really essential to most of what we want to do. The first section of the paper is dedicated to explicit examples of Lipschitz nonlinearities for which properness does not occur – there are functions $g \in Y$ for which $F^{-1}(g)$ contains a full half-line of functions in X. From such examples, we may obtain smooth nonlinearities with the same property, but we give no details.

The results in the paper indicate that in most directions, properness holds. What we mean by this is something similar to the fact that even when a map F does not have an invertible Jacobian $DF(u)$ at a point u, it may still have a subspace $L(u)$ on which the restriction of $DF(u)$ acts injectively – this is how bifurcation equations come up: they concentrate in a possibly small subspace $S(u)$ transversal to $L(u)$ the difficulties which are not resolved by the linearization at u. What we shall see is that the domain X of F splits into special (nonlinear) surfaces of finite dimension k, on which properness may break down, which have for tangent spaces the subspaces $S(u)$ when u is critical, but on transversal directions to these surfaces, the horizontal affine subspaces, properness is always present. As we shall see, counting or computing solutions of the differential equation $F(u) = g$ boils down to a finite-dimensional problem, simplifying both (abstract) analysis and numerics.

1.4. A related numerical algorithm to solve the PDE $F(u) = g$

Smiley and Chun ([12],[13]) showed that an analogous global Lyapunov–Schmidt decomposition exists for appropriate non-autonomous nonlinearities $f(x, u(x))$ and emphasized its relevance for numerical analysis: one might solve $F(u) = g$ by restricting F to the fiber α_g containing $F^{-1}(g)$. In [4], this project is accomplished: given a right-hand side g, one first obtains numerically a point in α_g, which in the Ambrosetti–Prodi case is a curve, and then proceeds to search for solutions by moving along it. The algorithm is sufficiently robust to handle more flexible nonlinearities: it does not require convexity of f or properness of F, and the range of f' may include other (finite) sets of eigenvalues of $-\Delta_D$.

1.5. Lipschitz nonlinearities

Now, what happens when f is not smooth anymore, but, say, Lipschitz? In particular, this is the scenario considered in the proof of the so-called one-dimensional Lazer-McKenna conjecture ([9], [10]) by Costa, Figueiredo and Srikanth in [3]. We state the result, for the reader's convenience. Let $X = H^2([0,\pi]) \cap H_0^1([0,\pi])$ be the Sobolev space of functions satisfying Dirichlet boundary conditions with square integrable second derivatives. Recall that $u \mapsto -u''$ acting on X to $Y = L^2([0,\pi])$ has eigenvalues $\lambda_k = k^2$, $k = 1, 2, \dots$ with corresponding eigenfunctions $\sin(kx)$. Take $f : \mathbb{R} \to \mathbb{R}$, a strictly convex smooth function f with asymptotic values a and b for its derivative f' satisfying

$$a < 1, \quad \lambda_k = k^2 < b < (k+1)^2 = \lambda_{k+1}.$$

Then the equation $F(u) = -u'' - f(u) = -t \sin x$, $u(0) = u(\pi) = 0$, has exactly $2k$ solutions for $t > 0$ sufficiently large.

The argument in [3] considers the nonlinearity $\tilde{f}$ given by $\tilde{f}'(x) = a$ or b, depending if $x < 0$ or $x > 0$. The related operator $\tilde{F}(u) = -u'' - \tilde{f}(u)$ now requires some care: it stops being differentiable everywhere and the usual differentiable normal forms at regular points and folds break down.

In this paper, we show that when f is merely Lipschitz, for appropriate conditions on the boundary of Ω, the operator F is still flat, in the sense that the global Lyapunov–Schmidt decomposition still holds, and sheets and fibers are still available as graphs of Lipschitz functions. A word of caution: piecewise linear nonlinearities may yield continua of points on which the map F takes a unique value. Such sets necessarily lie in a single fiber. More, the numerical analysis of the PDE $F(u) = g$ presented in [4] is still valid, after minor modifications.

We take this material to be an intermediate step towards a more geometric description of the operators of Hamilton–Jacobi–Bellman type, as studied by Felmer, Quaas and Sirakov in [7].

2. Some cautionary examples

For a box $\Omega \subset \mathbb{R}^n$, we consider again the differential equation

$$F(u) = -\Delta u - f(u) = g, \quad u|_{\partial\Omega} = 0 \tag{2}$$

with a Lipschitz nonlinearity $f(x)$. Once f is piecewise linear, there may be whole (straight line) segments on the domain restricted to which F is actually constant. This happens already for the one-dimensional case. Let φ_1 be the (positive) ground state associated to eigenvalue λ_1 and take $a < \lambda_1 < b$. Suppose that Ran $\varphi_1 = [0, M]$ and define $f(x)$ to be continuous, with derivatives equal to a, λ_1 and b in the intervals $[-\infty, 0]$, $[0, M]$ and $[M, \infty]$: clearly, $F(t\varphi_1) = 0$, for $t \in [0, 1]$.

2.1. Nonlinearities f with derivatives taking two values, $n = 1$

In a similar vein, we now provide examples of segments on which F is constant for the nonlinearity $f(x) = ax$ or bx, for $x < 0$ or $x > 0$, with the property that, for special values of a and b, there are right-hand sides g ($g \equiv 0$ is an example) with the property that $F^{-1}(g)$ contains a (straight) half-line of solutions. In particular, the map $F : X = H^2(\Omega) \cap H_0^1(\Omega) \to L^2(\Omega)$ is not proper and the equation has solutions which are not isolated.

We begin with the one-dimensional case, $n = 1$. Set $\Omega = I = [0, \pi]$. Split I into k closed intervals $I_i, i = 1, \ldots, k$ joined at their ends, of two different lengths, $I_{\text{odd } i} = \beta$, $I_{\text{even } i} = \alpha$. In the figure, $k = 4$. The smallest eigenvalues for the

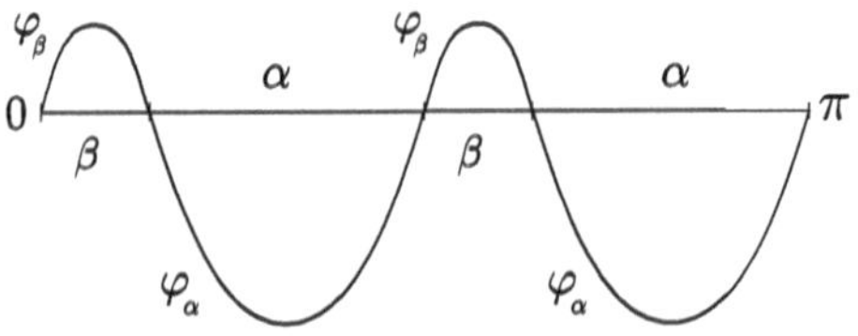

FIGURE 1. A half-line of solutions

operator $u \mapsto -u''$ with Dirichlet conditions in an interval of sizes β and α are respectively

$$\lambda_\beta = (\pi/\beta)^2 \quad \text{and} \quad \lambda_\alpha = (\pi/\alpha)^2 ,$$

with positive (normalized) eigenfunctions φ_β and φ_α. We set $b = \lambda_\beta$ and $a = \lambda_\alpha$ and construct a solution ψ by juxtaposing multiples of φ_β and φ_α as shown in the figure. On I_1, one may take $p\varphi_\beta$, for arbitrary $p > 0$. On I_2, the (negative) multiple of φ_α is determined by matching the first derivative – recall that $\psi \in X = H^2(I) \cap H_0^1(I)$, so ψ' is absolutely continuous. The procedure extends to the remaining intervals in a unique fashion. We have to make sure that the total length of the intervals I_i equals π. Thus, for example, in the simplest case $k = 2$, we must have

$$\beta + \alpha = \pi \quad \Longleftrightarrow \left(1/\sqrt{b}\right) + \left(1/\sqrt{a}\right) = 1.$$

For any value $a \in (1,4)$, there is a (unique) b, which turns out to be in $(4,\infty)$ which solves this equation. Said differently: any interval $[a,b]$ containing $\lambda_2 = 4$ for which a, b are not eigenvalues of the free problem admits a half-line of solutions of the equation above. For different numbers of intervals, one shows half-lines of solutions for any $a \in (\lambda_k, \lambda_{k+1})$ and appropriate $b \in (\lambda_{k+1}, \infty)$.

Alas, the only situation for which this argument does not provide a half-line of solutions $F^{-1}(0)$ is $a < \lambda_1$, the Ambrosetti–Prodi case. There are strong evidences that in this case there are no continua of solutions $F^{-1}(g)$, but we have no proof.

Clearly, one may replace $g \equiv 0$ by any g defined piecewise on intervals I_i as functions in the range of $u \mapsto -u'' - f(u)$ (Dirichlet conditions on I_i) acting on positive functions restricted to I_i. Thus, for k intervals, the set of such g is a vector subspace of $L^2(I)$ of codimension k. This construction ascertains that g is in the range of F (now considered in the full interval I), so that $g = F(u_0)$. By linearity on each interval I_i, adding a homogeneous solution $\psi \in F^{-1}(0)$ gives rise to $u_0 + \psi \in F^{-1}(g)$.

These ideas also suffice to prove that there is no nontrivial function in $[0,\pi]$ which is taken to 0 if $a < \lambda_1 < b$.

2.2. The case of arbitrary n

We now consider the case $n = 2$, the general situation being similar. We now have $\Omega = I \times I$ (rectangles would work also): just separate variables and proceed. Let $\psi(y)$ as before, solving $-\psi_{yy} - f(\psi) = 0$, where f is constructed from appropriate a and b, and let $\varphi(x) \geq 0$ be the ground state for Dirichlet conditions on I, so that $-\varphi_{xx} = \varphi(x)$. The product $\tilde{\psi}(x,y) = \varphi(x)\psi(y)$ satisfies

$$-\tilde{\psi}_{xx} - \tilde{\psi}_{yy} - f(\tilde{\psi}) = \tilde{\psi} + \varphi(x)(-\psi_{yy} - f(\psi)) = \tilde{\psi}$$

so that $\tilde{\psi}$ and its positive multiples solve

$$-u_{xx} - u_{yy} - \tilde{f}(u) = 0, \quad u|_{\partial\Omega} = 0,$$

for $\tilde{f}(x) = f(x) + x$.

3. Geometry of Lipschitz maps

Set $Y = L^2(\Omega)$ with inner product $\langle u, v\rangle_0 = \int_\Omega uv$ and norm $\|u\|_0$. Also let $X = H^2(\Omega) \cap H_0^1(\Omega)$, with inner product $\langle u, v\rangle = \langle -\Delta u\ ,\ -\Delta v\rangle_0$ and norm $\|u\|_2$.

We always consider sets Ω for which $-\Delta : X \subset Y \to Y$ is a self-adjoint isomorphism – we call such domains Ω *appropriate*. Also the same operator should have $C_0^\infty(\Omega)$ as a core, i.e., it is essentially self-adjoint in this domain. From the spectral theorem, there is an orthonormal basis of (Dirichlet) eigenfunctions $\varphi_i \in X$, $\|\varphi_i\|_0 = 1$, satisfying $-\Delta\varphi_i = \lambda_i\varphi_i$. Eigenfunctions associated to different eigenvalues are orthogonal with respect to both inner products. Concretely, one might take Ω to be a convex set or require $\partial\Omega$ to be $C^{1,1}$ ([12], [6]). Notice that, from standard results in spectral theory, operators

$$T : X \subset Y \to Y, \quad Tu = -\Delta u - qu,$$

for bounded real potentials q, are still self-adjoint with an orthonormal basis of eigenfunctions.

We assume that the nonlinearity $f : \mathbb{R} \to \mathbb{R}$ is Lipschitz, and f' takes values in an interval $[a, b]$ with the property that the bounds a and b are not eigenvalues λ_i. For this part of the paper, we make no assumptions about convexity for f. Notice that a and b do not have to be the asymptotic values of f', a degree of freedom which is convenient for numerical analysis.

For starters, $F : X \to Y$ given by $F(u) = -\Delta u - f(u)$ is a well defined map – it suffices to check that $f(u) \in Y$. This follows from the easy lemma below.

Lemma 1. *Say $f : \mathbb{R} \to \mathbb{R}$ is Lipschitz. Then the map $\hat{f} : Y = L^2(\Omega) \to Y$ given by $\hat{f}(u) = f \circ u$ is also well defined and Lipschitz with the same constant.*

Proof. Take first u, v continuous functions. Since f is M-Lipschitz, it is absolutely continuous, so that

$$|f(u(x))| = |f(0) + u(x) \int_0^1 f'(tu(x))dt\ | \le |f(0)| + M|u(x)|, x \in \Omega,$$

and, since Ω is bounded, we have $f(u) \in Y$. Similarly, applying the fundamental theorem of calculus to the function $\varphi(t) = f(tu(x) + (1-t)v(x))$, one obtains

$$\|f(u(x)) - f(v(x))\|_0 \le \int_0^1 |f'(tu(x) + (1-t)v(x))|dt\ \|u(x) - v(x)\|_0.$$

Now take Cauchy sequences of continuous functions converging to arbitrary functions in Y: the estimates above extend to the required L^2 estimates. □

3.1. The main result: $F_v : W_X \to W_Y$ is a homeomorphism

We now describe an orthogonal decomposition of X and Y. Take $\Omega \subset \mathbb{R}^n$ to be a bounded appropriate domain and let $\Lambda_f = \{\lambda_i\}_{i\in I}$ be the set of eigenvalues λ_i in (a, b). The *vertical subspaces* $V_X = V_Y$ equal the invariant subspace associated to Λ_f and $V_X \subset X, V_Y \subset Y$. The *horizontal* subspaces are $W_X = V_X^\perp \subset X$ and $W_Y = V_Y^\perp \subset Y$ where orthogonality takes into account the (different) inner products in X

and Y. These induce orthogonal decompositions $X = W_X \oplus V_X, \quad Y = W_Y \oplus V_Y$ and corresponding orthogonal projections P_Y and Q_Y from Y to W_Y and V_Y. Finally, we consider *affine horizontal subspaces* in X, which are sets of the form $x + W_X$, for a fixed x, and *affine vertical subspaces* in Y, of the form $y + V_Y$.

We need a label for this construction: a nonlinearity f induces an *I-decomposition* $X = W_X \oplus V_X$, $Y = W_Y \oplus V_Y$ associated to bounds a and b.

Theorem 1. *Let Ω be an appropriate domain, $f : \mathbb{R} \to \mathbb{R}$ Lipschitz,* $\operatorname{Ran} f' \subset [a, b]$, *where a and b are not eigenvalues λ_i. and $X = W_X \oplus V_X$, $Y = W_Y \oplus V_Y$ be the I-decomposition specified above. For $v \in V_X$, let $F_v : W_X \to W_Y$ be the horizontal projection of the restriction of F to the affine subspace $v + W_X$, $F_v(w) = P_Y F(w + v)$. Then F_v is a bi-Lipschitz homeomorphism. The Lipschitz constants for F_v and F_v^{-1} are independent of v.*

Proof. For $\gamma = (a + b)/2$, set $\tilde{f}(x) = f(x) - \gamma x$. Then $T : W_X \to W_Y$ given by $u \to -\Delta u - \gamma u$ is well defined and invertible, with eigenvalues $\lambda_j - \gamma$ with $j \notin I$. Let $\lambda_m - \gamma$ be the eigenvalue of T of smallest absolute value: clearly,

$$||T^{-1}|| = |\lambda_m - \gamma|^{-1} < |a - \gamma|^{-1} = (b - \gamma)^{-1}$$

For $u = w + v, w \in W_X, v \in V_X$, we have

$$F_v(w) = P_Y[-\Delta(w + v) - f(w + v)] = Tw - P_Y \tilde{f}(w + v).$$

The composition $F_v \circ T^{-1} : W_Y \to W_Y$ is of the form $I - K_v$, where $K_v(w) = P_Y \tilde{f}(T^{-1}w + v)$. We show that $K_v : W_Y \to W_Y$ is a contraction with constant uniformly bounded away from 1. For $w, \tilde{w} \in W_Y \subset L^2(\Omega)$,

$$\begin{aligned} \|K_v(w) - K_v(\tilde{w})\|_0 \leq & \ \|\tilde{f}(T^{-1}w + v) - \tilde{f}(T^{-1}\tilde{w} + v)\|_0 \\ \leq (b - \gamma)\|T^{-1}(w - \tilde{w})\|_0 \leq & \ \frac{b - \gamma}{|\lambda_m - \gamma|} \ \|w - \tilde{w}\|_0 = c \ \|w - \tilde{w}\|_0. \end{aligned}$$

Since b is not an eigenvalue λ_j, the Lipschitz constant c is uniformly bounded away from 1. From the Banach contraction theorem, $F_v : W_X \to W_Y$ is a homeomorphism. From the standard arguments bounding iterations by convergent geometric progressions, $(I - K_v)^{-1}$ is also Lipschitz, with constant $(1 - c)^{-1}$, hence independent of v. □

In particular, if $[a, b]$ does not contain eigenvalues λ_i, the result above recovers the Dolph–Hammerstein theorem for Lipschitz nonlinearities ([5], [8]).

3.2. Sheets and fibers make sense for Lipschitz nonlinearities

We are ready to extend to the Lipschitz context the global Lyapunov–Schmidt decomposition which is known for the smooth case, from the works of Berger and Podolak and Smiley. Consider the following diagram.

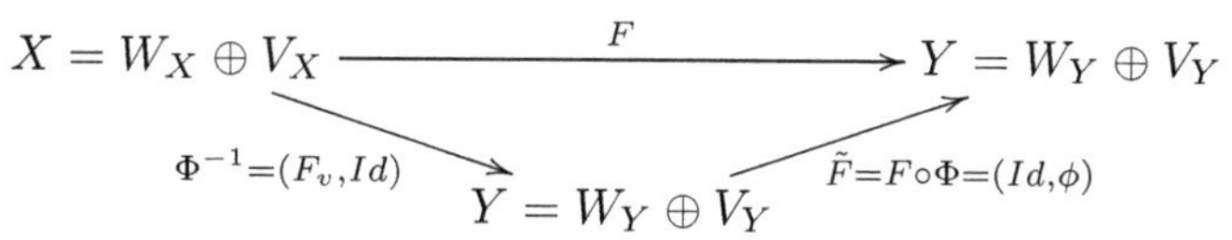

Thus the change of variables Φ yields $\tilde{F}(w,v) = F \circ \Phi(w,v) = (w, \phi(w,v))$, from which we will derive some convenient geometric properties. We first clarify a technicality: Φ is indeed a global change of variables in the Lipschitz category. In $W_X \oplus V_X$, we use the norm obtained by adding the norms in each coordinate.

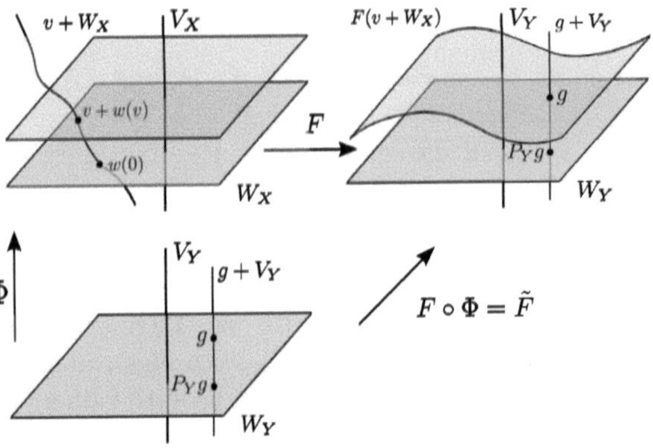

FIGURE 2. The change of variables

Proposition 1. *The map* $\Phi = ((F_v)^{-1}, Id) : Y = W_Y \oplus V_Y \to X = W_X \oplus V_X$ *is a bi-Lipschitz homeomorphism.*

Proof. The invertibility of Φ follows from the previous theorem. We use some elementary facts. The identity map $Id : (V_X, \|.\|_2) \to (V_Y, \|.\|_0)$ between normed spaces of the same finite dimension is bi-Lipschitz. Also, since W_X and V_X are orthogonal (in L^2 and H^2), $\|w\|_2 + \|v\|_2 \leq 2\|w \pm v\|_2$ for $w \in W_X$ and $v \in V_X$. To show that Φ^{-1} is Lipschitz, take $w+v, \tilde{w}+\tilde{v} \in X$. For appropriate constants C, $\tilde{C}$,

$$\begin{aligned}
&\|\Phi^{-1}(w+v) - \Phi^{-1}(\tilde{w}+\tilde{v})\|_0 = \|F_v(w) - F_{\tilde{v}}(\tilde{w})\|_0 + \|v - \tilde{v}\|_0 \\
&\quad \leq \| - \Delta(w - \tilde{w}) - P_Y(f(w+v) - f(\tilde{w}+\tilde{v}))\|_0 + C\|v - \tilde{v}\|_2 \\
&\quad \leq \|w - \tilde{w}\|_2 + \|f(w+v) - f(\tilde{w}+\tilde{v})\|_0 + C\|v-\tilde{v}\|_2 \leq \tilde{C}\|w + v - (\tilde{w}+\tilde{v})\|_2,
\end{aligned}$$

where the last inequality follows from Lemma 1.

We obtain a Lipschitz estimate for $\Phi = ((F_v)^{-1}, Id) : W_Y \oplus V_Y \to W_X \oplus V_X$. Take $z + v, \tilde{z} + \tilde{v} \in Y = W_Y \oplus V_Y$. Then

$$\|\Phi(z+v) - \Phi(\tilde{z}+\tilde{v})\|_2 \leq \|F_v^{-1}(z) - F_{\tilde{v}}^{-1}(z)\|_2 + \|F_{\tilde{v}}^{-1}(z) - F_{\tilde{v}}^{-1}(\tilde{z})\|_2 + \|v - \tilde{v}\|_2$$

Again from finite dimensionality of $V_X = V_Y$, there is an estimate of the form $\|v - \tilde{v}\|_2 \leq C\|v - \tilde{v}\|_0$. We also have $\|F_{\tilde{v}}^{-1}(z) - F_{\tilde{v}}^{-1}(\tilde{z})\|_2 \leq C\|z - \tilde{z}\|_0$ from the proof of the previous theorem. The first term is handled in a similar fashion. □

The picture should help putting pieces together. Here, $\dim V_X = \dim V_Y = 1$, as in the Ambrosetti–Prodi theorem: the convex span of Ranf' (f is Lipschitz!) contains only the eigenvalue λ_1. The map F takes an affine horizontal subspace

$v + W_X$ to a sheet, and the inverse of the vertical affine subspace $g + V_Y$ is a fiber, which crosses W_X at $w(0)$ and $v + W_X$ at $v + w(v)$, in the notation of the proof above. Clearly, sheet and fiber are graphs, as stated above. The change of variables Φ preserves horizontal affine subspaces and $\tilde{F}$ preserves affine vertical subspaces.

We are ready to prove the fundamental geometric property of such $F : X \to Y$: there are uniformly flat sheets and uniformly steep fibers.

Proposition 2. *Let $F : X = W_X \oplus V_X \to Y = W_Y \oplus V_Y$ with the hypotheses given in the beginning of the section. The image of each horizontal affine space $v + W_X \subset X$ under F is the graph of a Lipschitz function $\sigma_v : W_Y \to V_Y$. Similarly, the inverse of each vertical affine subspace $g + V_Y \subset Y$ under F is the graph of a Lipschitz function $\alpha_g : V_X \to W_X$. The Lipschitz constant can be taken to be the same, for all $v \in V_X$, $g \in W_Y$.*

Proof. We prove the result for fibers $F^{-1}(g + V_Y)$: the statement for sheets $F(v + W_X)$ is easier. Clearly, $\tilde{F}^{-1}(g + V_Y) \subset g + V_Y$, which is taken by the change of variables Φ to a set of the form $\alpha_g = \{(F_v)^{-1}(g) + v, v \in V_X\} \subset X$. From the theorem, for every $v \in V_X$, there is a unique $w(v) \in W_X$ for which $P_Y F(w(v)+v) = g$ – thus $F^{-1}(g + V_Y) = \{(F_v)^{-1}(g) + v, v \in V_X\}$. Said differently, $w(v) + v \in \alpha_g$: the set $\{(w(v), v), v \in V_X\} \subset X$ is the graph of a Lipschitz map.

The uniformity (on g) of the Lipschitz constant of the maps $v \mapsto w(v)$ is responsible for the uniform steepness of the fibers. □

In opposition to the arguments in [2], [13] and [4] for the smooth case, the geometric statements follow without recourse to implicit function theorems. Notice also that the uniform flatness of sheets and steepness of fibers are a counterpart to (differential) transversality between fibers and horizontal affine spaces in the domain and between sheets and vertical affine spaces in the counterdomain.

The restriction of F to horizontal affine subspaces is injective but the restriction to fibers α is not. In particular, in the standard Ambrosetti–Prodi case, F restricted to each fiber is simply the map $x \in \mathbb{R} \mapsto -x^2 \in \mathbb{R}$, after global changes of variables. The theorem becomes evident from this fact, first proved in [2].

Vertical lines may be taken by F to the horizontal plane, indicating yet another relevant transversality property of the fibers. To see this, take $\Omega = [-\pi/2, \pi/2]$ and $F(u) = -u'' - f(u)$, so that $\lambda_2 = 4$ and the corresponding eigenvector φ_2 is odd. Set $a = 3$ and $b = 5$, split $X = W_X \oplus \langle \varphi_2 \rangle$ and take $f(x) = e(x) + 4x$, where $e(x)$ is even (convexity is not necessary!) and $\mathrm{Ran} f' = (a, b)$. By symmetry, we have

$$\langle\, F(t\varphi_2)\,,\ \varphi_2\,\rangle = \int_\Omega (-t\,\Delta\varphi_2 - e(t\varphi_2) - 4t\,\varphi_2)\,\varphi_2 = 0.$$

4. Geometry and numerics

The statements in the previous section are exactly what we need to mimic the numerical algorithms in [4] for solving $F(u) = g$, i.e., the differential equation (2).

This section emphasizes the points where some alterations are needed, but most details common to the smooth and Lipschitz scenarios, which are provided in [4], will not be presented.

4.1. Finding the right fiber

To solve $F(u) = g$, first find any point $u = w + v \in \alpha_g$, the fiber associated to g. Said differently, find u so that $P_Y F(u) = P_Y g$. In order to do this, notice that, from the results of the previous section, each fiber α_g intersects each horizontal affine space $v + W_X$ at a single point. Thus each horizontal point $P_Y g \in W_Y$ corresponds to a unique $w(g) \in W_X$ for which $F_v(w(g)) = P_Y g$.

Said differently, each bi-Lipschitz map $F_v : W_X \to W_Y$ takes a point $w(g)$ in the fiber α_g to a point $P_Y g$, which is the only point in the vertical affine space $g + V_Y$ in the horizontal subspace W_Y. In words: there is a bi-Lipschitz map between the set of all fibers (represented by points in W_X) to the set of vertical affine spaces (represented by points in W_Y).

Now the good (numerical) news: to invert each map F_v, simply invert the (bi-Lipschitz) homeomorphism $F_v \circ T^{-1} = I - K$, where K is a contraction, as shown in the proof of Theorem 1! So the approximation of $F_v^{-1}(P_Y G)$ is amenable to standard numerical algorithms. In the differentiable case, we could do somehow better: we could invert by continuation where local steps are given by Newton iterations. In the strictly Lipschitz context, we lose quadratic convergence – some acceleration techniques are still available, but we provide no details.

4.2. Moving along a fiber

Once a point in α_g is identified, we need to learn to walk along the fiber, or more precisely, we have to compute the point in the fiber with a given height $v \in V_X$. This in turn is the main piece of information for a finite-dimensional inversion algorithm for the restriction $F : \alpha_g \to g + V_Y$, which is done by continuation starting from a given point in the fiber $u \in \alpha_g$.

Clearly small perturbations of a point $u_1 \in \alpha_g$ leave the fiber. But the previous algorithm – more precisely, the inversion of each map F_v – makes it possible to change $u_1 = w_1 + v_1$ to a point $\tilde{u} = \tilde{w} + v_2$ and then using $\tilde{u}$ as a starting point to solve $F_{v_2}(w_2) = P_Y g$, giving rise to a point $u_2 = w_2 + v_2 \in \alpha_g$.

4.3. Stability of the decomposition, a numerical necessity

The uniformity on the flatness of sheets and the steepness of fibers has a relevant consequence for numerics, which has been detailed in [4]. In a nutshell, whatever algorithm we use requires an approximation for the projection $P_Y : Y \to Y$. This is usually accomplished by computing (approximate) eigenfunctions associated to the eigenvalues of $-\Delta_D$ in the interval (a, b). A concrete possibility is to approximate functions by finite elements. The upshot is that the numerics handles an approximation $\tilde{V}_X = \tilde{V}_Y$ of $V_X = V_Y$ and its orthogonal complement. However, from the uniformities, such spaces, for sufficiently close approximations, still induce global Lyapunov–Schmidt decompositions giving rise to sheets and fibers, for which the algorithms described in the previous paragraphs hold and provide robust approximations to the real answers.

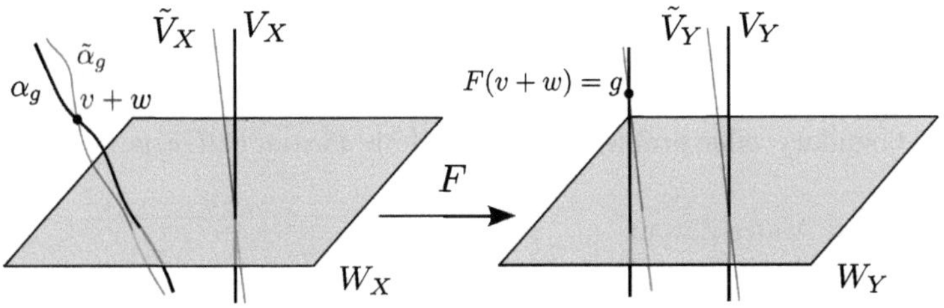

FIGURE 3. Irrelevant perturbations for numerical purposes

In the figure, we sketch fibers and their perturbations; there is an analogous reasoning for sheets. In the figure, $\tilde{V}_Y$ changes slightly the affine vertical subspace through the point g, which in turn, when inverted, gives rise to a slightly different fiber $\tilde{\alpha}_g$. Still, $\tilde{\alpha}_g$ is a Lipschitz graph of a function from $\tilde{V}_X$ to $\tilde{W}_X = (\tilde{V}_X)^{\perp}$. The (geometric) thing to notice is the fact that it is the steepness of the fibers which ascertain the robustness of the global Lyapunov–Schmidt decomposition.

References

[1] A. Ambrosetti and G. Prodi, On the inversion of some differentiable mappings with singularities between Banach spaces, Ann. Mat. Pura Appl. 93 (1972) 231–246.

[2] M.S. Berger and E. Podolak, On the solutions of a nonlinear Dirichlet problem, Indiana Univ. Math. J. 24 (1974) 837–846.

[3] D.G. Costa, D.G. Figueiredo and P.N. Srikanth, The exact number of solutions for a class of ordinary differential equations through Morse index computation, J. Diff. Eqns. 96 (1992) 185–199.

[4] J.T. Cal Neto and C. Tomei, Numerical analysis of semilinear elliptic equations with finite spectral interaction, J. Math. Anal. Appl. 395 (2012) 63–77.

[5] C.L. Dolph, Nonlinear integral equations of the Hammerstein type, Trans. AMS 66 (1949) 289–307.

[6] P. Grisvard, Elliptic problems in nonsmooth domains, Pitman, Boston, 1985.

[7] P. Felmer, A. Quaas and B. Sirakov, Landesman–Lazer type results for second order Hamilton–Jacobi–Bellman equations. J. Funct. Anal. 258 (2010) 4154–4182.

[8] A. Hammerstein, Nichtlineare Integralgleichungen nebst Anwendungen, Acta Math. 54 (1929) 117–176.

[9] A.C. Lazer and P.J. McKenna, On a conjecture related to the number of solutions of a nonlinear Dirichlet problem, Proc. R. Soc. Edinb. 95A (1983) 275–283.

[10] A.C. Lazer and P.J. McKenna, On the number of solutions of a nonlinear Dirichlet problem, J. Math. Anal. Appl. 84 (1981) 282–294.

[11] A. Manes and A.M. Micheletti, Un'estensione della teoria variazionale classica degli autovalori per operatori ellittici del secondo ordine, Boll. Unione Mat. Ital. 7 (1973) 285–301.

[12] G. Savaré, Regularity results for elliptic equations in Lipschitz domains, J. Funct. Anal. 152 (1998) 176–201.

[13] M.W. Smiley and C. Chun, Approximation of the bifurcation equation function for elliptic boundary value problems, Numer. Meth. Partial Diff. Eqs. 16 (2000) 194–213.

Carlos Tomei and André Zaccur
Departamento de Matemática, PUC-Rio
R. Mq. S. Vicente 225
Rio de Janeiro 22451-900, Brazil
e-mail: `tomei@mat.puc-rio.br`

Progress in Nonlinear Differential Equations
and Their Applications, Vol. 85, 457–466

The Backward λ-Lemma and Morse Filtrations

Joa Weber

Dedicated to Bernhard Ruf on the occasion of his 60th birthday

Abstract. Consider the infinite-dimensional dynamical system provided by the (forward) heat semi-flow on the loop space of a closed Riemannian manifold M. We use the recently discovered backward λ-Lemma and elements of Conley theory to construct a Morse filtration of the loop space whose cellular filtration complex represents the Morse complex associated to the downward L^2-gradient of the classical action functional. This paper is a survey. Details and proofs will be given in [6].

Mathematics Subject Classification (2010). 58-02 (Primary); 58B05, 35K91 (Secondary).

Keywords. Heat flow, Loop space, Morse filtration, Conley pair.

1. Introduction

Consider a closed smooth manifold M of dimension $n \geq 1$ equipped with a Riemannian metric and the Levi-Civita connection ∇. Pick a smooth function $V : S^1 \times M$, called the potential energy, and set $V_t(q) := V(t,q)$. Here and throughout we identify $S^1 = \mathbb{R}/\mathbb{Z}$.

For smooth maps $\mathbb{R} \times S^1 \to M : (s,t) \mapsto u(s,t)$ consider the *heat equation*

$$\partial_s u - \nabla_t \partial_t u - \nabla V_t(u) = 0. \tag{1}$$

It corresponds to the downward L^2-gradient equation of the *action* functional given by

$$\mathcal{S}_V(\gamma) = \int_0^1 \left(\frac{1}{2} |\dot{\gamma}(t)|^2 - V(t, \gamma(t)) \right) dt$$

for any element γ of the *free loop space* $\Lambda M := W^{1,2}(S^1, M)$ consisting of absolutely continuous loops in M whose derivative is square integrable. The critical

Financial support: FAPESP grants 2011/01830-1 and 2012/11634-8.

points of $\mathcal{S}_V$ are the closed orbits of the Euler–Lagrange flow associated to the mechanical Lagrangian given by kinetic minus potential energy, that is the solutions $x \in \Lambda M$ of the ODE $-\nabla_t \dot{x} - \nabla V_t(x) = 0$. For constant potential V these are the closed geodesics. *Throughout this paper* we fix a regular value a of $\mathcal{S}_V$ and assume that the Morse–Smale condition holds true below level a. Consider the sublevel set

$$\Lambda^a M := \{\gamma \in W^{1,2}(S^1, M) \mid \mathcal{S}_V(\gamma) < a\}.$$

In this case the action is a Morse function on $\Lambda^a M$ and the set of solutions to (1) that converge to critical points $x^\pm \in \Lambda^a M$, as $s \to \pm\infty$, carries the structure of a smooth manifold whose dimension is given by the Morse index difference $\mathrm{ind}_V(x^-) - \mathrm{ind}_V(x^+)$. Moreover, the set Crit of critical points of $\mathcal{S}_V$ in $\Lambda^a M$ is finite. By $m = m(a)$ we denote the maximal Morse index among them. By $\mathrm{Crit}_k \subset \mathrm{Crit}$ we denote the critical points of Morse index k. For each $x \in \mathrm{Crit}$ pick an orientation of the subspace E_x of the Hilbert space

$$X := T_x \Lambda M = W^{1,2}(S^1, x^*TM)$$

which is spanned by the eigenvectors corresponding to negative eigenvalues of the Hessian of $\mathcal{S}_V$ at x. (The dimension of E_x is finite and called the *Morse index* of x.)

Heat flow homology [4]. By definition the *Morse chain groups* $\mathrm{CM}_k = \mathrm{CM}_k(\Lambda^a M, \mathcal{S}_V; \mathbb{Z})$ are the free abelian groups generated by the (perturbed) closed geodesics x of Morse index k and below level a, that is $\mathbb{Z}^{\mathrm{Crit}_k}$. Set $\mathrm{CM}_k = \{0\}$ in case of the empty set. The chosen orientations provide the *characteristic sign* $n_u \in \{\pm 1\}$ for each heat flow solution u of (1) between critical points of index difference one. Up to shift in the time variable s, there are only finitely many such u. Counting them with signs n_u provides the *Morse boundary operator* $\partial_k : \mathrm{CM}_k \to \mathrm{CM}_{k-1}$. By HM_k we denote the corresponding homology groups.

Main result: The natural isomorphism to singular homology [6]. The idea to use cellular filtrations to calculate Morse homology goes back at least to Milnor [3]. One needs to construct a cellular filtration $\mathcal{F}$ of $\Lambda^a M$ whose cellular filtration complex $(\mathrm{C}_*\mathcal{F}, \partial_*)$ precisely represents the Morse complex, up to natural identification. In this case we are done, since

$$\mathrm{HM}_k \equiv \mathrm{H}_*\left((\mathrm{C}_*\mathcal{F}, \partial_*)\right) \simeq \mathrm{H}_*(\Lambda^a M) \tag{2}$$

where the isomorphism is provided by algebraic topology given any cellular filtration of $\Lambda^a M$ (related to the Morse complex or not); see, e.g., [2].

2. Morse filtrations and Conley pairs

Definition 2.1 (Cellular filtration and homology). Assume $\mathcal{F} = (F_{-1} \subset F_0 \subset F_1 \subset \cdots \subset F_\mu = \Lambda^a M)$ is a nested sequence of open subsets of $\Lambda^a M$ such that relative singular homology $H_\ell(F_k, F_{k-1})$ is trivial for any elements $\ell \neq k$ of the set $\{0, 1, \ldots, \mu\}$ and where $F_{-1} := \emptyset$. In this case $\mathcal{F}$ is (a special case of) a *cellular filtration of* $\Lambda^a M$. For the algebraic topology used in this section, see, e.g., [2]. The *cellular chain complex* consists of the *cellular chain groups* $\mathrm{C}_k\mathcal{F} := \mathrm{H}_k(F_k, F_{k-1})$ together with the triple boundary operators $\partial_k : \mathrm{H}_k(F_k, F_{k-1}) \to \mathrm{H}_{k-1}(F_{k-1}, F_{k-2})$. A cellular filtration $\mathcal{F}$ is called a *Morse filtration*, if $\mathrm{C}_k\mathcal{F} = \mathrm{CM}_k$ for every k, that is each relative homology group $H_k(F_k, F_{k-1})$ is generated precisely by the critical points of Morse index k.

Remark 2.2. To establish (2) we need a) to construct a Morse filtration $\mathcal{F}$ of $\Lambda^a M$ and b) to show that the associated triple boundary operator counts heat flow lines according to their characteristic signs between critical points of index difference one. How to solve these two problems is known for flows; cf. [3] or [1, Thm. 2.11]. The solution to b) carries over to our semi-flow situation, essentially since the semi-flow turns into a flow when restricted to the (finite-dimensional) unstable manifolds. It remains to construct a Morse filtration $\mathcal{F}$ of $\Lambda^a M$.

The Abbondandolo–Majer construction for flows [1]. In their construction of a Morse filtration $\mathcal{F}'$ of $\Lambda^a M$ openness of the sets F'_k follows from openness of the time-T-map and the Morse property is a consequence of forward flow invariance of the open sets F'_k. Start by setting N_0 equal to the union of open *local* sublevel sets, one for each local minimum x_0. Set $F'_0 := N_0$. Next choose a small open ball about each index one critical point and denote their (disjoint) union by N'_1. Then take the union of F'_0 and the whole forward flow of N'_1 and call it $F'_1 := F'_0 \cup \varphi_{[0,\infty)} N'_1$. Similarly define F'_2 and $F'_3, \dots, F'_m$.

A construction for semi-flows using Conley pairs [6]. The Cauchy problem associated to the heat equation (1) for maps $[0,\infty) \to \Lambda^a M : s \mapsto u_s = u(s,\cdot)$ is well posed and leads to the continuous *semi-flow*

$$\varphi : [0,\infty) \times \Lambda^a M \to \Lambda^a M$$

called the *heat flow*. In fact φ is of class C^1 on $(0,\infty) \times \Lambda^a M$. A characteristic feature of the heat flow is its extremely regularizing nature, namely $\varphi_s \gamma \in C^\infty(S^1, M)$ whenever $\gamma \in \Lambda M$ and $s > 0$. Observe that the set of nonsmooth elements is dense[1] in ΛM. Hence φ_s is not an open map for $s > 0$ and the Abbondandolo–Majer method does not work. Instead we propose the following construction.

It is a very simple – but far reaching – observation that *by continuity of φ_s preimages of open sets are open.* Define N_0 as above. Observe that the preimage ${\varphi_T}^{-1} N_0$ is open and semi-flow invariant. Assume x_1 is a critical point of Morse index one. The (one-dimensional) unstable manifold $W^u(x_1)$ necessarily[2] intersects N_0. Consequently our preimage gets very close to x_1 for T very large, however, it never contains x_1. To get over the barrier x_1 assume we had an open neighborhood N_{x_1} of x_1 containing no other critical points and a closed subset $L_{x_1} \subset N_{x_1}$ which does not contain x_1. Assume further that L_{x_1} is semi-flow invariant in N_{x_1} and every element leaving N_{x_1} under the semi-flow necessarily runs through L_{x_1} first.

[1] Pick $\gamma \in \Lambda M$ and a nonsmooth $\xi \in W^{1,2}(S^1, x^*TM)$. For large integers j consider $\exp_\gamma(\frac{1}{j}\xi)$.

[2] By Palais–Smale and $\mathcal{S}_V$ being Morse $\gamma_\infty := \lim_{s\to\infty} \varphi_s \gamma$ always exists and lies in Crit. If $\gamma \in W^u(x_1)$ and $\gamma \neq x_1$, then $\gamma_\infty \in \mathrm{Crit}_0$ by Morse–Smale.

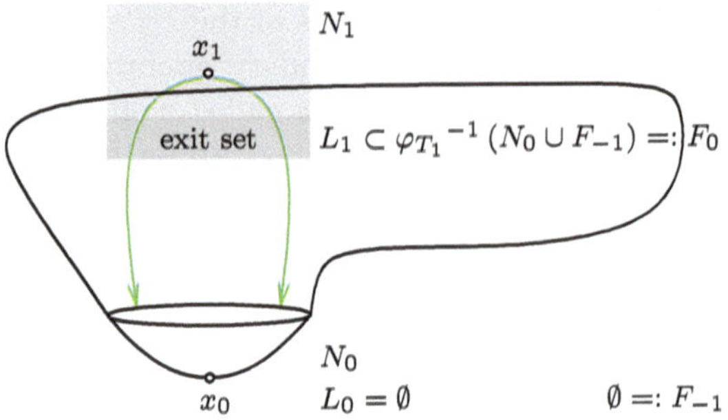

FIGURE 1. Morse filtration $\mathcal{F} = (\emptyset \subset F_0 \subset F_1 \subset \dots \subset F_m = \Lambda^a M)$

Such a pair (N_x, L_x) is called a *Conley pair* for $x \in \mathrm{Crit}$ and L_x is called an *exit set* for the *Conley set* N_x.

Pick $x \in \mathrm{Crit}$ and set $c := \mathcal{S}_V(x)$. For $\varepsilon > 0$ small and $\tau > 0$ large the sets

$$\begin{aligned} N_x &= N_x^{\varepsilon,\tau} := \left\{\gamma \in \Lambda^{c+\varepsilon} M \mid \mathcal{S}_V(\varphi_\tau \gamma) > c - \varepsilon\right\}_x \\ L_x &= L_x^{\varepsilon,\tau} := \{\gamma \in N_x \mid \mathcal{S}_V(\varphi_{2\tau}\gamma) \le c - \varepsilon\} \end{aligned} \tag{3}$$

form a Conley pair for x. Here $\{\dots\}_x$ indicates the path connected component of the set $\{\dots\}$ that contains x. By Theorem 3.2 d) below the sets N_x corresponding to different critical points x are pairwise disjoint. For $k \in \{0, \dots, m\}$ we define

$$N_k := \bigcup_{x \in \mathrm{Crit}_k} N_x, \qquad L_k := \bigcup_{x \in \mathrm{Crit}_k} L_x,$$

where $m = m(a)$ is the *maximal Morse index below level* a. Consider the preimages

$$F_k := {\varphi_{T_{k+1}}}^{-1}(N_k \cup F_{k-1}) \supset L_{k+1}, \qquad k = 0, \dots, m-1, \tag{4}$$

where the constant T_{k+1} is chosen sufficiently large[3] such that the inclusion holds true; see Figure 1. Note that if there are no critical points whose Morse index is k or $k+1$, then $F_k = F_{k-1}$ and $F_{k+1} = {\varphi_{T_{k+2}}}^{-1}(F_{k-1})$. Moreover, because there are no critical points in the complement of $N_m \cup F_{m-1}$ in $\Lambda^a M$, there is a constant T_{m+1} such that $\Lambda^a M$ is equal to $F_m := {\varphi_{T_{m+1}}}^{-1}(N_m \cup F_{m-1})$. Observe that each set F_k is open, because N_k and F_{k-1} are. Furthermore, although N_k is *not* semi-flow invariant the union $N_k \cup F_{k-1}$ *is*, because the exit set L_k of N_k is contained in F_{k-1}. Openness and semi-flow invariance heavily enter the calculation (5) in the proof of the Morse filtration property.

Morse filtration property. Constructing suitable homotopy equivalences and using excision one shows that

$$\mathrm{H}_\ell(F_k, F_{k-1}) \simeq \mathrm{H}_\ell(N_k, L_k) \simeq \bigoplus_{x \in \mathrm{Crit}_k} \mathrm{H}_\ell(N_x, L_x). \tag{5}$$

[3] Here Palais–Smale, Morse–Smale on neighborhoods, and $\mathcal{S}_V$ being bounded below enter.

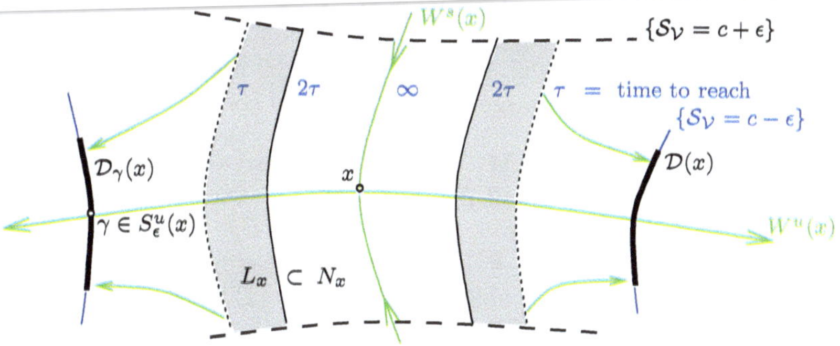

FIGURE 2. Conley pair (N_x, L_x) foliated by equal time disks ${\varphi_T}^{-1}\mathcal{D}_\gamma(x)$

Here the final step uses that N_k is a union of pairwise disjoint sets N_x. So in order to prove that the nested sequence $\mathcal{F}$ consisting of the open semi-flow invariant sets F_k defined by (4) is a Morse filtration of $\Lambda^a M$ – thereby concluding the proof of the main result (2) via Remark 2.2 – it remains to show that

$$\mathrm{H}_\ell(N_x, L_x) \simeq \mathrm{H}_\ell(\mathrm{D}^k, \partial\mathrm{D}^k) \simeq \begin{cases} \mathbb{Z}\,, & \ell = k, \\ 0\,, & \text{otherwise}, \end{cases} \tag{6}$$

for every $x \in \mathrm{Crit}_k$. To prove the first isomorphism was precisely the problem which inspired us to come up with the backward λ-Lemma in [5]: Since the part of N_x in the unstable manifold $W^u(x)$ is a k-disk bounded by the (relatively) closed annulus $L_x \cap W^u(x)$ it remains to deformation retract (N_x, L_x) onto its part in $W^u(x)$. A simple, but crucial, observation is that the semi-flow φ_s does the job on the *ascending disk*

$$W^s_\varepsilon(x) := W^s(x) \cap \Lambda^{c+\varepsilon}M = W^s(x) \cap N_x, \quad W^s(x) := \{\gamma \in \Lambda M \mid \varphi_s\gamma \overset{s\to\infty}{\longrightarrow} x\}.$$

Indeed it moves the elements of $W^s_\varepsilon(x)$ asymptotically to $x \in N_x \cap W^u(x)$, as $s \to \infty$. This fails on the complement $N_x \setminus W^s_\varepsilon(x)$. Note that $W^s_\varepsilon(x)$ is a C^1 graph over its tangent space, say X^+. The idea is to *foliate all of N_x by copies of $W^s_\varepsilon(x)$ (C^1 graphs over X^+), then extend φ_s artificially to all of N_x using the graph maps*; see (8) and Figure 4.

To understand the foliation structure assign to each point of N_x the time T at which it hits the level surface $\{\mathcal{S}_V = c-\varepsilon\}$; see Figure 2. This suggests that N_x is foliated by the equal time hypersurfaces ${\varphi_T}^{-1}\{\mathcal{S}_V = c-\varepsilon\}$ where $T \in (\tau, \infty)$. But for $T = \infty$ one obtains the codimension k ascending disk $W^s_\varepsilon(x)$. Because all leaves should be of the same codimension, consider the tubular neighborhood $\mathcal{D}(x) \to S^u_\varepsilon(x)$ associated to a (sufficiently small) radius r normal disk bundle of the descending sphere $S^u_\varepsilon(x) := W^u(x) \cap \{\mathcal{S}_V = c-\varepsilon\}$ in the Hilbert manifold $\{\mathcal{S}_V = c-\varepsilon\}$. Then each fiber $\mathcal{D}_\gamma(x)$ is a codimension k disk and so are the preimages $N_x \cap {\varphi_T}^{-1}\mathcal{D}_\gamma(x)$ which foliate N_x.

3. Backward λ-Lemma and stable foliations

Fix $x \in \mathrm{Crit}_k$ and set $c := \mathcal{S}_V(x)$. Because $N_x = N_x^{\varepsilon,\tau}$ fits into any neighborhood of x for $\varepsilon > 0$ small and $\tau > 0$ large, we will use local coordinates about $x \in \Lambda M$.

Local coordinates about $x \in \Lambda M$. The nonlinear part of the heat equation (1) determines a closed radius ρ_0 ball $\mathcal{B}_{\rho_0}$ about $0 \in X$ such that the following is true. Paths $s \mapsto u(s)$ in ΛM near x and $s \mapsto \xi(s)$ in $\mathcal{B}_{\rho_0}$ uniquely correspond to each other via the identity $u(s) = \exp_x \xi(s)$ pointwise for every $t \in S^1$. In the new coordinates ξ the Cauchy problem associated to (1) turns into the equivalent Cauchy problem

$$\zeta'(s) + A\zeta(s) = f(\zeta(s)), \qquad \zeta(0) = z \in \mathcal{B}_{\rho_0}, \tag{7}$$

for maps $\zeta : [0,T] \to \mathcal{B}_{\rho_0} \subset X$. Here $A = A_x$ is the Jacobi operator associated to the (perturbed) closed geodesic x. The semi-flow φ turns into the local semi-flow ϕ on $\mathcal{B}_{\rho_0} \subset X$. The nondegenerate critical point x corresponds to the hyperbolic fixed point 0 of ϕ. Furthermore, there is the orthogonal splitting

$$X := T_x\Lambda M \simeq T_xW^u(x) \oplus T_xW^s(x) =: X^- \oplus X^+.$$

Here X^- is of finite dimension $k = \mathrm{ind}_V(x)$ and consists of smooth loops along x. By $\pi_\pm : X \to X^\pm$ we denote the associated orthogonal projections. For coordinate representatives of global objects we shall use the global notation omitting x, for example $W^u(x)$ becomes W^u. By $\mathcal{S}$ we denote the representative of $\mathcal{S}_V$. Via a change of coordinates one achieves that locally near zero W^u is contained in X^-. By $\mathcal{B}_R^+$ we denote the closed ball of radius R about $0 \in X^+$. The *spectral gap* $d > 0$ is the distance between 0 and the spectrum of A_x.

Theorem 3.1 (Backward λ-Lemma, [5]). *Pick $\mu \in (0,d)$ and a hypersurface $\mathcal{D} \subset \mathcal{B}_{\rho_0}$ of the form $S_\varepsilon^u \times \mathcal{B}_\varkappa^+$. Then the following is true (see Figure 3). There is a ball $\mathcal{B}^+$ about $0 \in X^+$, a constant $T_0 > 0$, and a Lipschitz continuous map*

$$\begin{aligned} \mathcal{G} : (T_0,\infty) \times S_\varepsilon^u \times \mathcal{B}^+ &\to W^u \times \mathcal{B}^+ \subset \mathcal{B}_{\rho_0} \\ (T,\gamma,z_+) &\mapsto \left(G_\gamma^T(z_+), z_+\right) =: \mathcal{G}_\gamma^T(z_+) \end{aligned}$$

of class C^1. Each map $\mathcal{G}_\gamma^T : \mathcal{B}^+ \to X$ is bi-Lipschitz, a diffeomorphism onto its image, and $\mathcal{G}_\gamma^T(0) = \phi_{-T}\gamma =: \gamma_T$. The graph of G_γ^T consists of those $z \in \mathcal{B}_{\rho_0}$ which satisfy $\pi_+ z \in \mathcal{B}^+$ and reach the fiber $\mathcal{D}_\gamma = \{\gamma\} \times \mathcal{B}_\varkappa^+$ at time T, that is

$$\mathcal{G}_\gamma^T(\mathcal{B}^+) = {\phi_T}^{-1}\mathcal{D}_\gamma \cap \left(X^- \times \mathcal{B}^+\right).$$

Furthermore, the graph map $\mathcal{G}_\gamma^T$ converges uniformly, as $T \to \infty$, to the stable manifold graph map $\mathcal{G}^\infty$. More precisely, the estimates

$$\left\|\mathcal{G}_\gamma^T(z_+) - \mathcal{G}^\infty(z_+)\right\|_{W^{1,4}} \le e^{-T\frac{\mu}{16}}, \qquad \left\|d\mathcal{G}_\gamma^T(z_+)v\right\|_2 \le 2\left\|v\right\|_2,$$
$$\left\|d\mathcal{G}_\gamma^T(z_+)v - d\mathcal{G}^\infty(z_+)v\right\|_2 \le e^{-T\frac{\mu}{16}}\left\|v\right\|_2$$

hold true for all $T > T_0$, $\gamma \in S_\varepsilon^u$, $z_+ \in \mathcal{B}^+$, and v in the L^2 closure of X^+.

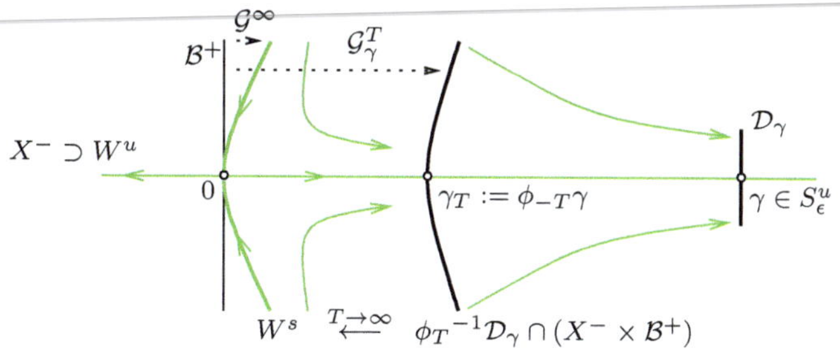

FIGURE 3. Backward λ-Lemma

Theorem 3.1 is based on the observation that the Cauchy problem for a heat flow line $\xi : [0,T] \to X$ with $\xi(0) = z$ is equivalent to a *mixed Cauchy problem* with data (T, γ, z_+). Namely, there is a unique heat flow line $\xi : [0,T] \to X$ with $\pi_+\xi(0) = z_+$ and $\pi_-\xi(T) = \gamma$.

That the (k-dimensional) unstable manifolds carry backward time information is evident from their definition. In contrast, Theorem 3.1 provides backward time information on *open* sets.

Stable foliation of Conley set. Theorem 3.1 foliates coordinate neighborhoods of x by (globally meaningless) codimension k disks. The next result provides global information in various directions. By definition the *descending disk* $W^u_\varepsilon(x)$ is given by $W^u(x) \cap \{\mathcal{S}_V > c - \varepsilon\}$.

Theorem 3.2 ([6]). *Given $\mu \in (0,d)$ there are constants $\varepsilon_0, \tau_0, r > 0$ such that the following is true. Assume $\tau > \tau_0$ and $\varepsilon \in (0, \varepsilon_0)$ and consider the radius r tubular neighborhood $\mathcal{D}(x) \to S^u_\varepsilon(x)$ defined in the paragraph preceding Section 3.*

a) *The Conley set $N_x = N_x^{\varepsilon,\tau}$ carries the structure of a C^0 foliation of codimension k. Its leaves are parametrized by the disk ${\varphi_\tau}^{-1}W^u_\varepsilon(x)$. It is of class C^1 away from the leaf over x, that is the ascending disk $W^s_\varepsilon(x)$. The other leaves are the disks*
$$N_x(\gamma_T) = \left({\varphi_T}^{-1}\mathcal{D}_\gamma(x) \cap \{\mathcal{S} < c + \varepsilon\}\right)_{\gamma_T}, \qquad \gamma_T := \varphi_{-T}\gamma,$$
where $T > \tau$ and $\gamma \in S^u_\varepsilon(x)$. Here $(\dots)_\gamma$ is the path connected component of γ.

b) *Leaves and semi-flow are compatible in the sense that*
$$z \in N_x(\gamma_T) \quad \Rightarrow \quad \varphi_\sigma z \in N_x(\varphi_\sigma \gamma_T), \quad \forall \sigma \in [0, T - \tau).$$

c) *The leaves converge uniformly to the ascending disk in the sense that*
$$\mathrm{dist}_{W^{1,2}}\left(N_x(\gamma_T), W^s_\varepsilon(x)\right) \le e^{-T\frac{\mu}{16}}$$
for all $T > \tau$ and $\gamma \in S^u_\varepsilon(x)$. Furthermore, if U is a neighborhood of $W^s_\varepsilon(x)$ in ΛM, then $N_x^{\varepsilon,\tau_*} \subset U$ for some constant τ_*.

d) *Assume U is a neighborhood of x in ΛM. Then there are constants ε_* and τ_* such that $N_x^{\varepsilon_*,\tau_*} \subset U$.*

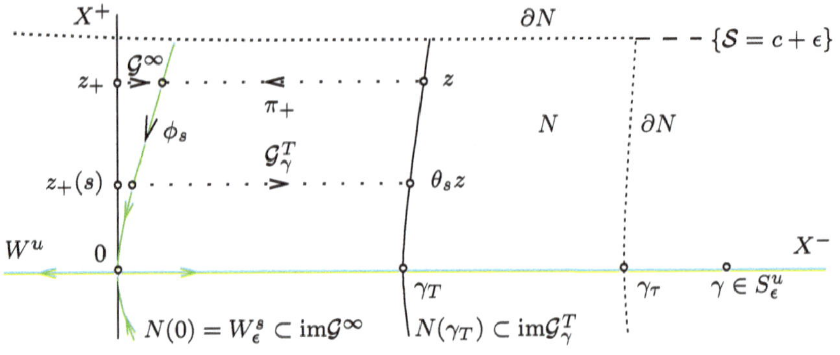

FIGURE 4. The induced flow θ_s on N

4. Strong deformation retract

Pick $x \in \mathrm{Crit}_k$. It remains to prove (6). If $k = 0$, then $L_x = \emptyset$ and $\{x\} = W^u(x) = N_x \cap W^u(x)$ is a strong deformation retract of $W^s_\varepsilon(x) = N_x$ where the retraction is provided by the semi-flow φ_s. So we are done. Now assume $k > 0$. Consider the local setup of Section 3 and denote the representative of N_x by N and similarly for other quantities. Fix $\rho_0 > 0$ so small that the only critical point in $\mathcal{B}_{\rho_0}$ is the origin of X.

Definition 4.1. By Theorem 3.2 each $z \in N$ lies on a leaf $N(\gamma_T)$ for some time $T > 0$ and some point γ in the descending disk S^u_ε where $\gamma_T := \phi_{-T}\gamma$. The continuous leaf preserving map $\theta : [0, \infty) \times N \to N$ defined by the composition of maps

$$\theta_s z := \mathcal{G}^T_\gamma \pi_+ \phi_s \mathcal{G}^\infty \pi_+ z \tag{8}$$

is called the *induced semi-flow on* N; see Figure 4. It is of class C^1 in $s \in (0, \infty)$.

That θ_s preserves the central leaf $N(0) = W^s_\varepsilon$ is due to the downward L^2-gradient nature of the heat equation. The proof for a general leaf $N(\gamma_T)$ turns out to be surprisingly complex although the idea is once more simple: Show that the map $s \mapsto \mathcal{S}(\theta_s z)$ strictly decreases whenever z lies in the (topological) boundary of a leaf. This implies preservation of leaves as follows. Firstly, note that θ is actually defined on a neighborhood of $N(\gamma_T)$ in $\mathcal{G}^T_\gamma(\mathcal{B}^+)$. Secondly, the (topological) boundary of a leaf lies on action level $c + \varepsilon$ whereas the leaf itself lies strictly below that level. Thus the induced semi-flow points inside along the boundary of each leaf – which is a disk by Theorem 3.2. So θ_s preserves leaves, thus N and L by Theorem 3.2. Moreover, it continuously deforms both topological spaces to their respective part in the unstable manifold and this concludes the proof of (6). Therefore $\mathcal{F}$ defined by (4) is indeed a Morse filtration for $\Lambda^a M$ and by Remark 2.2 this establishes the desired natural isomorphism (2).

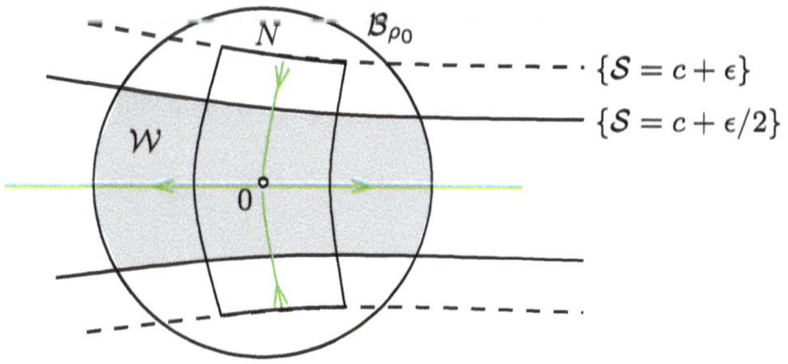

FIGURE 5. The neighborhood $\mathcal{W}$ of 0 used to define $\alpha > 0$

It remains to show that $\frac{d}{ds}\mathcal{S}(\theta_s z) < 0$ whenever z lies in the (topological) boundary of a leaf. Note that the L^2-gradient grad$\mathcal{S}$ is only defined on loops whose regularity is at least $W^{2,2}$. Consider the neighborhood $\mathcal{W} := \mathcal{B}_{\rho_0} \cap \{\mathcal{S} \leq c + \varepsilon/2\}$ of $0 \in X$ illustrated by Figure 5. By Palais–Smale the constant defined by

$$\alpha := \inf_{z \in (\mathcal{B}_{\rho_0} \cap W^{2,2}) \setminus \mathcal{W}} \|\mathrm{grad}\mathcal{S}(z)\|_2 > 0$$

is strictly positive. A rather technical argument, see [6], involving a long calculation which uses heavily the estimates provided by Theorem 3.1 shows that for all $\varepsilon > 0$ small and $\tau > 0$ large the following is true. If $T > \tau$ and $\gamma \in S^u_\varepsilon$, then

$$\begin{aligned} \frac{d}{ds}\mathcal{S}(\theta_s z) &= d\mathcal{S}|_{\theta_s z}\, d\mathcal{G}^T_\gamma|_{z_+(s)}\, \pi_+ \tfrac{d}{ds}\left(\phi_s \mathcal{G}^\infty \pi_+ z\right) \\ &= -\left\langle \mathrm{grad}\mathcal{S}|_{\theta_s z}, d\mathcal{G}^T_\gamma|_{z_+(s)}\pi_+ \mathrm{grad}\mathcal{S}|_{\phi_s q}\right\rangle_{L^2} \\ &\leq -\tfrac{1}{4}\alpha^2 \end{aligned}$$

for all $z \in \partial N(\gamma_T)$ and $s > 0$ small. It is precisely this calculation where we need convergence in $W^{1,4}$ and the extension to L^2 of the linearized graph map $d\mathcal{G}^T_\gamma(z_+)$ in Theorem 3.1. (The nonlinear part f of (1) maps $W^{1,4}$ into L^2.)

Acknowledgement

The author would like to thank Alberto Abbondandolo, Kai Cieliebak, and Klaus Mohnke for most useful discussions and comments.

References

[1] A. Abbondandolo and P. Majer, Lectures on the Morse complex for infinite dimensional manifolds, in *Morse theoretic methods in nonlinear analysis and in symplectic topology*, pp. 1–74, NATO Science Series II: Mathematics, Physics and Chemistry, P. Biran, O. Cornea, and F. Lalonde Eds, Springer, 2006.

[2] A. Dold, *Lectures on Algebraic Topology*, Grundlehren Math. Wiss. **200**, Springer-Verlag, Berlin, second edition 1980.

[3] J. Milnor, *Lectures on the h-Cobordism Theorem*, Princeton Univ. Press, 1965.

[4] J. Weber, Morse homology for the heat flow, *Math. Z.* 275 no. 1 (2013), 1–54. dx.doi.org/10.1007/s00209-012-1121-x.

[5] J. Weber, A backward λ-Lemma for the forward heat flow, Math. Ann. (2014). dx.doi.org/10.1007/s00208-014-1026-6.

[6] J. Weber, Stable foliations and the homology of the loop space. In preparation.

Joa Weber
IMECC
Instituto de Matemática, Estatística
e Computação Scientífica
Universidade Estadual de Campinas
Rua Sérgio Buarque de Holanda 651
Cidade Universitária "Zeferino Vaz"
Distr. Barão Geraldo
CEP 13083-859, Campinas-SP, Brasil
e-mail: `joa@math.sunysb.edu`